Ultrafiltration for Bioprocessing

Related Titles

An introduction to biotechnology
(ISBN 978-1-907568-28-2)

Fed-batch fermentation
(ISBN 978-1-907568-92-3)

Woodhead Publishing Series in Biomedicine: Number 29

Ultrafiltration for Bioprocessing

Development and Implementation of Robust Processes

Edited by

Herb Lutz

ELSEVIER

AMSTERDAM • BOSTON • HEIDELBERG • LONDON • NEW YORK • OXFORD
PARIS • SAN DIEGO • SAN FRANCISCO • SINGAPORE • SYDNEY • TOKYO

Woodhead Publishing is an imprint of Elsevier

WP WOODHEAD PUBLISHING

Woodhead Publishing is an imprint of Elsevier
80 High Street, Sawston, Cambridge, CB22 3HJ, UK
225 Wyman Street, Waltham, MA 02451, USA
Langford Lane, Kidlington, OX5 1GB, UK

British Library Cataloguing-in-Publication Data
A catalogue record for this book is available from the British Library

Library of Congress Control Number: 2014957102

ISBN 978-1-907568-46-6

For information on all Woodhead Publishing publications
visit our website at http://store.elsevier.com/

Typeset by Thomson Digital

Contents

List of figures

List of tables

About the contributors

Bala Raghunath As the global Director of Biomanufacturing Sciences Network, Bala leads a team of over 75 engineers and scientists responsible for the development, design and implementation of purification and separation process in the biopharmaceutical industry. Bala is a Chemical Engineer and has a B.Tech degree from the IIT, Mumbai (India), and MS and PhD degrees from the 'Center-of-Excellence for Membrane Technology at the University of Cincinnati, (USA). He has been an invited speaker at various technical conferences, has authored two patents and has also published several highly cited papers.

John Cyganowski John Cyganowski manages one of EMD Millipore's Process Development Sciences Team for North America. This field-based team of scientists and engineers, is dedicated to assisting EMD Millipore's customers in the development, optimization of separations technologies for biopharmaceutical manufacturing processes. John has specialized in Tangential Flow Filtration, Aseptic Processing, Viral Clearance and Large Scale Bioprocess Manufacturing over his 21 years with EMD Millipore. He has served as an instructor for the Tangential Flow Filtration section of the annual ASME Downstream BioProcessing and Economics course and is a member of the Parenteral Drug Association. John holds a Bachelor of Science Degree in Chemical Engineering from Northeastern University.

Joseph Parrella Joseph is an engineer in EMD Millipore's Biomanufacturing Engineering group. In his current role, Joseph manages a small group that supports the development, scale-up and implementation of tangential flow filtration and clarification technologies for biopharmaceutical processes. He has 10 years of experience supporting the development and manufacture of biopharmaceutical products and has presented and published on a variety of different technologies. Joseph holds a Bachelor of Science degree in Chemical Engineering from the University of California at Berkeley.

Pietro Perrone Pietro Perrone, P.E. works at EMD Millipore overseeing the engineering and design of Single-Use assemblies, systems and associated components for the pharmaceutical industry. He has managed the design and manufacture of custom-engineered stainless steel based filtration systems, chromatography systems and related process equipment with more than 20 years of purification/separation technology experience in process development and optimization, equipment scale-up and project management. Pietro has a Bachelor of Science Degree in Chemical Engineering from Tufts University and a Master of Science Degree in Biomedical Engineering and Biotechnology from the University of Massachusetts, Lowell. He is a member of the International Society for Pharmaceutical Engineering (ISPE) where his activities include Editorial Reviewer for Pharmaceutical Engineering Magazine, Chair of the Disposables Community of Practice and Industry Advisor to the Tufts University Student Chapter.

About the editor

As a Principal Consulting Engineer at EMD Millipore in the Biomanufacturing Sciences Network, Herbert currently focuses on the development, validation, scale-up and trouble-shooting of new downstream purification applications such as virus clearance, sterile filtration, clarification, tangential flow filtration, chromatographic purification and membrane adsorbers. Herbert is a frequent conference presentor, session chair, conference organizer and recognized thought leader. He has published several book chapters including the Membrane Separation Section of Perry's Chemical Engineer's Handbook, holds a filtration patent, is on the scientific advisory board for BioPharm International, and has published in the areas of scaling, ultra-filtration, sterile filtration, membrane adsorption, integrity testing and virus clearance. Herbert has degrees in chemistry and chemical engineering from UCSB and attended graduate school in Business and Chemical Engineering at MIT. He has worked in the area of separations and purification for 40 years. Mr. Lutz has taught membrane applications for Millipore, for the ASME Bioprocess course, and for the Society for Bioprocessing Professionals.

Preface

This book is intended as a teaching material and reference for practitioners in process development, transfer, validation, technology support, quality control, and operation within biomanufacturing firms. It is also intended as a reference for newcomers to biopharmaceutical manufacturing. It combines both academic and industrial perspectives with an eye toward useful practical information. Some mathematical treatment is provided to show from where certain conclusions and formulae originate. The intent here is not to dwell on the mathematical manipulations but on the resulting implications for what underlying mechanisms determine performance. Knowledge of the fundamentals enables one to extrapolate with confidence and trouble-shoot more efficiently.

As biopharmaceutical industry sales have grown worldwide with increasing numbers of people involved and regulatory scrutiny, the need for support material covering this universal unit operation has become apparent. General works on membranes such as Baker[1], Ho and Sirkar[2], Perry's Handbook[3] or general works on ultrafiltration such as Cheryan[4] and Zeman and Zydney[5] have omitted critical areas relevant to the industry such as validation and troubleshooting. Useful chapters have also appeared in various texts[6–9] but nothing as a comprehensive reference for the biopharmaceutical industry. It is hoped that this book will satisfy the need.

The book starts with the fundamentals. Chapter 1 is an overview intended to convey the terminology, principles of operation and an initial familiarity with the subject. This should give an overall perspective to read the subsequent in-depth chapters. Chapters 2–4 focus on membranes, modules and module performance. These are the fundamental building blocks of an ultrafiltration system and operation. Aspects of these chapters may be of more relevance to membrane suppliers. However, regulators are asking biomanufacturers to become more familiar with the raw materials used in their processes and what determines their performance.

Chapter 5 focuses on module configurations in a processing system. Chapters 6–8 consider the design of all the steps in an ultrafiltration process. Chapter 7 contains a number of case studies showing how ultrafiltration can be adapted to particular applications. There are a number of unique approaches here not found in any other published material.

Chapter 9 considers the hardware in more detail. Chapter 10 considers the steps involved in process development, process characterization and process validation. It contains elements of the quality-by-design approach regulators are promoting.

Chapter 11 provides a guide to troubleshooting. This is a unique chapter not found in other references. It contains a number of real-life problems that have arisen in technology transfer and manufacturing. Since speedy resolution of problems is required by quality and regulatory groups, this chapter can be invaluable to a practitioner.

Finally, Chapter 12 provides a summary and some thoughts on the direction of the field. Appendices are provided at the end and references provided throughout. Many experiments are governed by confidentiality agreements so no data can be shown without permission. Some data contained here have also been blinded. References to the literature and discussions with specific individuals are provided throughout. It is hoped that this book will serve as a valuable addition to the literature on the subject.

References

1. Baker RW. Membrane technology and applications. 2nd ed. Chichester, West Sussex, England: John Wiley & Sons; 2004.
2. Ho WS Winston, Sirkar K. Membrane handbook. New York, NY: Van Nostrand Reinhold; 1992.
3. Lutz H. Membrane separation processes. In: Perry's chemical engineers' handbook. 8th ed. New York, NY: McGraw-Hill; 2008.
4. Cheryan M. Ultrafiltration and microfiltration handbook. Lancaster, PA: Technomic Publishing Co. Inc; 1998.
5. Zeman LJ, Zydney AL. Microfiltration and ultrafiltration. New York, NY: Marcel Dekker; 1996.
6. Lutz H, Raghunath B. Ultrafiltration process design and implementation. In: Shukla AA, Etzel MR, Shishir Gadam, editors. Process scale in bioseparations for the biopharmaceutical industry. Boca Raton, FL: CRC Taylor & Francis; 2007.
7. Wang WK, editor. Membrane separations for biotechnology. 2nd ed. New York, NY: Marcel Dekker, Inc; 2001.
8. van Reis R, Zydney AL. Bioprocess membrane technology. J Membr Sci 2007;297:16–50.
9. EMD Millipore Corporation. Protein concentration and diafiltration by tangential flow filtration, Rev. C, TB032; 2003.

Acknowledgments

First, I would like to thank Marty Siwak for originally hiring me at Millipore and giving me a start in the membrane field. Marty's wisdom and encouragement has been valuable throughout the years. I thank my co-authors John Cyganowski, Joseph Parrella, Pietro Perrone and Bala Raghunath who assisted me in the challenge of completing this book. I also wish to thank my reviewers Elizabeth Goodrich, Pietro Perrone, Mischa Koslov, Gabriel Tkacik, Alex Xenopulos, Sarah Cummings and Willem Kools who substantially contributed to this work without the credit of authorship. I thank my publisher Glyn Jones for his understanding and encouragement during delays in the writing process. I also thank those who provided encouragement and assistance to continue working on this book in the face of other demands on time Willem Kools, Stephanie McGary, Mischa Koslov, Sarah Cummings and Mark Chisholm.

I would like to thank my other current and former colleagues at EMD Millipore and within the Biomanufacturing Support Network: Willem Kools, Joseph Silverberg, Elizabeth Goodrich, Jonathan Steen, Patrick Haughney, Gabriel Tkacik, Anthony Allegrezza, Leos Zeman, Vinay Goel, Ralph Kuriyel, Steve Pearl, Leon Mir and Gaston de los Reyes. Next my biomanufacturing collaboration customers: Robert van Reis, Bradley Wolk, Greg Blank, Joseph Shultz, Eva Gefroh, Chris Dowd and Matthew Westerby. Next my colleagues in academia and industry: William Eykamp, Andrew Zydney and George Belfort.

Finally, I would like to thank MaryEllen Ladd, my friends and family who provided encouragement and put up with my time and efforts over the years in completing this book.

Fundamentals

1

Herb Lutz
EMD Millipore, Biomanufacturing Sciences Network

1.1 Membrane pore sizes

A membrane can be idealized as a film that readily passes solvents and small solutes but retains large solutes above a particular size. UF (Ultrafiltration) membranes retain solutes with hydraulic diameters in the 5–150 nm range. This roughly corresponds to molecular weights in the 1–1000 kDa range covering most proteins, nucleic acids, nanoparticles, viruses, and some polymers. In the field of membranes, the term nanofiltration refers to membranes tighter than ultrafiltration but more open than reverse osmosis (Figure 1.1). However, in biotechnology, it has come to refer to virus-retaining membranes that fall in the open ultrafiltration (roughly 100 kDa) to tight microfiltration range (roughly 0.1 μm).

1.2 Applications

In a biopharmaceutical manufacturing process, biopharmaceutical products are expressed in a bioreactor (for mammalian cells) or fermentor (for bacterial or yeast cells). Ultrafiltration has been used to retain colloids and cell debris while passing the expressed product. In a similar way, ultrafiltration has been used after a refold pool to retain refold aggregates while passing the refolded product. Following clarification, ultrafiltration has been used to remove the largest contaminant – water – by retaining the product in a concentration step. This can be to reduce the size of subsequent steps, which may have to be sized based on the batch volume rather than on product mass. In addition, volume reduction facilitates transport and storage. This can provide flexibility to a manufacturing operation by decoupling the upstream from downstream operations.

Ultrafiltration is used to purify large solutes, such as vaccines, by retaining the desired product and passing through unwanted smaller components. This can include passing unreacted PEG or unconjugated polymers. Ultrafiltration is also used to retain unwanted viruses or aggregates while letting the desired product pass through. When the desired product and unwanted solutes are close in size, this is a challenging fractionation operation. This book will not cover virus ultrafiltration, or nanofiltration as it is frequently called.

The most common application is the use of ultrafiltration for final product formulation. This involves retaining the product while concentrating it to the desired target, and conducting a buffer exchange using a diafiltration process. Small contaminants and the old buffer components pass through the membrane.

Ultrafiltration for Bioprocessing

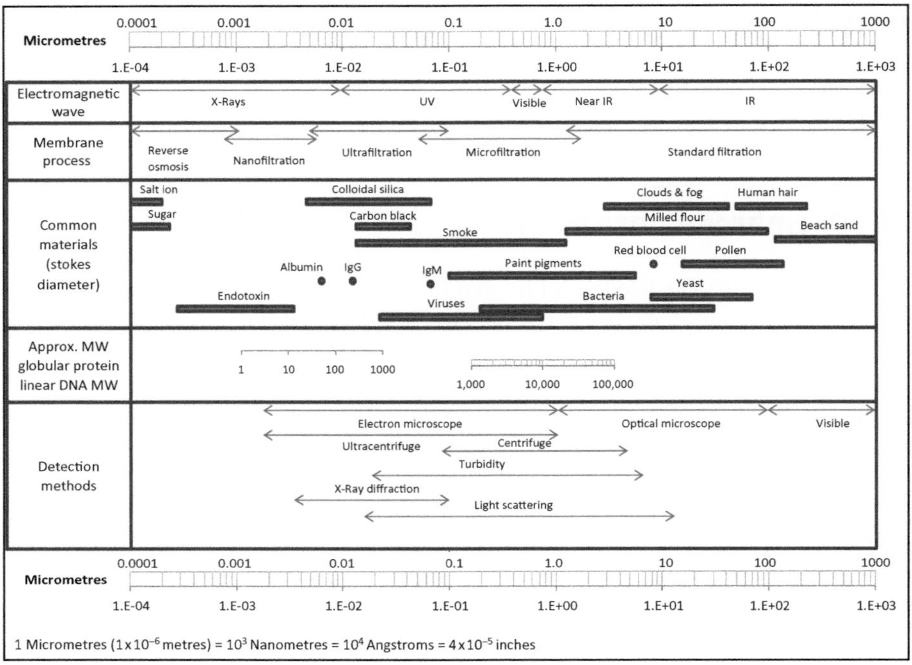

Figure 1.1 Ultrafiltration size range.

1.3 Modes of operation

Membranes are encased in modules for ease-of-operation. One could run the membrane in a static or dialysis mode where small buffer solutes can be exchanged by diffusion across the membrane (Figure 1.2). This is convenient at the bench scale but economical commercial operation requires bulk flow. UF can also be run in NFF (normal flow filtration) or TFF (tangential flow filtration) modes. NFF (Figure 1.2a) is the most common type of filtration. NFF mode involves passing the solvent through the filter under pressure where the fluid velocity is perpendicular (or normal) to the plane of the membrane. As the fluid passes through the membrane, it drags solutes with it to the membrane surface where they accumulate and cause the filter hydraulic resistance to increase. For high concentrations of retained solutes, NFF operation leads to relatively rapid plugging and low filter capacities. NFF is used at bench scale where low capacities pose less of a concern and the ease-of-use is convenient. NFF is also used for virus filtration.

TFF mode (Figure 1.2b) involves adding another fluid velocity component parallel to the plane of the membrane so the net solvent flow strikes the membrane at an angle. The presence of the tangential flow at the membrane surface facilitates backflow of solutes and prevents filter plugging. While there remains a region of high solute concentration at the membrane surface, steady-state operation is reached and TFF

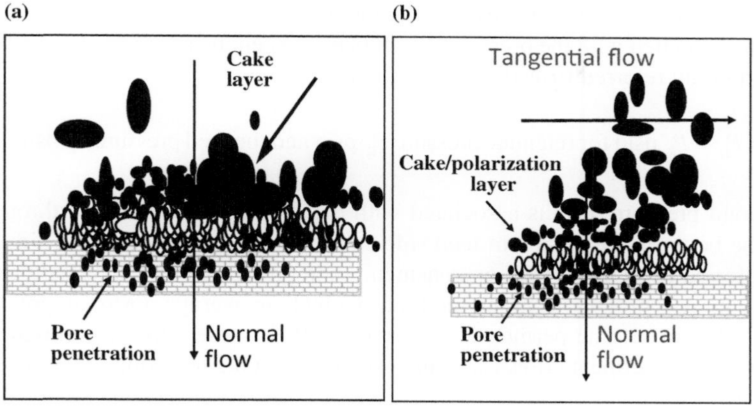

Figure 1.2 Operating modes.

operation shows very high filter capacities. TFF is used at manufacturing scale where capacity is important for economics but extra complexity is required to manage the tangential flow.

1.4 Module hydraulics

TFF modules have one entering feed stream with volumetric flowrate q_F and two streams leaving – the permeate (or filtrate) stream that has passed through the membrane and represents the normal flow component with volumetric flowrate q_P, and the retentate stream that has not passed through the membrane and represents the tangential flow component with volumetric flowrate q_R (Figure 1.3). Depending on the application, the product of interest may be retained in the retentate, passed through the membrane into the permeate stream, or separate recoverable products of value may lie in both the retentate and permeate. The ratio of the permeate flow to the feed flow is called the conversion. For dilute feeds, the conversion can be as high as 90%. For most applications with higher solute concentrations, conversions of 10–20% are common. This means that 80–90% of the feed flow winds up in the retentate as the tangential flow component.

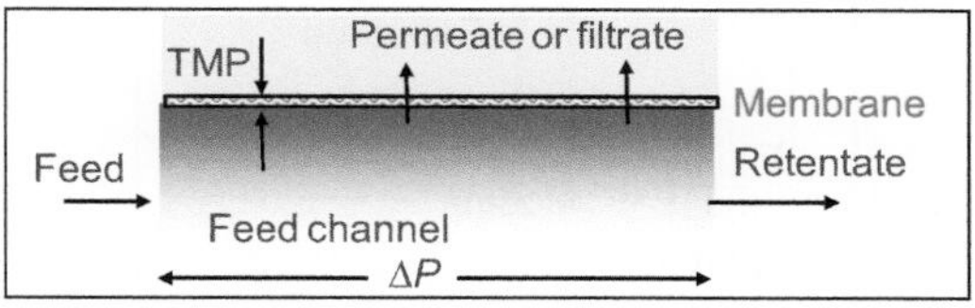

Figure 1.3 TFF module hydraulics.

Along with the two streams leaving the module are two pressure drops. The pressure drop associated with the pressure difference between the retentate P_R and the feed P_F is commonly referred to as the 'delta P' where

$$\Delta P = P_F - P_R \text{ psid for retentate pressure } P_R \text{ psig and the feed pressure } P_F \text{ psig} \quad (1.1)$$

A second pressure drop is associated with the driving force for flow through the membrane between the upstream feed side and the downstream permeate side, commonly referred to as the 'TMP' or transmembrane pressure. For a UF module where the upstream pressure varies along the feed channel, an average upstream pressure of $(P_F + P_R)/2$ is used. The permeate side pressure P_P is more uniform throughout the module since the permeate flows are much smaller than the feed side flows. The TMP is calculated as

$$\begin{aligned}\text{TMP} = (P_F + P_R)/2 - P_P \text{ psid for retentate pressure } P_R \text{ psig,}\\ \text{feed pressure } P_F \text{ psig, and permeate pressure } P_P \text{ psig}\end{aligned} \quad (1.2)$$

Solvent flow through the membrane is characterized by the flux or $J = q_p/A$, where A is the module membrane area. The flux is the flowrate per unit membrane area with common units of litres/hour-square metre or LMH. High flux means that the membrane module is very productive and can generate a large permeate volume in a small time with a small area. The units of flux can also be interpreted as a velocity of solvent through the membrane. A module with a flux of 36 LMH has a solvent velocity of 0.01 mm/s.

1.5 Basic system and operation

Figure 1.4 shows the components in a batch ultrafiltration system. During processing, the feed solution is contained within a feed tank and pumped by a feed pump into the module. Fluid passing through the membrane is diverted as permeate and fluid retained by the membrane is diverted as retentate. Typically, there is a retentate valve in the retentate line to provide some back pressure and help drive some of the fluid through the module. Depending on the application, the desired product of interest may

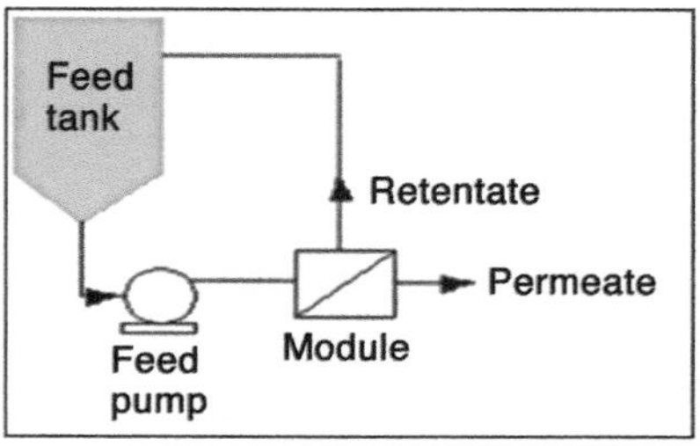

Figure 1.4 Batch UF system diagram.

Table 1.1 Ultrafiltration system-operating steps

Step	Description	Duration (h)
Preprocessing		
Installation	Install new modules or used modules into a holder	0.5
Flushing	Flush out preservatives and wet thoroughly	0.5
Integrity testing	Check for leaks	0.5
Equilibrate	Put modules in buffer	0.5
Processing		
Concentrate	Volume reduction	1
Diafilter	Buffer exchange	4
Postprocessing		
Recover product	Depolarize and drain system	0.5
Clean	Flush and incubate with cleaners	1
Verify cleaning	Check water permeability	0.5
Sanitize	Flush and incubate with sanitizers	1
Storage	Seal and store	0.5

be retained by the membrane and contained within the retentate stream, or may pass through the membrane and be contained within the permeate stream. It is common that after one pass through the module, the fluid has not been processed sufficiently to use in the next step. Rather, the retentate requires additional processing and is recycled back to the feed tank. The feed solution is repeatedly pumped across the UF modules until processing is complete.

The system may be plumbed in different configurations (Chapter 5). For example, diafiltration or buffer exchange takes place when a fresh diafiltration buffer is added to the tank during operation. The old buffer and other contaminants in the feed pass through the filter while the fresh buffer is added. Eventually the fresh buffer displaces the initial buffer while the product is retained by the membrane during the entire operation.

Complete operation of the ultrafiltration system requires multiple steps (Table 1.1). Operation of the UF system involves preprocessing steps to install, flush, integrity test, and equilibrate the modules (see Chapter 7). This is followed by feedstock processing (see Chapter 8). Postprocessing operation includes product recovery. It is most common for UF modules to be reused so postprocessing also includes cleaning, sanitization, and storage steps (see Chapter 9). The length of the preprocessing and postprocessing steps typically exceeds the processing time. It is common to prepare the system the day before processing. Single-use systems reduce the time, labour, and buffer requirements.

1.6 Process development and validation

Most of the processing steps (e.g., flushing, integrity testing, and cleaning) are routine and taken from vendor recommendations. Bioprocessors need to verify that these work properly for each molecule using scale-down test systems provided for the purpose by vendors. Bioprocessors need to specify processing steps and conditions, as well as demonstrate their performance. This can be relatively straightforward for the two main applications of batch concentration and diafiltration. These applications can be extremely robust and perform reliably and consistently. However, some applications (e.g., fractionation, clarification, and high final concentration) may require extra development (Chapter 7). One must be careful not to operate too close to an edge of failure without being able to monitor and control the process sufficiently.

1.7 Systems and troubleshooting

Most ultrafiltration systems are constructed using electropolished stainless steel components, designed for reuse. Recently, single-use systems have gained interest for small-scale processing. They may be manually operated, allow for automatic data storage, or be fully automated.

Improper attention to system design can cause the system to be unable to reach processing goals (Chapter 9). For example, a system that is not compact can have a large holdup volume that prevents one from achieving a concentration target. Poor mixing can limit buffer exchange during a diafiltration step. High retentate line pressure drops can prevent operation at the target feed flow and TMP specifications developed using a scale-down test system. Early chapters (1-5) draw connections between process performance and fundamental mechanisms. The troubleshooting chapter (Chapter 11) provides a useful summary guide from a manufacturing and quality perspective to determining root causes and correcting problems.

Membranes

2

Herb Lutz
EMD Millipore, Biomanufacturing Sciences Network

2.1 General properties

Ultrafiltration (UF) membranes are thin porous solids that perform separations by retaining and passing solutes primarily based on their size. Key membrane properties include retention capability, solvent permeability, chemical compatibility, mechanical strength for module fabrication and repeated use, consistency, and cost. They have nominal molecular weight limit (NMWL) ratings of 1–5000 kDa to retain solutes, such as macromolecules, nanoparticles, and colloids while passing smaller solutes and solvents. UF membranes, retain smaller solutes than microfiltration membranes and larger solutes than reverse osmosis membranes. Food and medical applications require extractables testing product and excipient, adsorption, particle shedding, and United States Pharmacopeia (USP) class VI toxicity testing. In addition, these applications require high consistency membranes with vendor quality programs (e.g., International Organization for Standardization (ISO), current good manufacturing practices (cGMP)). With reuse, membrane cost is significantly reduced (Table 2.1).

2.2 Materials and formation[1]

UF membranes are made by casting polymer films or by fusing particles together. UF membranes primarily consist of polymer films (polyethersulfone, regenerated cellulose, polysulfone, polyamide, polyacrylonitrile, or various fluoropolymers) formed by immersion casting on a web or as a composite cast on a microfiltration (MF) membrane. A multicomponent, high polymer concentration (5–25%) viscous solution (lacquer) is well-mixed, degassed, and filtered to remove particles or trapped air. The lacquer contains a non-solvent, which separates into droplets and becomes the pores of the UF membrane. Hollow fibres are formed by metering the lacquer through a die into an immersion bath. A flat sheet membrane is formed by metering the lacquer through a precisely controlled slot to coat a support web lying on a casting drum. Early UF immersion cast membranes cast on a solid web had thin surface retentive layers. Newer composite membranes use an MF membrane as the support web with the UF layer cast on top.[2] These composites demonstrate consistently high retention, higher strength in fabrication, and can be integrity tested using air diffusion in water.

The web coat or extruded fibre is fed into an immersion liquid bath (Figure 2.1). This causes the lacquer solvent to diffuse out of the lacquer coat (desolvation).

Table 2.1 **Membrane properties**

Material	Advantages	Disadvantages
Polyethersulfone (PES)	Resistance to temperature, Cl_2, pH, easy fabrication	Hydrophobic
Regenerated cellulose	Hydrophilic, low fouling	Sensitive to temperature, pH, Cl_2, microbial attack, mechanical creep
Polyamide		Sensitive to Cl_2, microbial attack
Polyvinylidene fluoride (PVDF)	Resistance to temperature, Cl_2, easy fabrication	Hydrophobic, coating sensitive to high pH
Inorganic	Resistance to temperature, Cl_2, pH, high pressure, solvents, long life	Cost, brittleness, high cross-flow rates

Then, solubility limits are reached (demixing) where the non-solvent phase nucleates and grows (spinoidal decomposition) and the solid polymer phase nucleates and grows (gelation). The polymer structure then coalesces into its final structure (phase transformation). The membrane structure is formed rapidly at non-equilibrium conditions, but the trajectory is commonly plotted on a ternary phase diagram. Foam or nodular structures form depending on the formation path. Asymmetry or skin structures can form depending on the kinetics of diffusive transport and nucleation. The key control parameters are the composition of the lacquer, composition of the immersion bath, coating or extrusion thickness, casting speed, bath temperature and bath flow velocity.

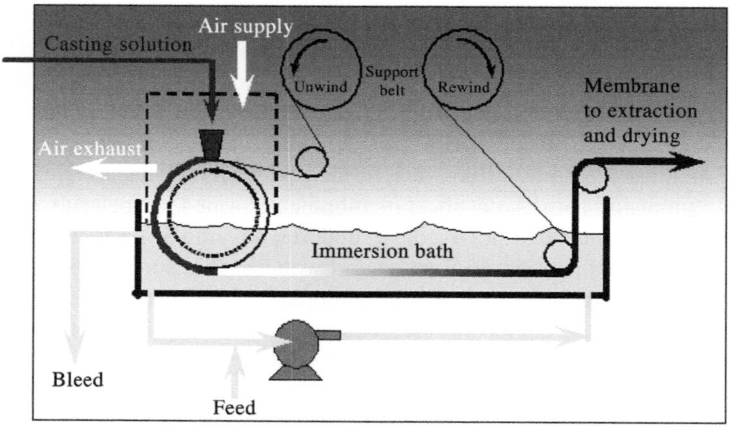

Figure 2.1 UF polymer membrane casting.
(Courtesy of EMD Millipore Corp.)

Residual solvent and anti-solvent are washed out of the polymeric membrane structure in separate extraction baths. Hydrophobic polymers are surface modified to render them hydrophilic and thereby reduce fouling, reduce product losses, and increase flux. Surface modification can include chemical grafting of hydrophilic groups to the base material surface or entangling hydrophilic polymers within the base material to coat the entire surface.

Inorganic UF membranes (alumina, glass and zirconia) are formed by sintering. Layers of spherical particles are deposited on a substrate and heated to fuse them together. Pores are formed in the gaps between the spherical particles. Successive layers of smaller and smaller particles can be deposited and sintered to make tighter membranes. These membranes find use in corrosive applications where their high cost can be justified.

2.3 Membrane structures

During immersion, the lacquer supported on a solid support web forms a surface skin where the solvent is rapidly extracted and a more open sub-structure where the solvent takes longer to diffuse out of the lacquer. Membranes formed this way have thin fragile skins with high permeability. They have relatively wide pore size distributions that limit solute separation. Their inability to handle back pressure limits reuse.

UF membrane flat sheet casting on a porous support web allows lacquer penetration into the porous structure depending on viscosity and surface tension. The resulting composite structure formed after immersion has a more uniform retention layer and stronger membrane (Figure 2.2).

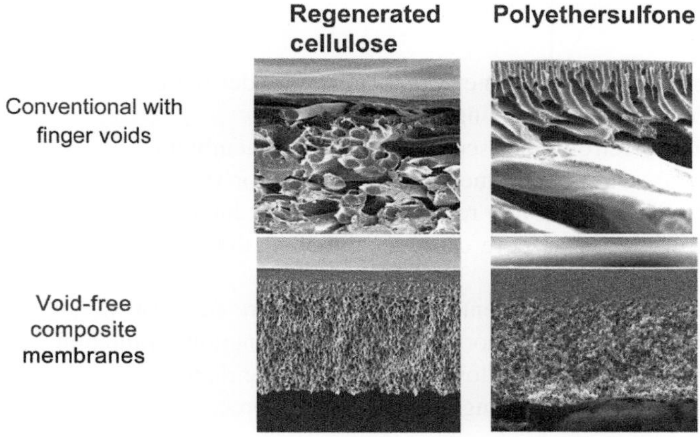

Figure 2.2 UF membrane structures.
(Courtesy of EMD Millipore Corp.)

UF membranes have an interconnected porous network that is sometimes described as a percolating foam structure. The 'bubbles' in the foam structure are formed by the non-solvent phase that becomes coated by the precipitating polymer. These foams show high interconnectedness with solvent permeability in both the normal and tangential directions. Voronai tessellation has been used to model these structures.

2.4 Solvent transport

Darcy's law[3] states that flow increases in proportion to the applied TMP driving force. A constant membrane permeability $L_p = J/\text{TMP}$ is a reflection of this law, where J is the LMH flux and TMP is the transmembrane pressure. For aqueous solvent systems, normal water permeability $\text{NWP} = L_p \cdot F(T)$ is commonly used where $F(T)$ is the temperature correction factor or ratio of fluid viscosity at a reference temperature of 23°C relative to the viscosity at the operating temperature (see Appendix A). This is approximated as about 2%/°C so a 28°C solvent has about 10% lower viscosity with a correction factor $F = 0.9$.

Permeability measurements are performed using membrane disks in holders with a screen support. These holders must be designed to use an accurately measured membrane area. Drilled metal support plates can block off part of the membrane area. Carbon black or dye staining is sometimes used to assess the effective membrane area in a holder. A porous support layer over the metal plate can aid in providing access to the complete membrane disk.

Flow through a porous medium can be modelled as flow through uniform parallel cylinders[3] (Hagen–Poiseuille flow):

$$J = \frac{n\pi r^4}{8\mu L}\text{TMP} \tag{2.1}$$

LMH for n cylinders per unit membrane area, cylinder radius r, solvent viscosity μ, membrane thickness L and driving force TMP.

One can use electron microscopy to count UF membrane pore density. It is also convenient to use measurable membrane porosity (or void volume) $\varepsilon = n\pi r^2$ so that $\text{NWP} = \varepsilon r^2 / 8\mu L$. Permeability remains constant for incompressible membranes and the strong dependence on pore diameter indicates that the larger pores have more flow.

Membrane pores are not identical but have a pore size distribution. This is commonly described using a log-normal probability function[4] (Equation 2.2). The log-normal distribution does not allow for negative pore diameters and provides a good description of actual pores arising from the casting process as counted using scanning electron microscopy.

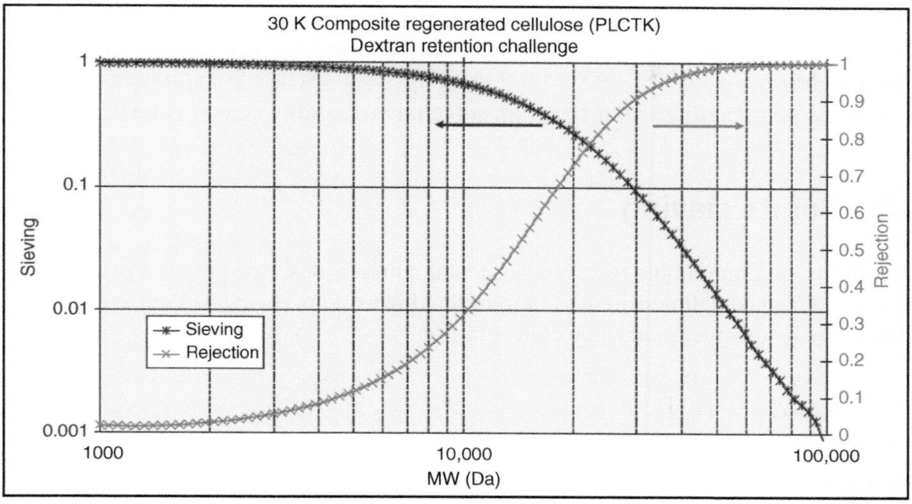

Figure 2.3 UF membrane sieving.[5]
(Courtesy of EMD Millipore Corp.)

$$J = \frac{\varepsilon \,\mathrm{TMP}}{8\mu L} \cdot \int_0^r r^2 p(r)\,dr \text{ for } p(r)dr = \frac{1}{r\sqrt{2\pi b}} \exp\left\{ -\frac{\left(\ln[r/\bar{r}] + b/2\right)^2}{2b} \right\} \tag{2.2}$$

where $\bar{r}$ is the measured mean pore radius, $b = \ln[1 + (\sigma/\bar{r})^2]$, σ is the measured pore radius standard deviation.

Rearranging the permeability expression yields a 'flow averaged pore radius' (Equation 2.3). Since one can measure membrane thickness, porosity, and NWP, this is a convenient form. Porosity or void volume is measured by comparing membrane dry and wet weights so ε = (wet weight − dry weight)/membrane volume/water density. As noted earlier, the larger pores take more of the flow so this flow average $\overline{r_{\mathrm{fa}}}$ will be larger than $\bar{r}$. Distributions with the same mean pore radius but increasing standard deviation will yield higher NWP membranes due to the presence of larger pores.

$$\overline{r_{\mathrm{fa}}} = \sqrt{\frac{8\mu L}{\varepsilon} \cdot \mathrm{NWP}} \tag{2.3}$$

Large defects or holes through membranes can be considered as a separate population from the log-normal distribution that add flow in parallel with membrane pores so that $\mathrm{NWP} = \pi(n_1 r_1^4 + n_2 r_2^4)/(8\mu L)$ where n_2 is the number of defects with radius r_2.

Flow through tight UF membrane pores is the dominant flow resistance. However, for some of the more open UF membranes, entrance effects can contribute additional pressure drop.

Process permeability will differ from solvent permeability (see Section 2.12).

2.5 Solute sieving

The ability of a membrane to pass a particular solute is described by its sieving coefficient s. This is the dimensionless ratio of the downstream permeate concentration c_P divided by its upstream feed side concentration c_F:

$$s = c_P / c_F \tag{2.4}$$

for permeate concentration c_P and feed side concentration c_F.

Solute retention, also referred to as rejection, is commonly described by $R = 1 - s$. A sieving coefficient of $s = 0.01$ means that the permeate concentration is 1% of that in the feed. This would also correspond to a retention of $R = 0.99$ or 99%. Figure 2.3 shows how a particular membrane passes a mixed dextran challenge. As the dextran solute molecular weight increases, the sieving coefficient decreases. Membrane retention is based on size, not molecular weight, but it is convenient to characterize membranes by a molecular weight because it is easier to measure than solute radius.

Measurements of UF membrane sieving for a particular solute can be influenced by adsorption of solutes causing changes in membrane pores, solute conformation, electrical charge, solute association, solute polarization, amount of processing, and solute size distribution. The impact of these effects on membrane selection is described further in Section 7.2. For consistent membrane characterization, it is useful to use a well-defined non-adsorbing solute system, such as mixed dextrans.[6] Dextrans are long linear molecules akin to the shape of nucleic acids rather than globular proteins. Figure 2.4 shows how the hydrodynamic or Stokes radius varies with solute molecular weight for a series of globular proteins,[7] linear DNA,[8] and linear dextrans.[9] This radius represents the size of an equivalent spherical particle with the same transport property, such as viscosity or diffusivity. One can use this radius as an indicator of how the particle might be retained by a membrane. The same hydraulic radius can represent solutes of different molecular weights due to the differences in conformation. The correlations can be used to extrapolate from the data. Globular solutes have a theoretical exponent of 1/3 while linear solutes have a theoretical exponent of 1/2.

Figure 2.5 compares sieving for different shaped solutes using their equivalent Stokes radius. For relatively large pore membranes, the different solutes match up well. However, as the solute size becomes larger relative to the membrane pores, the Stokes radius alone is not sufficient to characterize the solute size. In particular, the long straight dextran solutes find it easier to pass through the

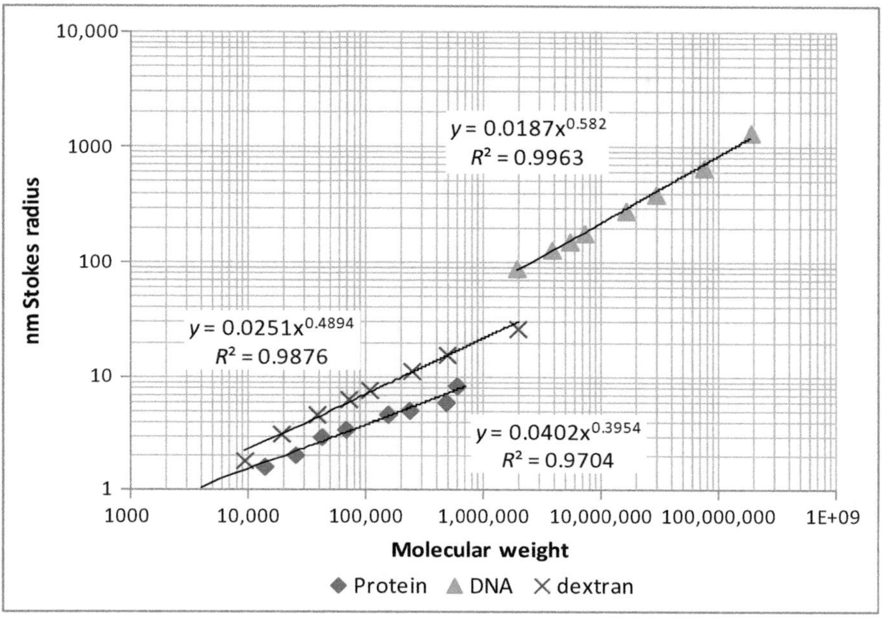

Figure 2.4 Solute sizes.

(a)

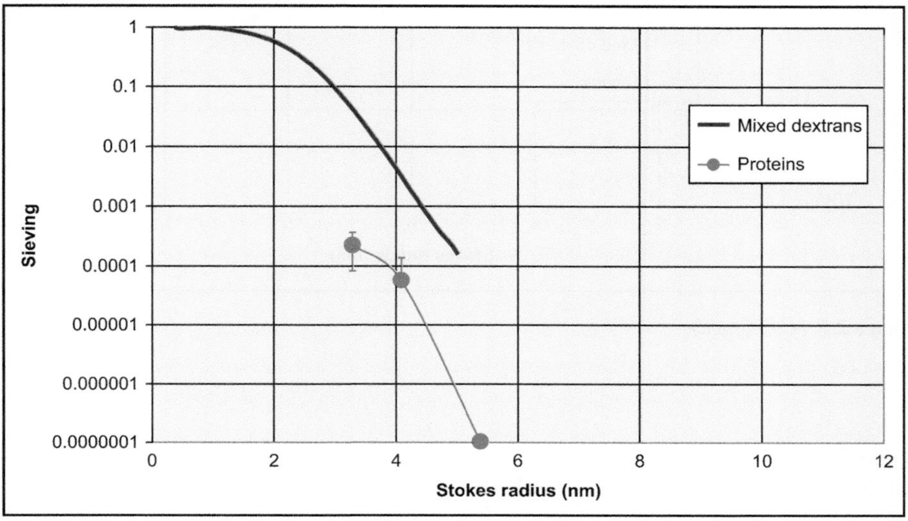

Figure 2.5 Solute retention.[11] (a) Tight membrane, (b) medium membrane, (c) open membrane.
(Courtesy of EMD Millipore Corp.)

(b)

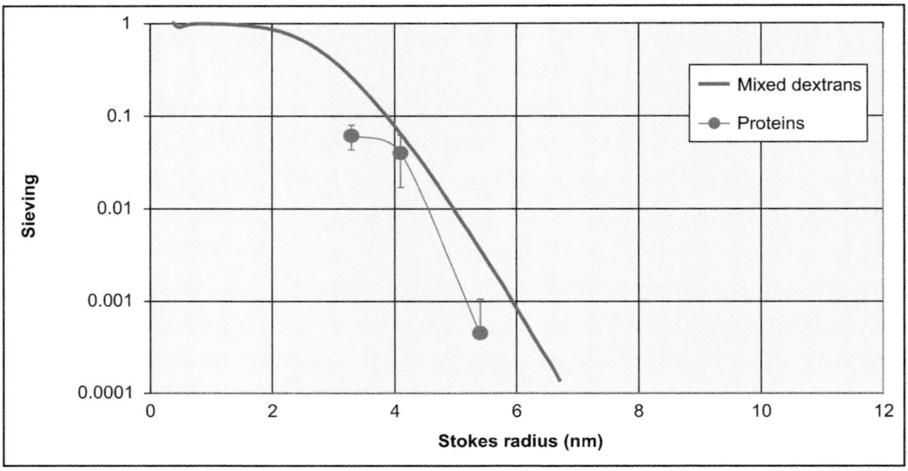

(c)

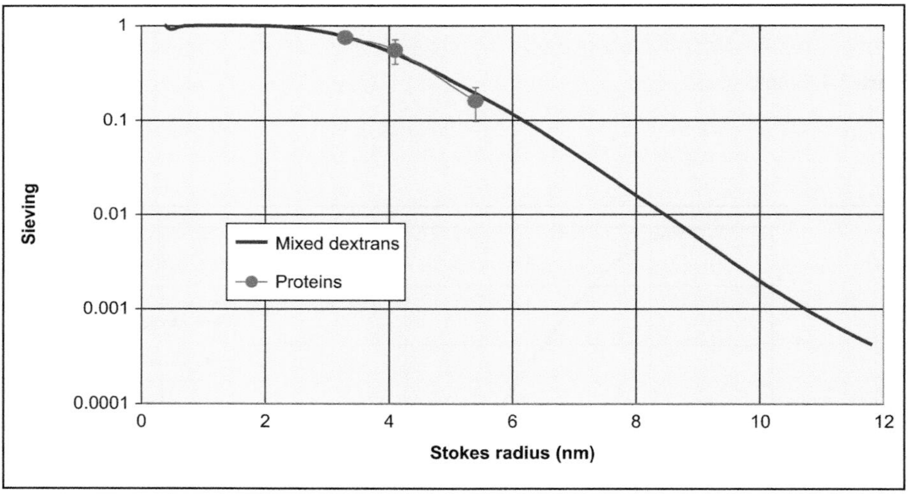

Figure 2.5 (*Continued*)

membranes than spherical proteins with the same Stokes radius. The velocity distribution at the surface of the membrane will apply a force on different sides of the solute to align the shape with the flow streamlines. This will align long, thin solutes (e.g. dextran, nucleic acids, and PEG) with the long diameter perpendicular to the pores and the small diameter entering the pores.[10] Solute conformation affects sieving.

2.6 Retention ratings

As described earlier, membranes are assigned an NMWL (nominal molecular weight limit) rating based on their retention. One can use the dextran molecular weight at 90% retention (10% sieving) or 'R90 value' as a consistent basis for assigning an NMWL. Using a different challenge solute and a different retention point will lead to a different rating. Not all manufacturers use the same basis so a 10 kDa membrane from one manufacturer can differ from that by another. Since pore size distributions vary between membrane families, the 'R99.9' value relevant for product yield can differ between membranes with the same R90. See Section 7.2 for more on membrane selection and factors that influence sieving.

Membrane sieving and solvent permeability are correlated for a particular membrane family through their mutual dependence on pore radius (Figure 2.6). For more open membranes, other physical tests measuring pore size, such as liquid–liquid porosimetry and bubble point, can be used.

Membrane properties can vary within a cast and between casts. Within a cast, the lacquer is the same but properties can vary 'across the web' or 'down the web' (along the length of the membrane sheet). Casts can be cut into separate rolls. Membrane coupon samples are typically taken at preselected points within each roll and tested to ensure they meet specifications. Figure 2.6 shows the variation in properties between separate rolls and separate casts. While water permeability variations typically have

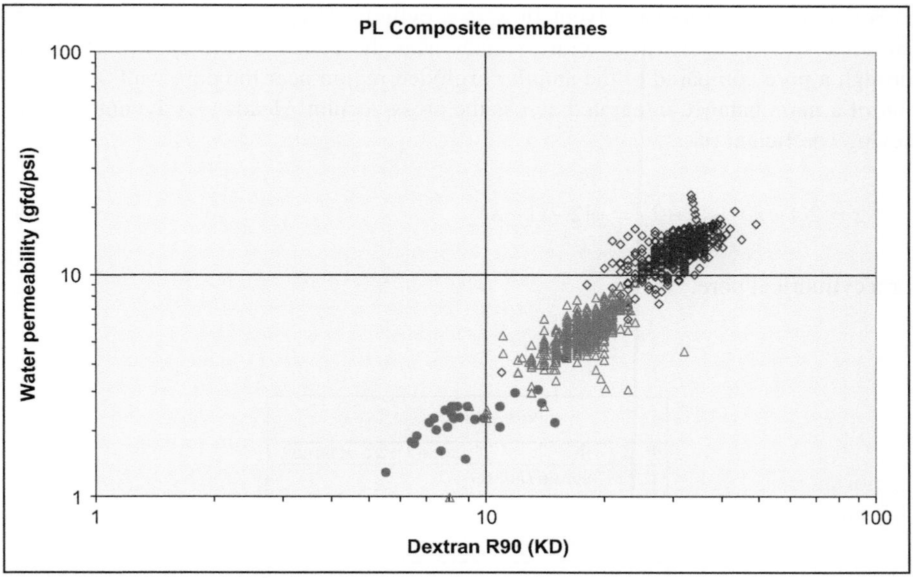

Figure 2.6 Lot-to-lot membrane variability.
(Courtesy of EMD Millipore Corp.)

little impact on process performance, the variation in retention properties can be significant.

Polarization causes elevated concentrations near the membrane surface and increases sieving. Small-scale stirred cell devices are commonly used for membrane characterization. They have poorer mass transfer properties than commercial modules and show higher passage (see Chapter 4).

2.7 Solute passage modelling

Pore shapes can be conveniently considered to fall in-between idealized cylindrical and slit geometries in terms of solute and pore wall interactions. The cylindrical pore has the greatest internal surface closest to the solute while the slit pore has the least. A hydrodynamic analysis can be used to characterize the sieving coefficient of idealized spherical solutes in cylindrical or slit pores. The centre of a spherical solute of radius r_s in a cylindrical pore with radius r_p can only fit in the middle of the pore out to a radius of $r = r_p - r_s$ compared to its ability to occupy any position in free solution.[12] As shown in Figure 2.7, this leads to an equilibrium partitioning of

$$\begin{aligned}&\phi = (1-\lambda)^2 \text{ for } \lambda = r_s/r_p \text{ for a cylindrical pore,}\\&\phi = (1-\lambda) \text{ for a slit pore with slit half width } r_p\end{aligned} \tag{2.5}$$

For laminar convective solvent flow in a cylinder (Hagen–Poiseuille flow[3]), the parabolic flow profile causes more solutes near the centre of the cylinder to flow through a pore compared to the annular excluded region near the pore wall. Application of a mass balance integrated across the cross-section[13] leads to a dynamic steric sieving coefficient of

$$s = 2(1-\lambda)^2 - (1-\lambda)^4 = \phi[2-(1-\lambda)^2 \tag{2.6}$$

for a cylindrical pore.

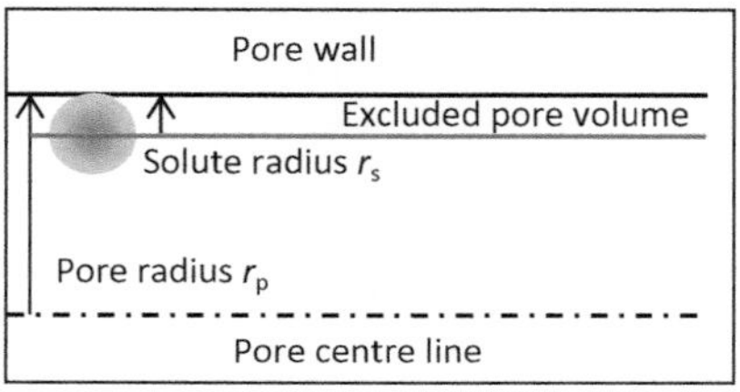

Figure 2.7 Slit pore sieving.

Further corrections can be made for interactions between the solute, wall, and hydrodynamic flow to yield the approximations

$$s \approx \phi[2-(1-\lambda)^2]\exp(-0.7146\lambda^2) \tag{2.7}$$

for a cylindrical pore, $s \approx \phi(3-\phi)^2[1/2-\lambda^2/6]$ for a slit pore.[14,15]

At low flows, solute passage by diffusion can be significant. The total solute flux through a pore is the sum of the convective and diffusive fluxes:

$$J_s = CK_c\frac{J}{\varepsilon} - K_d D_\infty \frac{dC}{dx} \tag{2.8}$$

where C is the g/L radially averaged solute concentration in the pore, K_c is the convective hindrance coefficient, J the LMH solute flux, ε the membrane porosity, D_∞ is the m^2/h solute free-solution binary diffusivity, K_d is the diffusive hindrance coefficient and x is the m distance along the membrane pore from the retentate upstream side to the permeate downstream side.

At the retentate upstream end ($x = 0$), $C_0 = \phi\, C_w$, and at the permeate downstream end ($x = L$), $C_L = \phi\, C_p$. Assuming steady-state in the pores (compared to upstream variations) and integrating yields

$$J_s = \phi K_c \frac{J}{\varepsilon}\left[\frac{C_w \exp(\mathrm{Pe}) - C_p}{\exp(\mathrm{Pe}) - 1}\right] \tag{2.9}$$

with membrane Peclet number $\mathrm{Pe} = [JL/D_\infty][\phi K_c/\varepsilon\phi K_d]$

Integration from $x = 0$ to x yields the dimensionless concentration profile

$$\frac{C}{C_w} = \phi\left[\frac{e^{\mathrm{Pe}\,x/L}(s-1) + e^{\mathrm{Pe}} - s}{e^{\mathrm{Pe}} - 1}\right] \text{for sieving coefficient } s = C_p/C_w \tag{2.10}$$

Asymptotic analysis yields

$$K_d \approx (1-\lambda)^{-13/2} \tag{2.11}$$

for a cylindrical pore, $K_d \approx (1 - 1.004\lambda + 0.418\lambda^3 + 0.21\lambda^4 - 0.619\lambda^5)$ for a slit pore.

The solute sieving coefficient becomes

$$s = C_p/C_w = s_\infty \exp(\mathrm{Pe})/[s_\infty + \exp(\mathrm{Pe}) - 1] \tag{2.12}$$

where $s_\infty = \phi K_c = 1 - \sigma_0$ is the asymptotic sieving coefficient for large Pe where convection dominates.

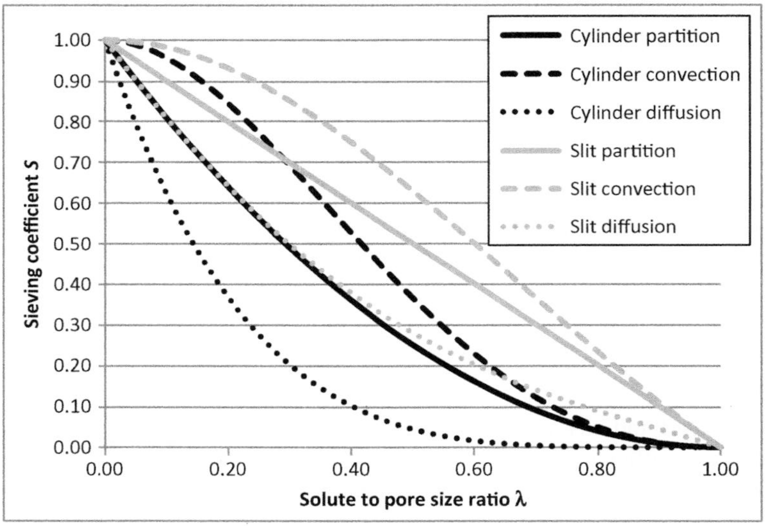

Figure 2.8 Ideal pore sieving.

The sieving coefficient for the idealized spherical solute in a slit or cylindrical pore geometries is shown in Figure 2.8. The partitioning curves represent a steady-state (no-flow) sieving, the convection curves a high flux sieving, and the diffusion a low flux sieving. These models show that even for an idealized single spherical solute and single cylindrical pore, the sieving is not a sharp 'S-shaped' curve between retained and passed solutes. For membranes with a log-normal pore size distribution, these sieving curves become even less sharp.

Average sieving across a membrane thickness from the wall to the permeate requires integration across the pore size distribution. Measured sieving coefficients from the bulk feed to the permeate stream requires accounting for polarization (Section 2.12). Solutes are dilute within pores but multicomponent interaction is required within the polarization layer where large solute polarization reduces the sieving of smaller solutes. Sieving along devices requires integration along the feed channel.

2.8 Solvent transport with electrical charges

Most membranes and other solids are negatively charged due to oxidation of surfaces forming COO- groups. Permeable charged solutes (electrolytes) will tend to have their positively charged cations associate closer to the membrane surface in a region referred to as the double layer. One can measure and derive relationships (see later) between the electrical current or solvent flow through a conductive pore and the electric field and pressure drop driving forces.[16] This shows that an electric field can induce flow (electro-osmosis) or a pressure drop, that fluid flow or a pressure drop can create an electric field (streaming potential) or a current (streaming current). As a result, the membrane permeability can be depressed.

Consider the electric potential $\Phi(r,x)$ described by Poisson's equation of electrostatics, where the axial electrical field is $E_x = -(\partial\Phi/\partial x)$ is constant:

$$\varepsilon_w \nabla^2\Phi = -\rho_e = -\sum_i z_i c_i \tag{2.13}$$

for pore liquid dielectric constant $\varepsilon_w = 6.375\times10^{-10}$ coulomb/Joule/m for water, ion charge density ρ_e, solute charge z_i, solute concentration c_i, axial coordinate x, and co-ordinate normal to the wall r.

At the wall, we can specify a potential Φ_w and take $\phi = \Phi - \Phi_w$ as the relative potential.

Charged solutes distribute in the electrical field according to the Boltzmann distribution. We can use the linearized Debye–Huckel approximation at moderate potentials to permit an analytical solution

$$c_i = c_{i0}\exp(-z_i F\phi/RT) \approx c_{i0}(1 - z_i F\phi/RT) \tag{2.14}$$

for Faraday constant F = 96,487 Coulombs/equiv, gas constant R = 1.987 cal/gmol/K, absolute temperature T in °K, and centre line ion concentration c_{i0}.

To solve for the radial electrical potential, combine equations (2.13 and 2.14) using the Debye length $\lambda^2 = \varepsilon_w RT/(F^2\sum_i z_i^2 c_{i0})$ and normalized potential $\psi = F\phi/RT$ to get $\lambda^2\nabla^2\psi = \psi - \Gamma$ with $\Gamma = -\sum_i z_i c_{i0}/\sum_i z_i^2 c_{i0}$. For a 0.1 M 1:1 salt in water, the Debye length λ = 1 nm while in more dilute 0.01 M solutions, this extends to 3 nm. This is comparable in magnitude to both the radii of UF pores and large solutes indicating it is relevant to UF processes. For a cylindrical pore the solution is $\psi = \Gamma[1 - I_0(r/\lambda)]$ for modified Bessel function of the first kind, of order zero I_0. The charge density also becomes $\rho_e = \sum_i z_i c_i = [\Gamma\varepsilon_w RT I_0(r/\lambda)]/F\lambda^2$. Relating this to the pore surface charge density q_2 using $\int_0^{r_p} \rho_e 2\pi r dr = 2\pi r_p q_2$ or $\Gamma = q_2 F\lambda/[\varepsilon_w RT I_1(r_p/\lambda)]$, and $\rho_e = [q_2 I_0(r/\lambda)]/[\lambda I_1(r_p/\lambda)]$. The radial electric field is $E_r = -(\partial\phi/\partial r) = (q_2/\varepsilon_w)\{[I_1(r/\lambda)]/[I_1(r_p/\lambda)]\}$.

Consider the average electric current flowing through the pore (x direction) $\langle i_x\rangle = \int_0^{r_p} i_x 2\pi r dr/(\pi r_p^2)$, where $i_x = F\sum_i z_i N_{ix}$ and the axial component flux is composed of convection, diffusion, and migration terms $N_{ix} = c_i v_x - D_i\nabla c_i - z_i u_i F c_i \nabla\Phi$ for ion mobility u_i. Here, the axial concentration is constant so there is no diffusive flux. The current becomes $i_x = \rho_e v_x - \kappa E_x$ for conductivity $\kappa = F^2\sum_i z_i^2 u_i c_i$.

Consider the momentum balance in the presence of an electric field

$$-\frac{dP}{dx} + \frac{\mu}{r}\frac{\partial}{\partial r}\left(r\frac{\partial v_x}{\partial r}\right) + E_x\rho_e = 0 \tag{2.15}$$

for liquid viscosity μ, axial velocity v_x, pressure P, and axial electric field E_x.

Combining (2.13) and (2.15), and integrating with boundary conditions $v_x(0) = 0$, $v_x(r_p) =$ finite, $\Phi(r_p) =$ finite, yields the velocity profile

$$v_z(r) = -\frac{dP}{dz}\left(\frac{r_p^2 - r^2}{4\mu}\right) + E_z \frac{\varepsilon_w}{\mu}\left(\Phi - \Phi_{r_p}\right) \tag{2.16}$$

Inserting this into the current flow equation and integrating over the pore cross-section, allows the total flow

$$\langle i_x \rangle = E_x \left\{ \kappa_{avg} + \frac{q_2^2}{\mu}\left[1 - \frac{I_0(r_p/\lambda) I_2(r_p/\lambda)}{I_1^2(r_p/\lambda)}\right]\right\} - \frac{dP}{dx}\frac{\lambda q_2}{\mu}\frac{I_2(r_p/\lambda)}{I_1(r_p/\lambda)} \tag{2.17}$$

where the average conductivity

$$\kappa_{avg} = \left\{F^2 \sum_i z_i^2 u_i c_{i0}\right\}\left\{1 + \frac{q_2 \lambda F}{\varepsilon RT}\left[\frac{2\lambda}{r_p} - \frac{1}{I_1(r_p/\lambda)}\right]\frac{\sum_i z_i^3 u_i c_{i0}}{\sum_i z_i^2 u_i c_{i0}}\right\}$$

$$\langle v_x \rangle = E_x \frac{\lambda q_2}{\mu}\frac{I_2(r_p/\lambda)}{I_1(r_p/\lambda)} - \frac{dP}{dx}\frac{r_p^2}{8\mu} \tag{2.18}$$

From equations (2.17 and 2.18), the electro-osmotic flow becomes $\langle v_x \rangle = E_x (\lambda q_2/\mu)[I_2(r_p/\lambda)/I_1(r_p/\lambda)]$, the streaming potential is $E_x = (\lambda q_2/\mu\kappa_{eff})(dP/dx)[I_2(r_p/\lambda)/I_1(r_p/\lambda)]$ and the streaming current is $\langle i_x \rangle = E_x 8\lambda q_2 \langle v_x \rangle [I_2(r_p/\lambda)/I_1(r_p/\lambda)]$. These can be used to experimentally determine charge density and pore radius. From equation (2.18), membrane permeability is depressed by charged membranes at low salt concentrations.[17]

2.9 Solute transport with electrical charges

Solutes can also have an effective size that is larger than what their steric size from NMR or crystal structure might imply. A charged solute has its electric field deformed when enclosed in a pore with a different pore wall dielectric coefficient than the solvent. This requires additional energy and reduces the likelihood of charged solutes from entering pores. In effect, the charge increases the effective size of the solute.[17] One expression developed using BSA[18] is

$$r_{eff} = r_s + 0.0137 z^2 / r_s \sqrt{I} \tag{2.19}$$

where r_s is the steric size in the absence of charge, I is the ionic strength and z is the net number of charged sites of the same charge on the molecule.

Most solutions have an ionic strength above isotonic (150 mM) where charge effects are passivated by the free salts. This makes the charge effect more limited in practice. For charged solutes and pore walls, opposite charges can increase sieving.[19]

2.10 Osmotic pressure

The presence of solutes can change solvent permeability. A retained solute will accumulate on the upstream membrane surface causing an osmotic pressure Π. Entropy is increased when the solvent dilutes high local concentrations. Osmotic pressure is the TMP required to overcome this effect. As a result, the driving force for flow must be modified. Membrane permeability may also be reduced. Solute adsorption (e.g., fouling) may coat the pore walls and reduce effective porosity, pore size, or add a cake layer resistance in series to reduce the membrane permeability from NWP to L_f. Solvent flow is then described by the osmotic model

$$J = L_f(\text{TMP} - \sigma_0 \Delta\Pi) \tag{2.20}$$

where σ_0 is the osmotic or Staverman reflection coefficient.[20]

For a totally retained solute $\sigma_0 = 1$ and for a totally passing solute $\sigma_0 = 0$. When TMP < Δ Π for a totally retained solute, reverse flow is experimentally observed showing that this is a driving force, not a resistance effect.

Figure 2.9 osmotic pressure data (taken with Osmometres) show pressures increase with concentration.[21–28] The smaller Dextran T10 shows higher pressures than the larger Dextran 500. Similarly, whey with lactalbumin and BSA show higher pressures than the larger IgG. Charged BSA at pH 7.4 shows higher pressures than at its pH 5.5 isoelectric point.[21]

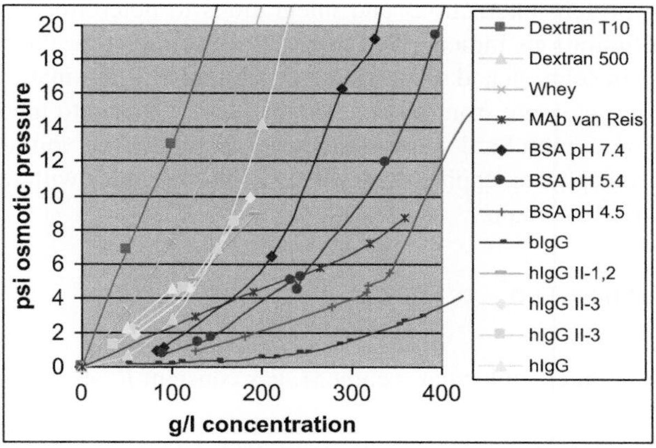

Figure 2.9 Osmotic pressures.

Table 2.2 Intermolecular forces

Type	Origin	Character
Covalent bond	Quantum mechanical	Attractive
Steric exclusion	Quantum mechanical	Repulsive
Charge–charge	Electrostatic	Repulsive
Charge–dipole	Electrostatic	Attractive
Dipole–dipole	Electrostatic	Attractive
Charge-induced dipole	Polarization	Attractive
Dipole-induced dipole	Polarization	Attractive
Dispersion	Polarization	Attractive
Hydrogen bond	Quantum mechanical	Varies

The osmotic pressure is a measure of the chemical potential of the solvent, taken here as water:[37]

$$\Pi = -\left(\mu_w - \mu_w^0\right) / v_w \tag{2.21}$$

where μ_w^0 is the chemical potential of pure water and v_w = 18 mL/gmole is the partial molar volume of water, generally taken at the pure water state.

The osmotic pressure arises from net intermolecular forces that resist the crowding of molecules associated with higher concentrations. Table 2.2 lists intermolecular forces.[29,30] Quantum mechanical forces create attractive covalent bonds, repulsive steric exclusion, and hydrogen bonding or hydrophobic forces arising from the water solvent that may be attractive or repulsive. Electrostatic forces create like charge repulsion or attractive charge dipole or dipole–dipole effects. Induced dipoles arise from external fields creating dipoles in an uncharged molecule. Magnetic forces are taken as negligible.

Simulations of protein interactions and integrations to determine average forces over multiple configurations indicate that steric exclusion, charge–charge repulsion moderated by ions in solution and dispersion attraction forces dominate.[22,31] One can consider the osmotic pressure from a suspension of hard spheres to model the steric exclusion force. For an inert hard sphere suspension that takes up volume but has no other forces between spheres, application of statistical mechanics with multibody interactions can be used to obtain[32]

$$\Pi = (CRT / M)\left(1+\phi+\phi^2-\phi^3\right)/(1-\phi)^3 \tag{2.22}$$

for solute concentration C, molecular weight M, gas constant R, absolute temperature T, and volume fraction ϕ.

This expression predicts that osmotic pressure increases with concentration, temperature, and volume fraction. It matches data with silica spheres in non-polar and comparable refractive index cyclohexane.[33] Molecular simulations predict a cubic centred solid phase[34] where

$$\Pi = (CRT/M)\begin{pmatrix} 2.558 + 0.125\beta + 0.176\beta^2 - 1.053\beta^3 + 2.819\beta^4 \\ -2.922\beta^5 + 1.118\beta^6 + 3(4-\beta)/\beta \end{pmatrix} \tag{2.23}$$

for $\beta = 4(1-\phi/0.74)$.

Simulations predict phase co-existance[35] at $\phi \sim 0.50$–0.55.

Concentration data can be converted to volume fraction using the specific volume. This is reported as $v_p = 0.745$ and 0.739 mL/g for γIgG,[42] and 0.72–0.73 for MAb.[37,38] Figure 2.10 shows that the polyclonal bovine IgG in 0.13 M NaCl osmotic pressure data matches the hard sphere model very well.[26] Polyclonal human IgG in 0.15 M NaCl and MAb osmotic pressures are higher than polyclonal bovine IgG. The van Reis MAb data were determined in a cassette, not an osmometre, which likely contributes to differences (see Chapter 3). Additional MAb data[37] showed $\Pi M/CRT \sim 0.47$, more consistent with the bovine IgG data.

The virial equation of state, applicable for dilute concentrations is[36]

$$\Pi = (CRT/M)\left(1 + B_2MC + B_3MC^2 + \ldots\right) \tag{2.24}$$

for virial coefficients B_2, B_3, and so on.

The second virial coefficient B_2 can be extracted from osmotic pressure, ultracentrifugation, and light scattering data[37] to determine sign, magnitude, and variation with test parameters. Large positive values of B_2 indicate molecular repulsion, while small or negative values indicate molecular attraction.

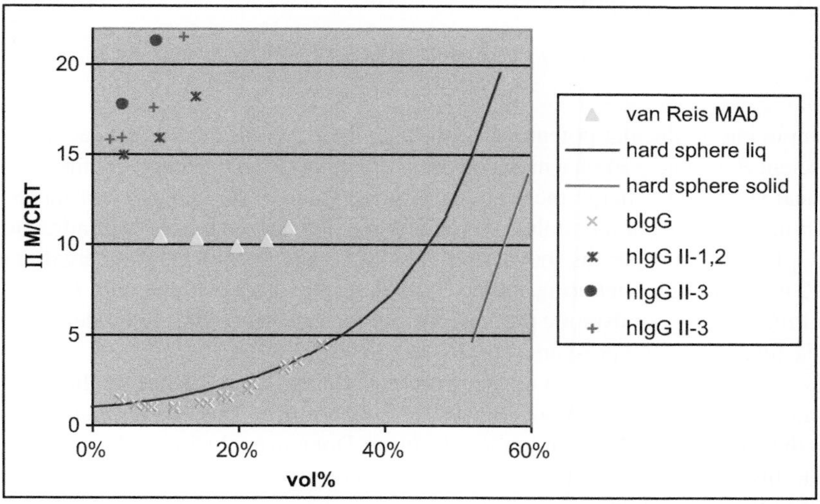

Figure 2.10 Hard sphere osmotic pressure model.

For a mixture of retained solutes i,

$$\Pi / RT = \sum_i (C_i / M_i) + \sum_i \left(B_{2i} C_i^2\right) + \ldots = C / \bar{M} + \bar{B}_2 C^2 + \ldots \tag{2.25}$$

where C is the total concentration of all the solutes, the overbar quantities represent averaged properties, the average second virial coefficient[39] $\bar{B}_2 = \sum_i \left(B_{2i} C_i^2\right) / C^2 = \sum_i \left(B_{2i} x_i^2\right)$

Plasma-derived hIgG at the pI shows association for <0.03 M NaCl and repulsion at 0.15 M. Repulsion increases for pH values above or below the pI.[27] For a 148 kDa MAb low ionic strength buffers show repulsion (positive slopes and B_2) and high ionic strength buffers show association (negative slopes and B_2).[37] This suggests electrostatic repulsion and polarization attraction. A series of MAbs differing only in the complementarity determining region (CDR) showed similar general behavior but one of the MAbs showed strong self-association.[38] It was inferred that the associating MAb showed electrostatic attraction between differently charged groups in the Fab region.

For general shapes and intermolecular forces, the McMillan–Meyer theory applies statistical mechanics by neglecting non-pair interactions to obtain the virial coefficients. They can be calculated using a configurational integral[36] that averages the pairwise solute interaction potential Φ_{ij}/kT over all the relative distances and orientations of protein molecules i and j.

$$B_2 = -\frac{1}{2V} \iint_V \{\exp(\Phi_{12} / kT) - 1\} dr_1 dr_2 \tag{2.26}$$

$$B_3 = -\frac{1}{3V} \int \iint_V \{\exp(\Phi_{12} / kT) - 1\}\{\exp(\Phi_{13} / kT) - 1\} \{\exp(\Phi_{23} / kT) - 1\} dr_1 dr_2 dr_3 \tag{2.27}$$

Simple intermolecular potentials allow analytical expressions for the second virial coefficient B_2 (Table 3.3) on a molecule rather than molar basis. Steric exclusion, nonspherical shape, and charge increase the positive value of B_2 indicative of molecular repulsion. Ion adsorption or adhesion reduces B_2, indicative of molecular attraction.

IgG_1 has the Y shape[27] as shown in Figure 2.11 with an equivalent sphere diameter of 10.2 nm from light scattering.[37] One could describe it as a cylinder with $L/d = 5$ and a resulting 25% higher osmotic pressure. Applying an oblate ellipsoid model produces an 80% higher osmotic pressure.[30]

The hard charged sphere in an electrolyte model in Table 2.3 has the hard sphere term along with additional repulsion and attraction terms. The first term corresponding to the electroneutrality requirement is due to Donnan exclusion and dominates the other terms. This effect creates a salt ion imbalance across the membrane.

For molecular volume v_p, spherical solute radius a, cylindrical solute length to diameter ratio L/d, ion charge z, electroneutrality parameter $\Sigma = \sum c_k z_k^2$, ionic strength

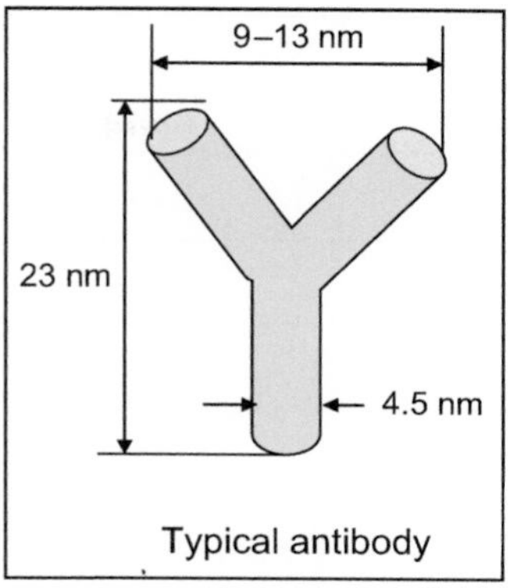

Figure 2.11 MAb dimensions.

parameter $\kappa^2 = (4\pi e^2 / \varepsilon kT)\Sigma$, parameter $\alpha = e^2\kappa / \varepsilon kT$, solvent dielectric κ, electron charge e, absolute temperature T, Boltzmann constant k.

2.11 Process permeability

Process permeability differs from solvent permeability. Figure 2.12 shows the flux vs. TMP curves for incompressible membrane (in pure solvent) and fouled membrane that has been in contact with a process solution and flushed. The fouled membrane has adsorbed components that reduce the effective pore size and lower the permeability. The unfouled permeability is restored by cleaning. Measuring the flux in the presence of retained protein further lowers the flux curve at high TMP and makes the flux insensitive to TMP changes at high values of TMP. Now the flux vs. TMP curve has a low TMP section called the 'linear region', a high TMP flat section called the 'polarized region', and an intermediate region where the curve bends called the 'knee' of the curve.

Figure 2.13 shows flux vs. TMP data for several membrane cutoffs and solutes. At low TMP, the curves mirror the linear solvent permeability lines, regardless of which protein is present, with the more open membranes showing higher slopes compared

Table 2.3 B_2 **models**

Model	Dimensionless potential Φ_{12}/kT	B_2
Hard sphere[40]	∞ inside radius a; 0 outside	$4v_p$
Hard cylinder[41]	∞ inside length $L \gg$ radius a; 0 outside	$(L/d)v_p$
Hard-charged sphere in electrolyte[40]	∞ inside; $(ze)^2\exp[-\kappa(r-a)]/\kappa r(1+\kappa a)$ outside	$4v_p + \frac{z^2}{2\Sigma} - \frac{z^4\alpha}{8(1+\kappa a)^2\Sigma}$

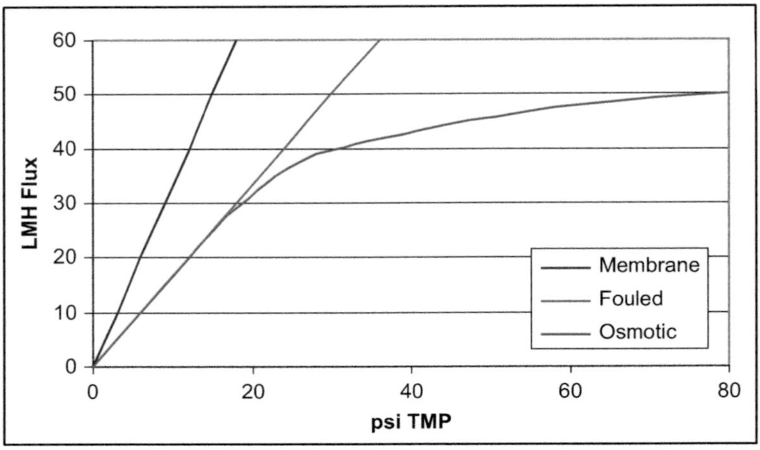

Figure 2.12 Process flux.

to the tighter membranes. At high TMP, the different membrane curves converge to the same constant flux value. The flux in this polarized region is only sensitive to the type of protein (here a MAb vs. a FAb), the cross-flow rate, and the feed concentration (data not shown). The flux values tend to be steady and reversible from one operating point to another.

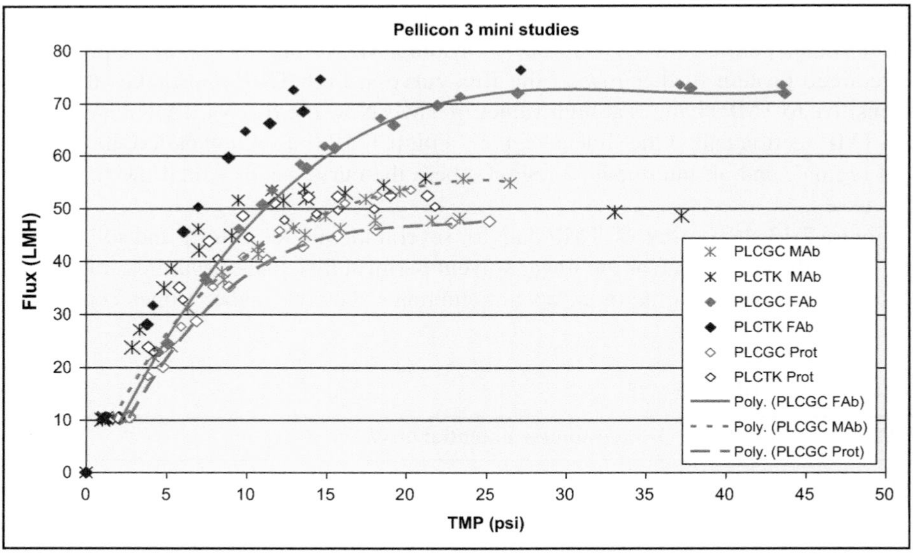

Figure 2.13 Process flux with different membranes (three membrane cutoffs, three proteins). (Courtesy of EMD Millipore Corp.)

2.12 Polarization

At high concentrations, the independence of flux on TMP is counter-intuitive and reflects the fundamental behavior of tangential flow filtration. In a module where there is a pressure drop from feed to retentate depending on cross-flow, one sees higher fluxes with pressure drop but not TMP. In the absence of membrane fouling, fluxes remain relatively constant over time and they are reversible as one moves from one operating point to another. Overall, this behavior is attributed to the presence of reversibly held high solute concentrations near the membrane surface called the polarization layer.

For a simple one-dimensional system at steady-state, a solute mass balance requires that the convective transport of solute to the membrane surface is balanced by the back transport of solute away from the membrane and the convective transport of solute through the membrane.[43] While multiple back transport mechanisms may play a role for different particle sizes,[44] for small nanometer-sized solutes diffusion dominates. We neglect transport along the membrane surface. For fluxes moving in a right-hand direction with the membrane at $x = 0$, the solute mass balance becomes

$$Jc - D\frac{dc}{dx} = Jc_P \text{ or } \frac{Jx}{D} = \ln\left[\frac{c - c_P}{c_W - c_P}\right] \text{or } \frac{Jx}{D} = \ln\left[\frac{c - c_P}{c_W - c_P}\right] \tag{2.28}$$

for $c = c_W$ wall concentration at $x = 0$.

For a completely retentive membrane ($c_P = 0$), a bulk concentration of c_B at the concentration boundary layer thickness $x = -\delta$, and a mass transfer coefficient of $k = D/\delta$, this simplifies to the film model

$$c_w = c_b \exp(J / k) \tag{2.29}$$

where the mass transfer coefficient, $k = k_0 v^n$, v = LMH cross-flow.

This expression suggests that the wall concentration (and polarization) is highest at high fluxes and low cross-flows while it is the lowest at low fluxes and high cross-flows. The amount of mass held up in the polarization layer can be calculated by integrating the concentration over the thickness of the boundary layer using equation (2.28). For a fully retained solute and a membrane area A, the additional mass held up in the boundary layer (above c_b) is

$$M / A = \int_{-\delta}^{0} (c - c_b)\, dx = \int_{-\delta}^{0} \left(c_w e^{Jx/D} - c_b\right) dx = \left[c_b e^{J/k}\right] \frac{\left[1 - e^{-J/k}\right]}{J / D} - c_b\delta = \frac{c_b D}{J} \left\{e^{J/k} - 1 - J / k\right\} \tag{2.30}$$

For values of J = 50 LMH, k = 25 LMH, $D = 4\times10^{-11}$ m^2/s,[45] c_b = 70 g/L, this predicts a mass of 0.9 g/m^2. Experimental measurements of the mass[46,47] are about 1.5 g/m^2.

For a fully retained solute ($\sigma_0 = 1$), equation (2.20) is $J = L_f(\mathrm{TMP} - \Pi_w)$ where Π_w is the osmotic pressure at the membrane wall. Consider this rearranged in terms of $\mathrm{TMP} = J/L_f + \Pi_w$. At low fluxes, the wall concentrations are small ($\Pi_w \ll J/L_f$) so TMP is linear with flux. At high fluxes, the wall concentrations are high ($\Pi_w \gg J/L_f$) so the TMP becomes independent of J and dependent on Π_w.

Empirically, the flux behaviour in the polarized region follows the TMP independent gel model:[44]

$$J = k_0 v^n \ln(c_g / c_b) \tag{2.31}$$

with parameters k_0, n, and the 'gel point' concentration c_g.

Since equation (2.29) follows a similar form, some have interpreted this to mean that the empirical parameter c_g represents a constant, maximum wall concentration corresponding to a protein 'gel' phase on the membrane surface. Solution studies at these concentrations show no phase change. Gel point parameter values are lower than solubility limits and much lower than concentrations implied by packing at the molar volume. See Chapter 5 for further discussion.

Polarization also has an impact on sieving. As described in Section 2.5, the membrane has an intrinsic sieving of $s_i = c_p/c_w$. With polarization raising the concentration at the wall, the observed sieving is $s_o = c_p/c_b$. For a single solute, observed sieving is equal to or less than intrinsic sieving depending on the extent of polarization. Using equation (2.28) with $c = c_b$ at $x = -\delta$, $-(J/k) = \ln[c_b - c_p / c_w - c_p]$. Dividing by c_p and substituting in the sieving definitions, and rearranging yields the sieving relation

$$1/s_o - 1 = (1/s_i - 1)e^{-J/k} \tag{2.32}$$

This can be plotted as $\ln(1/s - 1)$ vs. J to get a slope of $-1/k$ and an intercept of $\ln(1/s_i - 1)$.

For multicomponent systems, large solutes inhibit sieving of small solutes. The solute with the lowest diffusivity will partition itself in the polarization nearest the membrane wall and the solute with the highest diffusivity will be furthest from the wall. The presence of other solutes taking up volume near the wall will reduce concentrations of the target solute. This has impacts on membrane selection and operation as described in Chapter 7.

References

1. Zeman LJ, Andrew LZ. Microfiltration and ultrafiltration, principles and applications. New York, NY: Marcel Dekker; 1996.
2. DiLeo AJ, Allegrezza Jr. AE, Burke ET. Membrane, process, and system for isolating virus from solution. U.S. Patent 5,017,292, 1991.
3. Bird RB, Warren ES, Edwin NL. Transport phenomena. 2nd ed. New York, NY: Wiley & Sons; 2002.

4. Zydney AL, Aimar P, Meireles M, Pimbley JM, Belfort G. Use of the log-normal probability density function to analyze membrane pore size distributions: functional forms and discrepancies. J Membr Sci 1994;91:293.
5. Tkacik G. Personal communication 2003.
6. Tkacik G, Michaels S. A rejection profile test for ultrafiltration membrane & devices, Biotechnology 1991;9:941–46.
7. Erickson HP. Size and shape of protein molecules at the nanometer level determined by sedimentation, gel filtration, and electron microscopy. Biol Proceed Online 2009;11:32–51.
8. Pan S, Nguyen DA, Sridar T, Sunthar P, Prakash JR. Universal solvent quality crossover of the zero shear rate viscosity of semidilute DNA solutions. J Rheol 2014;58:339–68 arxiv:1112.3720v3 2013.
9. Armstrong JK, Wenby RB, Meiselman HJ, Fisher TC. The hydrodynamic radii of macromolecules and their effect on red blood cell aggregation. Biophys J 2004;87(6):4259–70.
10. Latulippe DR, Molek JR, Zydney AL. Importance of biopolymer molecular flexibility in ultrafiltration processes. Ind Eng Chem Res 2009;48(5):2395–403.
11. Xenopolous A. Personal communication 2014.
12. Ferry JD. Ultrafilter membranes and ultrafiltration. Chem Rev 1936;18:373.
13. Renkin EM. Filtration, diffusion, and molecular sieving through porous cellulosic membranes. J Gen Physiol 1955;38:225.
14. Deen WM. Hindered transport of large molecules in liquid-filled pores. AIChE J 1987;33:1409–1425.
15. Dechadilok P, Deen WM. Hindrance factors for diffusion and convection in pores. Ind Eng Chem Res 2006;45:6953–9.
16. Newman JS. Electrochemical systems. Englewood Cliffs, NJ: Prentice-Hall; 1991.
17. Pujar NS, Zydney AL. Pore size distribution effects on electrokinetic phenomena in semipermeable membranes. Ind Eng Chem Res 1974;33:2473.
18. Pujar NS, Zydney AL. Electrostatic effects on protein partitioning in size-exclusion chromatography and membrane ultrafiltration. J Chromatogr A 1998;796(2):229–38.
19. Deen WM, Smith FG III. Hindered diffusion of synthetic polyelectrolytes in charged microporous membranes. J Membr Sci 1982;12(2):217–37.
20. Kedem O, Katchalsky A. Thermodynamic analysis of permeability of biological membranes to non-electrolytes. Biochim Biophys Acta 1958;27:229.
21. Vilker VL, Colton CC, Smith KA. The osmotic pressure of concentrated protein solutions: effect of concentration and pH in saline solutions of bovine serum albumin. J Colloid Interf Sci 1981;79:548–66.
22. Vilker VL, Colton CC, Smith KA, Green DL. The osmotic pressure of concentrated protein and lipoprotein solutions and its significance to ultrafiltration. J Membr Sci 1984;20: 63–77.
23. Jonsson G. Boundary layer phenomena during ultrafiltration of dextran and whey protein solutions. Desalination 1984;51:61–77.
24. Da Costa AR, Fane AG. Pressure drop modelling in spacer-filled channels for ultrafiltration. J Membr Sci 1994;87:79.
25. van Reis R, Goodrich EM, Yson CL, Frautschy LN, Whitely R, Zydney AL. Constant C_{wall} ultrafiltration process control. J Membr Sci 1997;130:123–40.
26. Yousef MA, Datta R, Rogers VGJ. Free-solvent model of osmotic pressure revisited: application to concentrated IgG solution under physiological conditions. J Colloid Interf Sci 1998;197:108–18.
27. Oncley JL, Scatchard G, Brown A. Physical-chemical characteristics of certain of the proteins of normal human plasma. J Phys Chem 1947;51:184–98.

28. Ingerslev P, Andrek Larsen O, Lassen NA. Measurement of the colloid osmotic pressure in serum with Tybjaerg Hansen's osmometer. Scand J Clin Lab Invest 1966;18:431–6.
29. Israelachvilli J. Intermolecular and surface forces. 2nd ed. London, UK: Academic Press; 1991.
30. Hirschfelder JO, Curtiss CF, Byron Bird R. The molecular theory of gases and liquids. New York, NY: Wiley; 1954.
31. Lund M, Jonsson B. A mesoscopic model for protein-protein interactions in solution. Biophys J 2003;85:2940–7.
32. Carnahan NF, Starling KE. Equation of State for Nonattracting Rigid Spheres. J Chem. Phys 1969;51:635–6.
33. Vrij A, Jansen JW, Dhont JKG, Pathamanoharan C, Kops-Werkhoven MM, Fijnaut HM. Light scattering of colloidal dispersions in non-polar solvents at finite concentrations. Silica spheres as model particles for hard-sphere interactions. Faraday Disc Chem Soc 1983;76:19–36.
34. Hall KR. Another hard-sphere equation of state. J Chem Phys 1972;57:2252–4.
35. Russel WB, Saville DA, Schowalter WR. Colloidal dispersions. Cambridge, UK: Cambridge University Press; 1989.
36. Hill TL. An introduction to statistical thermodynamics. Reading, MA: Dover; 1960.
37. Salinas BA, Satish HA, Bishop SM, Harn N, Carpenter JF, Randolph TW. Understanding and modulating opalescence and viscosity in a monoclonal antibody formulation. J Pharm Sci 2009;99:82–93.
38. Liu J, Nguyen MDH, AndyaF J.D., Shire SJ. Reversible self-association increases the viscosity of a concentrated monoclonal antibody in aqueous solution. J Pharm Sci 2005;94:1928–40.
39. Cohn EJ, Edsall JT. Proteins, amino acids and peptides, ACS Monograph No. 90. New York, NY: Reichold; 1943. 428.
40. Tanford C. Physical chemistry of macromolecules. New York, NY: Wiley; 1961.
41. Zimm BH. Application of the methods of molecular distribution to solutions of large molecules. J Chem Phys 1946;14:164.
42. Pedersen KO. Ultracentrifugal studies on serum and serum fractions. Uppsala, Sweden: Almqvist ans Wiksells Boktryckeri AB; 1945.
43. Brian PLT. In: Merten U, editor. Desalination by reverse osmosis. Cambridge, MA: M.I.T. Press; 1966. p. 161–292.
44. Belfort G, Davis RH, Zydney AL. The behavior of suspensions and macromolecular solutions in crossflow microfiltration. J Membr Sci 1994;96:1.
45. Gosting LJ. Measurement and interpretation of diffusion coefficient for proteins. Adv Prot Chem 1956;11:430.
46. Vilker VL, Colton CC, Smith KA. Concentration polarization in protein ultrafiltration. Part I: an optical shadowgraph technique for measuring concentration profiles near a membrane solution interface. AIChE J 1984;27:632–7.
47. McDonough RM, Bauser H, Stroh N, Grauschoph U. Experimental in situ measurement of concentration polarization during ultra- and micro-filtration of bovine serum albumin and dextran blue solutions. J Membr Sci 1993;104:51–63.

Modules

3

Herb Lutz
EMD Millipore, Biomanufacturing Sciences Network

3.1 Stirred cell

Stirred cells use a punched cut flat sheet membrane disk inserted at the bottom of a reservoir vessel (Figure 3.1). An elastic O-ring provides a seal between the retentate reservoir and the permeate stream. One must be careful in practice to ensure this seal. A magnetic stir bar rests on top of the membrane and its rotation provides a tangential flow field on its surface. The vessel is pressurized to provide a transmembrane pressure (TMP) across the membrane disk. Various device sizes are available but working volumes as low as 10 mL are available so not much feedstock is required. Materials of construction permit the use of organic solvents. These modules are sometimes referred to as Amicon stirred cells after the company that first introduced them.

Much of the basic work on membrane characterization and modelling has employed a stirred cell. The flow field has also been well characterized (see Chapter 4). Stirred cells have a uniform mass transfer coefficient across the membrane surface. This uniformity is helpful in removing module effects from testing. One can then focus on membrane properties and uniform polarization effects. In addition, the TMP and cross-flow are independently controlled. However, the magnitude of the tangential flow is very small compared to commercial devices so polarization is very high. As a result, the flux and sieving results from stirred cell testing do not directly scale-up to larger commercial modules. One can however, see if membranes significantly foul in the feed using a stirred cell.

3.2 Hollow fibres and tubes

Hollow fibre modules are tubular devices containing fibre bundles (Figure 3.2). To make them, cast hollow fibre membranes are cut to a prescribed length, potted into end caps and sealed into a hollow fibre module. There is a lumen side (flowing along the inside of the fibres) and a shell side (flowing outside the fibre bundle) to the modules. Glue is commonly used to pot the end caps and provide the seal between the lumen and shell. The tight ultrafiltration (UF) skin can be on the inside, outside or both sides of the fibre.

For most tangential flow filtration (TFF) operations, feed flow is directed through the lumen side with the permeate withdrawn from the shell side (Figure 3.2). The small-diameter lumen promotes better back transport from the membrane surface to

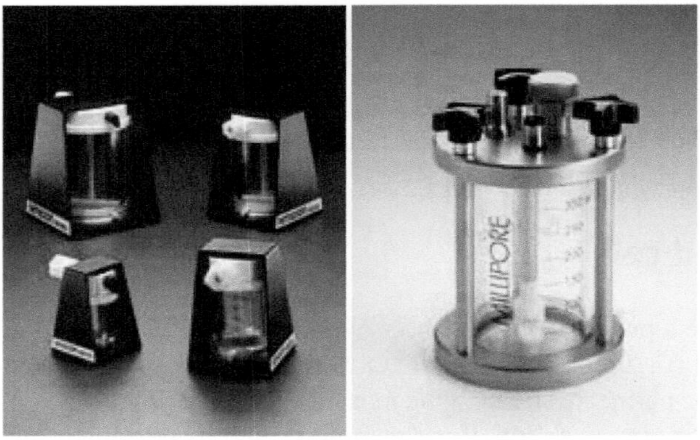

Figure 3.1 Stirred cells.
(Courtesy EMD Millipore Corp.)

reduce solute concentration at the membrane surface and increase fluxes. If feedstock is introduced on the shell side, there is minimal shear at the membrane surface to control polarization. However, there is a larger external surface area and the fibres are more resistant to high external pressures than internal pressures. These characteristics suggest that feed stocks in the polarized region of the flux curve are best run through the lumen and feed stocks in the linear region of the flux curve are best run through the shell.

A feed pump provides the cross-flow and a valve on the retentate provides back pressure. The permeate is generally kept at atmospheric pressure. The TMP and tangential flow rate decreases along the length of the fibres. One uses the module average $\overline{\text{TMP}} = (P_F + P_R)/2 - P_P$ and the module average cross-flow $\bar{Q}_F = (Q_F + Q_R)/2$

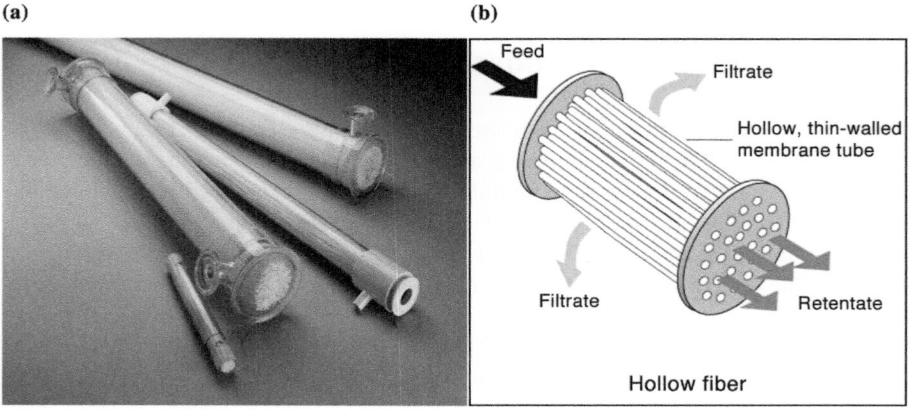

Figure 3.2 Hollow fibre modules: (a) module and (b) flow pattern.
(Courtesy EMD Millipore Corp.)

to characterize its operation. Changing either the feed flow or the retentate valve individually can affect the average cross-flow and the average TMP.

Hollow fibre modules have a convenient plug-and-play design for small-scale testing. Since they require more membrane area and feed flow than flat channel devices (see later section), they are not used as much in large-scale biopharmaceutical manufacturing. Some fibre scale-up is done by extending the fibre length instead of the number of fibres. Accurate scale-up should employ modules with the same fibre lengths and diameters.

3.3 Spirals

Spiral modules are tubular devices containing flat sheet membrane in a 'jelly roll' type configuration[1] (Figure 3.3). To make them, membranes are layered with a porous permeate channel spacer in-between to create a 'membrane sandwich'. The tight UF skin side faces away from the permeate channel. Three edges of the permeate channel are sealed together to create a 'leaf' and the fourth is sealed to a permeate collection tube containing drilled holes to allow flow from the permeate channel into the tube. Feed channel spacers are placed on either side. Additional leaves may be sealed to the collection tube for a multi-leaf spiral.[2] The leaves are then rolled and either inserted into a tubular shell (Figure 3.3a) for subsequent placement in housings or encased in their own housing (Figure 3.3b).

Feed flow is directed into the feed end of the module and flows axially through the feed spacer along the module length. It then exits out the retentate end of the module (Figure 3.3c). As the feed flows along the module length, some permeate flows radially through the membrane under a TMP driving force. It then flows through the permeate spacer channel in a spiral pattern, then into the permeate collection tube and axially out the retentate end of the module. Spiral modules placed in separate housings

(a) (b) (c)

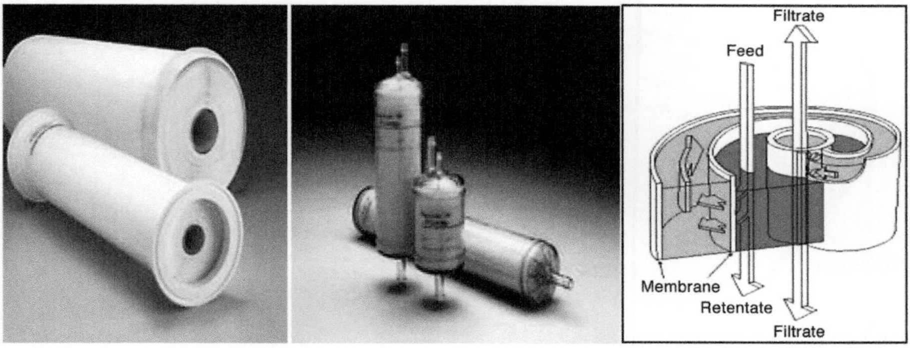

Figure 3.3 Spiral modules: (a) spiral, (b) encased spiral and (c) single-leaf flow pattern. (Courtesy EMD Millipore Corp.)

have an external seal to prevent feed channel bypass and an internal seal to separate the retentate and permeate flows.

Spiral modules are widely used in other industries (e.g., desalination, dairy and water treatment) but infrequently in biomanufacturing. The design has a low feed channel pressure drop and allows modules to be staged in series where the retentate from one module becomes the feed to the next module. This is critical for industries where pumping costs and high-pressure operation is critical. However, the unique permeate channel flow design creates pressure and flow profiles that are difficult to match in a scaled-down device that is critical to biomanufacturing.

3.4 Cassettes

Cassettes are rectangular devices containing flat sheet membranes layered with porous flow channel spacers (Figure 3.4). To make them, membranes are layered with a porous permeate channel spacer in-between to create a 'membrane sandwich'. The tight UF skin side faces away from the permeate channel. The membranes and permeate spacers have pre-cut holes that act as feed, retentate, and permeate flow channels. All

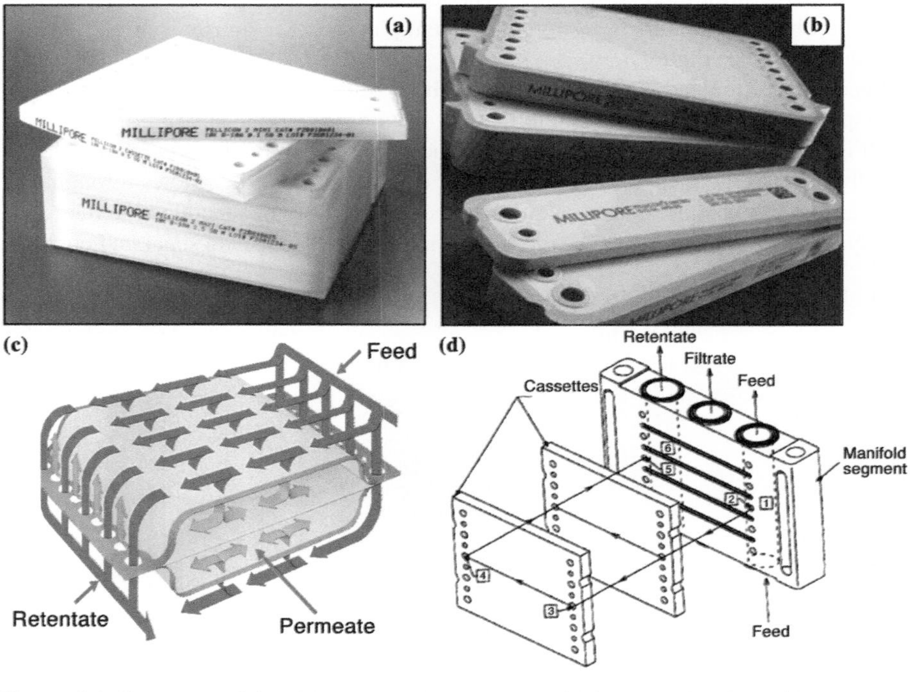

Figure 3.4 Cassette modules: (a) cassettes, (b) cassettes with integrated gaskets, (c) cassette flow pattern and (d) holder flow pattern.
(Courtesy EMD Millipore Corp.)

four edges of the permeate channel are sealed together to create a 'packet'. In addition, the circumference of the feed and retentate holes in the membrane and spacer are sealed to prevent flow from the permeate spacer channel into the hole. Multiple packets are assembled in a 'sandwich' interleaved with feed channel spacer containing pre-cut holes. Different area devices have a different number of packets. All four edges of the feed channel are sealed together and the circumference of the permeate holes in the membrane and spacer are sealed to prevent flow from the feed spacer channel into the hole. Multiple modules are placed into a cassette holder with interleaved sealing gaskets. Some modules may be encased in a housing with built in elastomeric seals to protect the membrane and avoid the need for separate gaskets. Cassette holders allow multiple cassettes to be operated in parallel from a common manifold. A compression seal using metal end-plates prevents bypass flow between the feed/retentate holes and the permeate holes.

Feed flow is directed from the holder into parallel feed holes on one side of the cassette. This flow proceeds from cassette to cassette in the holder along the feed channels created by the holes in the membranes and screens. A portion of this feed flow turns perpendicular and into each spacer-filled cassette feed channel. After traveling through each channel, the flow turns perpendicular and meets up with other retentate flows. This retentate flow proceeds from cassette to cassette in the holder along the retentate channels created by the holes in the membranes and screens. The retentate flow then enters the holder and proceeds out. Permeate flows through the membrane walls in each spacer-filled cassette feed channel and through the spacer-filled permeate channel. After traveling through each permeate channel, the flow turns perpendicular and meets up with other permeate flows. This permeate flow proceeds from cassette to cassette in the holder along the permeate channels created by the holes in the membranes and screens. The permeate flow then enters the holder and proceeds out.

Cassettes are widely used in biomanufacturing. The design allows for high-pressure drop feed channel spacers that increase permeate flux at low feed flow rates. This permits the use of compact systems with low membrane areas and small feed pumps to process highly valuable pharmaceutical products with minimal pump pass degradation and low holdup yield losses. In addition, the parallel channel cassette design permits the use of accurate scaled-down devices critical to biomanufacturing.[3]

3.5 Module characteristics

Commercial UF module characteristics are summarized in Table 3.1.

For biological manufacturing, cassettes are the most widely used module format with some fibre usage at smaller scale and for plugging feedstocks. Stirred cells are used for research and membrane characterization. Their poor mass transfer characteristics mean that their uses are limited to a lab.

Hollow fibre modules are a convenient plug-and-play format for small-scale operation. They do not utilize turbulence promoters in the feed channel and show poorer mass transport than spirals or cassettes. This translates to larger area

Table 3.1 **Commercial module comparison**

Module	Spiral	Fibre	Cassette
Description			
Feed channel	Spacer	Open	Tight screen
Typical feed flux LMH	700–5,000	500–18,000	200–400
Area m^2/module	0.1–35	0.001–5	0.05–2.5
ΔP psid/module	6–15	1–6	10–50
Performance			
Feed channel plugging sensitivity	Moderate	Low-mod	High
Working volume L/m^2	1.0	0.5	0.4
Holdup volume L/m^2	0.03	0.03	0.02
Ruggedness	Moderate	Low-mod	High
Ease-of-use	Moderate	High	Moderate
Scalability	Fair	Moderate	Good
Mass transfer coefficient	Moderate	Low	High
Cost \$/m^2	40–200	200–900	500–1,000
Vendors	EMD Millipore	EMD Millipore, GE, Spectrum	EMD Millipore, Pall, GE, Sartorius, Novasep
Key advantages		Ease-of-use, cost	High mass transfer, low pumping, scalability
Key disadvantages	Scalability, low mass transfer, high pumping	Low mass transfer, high pumping	Cost
Uses	Dairy, desalination	Bench scale preps.	Biotech. mfg.

requirements for commercial processing. In addition, fibres require higher feed flows to generate reasonable fluxes. At larger scales, fibre systems are limited by pumps available. Fibres with relatively open pores are common for vaccine production in small batches.

Spirals have been the preferred format for water and dairy industry application due to their lower feed channel pressure drops and reduced pumping energy requirements. Spirals use a relatively open screen or feed channel spacer that is not very efficient in promoting mass transfer. This means more area is required for commercial operation compared to cassettes with efficient turbulence promoting screens. It also means that less feed flow is needed to obtain good flux performance. As a result, cassette systems are more compact with less holdup volume and a smaller footprint to fit into expensive, environmentally controlled, manufacturing space. Higher yields are valuable for products valued at thousands of dollars per gram. Reduced bioproduct pumping also reduces potential product damage. These key attributes contribute to making cassettes the preferred format for most applications.

3.5.1 *Polarization control*

As described in Section 2.12, retained solutes accumulate in a reversibly held layer near the membrane surface. Maintaining a uniform state of polarization along the length of the feed channel is critical to some applications. For example, poor polarization can lead to irreversible cell and colloid buildup on the membrane surface and flux crashing for clarification applications. Polarization control is also critical for fractionating applications where closely sized solutes are separated between the retentate and permeate streams.

Polarization is most directly affected by the feed flow and the permeate flow. The permeate flow controls the rate at which solutes are brought to the membrane surface while the feed flow controls the rate of back transport. One could control these directly through the use of a two-pump system with a feed pump and a permeate pump. This approach is preferred for the polarization sensitive applications described earlier.

Since the permeate flow can be influenced by the TMP, it is common in most cases to manipulate feed flow and TMP to set polarization conditions. The TMP is set using a retentate valve.

3.5.2 *Series/parallel operation*

Modules are employed in series and in parallel configurations. Series operation creates a longer feed channel by directing the retentate flow from the first module into the feed flow of the next module. Longer feed channels enable a higher residence time within the module and more of the feed flow is converted into permeate flow. Properties vary along the length of the feed channel. Fibre modules require additional pumping so they need to be configured in series for scaling up.

Parallel operation keeps the same feed channel length while increasing the area. This linear scaling provides assurance of consistent performance between small- and large-scale operation. This is critical for the polarization-sensitive applications. Cassette assemblies of six high stacked holders containing 120 m^2 of area are in commercial operation. See Chapter 10 for more on manufacturing systems.

3.6 Module shear

Section 7.6 covers aggregation and denaturing during processing. Some studies have suggested that module shear may be a cause of protein damage. Fluid flowing past a solid surface, such as a membrane, has a zero tangent velocity at the surface. This creates a velocity gradient (or shear rate) perpendicular to the surface. Assume as a first-order approximation that the parabolic flow is quickly established and that flow through the membrane has no significant affect on the velocity distribution. Table 3.2 shows the velocity distributions for flow through a hollow fibre tube and through an open channel plate and frame channel. The shear rate is the velocity gradient perpendicular to the membrane surface as s^{-1}. Shear rates are constant along the channel length and maximum at the wall. The shear stress has units of pressure.

Table 3.2 Open channel flow[4]

Device	Fibre radius $r = 0$ to R	Plate and frame (slit flow) half slit height $x = 0$ to H
Tangent velocity	$v_z = \frac{\Delta PR^2}{4\mu L}\left[1-\left(\frac{r}{R}\right)^2\right]$	$v_z = \frac{\Delta PH^2}{2\mu L}\left[1-\left(\frac{x}{H}\right)^2\right]$
Shear rate	$\gamma = \frac{\Delta \mathrm{Pr}}{2\mu L} = \frac{4\langle v\rangle r}{R^2}$	$\gamma = \frac{\Delta Px}{\mu L} = \frac{3\langle v\rangle x}{H^2}$
Wall shear rate	$\gamma_w = \frac{\Delta PR}{2\mu L} = \frac{4\langle v\rangle}{R}$	$\gamma_w = \frac{\Delta PH}{\mu L} = \frac{3\langle v\rangle}{H}$
Wall shear stress	$\tau = -\mu\gamma = \frac{\Delta PR}{2L}$	$\tau = -\mu\gamma = \frac{\Delta PH}{L}$

The formulas for the wall shear rate in Table 3.2 imply that at a given feed channel pressure drop the shear rate and comparable feed channel dimensions, the shear rate in a fibre is less than that in a plate and frame module. However, fibres require higher feed flows and have lower conversion. This means that they have more pump passes than the plate and frame modules and can have more pump damage.

A semi-rigid globular polymer immersed in the fluid will experience slower fluid velocities on the side facing the solid surface and higher fluid velocities away from the surface. This will induce a mechanical torque on the polymer macromolecule causing it to (1) deform its conformation by stretching in the direction of flow as component chains experience shear, and (2) rotate or tumble as a polymer ensemble clockwise towards the solid surface.[5,6] Randomly coiling globular 1300 kDa polyisobutylene was observed to stretch by 40% under 600 s^{-1} shear rates.[7]

Protein three-dimensional structures have stabilizing intermolecular forces[8] (i.e., van der Waals, ionic and hydrogen bonding) that are expected to resist deformation. However, any changes that do occur can lead to a denatured protein with associated loss of biological activity.[9] Published studies on protein sensitivity to shear are varied. Some studies show a loss of enzymatic activity above a threshold value of 10^3 s^{-1} shear rates[10] over 30 min. These studies correlated activity loss to the product of shear rate and residence time averaged over the flow area perpendicular to the solid surface:

$$\langle\gamma\theta\rangle = \frac{\int_{\text{area}} \gamma\theta v_z\, dA}{\int_{\text{area}} v_z\, dA} \text{ or } \frac{8L}{3R} \text{ for tube flow} \tag{3.1}$$

Activity losses were seen above a threshold value of $\langle\gamma\theta\rangle > 10^5$. This is consistent with a denaturation reaction path limited by the step: protein + shear -> denatured intermediate.

Other studies show no loss in enzymatic activity at 10^5–10^6 s^{-1} shear rates[11] with some of the same enzymes, or no protein changes (by SEC, CD, scanning microcalorimetry, SDS-PAGE) at 10^5 s^{-1} shear rates[12,13] over hours of operation. Activity losses may increase or decrease with higher temperatures of 60 °C, and addition of glycerol to increase viscosity and shear stresses did not induce shear damage.[7] Differences in shear sensitivity results have been attributed to (1) air bubbles that can break up and increase their area under shear, shear increasing the turnover of protein molecules changing conformation at the air interface,[14] and (2) the presence of catalytic metal ions that can get greater access to shear-stressed proteins.[15]

If one replaces the cassette holder with a valve to generate a comparable pressure drop, aggregation tends to still be present. This implies that the aggregation is due to the pumping system alone. Gas–liquid interfaces can occur within piping systems as the result of gas effervescing out of solution. These can be visualized by replacing sections of pumps and piping with clear materials and using a strobe to 'freeze' the effect in place.[16] One can see white foam composed of many tiny bubbles that are localized in certain regions of the piping system. These can occur when there is localized low pressure as a result of hydrodynamics, and nucleation sites for formation and growth. As the fluid moves to regions of higher pressure, the bubbles collapse. Diafiltrate buffer stored under pressure to minimize bioburden has a higher concentration of dissolved gas and can show enhanced aggregation. The addition of nonionic surfactants, such as Tween or Triton, can minimize aggregation by migrating to the gas–liquid interface and blocking protein contact with the interface.

For screened channel TFF devices, such as cassettes or spirals, the screens generate complex flow patterns with velocity profiles that vary along the screened feed channel. 2D computer simulations show stable circulation patterns (no vortex shedding), and flow reversal as a circulating eddy behind the screen.[17] The associated shear stress τ ($= -\mu\gamma$) varies from 0 to 3.3 N/m^2 with corresponding shear rates of 0 to 3300 s^{-1}. Shear is zero where the screen fibres touch the wall and at the stagnation point behind a recirculating eddy. The overall maximum shear occurs opposite a screen fibre while local shear maxima occur due to eddies near the screen fibres.

Consider wall shear calculated using the open channel formula with channel height given by the screen thickness and velocity modified to account for the presence of solid spacers. Screen shear is calculated from the fluid mechanics solution of a fluid flowing past a cylinder-oriented perpendicular to flow:[18]

$$\overline{\gamma}_{\text{screen}} = 0.375\sqrt{\frac{\langle v\rangle \sin^3(\theta/2)}{\upsilon r_{\text{f}}}} = 2.23\sqrt{\langle v\rangle / r_{\text{f}}} \tag{3.2}$$

for $\theta = 90°$ angle between crossing screen fibres, water $\upsilon = \mu/\rho = 0.01$ cm^2/s, and screen fibre radius r_f.

Noting that $\langle v\rangle = J_F / 36{,}000a\varepsilon$ for average feed channel velocity $\langle v\rangle$ in cm/s, feed flux J_F in LMH, and feed channel area to membrane area a (dimensionless), expressions for wall and screen shear can be obtained for Pellicon cassettes with A and C screens as shown in Table 3.3. These calculations indicate that wall shear dominates screen shear for the simplified model. The 10^3 s^{-1} threshold shear value observed by some investigators occurs at feed flux rates of 310 LMH for the A screen and 530 LMH for the C screen.

Table 3.3 **Shear in screened channels**

Property		A screen	C screen
Fibre radius	r_f	0.011 cm	0.014 cm
Porosity	ε	0.63 (est.)	0.63
Half-height	H	0.021 cm	0.026 cm
Feed channel area/ membrane area	a	0.0014	0.0018
Average wall shear	$\bar{\gamma}_{\text{wall}}$	$160\langle v\rangle = 3.2J_F\ \text{s}^{-1}$	$120\langle v\rangle = 1.9J_F\ \text{s}^{-1}$
Average screen shear	$\bar{\gamma}_{\text{screen}}$	$28\sqrt{\langle v\rangle} = 3.9\sqrt{J_F}\text{s}^{-1}$	$24\sqrt{\langle v\rangle} = 3.0\sqrt{J_F}\text{s}^{-1}$

Consider a slit channel run comparably to the computer-simulated screened channel with wall shear rates from 0 to 3300 s^{-1}. For the channel with $\text{Re} = 171 (= \langle v\rangle d_h / \upsilon)$, similar screen dimensions as used in the simulation, υ = 0.01 cm^2/s, a C screen channel with H = 0.026 cm, and a screen channel hydraulic diameter of $d_h = 4H\{\varepsilon / [1 + 2(1-\varepsilon)H / r_f]\} = 0.033\text{cm}$ (= 4* cross-section available for flow/wetted channel perimeter), the velocity $\langle v\rangle$ must be 52 cm/s. For a slit channel with the same H and Re, $d_h = 4H = 0.092$ cm, $\langle v\rangle = 19\,\text{cm/s}$ and $\gamma = 2500\ \text{s}^{-1}$. If similar ratios apply to other Reynold's numbers and screen geometries, one can estimate wall shear rates in the screened channel as 0 to 1.3 times the shear rate in the slit channel. Staying with a maximum shear rate, this implies that the $10^3\ \text{s}^{-1}$ threshold shear value observed by some investigators occurs at feed flux rates of 240 LMH for the A screen and 410 LMH for the C screen (Figure 3.5).

For slit flow, the product of shear rate and residence time averaged over the flow area perpendicular to the solid surface is $\langle\gamma\theta\rangle_{1\text{pass}} = 9L/8H$. This term

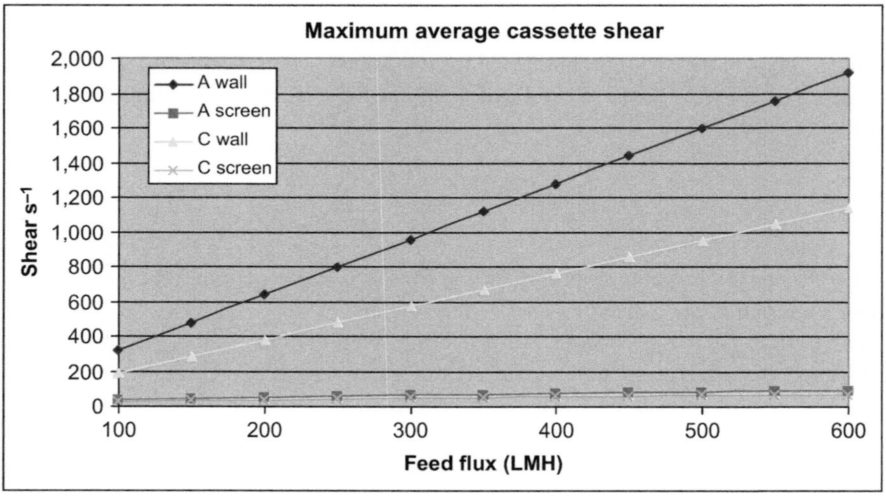

Figure 3.5 Maximum average cassette shear.

is 980 for the A screen and 730 for the C screen, significantly below a 10^5 threshold value. These values are for a single pass of protein through the cassette feed channels. The average number of pump passes during processing is $N_p = V_{pumped}/V_{Feed} = J_F At/V_{Feed} = J_F V_{Permeate} Ft/V_{Feed} Jt = [1+(N-1/X)](F/Y)$, where J is the average flux, Y is the average conversion (J/J_F) $\sim$ 0.10, F is the design safety factor $\sim$1.2, N is the number of diavolumes $\sim$ 12 and X is the initial volume reduction factor $\sim$ 30. This makes $N_p \sim 28$ and $\langle\gamma\theta\rangle_{process} = \langle\gamma\theta\rangle_{1pass} N_p = 28{,}000$ for the A screen and 20,000 for the C screen. These values are both below a 10^5 threshold value.

For the screened channel, examination of the simulated velocity profiles indicates the possibility of partitioning the flow into one zone with a narrower height $H' = H - r_f$ and a longer flow path $L' = \sqrt{H^2 + L^2}$. The other zone contains stable recirculation flow patterns both in front of and behind the screen fibres. Transport by diffusion across streamlines is considered negligible as in the slit flow case. Assuming a parabolic velocity distribution in zone one, the shear residence time term becomes $\langle\gamma\theta\rangle_{screen\,1\,pass} = 9L'/8H' = \langle\gamma\theta\rangle_{slit\,1\,pass}\,\psi$, where $\psi = \dfrac{\sqrt{(H/L)^2+1}}{1-r_f/H}$. Evaluation for a C screen gives $\psi = 2.4$, $\langle\gamma\theta\rangle_{screen\,1\,pass} = 1800$ and $\langle\gamma\theta\rangle_{screen\,process} = 49{,}000$. This remains below any 10^5 threshold value.

3.7 Comparison of flat sheet and tube modules

Chapter 4 explores the performance of different module types. Table 3.4 summarizes the performance relations for flat sheet and tube modules.

Table 3.4 Module performance summary

	Flat sheet	Tube
Cross-sectional area	$A_{flow} = n2hW$	$A_{flow} = n\pi R^2$
Membrane area	$A_{memb} = 2nWL$	$A_{memb} = 2\pi nRL$
Channel hydraulics	$\bar{v} = \dfrac{\Delta Ph^2}{3\mu L}$	$\bar{v} = \dfrac{\Delta PR^2}{4\mu L}$
Device hydraulics	$Q_F = \bar{v}A_{flow} = \dfrac{2n\Delta Ph^3 W}{3\mu L}$	$Q_F = \bar{v}A_{flow} = \dfrac{n\pi\Delta PR^4}{4\mu L}$
Mass transfer coefficient	$k = 1.86\left(\dfrac{\bar{v}D^2}{2h}\right)^{1/3}$	$k = 1.62\left(\dfrac{\bar{v}D^2}{2R}\right)^{1/3}$
Permeate flow	$Q_p = 3.72nWL = \left(\dfrac{\Delta PhD^2}{6\mu L}\right)^{1/3} \ln\left(\dfrac{C_g}{\bar{C}}\right)$	$Q_p = 1.62\pi nRL = \left(\dfrac{\Delta PRD^2}{\mu L}\right)^{1/3} \ln\left(\dfrac{C_g}{\bar{C}}\right)$

Table 3.5 **Module performance ratios (sheet/tube)**

Membrane areas	$A_{membs} / A_{membt} = 0.25(h / R)^{1/3}$
Feed flows	$Q_{Fs} / Q_{Ft} = 0.67(h / R)^{8/3}$

Consider a comparison of these modules using a common feed with properties (D, C_g, μ), the same (maximum) feed channel pressure drop ΔP, and the same feed channel length L. It is convenient to consider a common permeate flow between modules. On this basis, Table 3.5 shows the performance ratios of these modules. This comparison indicates that for common channel dimensions, it takes about four times more tube module area and 50% more feed flow to perform the separation compared to a flat sheet module. This implies larger systems with more membrane cost, more holdup and minimum working volume constraints, and more yield losses for tubes. It also implies more pump passes and potential protein degradation for tubes.

3.8 Other modules

Rigid tube modules have membrane cast on the inner walls of 1/2″–1″ paper or plastic porous tubes. Feed is pumped at high velocities through these tubes to create turbulent flow. Large pumps are required and holdup volumes can be significant. These tubes are used in the automobile paint industry and some dairy applications.

A variation on the hollow fibre module has been proposed where the fibres are wound in a helix configuration. This causes the flow to rotate in order to conserve angular momentum. The resulting fluid rotation within the fibre lumen is referred to as a Dean vortex.[19] These modules provide greater wall shear and flux while producing a larger pressure drop. They have not been commercialized.

Rotating cylinder modules employ a rotating cylinder placed inside of a stationary hollow cylinder with flat sheet membranes along both the inside and outside walls of the annulus. Rotation creates a secondary flow pattern in the annular space between them called Taylor vortices.[20] The resulting fluid instabilities control concentration polarization and generate high fluxes.[21,22] However, these systems do not scale up well. As the concentric cylinders are made larger in length or diameter, there is more mechanical stress on them and they start to distort under operation. They have been employed with microfiltration membranes at small scale for blood cell separation and washing.

Vibrating modules, such as the VSEP[23] (Vibratory Shear Enhanced Process), use flat sheet membranes mounted on a vibrating base. The oscillations independently control polarization and can generate high fluxes and reach very high solid concentrations. However, these systems also do not scale up well. They are used for concentrating sludges in the pulp and paper, chemical processing, oil and gas, and wastewater treatment industries.

Rotating disc modules[24] employ alternating rotating and static disks with membranes on their surfaces.[25] These can develop high shear separate from the TMP driving force and work with thick suspensions.

References

1. Gulf General Atomic, Inc. assignee, Bray DT. U.S. Patent 3,367,505, 1968.
2. Gulf General Atomic, Inc. assignee, Westmoreland JC. U.S. Patent 3,367,504, 1968.
3. van Reis R, Goodrich EM, Yson CL, Frautschy LN, Dzengeleski S, Lutz H. Linear scale ultrafiltration. Biotechnol Bioeng 1997;55:123–40.
4. Bird RB, Stewart WE, Lightfoot EN. Transport phenomena. 2nd ed. New York, NY: John Wiley; 2002.
5. Smith DE, Babcock HP, Chu S. Single-polymer dynamics in steady shear flow. Science 1999;283:1724–7.
6. Keller A, Odell JA. The extensibility of macromolecules in solution; a new focus for macromolecular science. Colloid Polym Sci 1985;263:181–201.
7. Cottrell FR, Merrill EW, Smith KA. Conformation of polyisobutylene in dilute solutions subject to a hydrodynamic shear field. J Polym Sci 1969;A-2(7):1415.
8. Creighton TE. Proteins. New York, NY: W.H. Freeman & Co.; 1984.
9. Charm SE, Wong BL. Shear effects on enzymes. Biotechnol Bioeng 1970;12:1103.
10. Thomas CR, Dunnill P. Action of shear on enzymes: studies with catalase and urease. Biotechnol Bioeng 1971;21:2279–302.
11. Maa YF, Hsu CC. Effect of high shear on proteins. Biotechnol Bioeng 1996;51:458–65.
12. Maa YF, Hsu CC. Protein denature by combined effect of shear and air-liquid interface. Biotechnol Bioeng 1997;54:503–12.
13. Virkar PD, Narendranathan TJ, Hoare M, Dunnill P. Studies on the effect of shear on globular proteins: extensions to high shear fields and to pumps. Biotechnol Bioeng 1981;23:425–9.
14. Matthew T, Middleman S. Shear deformation effects in enzyme catalysis: preliminary experimental results on lactic dehydrogenase. Biotechnol Bioeng 1978;20:605–10.
15. Matthew T, Middleman S. Shear deformation effects in enzyme catalysis: metal ion effect in the shear inactivation of urease. J Biophys Soc 1978;23:121–8.
16. Wolk BM. Personal communication 1997.
17. Schwinge J, Wiley DE, Fletcher DF. Simulation of the flow around spacer elements in a small channel. 1. Hydrodynamics. Ind Eng Chem Res 2002;41:2977–87.
18. Da Costa A, Fane AG, Wiley, DE. Spacer characterization and pressure drop modeling in spacer-filled channels for ultrafiltration. J Membr Sci 1994;87:79–98.
19. Dean WR. The streamline motion of fluid in a curved pipe. Phil Mag 1928;5(7):673–95.
20. Schlichting H. Boundary-layer theory. New York, NY: McGraw-Hill; 1955. pp. 154–162.
21. Hallstrom B, Lopez-Leiva M. Description of a rotating ultrafiltration module. Desalination 1978;24:339.
22. Holeschovsky UB, Cooney CL. Quantitative description of ultrafiltration in a rotating filtration device. AIChE J 1991;37:1219.
23. Culkin B. U.S. Patent 4,872,899, issued Oct. 10, 1989.
24. SpinTek.com.
25. Serra CA, Wiesner MR. A comparison of rotating and stationary membrane disk filters using computational fluid dynamics. J Membr Sci 2000;165:19–29.

Module performance

4

Herb Lutz
EMD Millipore, Biomanufacturing Sciences Network

4.1 Solute transport mechanisms

Module flux is often limited by the accumulation of solutes on the membrane surface. Solute transport from the bulk flow channel to the wall (membrane surface) is governed by convective flow that drags the solute along with it. The convective solute flux vector $\vec{J}$ (g/cm²/s) is described by $\vec{J} = c\vec{v}$, for solute concentration c (g/cm³) and the velocity vector $\vec{v}$ (cm/s).

Several mechanisms have been identified that govern solute back transport away from the wall: Brownian diffusion, shear-induced diffusion, and inertial lift. Transport by Brownian diffusion is the result of solute–solvent collisions that lead to random motion, some in the direction away from the membrane. The Brownian diffusion solute flux vector $\vec{J}$ (g/cm²/s) is described by $\vec{J} = -D\vec{\nabla}c$, for Brownian diffusivity D (cm²/s), solute concentration c (g/cm³) and the gradient vector operator $\vec{\nabla}$ (1/cm).

Transport by shear-induced diffusion[1] is the result of solute–solute collisions that lead to random motion, some in the direction away from the wall. The diffusive solute flux vector $\vec{J}$ (g/cm²/s) is described by $\vec{J} = -D_s\vec{\nabla}c$, for shear-induced diffusivity D_s (cm²/s). The shear-induced diffusivity (cm²/s) is described by $D_s = r_s^2\gamma_w f(\phi)$ for solute radius r_s (cm), wall shear rate $\gamma_w = (\partial v_x / \partial y)|_{wall}$ (s^{-1}) for v_x tangent to the wall and y perpendicular to the wall, and dimensionless function f of the particle volume fraction ϕ.

Transport by inertial lift[2] is the result of a non-uniform flow field around the solute particle that creates a net force on the solute. Fluid moves slower as it nears a surface. This leads to higher velocities on the side of the particle away from the surface. Higher velocities lead to lower pressures. The net result is that the particle sees a pressure difference pushing it away from the wall, similar to the lift on an airplane wing. The velocity difference also induces a torque on the particle causing it to rotate. The resulting inert lift solute flux vector $\vec{J}$ (g/cm²/s) is described by $\vec{J} = \psi r_s^3 \gamma_w^2 / \upsilon$ for constant ψ.

The relative magnitude of these back transport mechanisms can be compared. Consider a steady-state solvent flux that just balances the convective solute flux toward the wall for each mechanism. Calculations[3] were made for a suspension of uniform spherical particles with density of 1 g/cm³ and bulk concentration of 1% volume flowing in room temperature water with viscosity of 1 cP along a porous channel with length 0.3 m and a wall shear of 2500 s^{-1}. It is also assumed that the maximum particle packing density at the wall is 60% corresponding to a solid of close packed spheres. As shown in Figure 4.1, Brownian diffusion dominates (with highest flux) for small particles (<0.1 μm), shear-induced diffusion dominates for medium-sized particles (1–10 mm) and inertial lift dominates for large particles (>100 μm). Proteins, nucleic acids, viruses, and other solutes typical of ultrafiltration processing have diameters on

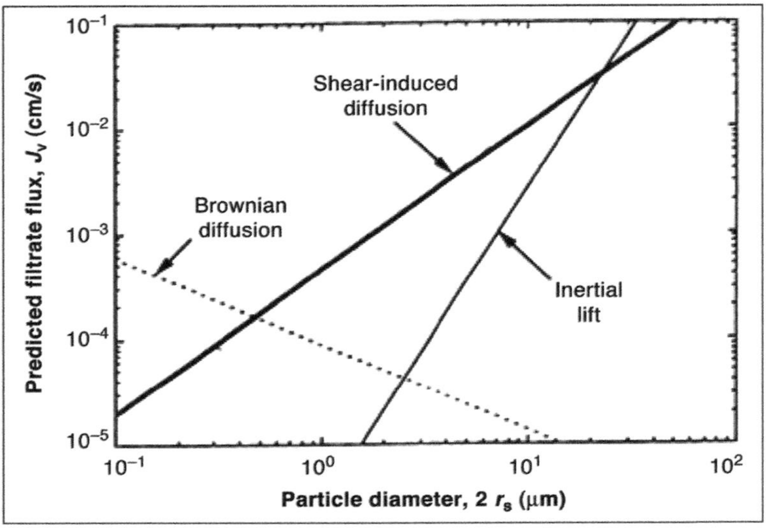

Figure 4.1 Solute back-transport magnitudes.[3]
(Courtesy of Journal of Membrane Science.)

the order of nanometers (Figure 1.1). This means back-transport from an ultrafiltration membrane is governed primarily by Brownian diffusion.

Back transport mechanisms limit solute build up at the wall. Solutes can also move tangential to the wall (along the membrane surface) by fluid convection, Brownian diffusion, or motion at the membrane surface. Convective flow in the fluid is accommodated in the mass balance as $J = cv$ where the velocity is taken tangent to the membrane surface. Brownian diffusion tangent to the surface is generally neglected compared to other mechanisms. Motion at the wall has been modelled as spheres rolling along an irregular surface.[4]

4.2 Velocities and friction factors

Fluid velocity $\vec{v}$ is a vector quantity with magnitude and direction in three dimensions. The velocity vector must satisfy the steady-state mass balance[4,5] (continuity equation) $\vec{\nabla} \cdot \vec{v} = 0$ for an incompressible fluid. The velocity vector must also satisfy steady-state momentum balances[5,6] in each of three dimensions: $\vec{\nabla} \cdot \rho\vec{v}\vec{v} + \vec{\nabla} p + \vec{\nabla} \cdot \mu\vec{v} = \vec{0}$ where body fields (e.g., gravity and electromagnetic) are neglected, ρ is the g/cm^3 mass density, $\vec{v}\vec{v}$ is a dyadic product, p is the scalar pressure field and μ is the g/cm/s (poise) viscosity. Boundary conditions are $\vec{v} = \vec{0}$ at an impermeable wall, the velocity component normal to the wall is v_w for a permeable membrane surface and $\vec{v} = \vec{v}_b$ or bulk values far away from the wall in free solution.

Solving the set of four coupled non-linear partial differential equations (total mass, momentum in each of three dimensions) with boundary conditions is complex. Some insight into the form of the velocity profiles can be obtained using dimensionless

Table 4.1 Module friction factors

Module	Reynolds number	Friction factor f Re
Stirred cell	$\omega R^2/\upsilon$	$\frac{\pi}{2}\rho R^5\omega^2 G'(0)\mathrm{Re}^{-1/2}$
Slit	$\bar{v}_{x0}h/\upsilon$	$24\left[1-\frac{27\mathrm{Re}_w}{35}\right]\left[1-\frac{\mathrm{Re}_w x}{\mathrm{Re}\, h}\right]$
Tube	$\bar{v}_{z0}2R/\upsilon$	$16\left[1-\frac{2\mathrm{Re}_w z}{\mathrm{Re}\, R}\right]$
Cassette	$\bar{v}_{x0}d_h/\upsilon$*	Constant

* See Section 4.13 for a more complete description of d_h.

variables. Dividing each linear dimension with a single constant characteristic length L_{char}, each velocity with a single constant characteristic velocity V_{char}, and each parameter by its bulk value far from the membrane allows one to rewrite the equations in terms of corresponding dimensionless variables designed with a ~ as $\tilde{\vec{\nabla}} \cdot \tilde{\vec{v}} = 0$ and $\mathrm{Re}\tilde{\vec{\nabla}} \cdot \tilde{\rho}\tilde{\vec{v}}\tilde{\vec{v}} + \tilde{\vec{\nabla}}\tilde{p} + \tilde{\vec{\nabla}} \cdot \tilde{\mu}\tilde{\vec{v}} = \vec{0}$ for a characteristic viscous pressure $p_{char} = \rho V^2_{char}/2$, and dimensionless Reynolds number $\mathrm{Re} = V_{char}L_{char}/\upsilon$ where the fluid parameters are evaluated at the bulk conditions. This implies that the dimensionless velocities can be expressed by the function $\tilde{\vec{v}} = f\left(\tilde{\vec{r}}, \mathrm{Re}, \tilde{\rho}, \tilde{\mu}, \tilde{v}_w\right)$. Here it is presumed that concentration gradients do not have a significant impact on the fluid properties $\tilde{\rho}, \tilde{\mu}$ and that in turn this does not have a significant impact on the velocity profiles.

For a stationary wall, the fluid force on the wall must balance the pressure drop as $P_0 - P_L = F/A = \tau_w/A = -(\mu_w/A)(\partial\bar{v}_x/\partial y)\big|_{wall}$ where F is the total force g/cm/s^2 integrated over the wall area A, fluid shear at the wall τ_w g/cm/s^2, fluid viscosity at the wall μ_w g/cm/s, fluid velocity tangential to the surface v_x cm/s and direction perpendicular to the surface y cm. This can be evaluated from the steady-state fluid velocity profile. The dimensionless average shear by the Fanning friction factor[5,6] is $f = [\mu_w/(\rho_b\bar{v}^2/2)](\partial\bar{v}_x/\partial y)\big|_{wall} = f(\mathrm{Re})$. (The Moody friction factor $f_m = 4f$ Fanning friction factor). This can be written in terms of pressure drop using the relationship between pressure drop and shear described earlier $f = \Delta PA/(\rho_b\bar{v}^2/2)$. This form allows for calculation from experimental data. One empirical form widely used for correlation of non-porous transport with constant properties is $f = a\mathrm{Re}^b\left(L_{char}/L\right)^e$ where L is the length of the feed channel. A summary of friction factors for different devices is shown in Table 4.1. Increasing suction causes the friction factor to increase.

4.3 Concentrations and mass transfer coefficients

A steady-state solute concentration profile is described by a component mass balance[5,6] with no chemical reaction: $\vec{\nabla} \cdot c\vec{v} + \vec{\nabla} \cdot D\vec{\nabla}c = 0$. In this partial differential equation, Brownian diffusion is the only back-transport mechanism for small solutes, c is the g/cm^3 concentration, D is the cm^2/s Brownian diffusivity, $\vec{v}$ is the cm/s velocity vector field and $\vec{\nabla}$ is the cm^{-1} gradient vector operator. Boundary conditions are $c = c_w$ at the wall and $c = c_b$ far away from the wall in free solution. This equation must be solved with the velocity equations.

Solving the set of five coupled non-linear partial differential equations (solute concentration, total mass, momentum in each of three dimensions) with boundary conditions is complex. As earlier, dividing each linear dimension with a single constant characteristic length L_{char}, each velocity with a single constant characteristic velocity V_{char}, each parameter by its bulk value far from the membrane and the concentration by the difference $c_w - c_b$ allows one to rewrite the equations as in terms of dimensionless variables designated with a ~ on top $\tilde{\vec{\nabla}} \cdot \tilde{\vec{v}} = 0$, $\mathrm{Re}\,\tilde{\vec{\nabla}} \cdot \tilde{\rho}\tilde{\vec{v}}\tilde{\vec{v}} + \tilde{\vec{\nabla}}\tilde{p} + \tilde{\vec{\nabla}} \cdot \tilde{\mu}\tilde{\vec{v}} = \vec{0}$ and $\mathrm{Re}\,\mathrm{Sc}\,\tilde{\vec{\nabla}} \cdot \tilde{c}\tilde{\vec{v}} + \tilde{\vec{\nabla}} \cdot \tilde{D}\tilde{\vec{\nabla}}\tilde{c} = 0$, for a characteristic viscous pressure $p_{char} = \mu\, V_{char}/L_{char}$, dimensionless Reynolds number $\mathrm{Re} = V_{char} L_{char} / \upsilon$, and dimensionless Schmidt number $\mathrm{Sc} = \upsilon / D$ where the fluid parameters are evaluated at the bulk conditions. This implies that the dimensionless velocities can be expressed by the function $\tilde{\vec{v}} = f\left(\tilde{\vec{r}}, \mathrm{Re}, \tilde{\rho}, \tilde{\mu}, \tilde{v}_w\right)$, and the dimensionless concentration by the function $\tilde{c} = f\left(\tilde{\vec{r}}, \mathrm{Re}, \mathrm{Sc}, \tilde{\rho}, \tilde{\mu}, \tilde{D}, \tilde{v}_w\right)$. If concentration gradients significantly distort the velocity profiles, these will need to be solved together. Otherwise, the velocity profiles can be solved separately and inserted into the mass balance equation to solve for the concentration profile.

The mass transfer coefficient k is a convenient way to express solute transport. It enters into the gel flux model $J = k \ln\left(c_g / c\right)$ applicable to the polarized region of the flux curve. The value of the mass transfer coefficient is obtained from the concentration gradient normal to the wall (membrane surface)[5,6]: $k = \frac{D_w}{\left(c_b - c_w\right)} \left.\frac{\partial c}{\partial z}\right|_{wall}$ for cm/s mass transfer coefficient k, cm^2/s diffusivity at the wall D_w, g/cm^3 concentration c, cm direction normal to the wall z, g/cm^3 wall concentration c_w, and g/cm^3 bulk concentration far away from the wall in free solution c_b. Solving for k requires knowing the concentration gradient at the wall. This can be obtained from the steady-state concentration profile in the fluid. The dimensionless mass transfer coefficient or Sherwood number[5,6] by $\mathrm{Sh} = \frac{kL_{char}}{D_b} = -\tilde{D}\left.\frac{\partial \tilde{c}}{\partial \tilde{z}}\right|_{wall} = f\left(\tilde{\vec{r}}, \mathrm{Re}, \mathrm{Sc}, \tilde{\rho}, \tilde{\mu}, \tilde{D}, \tilde{v}_w\right)$. One empirical form widely used for correlation of non-porous transport with constant properties is $\mathrm{Sh} = a\,\mathrm{Re}^b\,\mathrm{Sc}^c\left(L_{char} / L\right)^e$ where L is the length of the feed channel. A summary of Sherwood numbers for laminar flow in different devices is shown in Table 4.2.

Constant property mass transfer problems are analogous to comparable heat transfer problems where one replaces the variables (c, D, Sc, Sh) with the variables (T, α, Pr, Nu), where α is the thermal diffusivity, Pr is the Prandtl number and Nu is the Nusselt number. The dimensionless equations are identical. This means that the solutions to either problem are interchangeable.

The derivations can get complex and definitions of terms and constants, such as hydraulic radius, friction factors and Reynolds numbers, can vary among authors.

Multi-component systems involve interactions between different solutes in the polarization layer.

Table 4.2 Sherwood numbers

Module	Reynolds number	Sherwood number	References
Stirred cell	$\omega R^2/\upsilon$	$0.29\,\mathrm{Re}^{1/2}\,\mathrm{Sc}^{1/3}$	Smith[7]
Slit	$\bar{v}_{x0}h/\upsilon$	$1.86\,\mathrm{Re}^{1/3}\,\mathrm{Sc}^{1/3}(d_h/L)^{1/3}$	Leveque[8]
Tube	$\bar{v}_{z0}2R/\upsilon$	$1.62\,\mathrm{Re}^{1/3}\,\mathrm{Sc}^{1/3}\,(d_h/L)^{1/2}$	Leveque[8]
Cassette	$\bar{v}_{x0}d_h/\upsilon$*	$0.664\,\mathrm{Re}^{1/2}\,\mathrm{Sc}^{1/3}\,(d_h/L_s)^{1/2}$	Da Costa[9]

Sc = υ/D.
* See Section 4.13 for a description of d_h and L_s for cassettes.

4.4 Stirred cell velocities and torque

The stirred cell is widely used as a simple low-volume system to study membrane transport. It has the virtue that the mass transfer coefficient is uniform over the surface so variations are due to the membrane non-homogeneity and not the test device.

Consider rotating flow at an angular rotation rate of ω above a stagnant disk of radius R where the flow field is three-dimensional with velocity components in the radial v_r, tangential v_φ and normal v_z directions (Figure 4.2). For constant fluid density

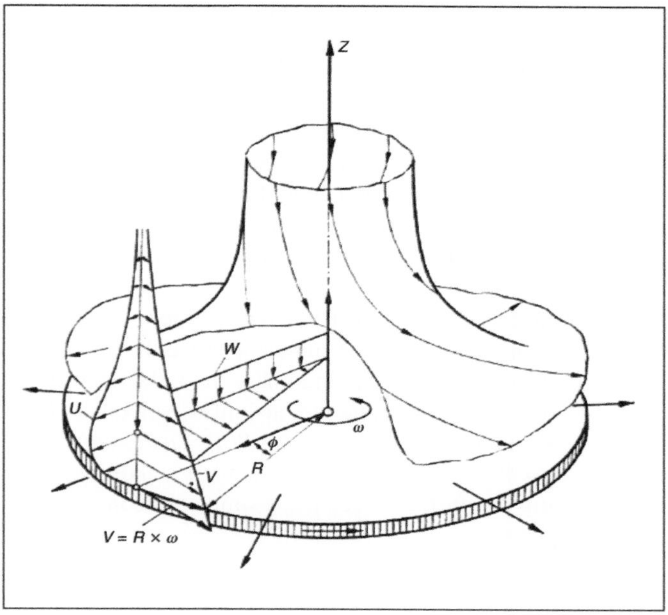

Figure 4.2 Stirred cell velocity profiles.[10]
(Courtesy of McGraw-Hill Education.)

ρ and viscosity μ, the flow field must satisfy the steady-state mass and momentum balances (Navier–Stokes equations) in cylindrical co-ordinates. Rotational symmetry in φ means all derivatives in φ are zero and the equations simplify to

1. Continuity $\dfrac{\partial v_r}{\partial r}+\dfrac{v_r}{r}+\dfrac{\partial v_z}{\partial z}=0$

2. r momentum $v_r\dfrac{\partial v_r}{\partial r}+v_z\dfrac{\partial v_r}{\partial z}-\dfrac{v_\varphi^2}{r}=-\dfrac{1}{\rho}\dfrac{\partial p}{\partial r}+\upsilon\left[\dfrac{\partial^2 v_r}{\partial r^2}+\dfrac{\partial}{\partial r}\left(\dfrac{v_r}{r}\right)+\dfrac{\partial^2 v_r}{\partial z^2}\right]$

3. φ momentum $v_r\dfrac{\partial v_\varphi}{\partial r}+v_z\dfrac{\partial v_\varphi}{\partial z}+\dfrac{v_r v_\varphi}{r}=\upsilon\left[\dfrac{\partial^2 v_\varphi}{\partial r^2}+\dfrac{\partial}{\partial r}\left(\dfrac{v_\varphi}{r}\right)+\dfrac{\partial^2 v_\varphi}{\partial z^2}\right]$

4. z momentum $v_r\dfrac{\partial v_z}{\partial r}+v_z\dfrac{\partial v_z}{\partial z}=-\dfrac{1}{\rho}\dfrac{\partial p}{\partial z}+\upsilon\left[\dfrac{\partial^2 v_z}{\partial r^2}+\dfrac{1}{r}\dfrac{\partial v_z}{\partial r}+\dfrac{\partial^2 v_z}{\partial z^2}\right]$

5. boundary conditions at $z = 0$ (wall), $v_r=0,\ v_\varphi=r\omega,\ v_z=v_w$ (<0 for suction or permeate flow)
6. boundary conditions at $z\to\infty$ (far from the surface), $v_r=0,\ v_\varphi=0$.

(Circumferential reservoir vessel wall and disk edge effects are neglected.)

The streamlines in Figure 4.2 show a uniform rotational flow with radial fluid flow outward at the disk surface drawing more fluid down from a shrinking core above the disk. In the presence of an enclosed cylindrical shell, this sets up a circulation pattern with fluid rising at the cylinder wall and falling in the centre. Following von Kármán,[11] it is convenient to define a momentum boundary layer thickness δ in the z direction over which the velocity changes occur. Consider a fluid volume element $dV = \delta dr ds$ at a distance r from the centre with width ds and rotating with the disk. The radial centrifugal force on this volume element $\rho r\omega^2 dV$ is balanced by a radial component of the wall shear $\tau_w \sin\theta\, dA$ acting on the surface $dA = dr ds$, where the angle θ is formed between the direction of the streamline and the radial direction. Also, for a Newtonian fluid, the circumferential component of the wall shear must be proportional to the gradient of the circumferential velocity v_φ at the wall so $\tau_w\cos\theta \sim \mu r\omega/\delta$. Satisfying both conditions requires the boundary thickness in the z direction to scale as $\delta\sim\sqrt{\tan\theta v/\omega}$.

The variables can then be non-dimensionalized. The z co-ordinate becomes $\zeta=z/\delta=z\sqrt{\omega/\upsilon}$. It is also useful to non-dimensionalize the angular velocity with $r\omega$ so that $G(\zeta)=v_\varphi/(r\omega)\sim 1$. Substituting these into the Navier–Stokes equations then shows it is convenient to define dimensionless velocities $F(\zeta)=v_r/(r\omega)$, $H(\zeta)=v_z/\sqrt{\upsilon\omega}$ and a dimensionless pressure of $P(\zeta)=p/(\rho\upsilon\omega)$.

Substituting these into the four balances and boundary conditions, and neglecting the smaller terms, the original problem becomes a simpler one involving solving four ordinary differential equations:

1. Continuity $2F+H'=0$
2. r-momentum $F^2-G^2+HF'-F''=0$
3. φ-momentum $2GF+HG'-G''=0$
4. z-momentum $P'+HH'-H''=0$

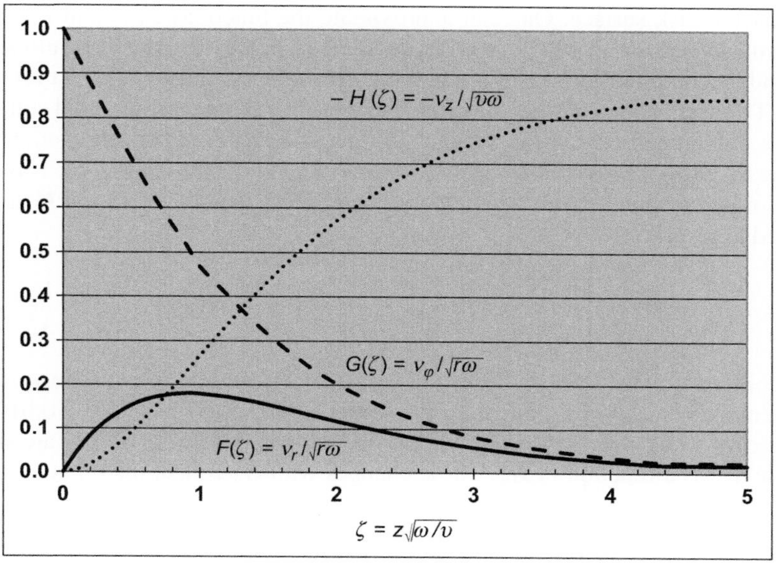

Figure 4.3 Stirred cell velocities[10] with $v_w = 0$.

5. boundary conditions at $\zeta = 0, F = 0, G = 1, H = v_w/\sqrt{\upsilon\omega}$
6. boundary conditions as $\zeta \to \infty, F = 0, G = 0$.

Numerical integration of this system[11,12] with $v_w = 0$ produces the dimensionless velocities shown in Figure 4.3. The negative dimensionless axial velocity H represents fluid drawn downward to replace the fluid thrown outward by centrifugal force at the disk. The axial velocity increases as one moves away from the disk and the net downward flow is $Q(z \to \infty) = v_z(\infty)\pi R^2 = \sqrt{\upsilon\omega}H(\infty)\pi R^2$ where the numerical solution[11] shows $H(\infty) = 0.886$. The momentum boundary layer thickness can be taken as the point $\varsigma = 4.5$ where the tangential flow is 98% of its bulk rotational flow value.

The dimensionless radial velocity F is zero at the disk surface and far away from the disk, with a positive maximum in-between showing outward radial flow. The dimensionless tangential velocity G is highest at the disk surface and decays as one moves away into free solution.

Including suction at the disk surface[7,13] involves changing the boundary condition $H(0) = H_w = v_w/\sqrt{\upsilon\omega} < 0$. Increasing suction makes the dimensionless axial velocity H more uniform with distance z from the disk surface so $H(z) \approx v_w/\sqrt{\upsilon\omega} < 0$ for all z. More of the inflow goes through the porous disk and not radially outward. Increasing suction also decreases the radial velocity F with the maximum occurring closer to the disk. The tangential profile G decreases with suction, thinning the boundary layer.

The concentration profile changes in a region very close to the surface known as the concentration boundary layer. As a result, it is important to characterize the velocity field near the disk surface. One can approximate the functions F, G and H using a polynomial expansion around $\zeta = 0$. The coefficients in these expansions must satisfy the boundary conditions at $\zeta = 0, F = 0, G = 1, H = H_w$. Initially consider the case of $H_w = 0$. Then the expansions are substituted into each of the four differential equations to give a polynomial expression for each in powers of ζ. Since each equation must hold true for any ζ, each coefficient of ζ in each equation must equal zero. This gives a series of relationships involving the polynomial coefficients of F, G and H that can be solved to give

$$F(\zeta) = a_1\zeta - \frac{1}{2}\zeta^2 - \frac{1}{3}b_1\zeta^3 - \cdots, G(\zeta) = 1 + b_1\zeta + \frac{1}{2}\zeta^3 + \cdots, H(\zeta) = -a_1\zeta^2 + \frac{1}{3}\zeta^3 + \cdots$$

where constants a and b are yet to be determined. Comparison of these polynomials with the numerical solution then yields the values of[14] $a_1 = 0.510$ and $b_1 = -0.616$. For the case of $v_w \neq 0$, $H_w = H(0) = v_w / \sqrt{\upsilon\omega} < 0$, the parameters[7] a_1 and b_1 are shown in Figure 4.4 as a function of H_w. These changes with H_w reflect the thinning of the boundary layer. At large values of H_w, it was observed that $H(\varsigma) \approx H_w$. Eliminating F from equations (4.1) and (4.3), one can solve for $G(\varsigma) = \exp(H_w\varsigma)$ with $G'(0) = H_w$.

The total fluid torque[4] on the disk is $T = \int r\tau\, dA = \int_0^R r\left[\mu r\omega G'(0)\sqrt{\omega\upsilon}\right] 2\pi r\, dr = \frac{\pi}{2}\rho R^5\omega^2 G'(0)\mathrm{Re}^{-1/2}$ for Reynolds number $\mathrm{Re} = \omega R^2/\upsilon$ and fluid contact on just one side of the disk. For turbulent flow with a $v^{1/7}$ velocity profile, the torque[4] is $T = 0.073\rho R^5\omega^2\,\mathrm{Re}^{-1/5}$. These expressions compare well with experimental data[7,10] with a transition from laminar to turbulent flow around Re = 6×10^4. The addition of suction increases torque.[7]

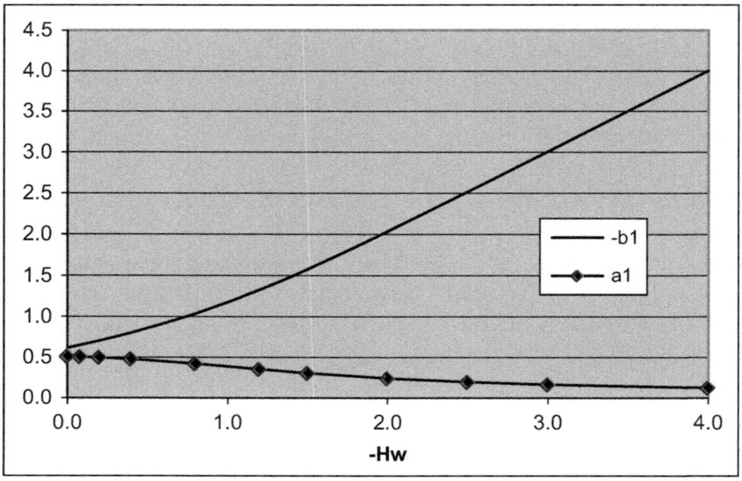

Figure 4.4 Stirred cell velocity parameters.[7]

4.5 Stirred cell concentrations and mass transfer coefficient

The steady-state solute balance in cylindrical co-ordinates with rotational symmetry (in φ) is

$$v_r \frac{\partial c}{\partial r} + v_z \frac{\partial c}{\partial z} = \frac{1}{r}\frac{\partial}{\partial r}\left(rD \frac{\partial c}{\partial r} \right) + \frac{\partial}{\partial z}\left(D \frac{\partial c}{\partial z} \right)$$

where at the wall $v_w c_w (1-s) = D \left. \frac{\partial c}{\partial z} \right|_w$; far from the surface as $z \to \infty$, $c = c_b$; at the centreline $r = 0$, $\partial c / \partial r = 0$ and far from the centreline as $r \to \infty$, $\partial c / \partial r = 0$.

Following Levich,[15] the radial boundary conditions show no gradient at the centre of the disk or far away from the centre of the disk. This implies that there is no driving force for variation in the radial direction. Consider concentration as a function of only the normal direction $c(z)$. In terms of $\zeta = z\sqrt{\omega/\upsilon}$, the solute balance becomes $\upsilon H(\zeta)\frac{dc}{d\zeta} = \frac{d}{d\zeta}\left(D \frac{dc}{d\zeta} \right)$. Integrating and applying the boundary conditions in z yields $c(\zeta) = c_w + (c_b - c_w) \int_0^{\zeta} \exp\left(\mathrm{Sc} \int_0^{\zeta} H(\zeta) d\zeta \right) d\zeta \Big/ \int_0^{\infty} \exp\left(\mathrm{Sc} \int_0^{\zeta} H(\zeta) d\zeta \right) d\zeta.$

The mass transfer coefficient is $k = \frac{D\sqrt{\omega/\upsilon}}{(c_b - c_w)} \left. \frac{\partial c}{\partial \zeta} \right|_{\zeta=0}$ and the Sherwood number is

$\mathrm{Sh} = \frac{kd}{D} = \frac{\left(d\sqrt{\omega/\upsilon} \right)}{(c_b - c_w)} \left. \frac{\partial c}{\partial \zeta} \right|_{\zeta=0} = \mathrm{Re}^{1/2} \Big/ \int_0^{\infty} \exp\left(\mathrm{Sc} \int_0^{\zeta} H(\zeta) d\zeta \right) d\zeta$ for Reynolds number $\mathrm{Re} = d^2\omega/\upsilon$ with characteristic length L_{char} = disk diameter $d = 2R$ and characteristic tangential velocity $V_{char} = d\omega$.

For the large solutes retained by ultrafilters, $D \sim 10^{-10}$ m^2/s compared to $\upsilon \sim 10^{-6}$ m^2/s for water. This means that $\mathrm{Sc} = \upsilon/D \sim 10^4 \gg 1$. An order of magnitude analysis indicates that for large Sc, the momentum boundary layer is much larger than the concentration boundary layer[5,6] $\delta \gg \delta_c$. This means that the velocity profiles over the concentration boundary layer (near the surface) can be well approximated by their asymptotic values near the wall $H(\zeta) = -a\zeta^2 - H_w$. For small H_w the integral becomes[15] $\int_0^{\infty} \exp\left(-\mathrm{Sc}\left[a\zeta^3/3 \right] \right) d\zeta = \frac{\Gamma(1/3)}{3(a\mathrm{Sc}/3)^{1/3}} = 1.596\,\mathrm{Sc}^{-1/3}$ and $\mathrm{Sh} = 0.626\,\mathrm{Re}^{1/2}\,\mathrm{Sc}^{1/3}$.

Smith[7] found that the liquid phase Sherwood number for benzoic acid dissolution in a custom spinning disk device correlated as $\mathrm{Sh} = A\,\mathrm{Re}^b\,\mathrm{Sc}^{1/3}\,\psi(\text{lengths})\,\phi(\mathrm{Sh}_w)$ for $\mathrm{Sh} = kb/D$, $\mathrm{Re} = \omega b^2/\upsilon$, $\mathrm{Sh}_w = 2P_m\sqrt{\upsilon/\omega}/D$ (~ratio of permeability to mass transfer coefficient) for vessel diameter b and membrane permeability P_m. The correlated values for A and c varied smoothly from asymptotes with $A = 0.285$ and $c = 0.567$ for laminar flow (Re < 32,000) and $A = 0.0443$ and $c = 0.746$ for turbulent flow. The exponent $c = 0.567$ for laminar flow matches the derived value of $c = 0.5$ within 15%. A mathematical simulation using the equations for flow was used to derive length corrections based on the ratio of the vessel diameter/disk diameter and the ratio of the disk separation distance/ disk diameter. These corrections ranged from 0 to 25%.

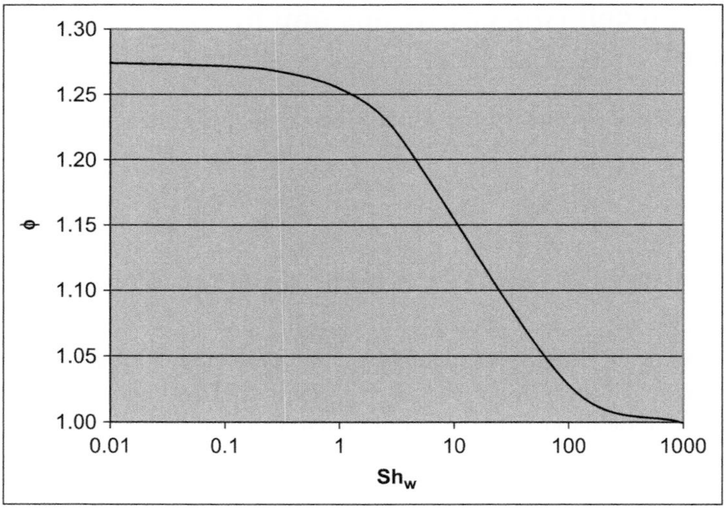

Figure 4.5 Sh_w correction factor.

The mathematical simulation was also used to derive a correction for membrane permeability $\phi = 1.1374 + 0.1632x - 0.0027x^2 - 0.0235x^3$ for $x = \tanh[1.1086 - \log_{10}(Sh_w)]$. An infinite Sh_w corresponds to the constant wall concentration or impermeable wall base case. Smaller Sh_w corresponds to a finite v_w and leads to higher transport (Figure 4.5).

Agreement with the Smith correlation was good with some benzoic acid and salt data[15,16] when correcting for different test apparatus lengths. Mitchell[17] correlated $Sh = 0.373 Re^{0.557} Sc^{1/3}$ using a polarographic technique. However, other salt data[18] indicated comparable exponential dependencies but the constant term A was only 30% as large after correction for lengths. The correlated constant term $A = 0.285$ for laminar flow with the derived result $A = 0.626$ indicates that the length corrections may be significant in matching experimental conditions with the flow modelling assumptions. Kozinski[18] found that the liquid phase Sherwood number for benzoic acid dissolution closely matched the derived Sherwood number as cited for other investigators.[19] Deviations below predictions at lower spinning rates were attributed to poor mixing and edge effects.

BSA stirred cell data involves parameters D and μ changing with concentration and the permeability correction for $v_w \neq 0$. Opong[20] used a stirred cell containing a permeable membrane to correlate as $Sh = 0.23 Re^{0.567} Sc^{1/3}$. The exponents reasonably match predictions but the constant term is lower than the 0.626 predicted value and 17% below that measured by Smith.[7] This is attributed to edge effects and property variations. Kozinski[18] used a custom spinning disk device containing a permeable membrane to correlate $Sh = A Re^{1/2} Sc^{1/3} (\bar{D}^2 / \bar{\mu})^{1/3} (\tau c_w / c_b)^{1/3}$ where the averaging is between the wall and bulk values. Fluxes predicted using an osmotic model matched the data reasonably well given uncertainties in the viscosity and diffusivity dependencies on concentration.

4.6 Slit channel velocities and friction factor

Slit channel flow can be used to model plate and frame modules where spacers are used to hold the membranes apart (Figure 4.6). One can also make an analogy to cassettes with very coarse screens. For spirals where the feed channel height is small compared to the radial distance from the centreline permeate tube, curvature in the feed channel can be neglected. Additional discussion of screened channels in cassettes will be considered later (Section 4.11).

For an infinite slit with internal flow, the flow field is two-dimensional with an x direction moving down the flow path and a y direction moving towards the membrane wall. The velocity components are v_x, tangent to the wall, and v_y normal to the wall. For constant fluid density ρ and viscosity μ, pressures and flows must satisfy the steady-state mass and momentum balances in rectangular co-ordinates (Navier–Stokes equations):

1. Continuity: $\dfrac{\partial v_x}{\partial x}+\dfrac{\partial v_y}{\partial y}=0$
2. x-momentum: $v_x\dfrac{\partial v_x}{\partial x}+v_y\dfrac{\partial v_x}{\partial y}=-\dfrac{1}{\rho}\dfrac{\partial p}{\partial x}+\upsilon\left[\dfrac{\partial^2 v_x}{\partial x^2}+\dfrac{\partial^2 v_x}{\partial y^2}\right]$
3. y-momentum: $v_x\dfrac{\partial v_y}{\partial x}+v_y\dfrac{\partial v_y}{\partial y}=-\dfrac{1}{\rho}\dfrac{\partial p}{\partial y}+\upsilon\left[\dfrac{\partial^2 v_y}{\partial x^2}+\dfrac{\partial^2 v_y}{\partial y^2}\right]$
4. boundary conditions at $y = 0$ (centreline) $v_y=0,\ \dfrac{\partial v_x}{\partial y}$ by symmetry
5. boundary conditions at $y = h$ (wall) $v_y=v_w$ and $v_x=0$ no slip.

4.6.1 Slit developed flow – no permeate

Figure 4.7 shows how the measured[21] axial flow velocity profiles $v_x(y)$ develop as one moves down the feed channel for $v_w = 0$. For downstream, the velocity profile is uniform and parabolic. At this point $v_y = 0$ so the Navier–Stokes equations reduce to $0=-(1/\rho)(\mathrm{d}p/\mathrm{d}x)+\upsilon(\mathrm{d}^2 v_x/\mathrm{d}y^2)$. Integration for constant $\mathrm{d}p/\mathrm{d}x$ yields $v_x(y)=1.5\overline{v}_{x0}[1-(y/h)^2]$ describing Poiseuille flow where $\overline{v}_{x0}=(-h^2/3\mu)(\mathrm{d}p/\mathrm{d}x)$. One basis for choosing a characteristic length is the hydraulic diameter d_h= 4(cross-sectional area for flow)/(wetted perimeter). For the slit geometry this is $d_h = (2hW)/(2h + 2W)\sim 4h$ for $W \gg h$. Picking a Reynolds number

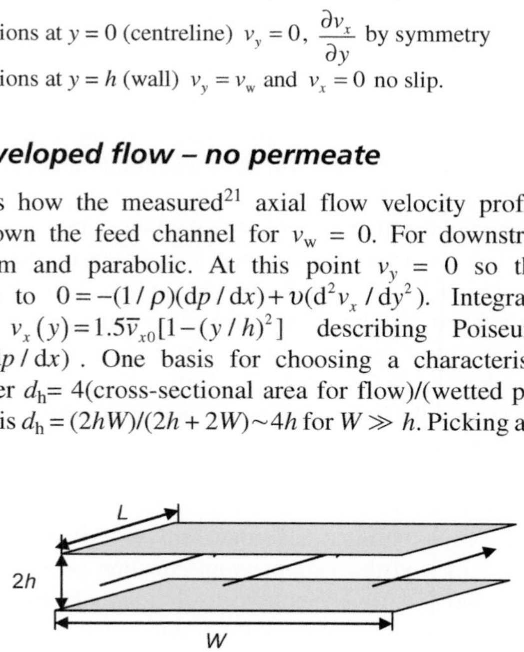

Figure 4.6 Slit channel flow.

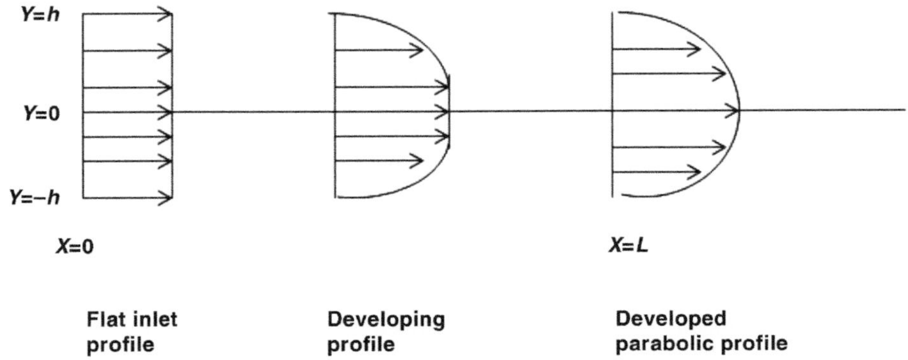

Figure 4.7 Developing velocity profile.

on this basis yields $\mathrm{Re} = 4h\overline{v}_{x0}/\nu$. The friction factor can then be written as $f = \dfrac{\mu}{\rho\overline{v}_x^2/2} \dfrac{\partial v_x}{\partial y}\Big|_{\text{wall}} = \dfrac{2\nu}{\overline{v}_x^2}\dfrac{3\overline{v}_x}{h} = 24/\mathrm{Re}$.

4.6.2 Slit developing flow – no permeate

At the inlet, the velocity profile is dominated by inlet conditions, taken here as a uniform velocity $\overline{v}_{x0}$. As the fluid moves along the channel, the influence of the channel walls on the flow grows out from the walls. This region is called the momentum boundary layer thickness δ. An order of magnitude analysis of the Navier–Stokes equations using the boundary layer concept[10] involves noting that changes in the y direction occur over a distance δ while changes in the x direction occur over a much larger characteristic distance x. In the momentum balances, $\delta \ll x$ implies that $(\partial^2 v_x/\partial x^2) \ll (\partial^2 v_x/\partial y^2)$ and $(\partial^2 v_y/\partial x^2) \ll (\partial^2 v_y/\partial y^2)$ so we can neglect the first terms. The conservation equation implies that changes in v_y scale according to $(\overline{v}_{x0}/x) \approx (v_y/\delta)$. Then the x-momentum inertia terms on the left-hand side scale as $\overline{v}_{x0}^2/x$ while the viscous terms on the right-hand side scale as $\nu\overline{v}_{x0}/\delta^2$. For both these terms to be significant and therefore of the same magnitude, the dimensionless boundary layer thickness must scale as $(\delta/x) \approx \sqrt{(\nu/x\overline{v}_{x0})}$. For a Reynolds number $\mathrm{Re}_x = x\overline{v}_{x0}/\upsilon_b$, then $\delta(x) \approx x\,\mathrm{Re}_x^{-1/2}$ where $\overline{v}_{x0}$ is the average cross-sectional flow in the x direction at $x = 0$, υ_b is the bulk solution kinematic viscosity.

Blausius[22] used this boundary layer thickness to scale the y direction. Assume that the x-velocity is only a function of the scaled variable $v_x(\eta) = \overline{v}_{x0}f(\eta)$ where $\eta = y/\delta(x)$. Then the stream function ψ defined by $v_x = \partial\psi/\partial y$ and $v_y = -\partial\psi/\partial x$ so it satisfies the continuity equation must have the form $\psi = \sqrt{\nu x\overline{v}_{x0}}\,f(\eta)$. This makes $v_y = 0.5\sqrt{\nu\overline{v}_{x0}/x}\,(\eta f' - f)$ and the y-momentum equation becomes $ff'' + 2f''' = 0$ with $f(0) = 0, f'(0) = 0$ and $f'(\infty) = 1$. Numerical solution by Horwath[23] indicates that $v_x(\eta = 5) = 0.99\ \overline{v}_{x0}$. Nikuradse[24] confirmed the results experimentally.

Eventually, the boundary layers from each wall meet each other. At that point the flow becomes 'fully developed' and independent of the original inlet velocity profile.

The channel length required to reach 99% of the fully developed velocity profile is called the 'entrance length'. For $v_w = 0$, the entrance length is[21] $L_e = (1.25 + 0.088\,\text{Re})h$ for. At the transition from laminar to turbulent flow (Re~2500), $L_e / h = 221$. Permeate flows ($v_w > 0$) increase the entrance length[25] and in the extreme may prevent achievement of the fully developed profile. The fully developed flow analysis applies for (1) long channels, (2) slow inlet velocities and (3) slow permeate velocities.

4.6.3 Slit developing flow – constant permeate

There have been quite a number of analyses for fully developed laminar flow in slits and tubes,[25–36] each making various simplifications to the Navier–Stokes equations and boundary conditions to permit an analytical solution. Here we first follow the classic treatment of Berman,[26] which assumes a constant wall velocity v_w along the entire length of the slit. This assumption is reasonable for the polarized region of the flux curve where fluxes are independent of TMP and vary slowly with changes in cross-flow. It is probably not applicable for the linear region of the flux curve or for pure solvent flow.

Natural scaling variables are h for the y direction and $\overline{v}_{x0}$, the average or superficial inlet x-velocity. Consider the form $v_x(x,y) = \overline{v}_{x0} h f(x) g'(y)$. Satisfying the continuity equation and boundary conditions leads to $v_y(x,y) = v_w g$ and $v_x(x,y) = \overline{v}_{x0} h[1-(v_w x / \overline{v}_0 h)]g'$ (where $g(0) = 0$ and $g(h) = 1$ with no loss of generality). Substituting these x and y velocity expressions into the y- and x-momentum equations, respectively $-(1/\rho)(\partial p/\partial y) = v_w^2 g g' - v_w \upsilon g''$ and $-(1/\rho)(\partial p/\partial x) = \overline{v}_{x0}[1-(v_w x/\overline{v}_0 h)]\{-\upsilon h g''' - v_w[(g')^2 - g g'']\}$. Noting that the y-momentum balance is a function of y only, taking its x-derivative yields $\partial^2 p/\partial x \partial y = 0$. Similarly, taking the y-derivative of the x-momentum balance and applying $\partial^2 p/\partial x \partial y = 0$ yields $0 = \overline{v}_{x0}[1-(v_w x/\overline{v}_0 h)](\partial\{-\upsilon h g''' - v_w[(g')^2 - g g'']\}/\partial y)$. This requires $h^3 g''' + h^2 \text{Re}_w[(g')^2 - g g''] = k$, where k is a constant and $\text{Re}_w = v_w h/\upsilon$ is a wall Reynolds number. $\text{Re}_w > 0$ for permeate flow. The boundary conditions are $g(0) = g'(1) = g''(0) = 0$, $g(1) = 1$.

This third-order ordinary-linear equation defining $g(y)$ requires a perturbation solution in the parameter Re_w. It is convenient to consider $g(\lambda, \text{Re}_w)$ where $\lambda = y/h$. Expanding g in a power series of Re_w where the coefficients are functions of λ, and expanding k in a power series of Re_w

$$g(\lambda, \text{Re}_w) = g_0(\lambda) + g_1(\lambda)\text{Re}_w + g_2(\lambda)\text{Re}_w^2 + \cdots \quad \text{and} \quad k(\text{Re}_w) = k_0 + k_1 \text{Re}_w + k_2 \text{Re}_w^2 + \cdots$$

Substituting both series in the differential equation, and collecting terms of same power of Re_w gives ordinary differential equations for the coefficients in λ that can be solved with the boundary conditions to give $k(\text{Re}_w) \approx -3 + 81\text{Re}_w/35$ and $g(\lambda, \text{Re}_w) \approx \lambda(3-\lambda^2)/2 + \text{Re}_w(3\lambda^3 - 2\lambda - \lambda^7)/280$ for the first two terms. This makes the velocities $v_x[(x/h), \lambda, \text{Re}_w] = (3\overline{v}_0/2)[1-(v_w x/\overline{v}_0 h)](1-\lambda^2)[1-(\text{Re}_w/420)(2-7\lambda^2-7\lambda^4)]$ and $v_y(\lambda, \text{Re}_w) = (v_w \lambda/2)[(3-\lambda^2) - \text{Re}_w(2-3\lambda^2+\lambda^6)/140]$. After a channel length of $x = \overline{v}_{x0} h / v_w$, all the fluid entering the slit will have been drawn out of the channel walls and the solution is no longer valid. The x-velocity retains a

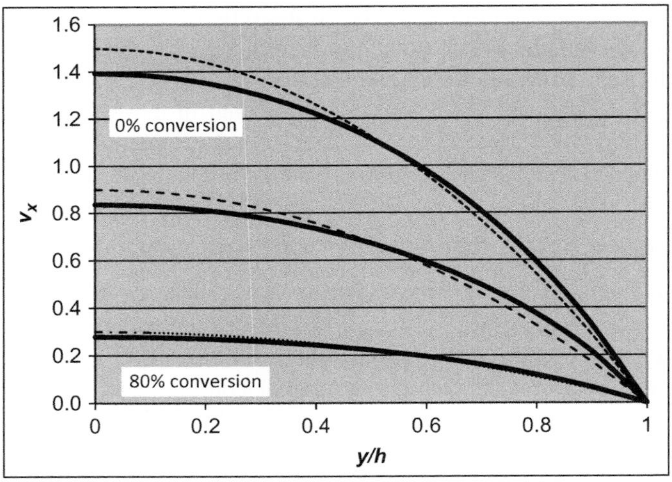

Figure 4.8 Slit tangential velocity profile $v_x/\bar{v}_0$ vs. radial position at different conversions (solid line $\mathrm{Re_w} = -10$, dotted line $= -0.1$).

roughly parabolic profile in λ as the fluid flows down the channel and gets converted to permeate (Figure 4.8).

Substituting the velocity expressions back in the x-momentum balance $-\frac{1}{\rho}\frac{\partial p}{\partial x} = -\frac{\upsilon \bar{v}_{x0} k}{h^2}\left[1 - \frac{v_w x}{\bar{v}_{x0} h}\right]$ and integrating $\frac{p(0) - p(L)}{\rho \bar{v}_{x0}^2/2} = \left[\frac{24}{\mathrm{Re}} - \frac{648\,\mathrm{Re_w}}{35\,\mathrm{Re}}\right]\left[1 - \frac{2\,\mathrm{Re_w}\,x}{\mathrm{Re}\,h}\right]\left[\frac{x}{h}\right]$ or $f\,\mathrm{Re} = 24\left[1 - \frac{27\,\mathrm{Re_w}}{35}\right]\left[1 - \frac{2\,\mathrm{Re_w}\,x}{\mathrm{Re}\,h}\right]\left[\frac{x}{h}\right]$ for Fanning friction factor f. The pressure drop reverts to parabolic flow[5,6] in a solid wall slit for $\mathrm{Re_w} = 0$. For permeate flow where $\mathrm{Re_w} > 0$, the friction is lower.

The flow transitions from laminar to turbulent at a critical Reynolds number. From experimental observation,[23] a smooth entrance transitions at Re~4400 while an abrupt entrance transitions at Re~2500. Turbulent flow is characterized by very small diameter fluid eddies that arise spontaneously from fluid instabilities. The condition of the fluid is unaffected by the overall duct shape. The pressure drop in turbulent slit flow is the same function of the Reynolds number as found for tubes as long as the hydraulic diameter is used (Figure 4.9).

4.6.4 Slit developing flow with constant permeability

A recent solution of the Navier–Stokes equations by Tilton[35] allows for variable permeate flow. In this solution, Darcy's law is applied to the permeable wall ($y = h v_y = v_w = L_f p$), and a long channel length compared to height is assumed. The pressures p are transmembrane pressures relative to a constant permeate pressure (which can

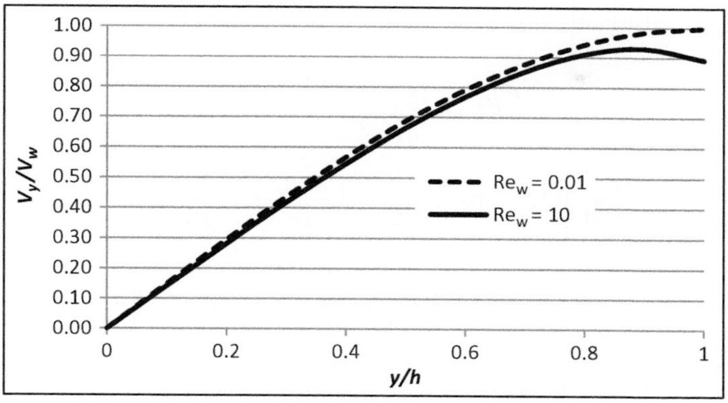

Figure 4.9 Slit normal velocity profile vs. radial position (solid line $Re_w = -10$, dotted line = -0.1).

be taken as 0 to make gauge pressures with no loss of generality). The order of magnitude analysis[10] on the Navier–Stokes equations with scaling magnitudes $x \sim L$, $y \sim h$, where $h/L \sim \varepsilon < <1$ and $v_x \sim \bar{v}_{x0}$ (average cross-flow velocity at $x = 0$). The continuity equation (4.1) then requires $v_y \sim \bar{v}_{x0} h / L = \bar{v}_{x0}\varepsilon$. The x-momentum equation (4.2) terms scale as $v_x \frac{\partial v_x}{\partial x} + v_y \frac{\partial v_x}{\partial y} \sim \frac{(\bar{v}_{x0})^2 \varepsilon}{h}$, $\upsilon \frac{\partial^2 v_x}{\partial x^2} \sim \frac{\upsilon \bar{v}_{x0} \varepsilon^2}{h^2} \ll \upsilon \frac{\partial^2 v_x}{\partial y^2} \sim \frac{\upsilon \bar{v}_{x0}}{h^2}$. Since this system reduces to Hagen–Poiseuille flow with $v_w = 0$, the pressure term must be significant and scales as $(1/\rho)(\partial p / \partial x) \sim (p / \rho L) \sim (\upsilon \bar{v}_{x0} / h^2)$. Then p scales as $p \sim (\mu \bar{v}_{x0} / \varepsilon h)$ and the x-momentum equation simplifies to $\partial p / \partial y = \mu(\partial^2 v_y / \partial y^2)$. The y-momentum equation (4.3) terms scale as $v_x(\partial v_y / \partial x) + v_y(\partial v_y / \partial y) \sim \{[(\bar{v}_{x0})^2 \varepsilon^2] / h\}$, $\upsilon(\partial^2 v_y / \partial x^2) \sim \upsilon(\bar{v}_{x0}\varepsilon^3 / h^2)$, $\upsilon(\partial^2 v_y / \partial y^2) \sim \upsilon(\bar{v}_{x0}\varepsilon / h^2)$ and $(1/\rho)(\partial p / \partial y) \sim (\upsilon \bar{v}_{x0} / \varepsilon h^2)$. This means all the terms in the y-momentum equation except the pressure term are negligible so it reduces to just $\partial p / \partial y \sim 0$. Boundary condition (4.5) scales as $v_y \sim \bar{v}_{x0}\varepsilon = L_f p \sim L_f \mu \bar{v}_{x0} / \varepsilon h$ or $\varepsilon^2 \sim L_f \mu / h$. This means that permeability is directly related to the scaling parameter ε.

The Tilton solution proceeds by substituting non-dimensional variables $x' = x/L$, $y' = y/h$, $v_{x'} = v_x / \bar{v}_{x0}$, $v_{y'} = v_y / \bar{v}_{x0}\varepsilon$, $p' \sim (\mu \bar{v}_{x0} / \varepsilon h)$ along with the Reynolds number ($\mathrm{Re} = 2h\bar{v}_0 / \upsilon$) in the Navier–Stokes equations. Then, a regular perturbation solution[36] in the parameter ε is applied where $v_x = v_{x0} + \varepsilon v_{x1}$, $v_y = v_{y0} + \varepsilon v_{y1}$ and $p = p_0 + \varepsilon p_1$. The modified Navier–Stokes equations and boundary conditions should be satisfied for any value of the small parameter ε. Looking first at a solution with $\varepsilon = 0$ using the technique of separation of variables[37] involves separating the functions into products of functions, each with a different variable: $v_x(x, y) = v_x^x(x) v_x^y(y)$, $v_y(x, y) = v_y^x(x) v_y^y(y)$, $p(x, y) = p^x(x) p^y(y)$. Solution with the boundary conditions yields the zeroth-order terms. Substitution of some of these zero-order terms back into the Navier–Stokes equations and solving for the first-order terms in ε using the separation of variables method then yields the ε-order solution.

The non-dimensional Tilton solution for the zeroth- and first-order terms in ε is

$$v_x / \bar{v}_{x0} = \frac{\mathrm{Re}}{12}(3y - y^3)\left[\frac{d^2 p_0}{dx^2} + \varepsilon\frac{d^2 p_1}{dx^2}\right] + \frac{\mathrm{Re}^3}{20160}(322y - 105y^3 - y^7)\left[\left(\frac{d^2 p_0}{dx^2}\right)^2 + \frac{dp_0}{dx}\frac{d^3 p_0}{dx^3}\right]$$

$$v_y / \bar{v}_{x0} = -\frac{\mathrm{Re}}{4}(1 - y^2)\left[\frac{dp_0}{dx} + \varepsilon\frac{dp_1}{dx}\right] - \frac{\mathrm{Re}^3}{2880}(46 - 45y^2 - y^6)\left[\frac{dp_0}{dx}\frac{d^2 p_0}{dx^2}\right]$$

$$\bar{v}_{x0} = \frac{-\mathrm{Re}}{6}\left[\frac{dp_0}{dx} + \varepsilon\frac{dp_1}{dx}\right] - \frac{3\mathrm{Re}^3}{280}\left[\frac{dp_0}{dx}\frac{d^3 p_0}{dx^3}\right].$$

$$p / (\rho \bar{v}_{x0}^2) = p_0 + \varepsilon p_1, \text{where } p_0 = -\frac{2\sqrt{3}}{\varepsilon \mathrm{Re}}\sinh\left(\sqrt{3}\varepsilon x\right) + \pi\cosh\left(\sqrt{3}\varepsilon x\right)$$

$$p_1 = -\frac{9\sqrt{3}\,\mathrm{Re}\,\pi}{70}\left[\sinh\left(\sqrt{3}\varepsilon x\right) - 2\sinh\left(2\sqrt{3}\varepsilon x\right)\right] + \frac{1}{20\varepsilon}[12 + (\varepsilon \mathrm{Re}\,\pi)^2]\left[\cosh\left(\sqrt{3}\varepsilon x\right) - \cosh\left(2\sqrt{3}\varepsilon x\right)\pi = p(x{=}0)/(\rho \bar{v}_{x0}^2)\right]$$

For a dead-ended system, a different boundary condition is applied to obtain different expressions for p_0 and p_1.

4.6.5 Starling flow

When the pressure drop in the feed channel is high enough, the retentate pressure can drop below the permeate pressure or the retentate osmotic pressure. This can lead to an unusual flow pattern known as 'Starling flow' with flow from the feed channel into the permeate channel over an initial feed channel length and a reverse flow from the permeate channel into the feed channel over the remaining module channel length. The flow modules described here do not include describing flow in the permeate channel and do not account for this effect. This effect is pronounced with low TMP, high cross-flow, and low permeability membranes. Reverse flow can foul the membrane from the back side and stress the membrane module seals to create defects.

4.7 Slit channel concentrations and mass transfer coefficient

The concentration profile must satisfy the steady-state solute balance

$$v_y\frac{\partial c}{\partial y} + v_x\frac{\partial c}{\partial x} = \frac{\partial}{\partial x}\left(D\frac{\partial c}{\partial x}\right) + \frac{\partial}{\partial y}\left(D\frac{\partial c}{\partial y}\right)$$

for boundary conditions at (wall) $c(x, \pm h) = c_w$ and at the centreline $y = 0$, $\frac{\partial c}{\partial y} = 0$, while at the inlet $c(0,y) = c_0$.

The concentration profile in the channel develops in a similar manner to the axial velocity profile shown in Figure 4.7. At the inlet, the concentration profile is dominated by inlet conditions, taken here as the uniform concentration c_0. As the fluid moves along the channel, the influence of the channel walls on the concentration grows out from the walls. This region is now called the concentration boundary layer thickness $\delta_c(x)$. Eventually, the concentration boundary layers from each wall meet each other. At that point the profile becomes 'fully developed' and independent of the original inlet concentration profile.

4.7.1 Slit developing concentration with developed flow

For fully developed parabolic flow with a non-porous wall, the y velocity is zero. For a long slit, changes in x are small so axial diffusion is neglected compared to radial diffusion. Non-dimensionalizing terms[25] using $\theta = \frac{c - c_w}{c_0 - c_w}$, $\zeta = \frac{Dx}{4\bar{v}_{x0}h^2} = \frac{x/h}{\text{Re Sc}}$ and $\eta = \frac{y}{h}$, the mass balance becomes $\frac{3}{8}\left[1-\eta^2\right]\frac{\partial \theta}{\partial \zeta} = \frac{\partial^2 \theta}{\partial \eta^2}$ with $\theta(\zeta \pm 1) = 0$ at the walls, at the centreline $\frac{\partial \theta}{\partial \eta} = 0$, at the inlet $\theta(0,\eta) = 1$. This Graetz problem[38] is solved using separation of variables.[39] This assumes a trial solution $\theta(\xi,\eta) = X(\xi)Y(\eta)$. Substitution requires $\frac{3}{8X}\frac{dX}{\partial \zeta} = \frac{1}{\left[1-\eta^2\right]Y}\frac{d^2Y}{\partial \eta^2} = -\lambda_n^2$ for constant λ_n so the solution becomes $\theta(\xi,\eta) = \sum_{n=1}^{n=\infty} C_n Y_n(\eta)\exp(-8\lambda_n^2 \varsigma/3)$ where λ_n is the nth eigenvalue in $\frac{d^2Y_n}{\partial \eta^2} = -\lambda_n^2 \left[1-\eta^2\right]Y_n$ with boundary conditions $Y_n(1) = 0$, $Y_n'(0) = 0$. This is known as a Sturm–Liouville problem.[39] The coefficients C_n are derived by applying the boundary condition at $\zeta = 0$, multiplying each side by $\left[1-\eta^2\right]Y_n d\eta$ and integrating to get $C_n = \frac{\int_{\eta=0}^{\eta=1}\left[1-\eta^2\right]Y_n\, d\eta}{\int_{\eta=0}^{\eta=1}\left[1-\eta^2\right]Y_n^2\, d\eta} = \frac{-2}{\lambda_n(\partial Y_n/\partial \lambda_n)_{\eta=1}}$. The functions Y_n are obtained by the method of Frobenius[40] using $Y_n(\eta) = \sum_{i=0}^{i=\infty} b_{ni}\eta^i$ with the coefficient recursion formula $b_{ni} = 0$ if $i < 0$, $b_{n0} = 1$ and $b_{ni} = -\lambda_n^2(b_{n,i-2} - b_{n,i-4})/[n(n-1)]$. A trial and error procedure was used[41] to obtain the values in Table 4.3.

The mass transfer coefficient is obtained from the concentration profile using $k = \frac{D\partial c/\partial y}{(c_w - c_m)} = \frac{D\partial\theta/\partial\eta}{h\theta_m}$ or $Sh_m = kh/D = [(\partial\theta/\partial\eta)/\theta_m]$, where m designates the mean concentration in the channel. Figure 4.10 shows the Sherwood number based on the mean slit concentration vs. the scaled feed channel parameter. The value decreases as the concentration boundary layer grows and eventually reaches a constant value

Table 4.3 **Non-porous slit eigenvalues and derivatives**[37,41]

n	λ_n	$(\partial Y_n/\partial\lambda_n)_{\eta=1}$	$(\partial Y_n/\partial\eta)_{\eta=1}$
1	1.68160	−0.99044	−1.42916
2	5.66986	1.17911	3.80707
3	9.66824	−1.28625	−5.92024
4	13.66766	1.36202	7.89254

when the boundary layer is fully developed. It has the highest value when the boundary layer is the thinnest.

The infinite series solution converges slowly. Over the initial region, one can consider a similarity solution based on concentration boundary layer thickness scaling. Define variable $\chi = 1-\eta$ as the distance from the bottom wall. Then the mass balance becomes $\frac{3}{8}\left[2\chi-\chi^2\right]\frac{\partial\theta}{\partial\zeta}=\frac{\partial^2\theta}{\partial\chi^2}$ where $\theta=\frac{c-c_w}{c_0-c_w}$, with $\theta(\zeta,0) = 0$ at the wall, $\theta(\zeta,\infty) = 1$ away from the wall, and at the inlet $\theta(0,\chi) = 1$. Consider a concentration boundary layer thickness δ_c over which concentration changes in the y- or χ-direction occur. Applying this to the mass balance shows $\chi^2 \ll \chi$ so the velocity profile uses the Leveque[42] approximation by its slope at the wall and $\delta_c/\xi \approx 1/\delta_c^2$ so the boundary layer thickness grows as $\delta_c \approx \xi^{1/3}$. We can then seek a similarity solution for $\theta(s)$ where $s=\chi/\delta_c(\xi)$ where $\frac{\partial\theta}{\partial\zeta}=\frac{d\theta}{ds}\frac{\partial s}{\partial\zeta}=\frac{d\theta}{ds}\frac{-\chi\delta_c'}{\delta_c^2}$, $\frac{\partial^2\theta}{\partial\chi^2}=\frac{d^2\theta}{ds^2}\frac{1}{\delta_c^2}$, so $\frac{d^2\theta}{ds^2}+\frac{6}{8}\frac{d\theta}{ds}s^2\delta_c^2\delta_c'=0$. For this to be true $\delta_c^2\delta_c'=(\delta_c^3)'/3=\alpha$ constant or $\delta_c=3\alpha\xi^{1/3}$ for $\delta_c(0)=0$. The mass balance $\frac{d^2\theta}{ds^2}+\frac{3\alpha}{4}s^2\frac{d\theta}{ds}=0$ can then be solved using an integration factor to $\theta(s)=\frac{3}{\Gamma(1/3)}\int_s^\infty e^{-t^3}\,dt$ with $\mathrm{Sh_m}=\frac{kh}{D}=-\frac{\partial\theta}{\partial\eta}=1.86\,\mathrm{Re}^{1/3}\,\mathrm{Sc}^{1/3}(x/L)^{1/3}$.

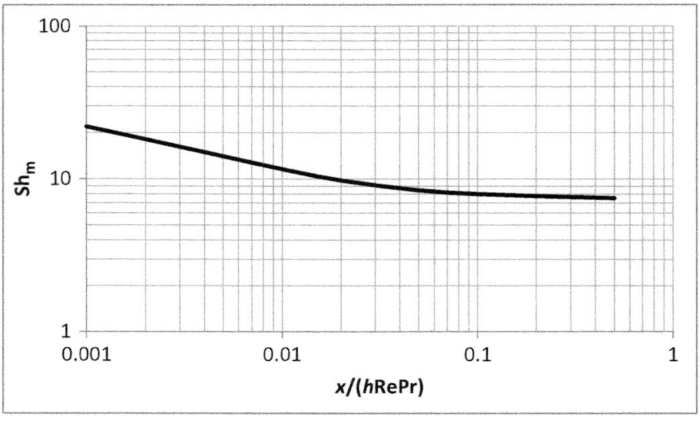

Figure 4.10 Sherwood number developed flow.

4.7.2 Slit developing concentration with developing flow

Consider a concentration boundary layer thickness δ_c over which concentration changes in the y-direction occur. Applied to the mass balance, this means that the y inertial terms on the left-hand side must scale as the y-diffusion term $\frac{v_y}{\delta_c} \sim \frac{\bar{v}_{x0}\delta}{x\delta_c} \sim \frac{D}{\delta_c^2}$. Substituting the earlier solution on developing flow, $\left(\frac{\delta}{x}\right) \approx \sqrt{\frac{v}{x\bar{v}_{x0}}}$ shows that $\frac{\delta}{\delta_c} \sim \frac{D}{\upsilon} \sim \frac{1}{\mathrm{Sc}}$ where Sc is the Schmidt number. For the large solutes retained by ultrafilters, $D \sim 10^{-10}$ m^2/s compared to $\upsilon \sim 10^{-6}$ m^2/s for water. This means that $\mathrm{Sc} = \upsilon/D \sim 10^4 \gg 1$ and the momentum boundary layer grows much larger and faster than the concentration boundary layer[5,6] $\delta \gg \delta_c$. This means that the velocity profiles over the concentration boundary layer (near the surface) can be well approximated by their asymptotic values for large λ. In addition, confining concentration variations to a small boundary layer means $(\partial/\partial x)[D(\partial c/\partial x)] \ll (\partial/\partial y)[D(\partial c/\partial y)]$ so axial diffusion can be neglected.

Application of the scaled variable $\eta = y/\delta(x)$ to the mass balance results in the ordinary differential equation $\theta'' + \mathrm{Sc}\, f\theta'/2 = 0$ where $\theta = (c - c_w)/(c_0 - c_w)$ and boundary conditions $\theta(0) = 1$, $\theta(\infty) = 0$. The solution first given by Pohlhausen[43] can be written as $\theta(\eta, \mathrm{Sc}) = \int_\eta^\infty [f''(\xi)]^{\mathrm{Sc}} d\xi \,/\, \int_0^\infty [f''(\xi)]^{\mathrm{Sc}} d\xi$ with mass transfer at the wall given by[10] $\mathrm{Sh} = \frac{kh}{D} = \frac{\partial\theta/\partial\eta}{\theta_m} = 0.339\,\mathrm{Sc}^{1/3}\,\mathrm{Re}^{1/2}$ for large Sc. Note that this solution differs from the developed flow solution because of its dependence on Re to the 1/2 rather than 1/3 power.

4.7.3 Slit developing concentration with developed flow

Recognizing that for typical ultrafiltration operation with a flux of 40 LMH, channel height of 1 mm, and kinetic viscosity of 10^{-6} m^2/s, $\mathrm{Re}_w = hv_w/\upsilon_b \sim 0.01$, small. Following Sherwood,[44] the steady-state mass balance is $\frac{\partial(v_x c)}{\partial x} + \frac{\partial(v_y c)}{\partial y} = \frac{\partial}{\partial y}\left(D\frac{\partial c}{\partial y}\right)$ with simplified Berman velocity profiles $v_x\left(\frac{x}{h}, \lambda\right) \sim 1.5\bar{v}_0\left[1 - \frac{v_w x}{\bar{v}_0 h}\right](1-\lambda^2)$, $v_y(\lambda) \sim v_w[3 - \lambda^2]$. Non-dimensionalizing using $\xi = x/h$, $\lambda = y/h$, $\theta = c/c_0$, velocities normalized with $\bar{v}_{x0}$, leads to the non-dimensional equation $\frac{\partial(\tilde{v}_x\theta)}{\partial\xi} = \sigma\frac{\partial}{\partial\lambda}\left(-\hat{v}_y\theta + \alpha\frac{\partial\theta}{\partial\lambda}\right)$ where $\gamma = v_w/\bar{v}_{x0}$ conversion and $\alpha = D/v_w h$ and boundary conditions $\theta(0,\lambda) = 1$, $\theta(\xi, 1) = \alpha\frac{\partial\theta}{\partial\lambda}$ and $\frac{\partial\theta}{\partial\lambda} = 0$ at $\lambda = 0$. As before, application of separation of variables leads to the solution $\theta(\xi,\lambda) = \sum_{n=0}^{n=\infty} A_n Y_n(\lambda)(1-\sigma\xi)^{2\beta_n/3-1}$ where β_n is the nth eigenvalue in $\frac{d}{d\lambda}\left[\hat{v}_y Y_n - \alpha\frac{\partial Y_n}{\partial\lambda}\right] = \beta_n\left[1-\lambda^2\right]Y_n$ with boundary conditions $\alpha Y'_n(1) = Y_n$, $Y_n'(0) = 0$. The functions Y_n obtained by the method of Frobenius[40] using $Y_n(\eta) = \sum_{i=0}^{i=\infty} a_{ni}\eta^i$.

Sherwood was able to plot the results as a polarization index involving the mean or 'mixing cup' concentration, $\Gamma = (c_w - c_m)/c_m = (1-\gamma L)c(L,h) - 1$, vs. the group $\kappa = \gamma h/3\alpha^2 = (v_w h/D)^3 (x/h\text{Re Sc})/3$ with values for different α generally falling on a common curve. This result shows the new dependence of the solution on the term with v_w. Dresner[45,46] developed analytical approximations to this solution as $\Gamma \approx 1.536\kappa^{1/3}$ for small $\kappa < 0.02$, $\Gamma \approx \kappa + 5\left[1-\exp(-\sqrt{\kappa/3})\right]$ for $\kappa > 0.02$ and $\Gamma \approx 1/(3\alpha^2)$ far downstream. Maximum analytical expression approximation errors are <15%. Comparison[32] of average module flux predictions with experiments showed agreement within 5%. Further analysis indicates that developing flow causes greater polarization than developed flow with permeate flow present in both cases.[45]

4.8 Tube velocities and friction factor

Tube flow through a cylindrical channel corresponds to hollow fibres or tubular membrane modules with feed flow through the lumen (Figure 4.11).

The tube flow field is two-dimensional with velocity components in the radial v_r, and normal v_z directions. For constant fluid density ρ and viscosity μ, pressures and flows must satisfy the steady-state mass and momentum balances in cylindrical co-ordinates with rotational symmetry (in φ)

1. Continuity $\frac{1}{r}\frac{\partial(rv_r)}{\partial r} + \frac{\partial v_z}{\partial z} = 0$
2. r momentum $v_r\frac{\partial v_r}{\partial r} + v_z\frac{\partial v_r}{\partial z} = -\frac{1}{\rho}\frac{\partial p}{\partial r} + \upsilon\left[\frac{\partial}{\partial r}\left(\frac{1}{r}\frac{\partial r v_r}{\partial r}\right) + \frac{\partial^2 v_r}{\partial z^2}\right]$
3. z momentum $v_r\frac{\partial v_z}{\partial r} + v_z\frac{\partial v_z}{\partial z} = -\frac{1}{\rho}\frac{\partial p}{\partial z} + \upsilon\left[\frac{1}{r}\frac{\partial}{\partial r}\left(r\frac{\partial v_z}{\partial r}\right) + \frac{\partial^2 v_z}{\partial z^2}\right]$
4. boundary conditions at $r = R$ (wall) $v_r = v_w$ and $v_z = 0$
5. boundary conditions at $r = 0$ (centreline) $v_r = 0$, and $\frac{\partial v_z}{\partial r} = 0$.

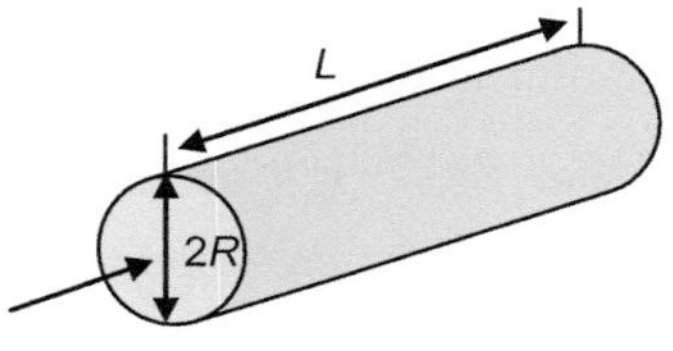

Figure 4.11 Cylindrical channel flow.

4.8.1 Tube developed flow – no permeate

As with slit flow, far downstream the velocity profile is uniform and parabolic with $v_r = 0$ so the Navier–Stokes equations reduce to $0 = -\frac{1}{\rho}\frac{\partial p}{\partial z} + \frac{\nu}{r}\frac{\partial}{\partial r}\left(r\frac{\partial v_z}{\partial r}\right)$. Integration for constant $\frac{dp}{dz}$ yields $v_z(r) = 2\bar{v}_z\left[1 - \left(\frac{r}{R}\right)^2\right]$ describing Poiseuille flow where $\bar{v}_z = \frac{-R^2}{8\mu}\frac{dp}{dz}$. The hydraulic diameter for the tube geometry is $d_h = 4(\pi R^2)/(2\pi R) \sim 2R$. Picking a Reynolds number on this basis yields $\mathrm{Re} = 2R\bar{v}_z / \nu$. The friction factor can then be written as $f = \frac{\mu}{\rho\bar{v}_z^2/2}\left.\frac{\partial v_z}{\partial r}\right|_{\text{wall}} = \frac{2\nu}{\bar{v}_z^2}\frac{4\bar{v}_z}{R} = 16/\mathrm{Re}$.

As with slit channel flow, at the inlet of the tube the velocity profiles are dominated by inlet conditions. After flowing through a distance referred to as the 'entrance length', the velocity profiles approach fully developed profiles. For $\mathrm{Re_w} = 0$, the entrance length required to reach 99% of the fully developed velocity profile is[47] $L_e = (1.18 + 0.112\,\mathrm{Re})h$ for $\mathrm{Re} = v_z D/\nu$. Large permeate flows ($\mathrm{Re_w} \ll 0$) increase the entrance length.[23]

Following the earlier analysis for slit flow,[19] consider the separation of variables solution $v_z(z,r) = \bar{v}_{z0} R f(z) g'(\lambda)$, where $\bar{v}_{z0}$ is the average cross-sectional flow at $z = 0$ and $\lambda = (r/R)$. This yields $\bar{v}_z(z) = \bar{v}_{z0} - (2v_w z / R)$ with velocity profiles $v_r(r) = 2v_w\left[\left(\frac{r}{R}\right) - \frac{1}{2}\left(\frac{r}{R}\right)^3\right]$, $v_z(r,z) = 2\left[\bar{v}_{z0} - \frac{2v_w z}{R}\right]\left[1 - \left(\frac{r}{R}\right)^2\right]$, pressure profile $\Delta p = \frac{8\mu\bar{v}_{z0}L}{R^2}\left(1 - \frac{2v_w L}{\bar{v}_{z0}R}\right)$ and Fanning friction factor $f = \left(\frac{p_L - p_0}{\rho\bar{v}_{z0}^2/2}\right)\left(\frac{\pi R^2}{2\pi RL}\right) = \frac{16}{\mathrm{Re}}\left(1 - \frac{2\,\mathrm{Re_w}\,L}{\mathrm{Re}\,R}\right)$. For $\mathrm{Re_w} = 0$, this reverts to Hagen–Poiseuille flow.[6,10]

The transition from laminar to turbulent flow occurs at Re = 4000 in the porous tube vs. non-porous at 2100.

The corresponding non-dimensional Tilton solution is

$$v_r / \bar{v}_{z0} = \frac{\mathrm{Re}}{32}(2r - r^3)\left[\frac{d^2 p_0}{dz^2} + \varepsilon\frac{d^2 p_1}{dz^2}\right] + \frac{\mathrm{Re}^3}{73728}(58r - 36r^3 + 6r^5 - r^7)\left[\left(\frac{d^2 p_0}{dz^2}\right)^2 + \frac{dp_0}{dz}\frac{d^3 p_0}{dz^3}\right]$$

$$v_z / \bar{v}_{z0} = -\frac{\mathrm{Re}}{8}(1 - r^2)\left[\frac{dp_0}{dz} + \varepsilon\frac{dp_1}{dz}\right] - \frac{\mathrm{Re}^3}{18{,}432}(29 - 36r^2 + 9r^4 - 2r^6)\left[\frac{dp_0}{dz}\frac{d^2 p_0}{dz^2}\right]$$

$$\bar{v}_z / \bar{v}_{z0} = -\frac{\mathrm{Re}}{16}\left[\frac{dp_0}{dz} + \varepsilon\frac{dp_1}{dz}\right] - \frac{3\mathrm{Re}^3}{4096}\left[\frac{dp_0}{dz}\frac{d^2 p_0}{dz^2}\right]$$

$$p/(\rho \overline{v}_{z0}^{2}) = p_0 + \varepsilon p_1, \text{where } p_0 = -\frac{4}{\varepsilon \mathrm{Re}}\sinh(4\varepsilon z) + \pi \cosh(4\varepsilon z)$$

$$p_1 = -\frac{\mathrm{Re}\,\pi}{4}[\sinh(4\varepsilon z) - 2\sinh(8\varepsilon z)] + \frac{1}{16\varepsilon}[16 + (\varepsilon \mathrm{Re}\,\pi)^2][\cosh(4\varepsilon z) - \cosh(8\varepsilon z)] \text{ and } \pi = p/(\rho \overline{v}_{z0}^{2})$$

For a dead-ended system, different boundary conditions are applied to get

$$p_0 = -\frac{4}{\varepsilon \mathrm{Re}}\sinh(4\varepsilon z) + \frac{4\coth(4\varepsilon L)}{\varepsilon \mathrm{Re}}\cosh(4\varepsilon z)$$

$$p_1 = -\frac{\coth(4\varepsilon L)}{\varepsilon}\left[\sinh(4\varepsilon z) - 2\sinh(8\varepsilon z) + A\cosh(4ez) = \frac{1}{\varepsilon}[\coth^2(4eL)+1]\cosh(8ez)\right],$$

$$A = \frac{\coth(4\varepsilon L)}{\sinh(4\varepsilon L)}\left[[\cosh(4\varepsilon L) - \cosh(8\varepsilon L)] + \frac{\sinh(8\varepsilon L)}{2\sinh(4\varepsilon L)\varepsilon}[\coth^2(4eL)+1]\right],$$

where L is the non-dimensional length of the filter feed channel.

Experimental measurement of friction coefficients for stainless porous tube friction factors shows good correlation with the theoretical values.

4.9 Tube concentrations and mass transfer coefficient

The steady-state solute balance in cylindrical co-ordinates with rotational symmetry (in φ) is

$$v_r\frac{\partial c}{\partial r} + v_z\frac{\partial c}{\partial z} = D\left[\frac{1}{r}\frac{\partial}{\partial r}\left(r\frac{\partial c}{\partial r}\right) + \frac{\partial^2 c}{\partial z^2}\right]$$

for boundary conditions at $r = R$ (wall) $c = c_w$ and at the centreline $c = c_b$.

For fully developed parabolic flow with a non-porous wall, the radial velocity is zero. For a long tube, changes in x are small so axial diffusion is neglected compared to radial diffusion. Non-dimensionalizing terms[25] using $\theta = \frac{c - c_w}{c_0 - c_w}$, $\zeta = \frac{Dx}{2R\overline{v}_{x0}} = \frac{x/R}{\mathrm{Re\,Sc}}$ and $\eta = \frac{r}{R}$, the mass balance becomes $\left[1-\eta^2\right]\frac{\partial\theta}{\partial\zeta} = \frac{1}{\eta}\frac{\partial}{\partial\eta}\left(\eta\frac{\partial\theta}{\partial\eta}\right)$ with $\theta(\zeta \pm 1) = 0$ and at the centreline $\partial\theta/\partial\eta = 0$ while at the inlet $\theta(0,\eta) = 1$. This is known as a Graetz–Nusselt equation,[26,27] which is solved using separation of variables. This assumes a trial solution $\theta(\xi,\eta) = X(\xi)Y(\eta)$. Substitution requires $\frac{1}{X}\frac{dX}{d\zeta} = \frac{1}{\left[1-\eta^2\right]Y}\left[\frac{d^2Y}{d\eta^2} + \frac{1}{\eta}\frac{dY}{d\eta}\right] = -\lambda_n^2$ for constant λ_n so the solution becomes $\theta(\xi,\eta) = \sum_{n=1}^{n=\infty} C_n Y_n(\eta)\exp(-\lambda_n^2\varsigma)$ where

Table 4.4 Non-porous tube eigenvalues and derivatives[25]

n	λ_n	$(\partial Y_n/\partial \lambda_n)_{\eta=1}$	$(\partial Y_n/\partial \eta)_{\eta=1}$
1	2.70436	−0.50089	−1.01430
2	6.67903	0.37146	1.34924
3	10.67337	−0.31826	−1.57231
4	14.67107	0.28648	1.74600

λ_n is the nth eigenvalue in $\frac{d^2Y}{d\eta^2}+\frac{1}{\eta}\frac{dY}{d\eta}=-\lambda_n^2\left[1-\eta^2\right]Y$ with boundary conditions $Y(1) = 0$, $Y'(0) = 0$. The coefficients C_n are derived by applying the boundary condition at $\zeta = 0$, multiplying each side by $\left[1-\eta^2\right]Y_n\eta d\eta$ and integrating to get $C_n = \frac{\int_{\eta=0}^{\eta=1}\left[1-\eta^2\right]Y_n\eta\, d\eta}{\int_{\eta=0}^{\eta=1}\left[1-\eta^2\right]Y_n^2\eta\, d\eta} = \frac{-2}{\lambda_n(\partial Y_n/\partial\lambda_n)_{\eta=1}}$. The functions Y_n are obtained by the method of Frobenius using $Y_n(\eta)=\sum_{i=0}^{i=\infty}a_{ni}\eta^i$ with the coefficient recursion formula $a_{ni} = 0$ if $i < 0$, $a_{n0} = 1$ and $a_{ni} = -\lambda_n^2(a_{n,i-2}-a_{n,i-4})/i^2$. A trial and error procedure was used to obtain the values in Table 4.4. These values can be used to determine the Sherwood number as $\mathrm{Sh}=\frac{k2R}{D_b}=\frac{\partial\theta}{\partial\eta}\Big|_{\text{wall}}=-\tilde{D}\sum_{n=1}^{n=\infty}C_nY_n{}'(\eta)\exp(-\lambda_n^2\varsigma)$ -> 1.86

Figure 4.12 shows the evolution of the concentration.

The infinite series solution just presented is slowly convergent and does not lend itself easily to physical insights about the solution. Consider the Leveque[42] solution associated with short entrance lengths and thin concentration boundary layers. Letting

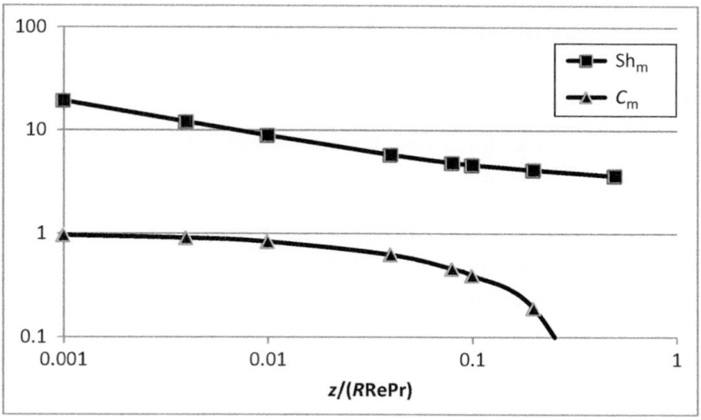

Figure 4.12 Sherwood number.

$n = 1-\eta$ where $n \ll 1$ and simplifying the mass balance yields $2n(\partial\theta/\partial\zeta) = \partial^2\theta/\partial n^2$. Order of magnitude analysis indicates that $\delta_c/\zeta \sim 1/\delta_c$ or $\delta_c \sim \zeta^{1/3}$ for the concentration boundary length scaling. Following Deen, we consider a similarity solution $\theta(p)$ where $p = n/g(\zeta)$. Then $\partial\theta/\partial\zeta = (d\theta/dp)(\partial p/\partial\zeta) = \theta'(-ng^{-2}g')$ and $\frac{\partial^2\theta}{\partial n^2} = \frac{\partial}{\partial n}\left(\frac{d\theta}{dp}\frac{\partial p}{\partial n}\right) = \frac{\partial}{\partial n}(\theta'/g) = \theta''/g^2$. Substituting in the mass balance and substituting for n yields $\theta'' + 2p^2(g^2g')\theta' = 0$ where $g^2g' = \alpha$ constant is required to make $\theta(p)$. Using $g(0) = 0$, the equation for $g(\xi) = \left(\frac{9}{2}\xi\right)^{1/3}$ where $\alpha = 3/2$ for convenience. This matches the expected boundary length scaling. The equation for concentration becomes $\theta(p) = \frac{3}{\Gamma(1/3)}\int_p^\infty \exp(-t^3)dt$ where $\Gamma(1/3) \sim 2.67894$ with Sherwood number $\mathrm{Sh} = \frac{k2R}{D} = \frac{-2(\partial\theta/\partial n)_{n=0}}{\theta_w - \theta_b} \approx \frac{6}{\Gamma(1/3)g(\xi)} = 1.357\,\mathrm{Re}^{1/3}\,\mathrm{Sc}^{1/3}(R/z)^{1/3}$.

4.9.1 Tube with developing flow and permeate

Using the same approach as described earlier for flat plates, Sherwood[44] used the steady-state mass balance $\frac{\partial(v_x c)}{\partial x} + \frac{\partial(v_y c)}{\partial y} = \frac{\partial}{\partial y}\left(D\frac{\partial c}{\partial y}\right)$ with simplified Yuan velocity profiles[27] to develop a non-dimensional balance. Separation of variables led to a solution that could be conveniently plotted in terms of a polarization index $\Gamma = (c_w - c_m)/c_m = (1-2\gamma L)c(z,R) - 1$ vs. the parameter $\kappa = 2\gamma z/3R\alpha^2$ with values for different α falling on a common curve. The Dresner slit geometry solutions can be adapted for tube flow by recognizing that curvature effects in the velocity profile are negligible and that the tube velocity is 4/3 that of the flat plate[44] so one can substitute $\kappa_{\text{tube}} = \gamma z/4R\alpha^2$ in his approximations.

4.10 Cassette velocities and friction factor

4.10.1 Simulation

A 3D flow simulation model for a cassette was developed using COMSOL computational fluid dynamics software.[48] As shown in Figure 4.13, the simulation volume consisted of the screen height (y axis) with two twill weave sequences in the flow direction (z axis) and two twill weave sequences in the x axis direction. Laminar flow and Newtonian viscosity models were used. No allowances were made for permeate flow, screen intrusion into the membrane, shear at the membrane surface, or membrane deformation.

The simulation feed flow was 8 LMM. For water flow, the z direction velocities can vary sixfold across the channel. The pressure field varies slightly across the feed channel (x-direction) and shows a relatively uniform decline along the length of the feed channel (z-direction).[48]

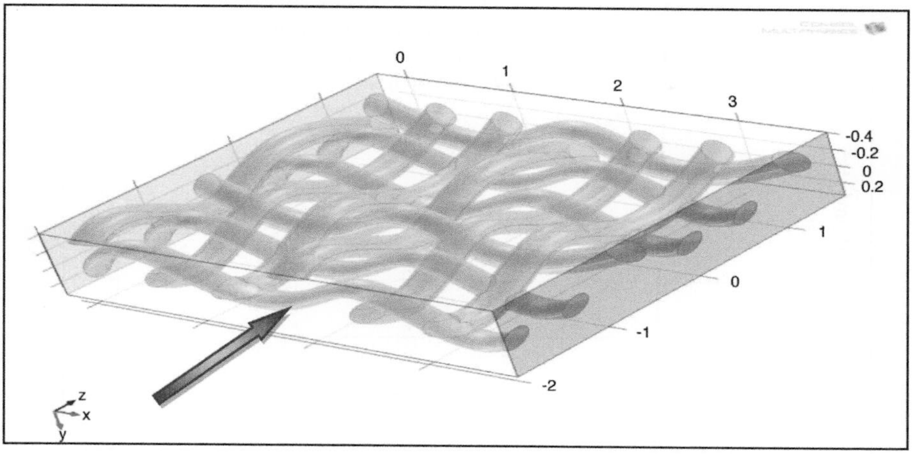

Figure 4.13 3D model of 2/1 twill weave D screen. The arrow shows the inlet feed flow direction. Dimensions in millimetres.
(Courtesy of EMD Millipore Corp.)

Screens are generally inserted in cassette feed channels to enhance back transport of retained solutes away from the membrane surface. This increases fluxes and reduces membrane area requirements. Figure 4.14 shows a 2D cross-section with flow streamlines through the screened channel fibres. The screen fibres force the fluid to

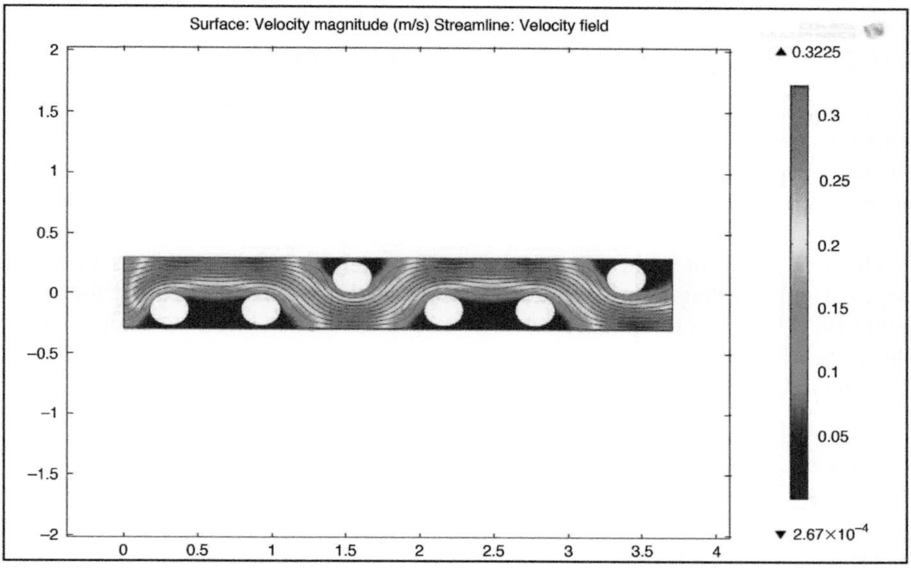

Figure 4.14 Fluid streamlines in the screened channel.
(Courtesy of EMD Millipore Corp.)

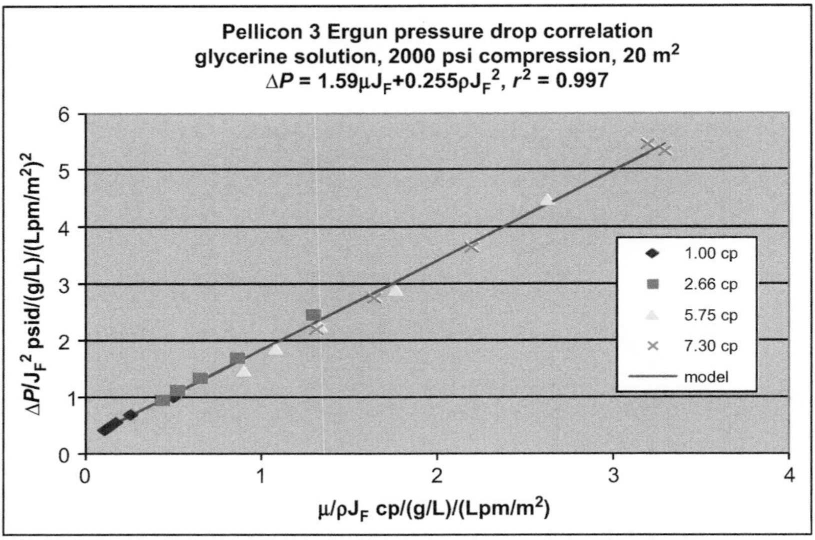

Figure 4.15 Feed channel pressure drop simulation values and data.

follow a zigzag path. This induces circular eddies perpendicular to the main flow path by conservation of angular momentum. In a curved pipe, these eddies are referred to as Dean vortices.[49] Tighter screens require the fluid to make tighter turns and induce faster rotation. This increases both the pressure drop and the mass transfer coefficient.

4.10.2 Cassette pressure drop and friction factor

The Ergun equation[5] (shown in Figure 4.15 for Pellicon C screen ultrafiltration cassettes) was used to calculate pressure drop for a given viscosity and feed flow rate. The Ergun equation embodies both form drag (the frictionless conversion of kinetic energy (ρv^2) to pressure changes[5] and skin drag (fluid friction moving parallel to a surface converted to heat calculated as multiples of $\rho v^2 f$ for friction factors f). It shows excellent statistical fit to the data of $r^2 = 0.997$ over the 1.0–7.3 cp and 2–9 Lpm/m² ranges. The form drag term (intercept term) is small and can be neglected for most applications (medium flows and moderate viscosities) leaving just the skin drag term. This correlation has been confirmed for retained MAbs at high viscosities as well.

Using the definition of the friction factor, the skin drag term becomes

$$f = \frac{p_L - p_0}{\rho \overline{v}_{z0}^2/2} = \frac{K\mu J_F}{\rho \overline{v}_{z0}^2/2} = \frac{2K\nu A_{xs}}{\overline{v}_{z0} A_{memb}} = \frac{2KA_{xs}H_{char}}{\mathrm{Re}\, A_{memb}}$$

or $f\,\mathrm{Re} =$ constant, where A_{xs} is the screen feed channel cross-sectional area for flow, A_{memb} is the membrane area, H_{char} is a characteristic height of the screen channel, Re is the Reynolds number.

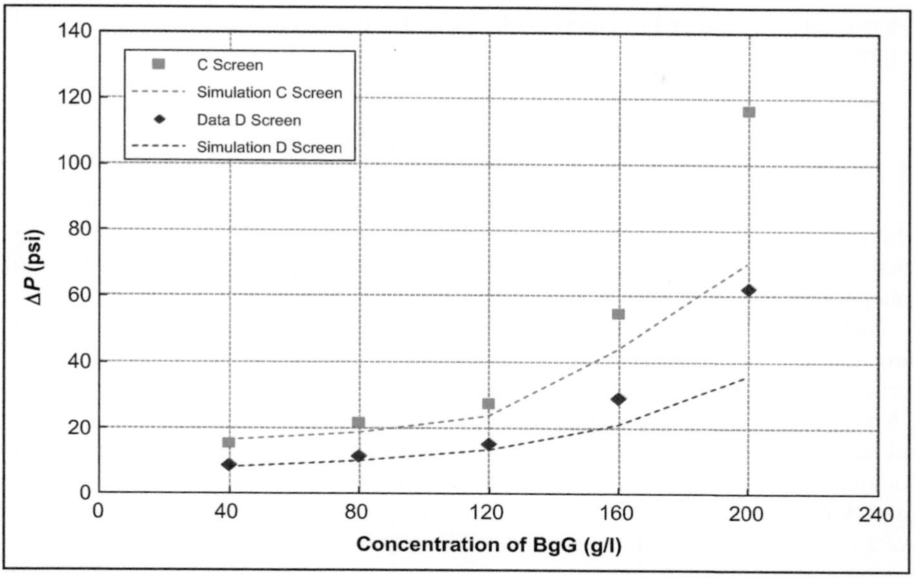

Figure 4.16 Feed channel pressure drop simulation values and data. (Courtesy of EMD Millipore Corp.)

The hydraulic diameter is a useful characteristic length scale. For a screened cassette channel, it can be calculated[9] as $d_h = \dfrac{4e}{2/h+(1-e)s}$ where e = spacer porosity or void fraction, $2h$ = spacer height, s = spacer specific surface area per unit volume. Another characteristic length is the mesh spacing L_s. One can consider flow and concentration boundary layers growing within L_s and then starting over at the next spacer.

A comparison of simulation pressure drops and experimental data for C and D feed screens shows excellent agreement at low feed concentrations (Figure 4.16). This shows that the Navier–Stokes equations and boundary conditions can accurately predict the pressure drop through the complex geometry of a screened channel. As shown in Figure 4.15, this includes flow phenomena, such as stagnant points where the flow is zero, and the presence of eddies, which have reverse flow.

The divergence between the simulation results and experimental data at higher protein concentrations could be explained by (1) an inaccurate viscosity model, (2) cassette channel inlet–outlet effects or (3) polarization effects. Viscosity measurements vary among viscometers and may show shear-thinning or non-Newtonian effects not included here. Inlet–outlet effects are due to experimental cassette pressure drops in the holder and cassette flow distribution to the various feed channels. Polarization effects are due to experimental concentrations near the membrane surface that increase viscosity and alter velocity profiles.

A macroscopic analysis of channel hydraulics can be used to see if pressure drop can be related to average shear. From a steady-state macroscopic mechanical energy balance[2] (isothermal, non-compressible fluid, no pump and no gravitational head height)

$$\Delta P = -\Delta \frac{\rho}{2} \frac{\langle \overline{v}^3 \rangle}{\langle \overline{v} \rangle} - \frac{\rho \langle \overline{v} \rangle^2 L(f_{\text{formscreen}} + f_{\text{skinscreen}} + f_{\text{skinwall}})}{2R_{\text{h}}} \tag{4.2}$$

where $\langle \overline{v} \rangle$ is a time and cross-section averaged velocity, f is a generalized Fanning friction factor averaged along the channel and R_{h} the hydraulic radius (cross-section area/perimeter).

These terms arise from kinetic energy changes associated with changes in fluid direction, form drag on the screen (arising from pressure the field acting on the wall surfaces), and skin drag (arising from fluid shear at the wall surfaces and screen for a Newtonian fluid). Evaluation of detailed expressions for each term, assuming no interaction effects, and comparison to experimental data showed good agreement dominated by form drag on the screen and kinetic energy changes, not from shear related skin drag effects ($a < 5\%$ effect).[9] This analysis indicates that one cannot obtain average shear values in a screened channel directly from pressure drop measurements as in an open channel.

4.11 Cassette concentrations and mass transfer coefficient

Given the complex flow in the screened feed channel, detailed 3D modelling of the concentration profiles with permeate flow and polarization phenomena requires high simulation resolution and modelling capabilities. A 2D simulation by Schwinge[8] showed variations in concentration across the channel and variations in the mass transfer coefficients along the membrane surface.

Using a larger scale of resolution of 0.5 cm slices in the z direction, corresponding to several mesh lengths, these non-uniformities within the spacer mesh average out. Within this slice, one can consider an average pressure, mixing cup concentration, wall concentration, mass transfer coefficient, and flux. Mixing cup concentration, feed flux, and pressures are calculated using mass and momentum balances. Flux calculation by the osmotic model requires an osmotic pressure model,[48] mass transfer coefficient model (velocity dependence), viscosity model (concentration dependence) and polarization model (standard film model). Non-linearities require a trial and error solution procedure.

Application of the model to data sets indicates that quadratic osmotic pressure models, mass transfer coefficient velocity dependence of a 0.5 power, viscosity exponential dependence on concentration, a pressure drop model proportional to viscosity and feed flow, and the standard film model together can describe the data. The 0.5 power dependence of the mass transfer coefficient is consistent with Da Costa[9] and a fluid mechanics model that considers simultaneously growing velocity and concentration boundary layers over each mesh distance (Section 4.7). The model can predict the

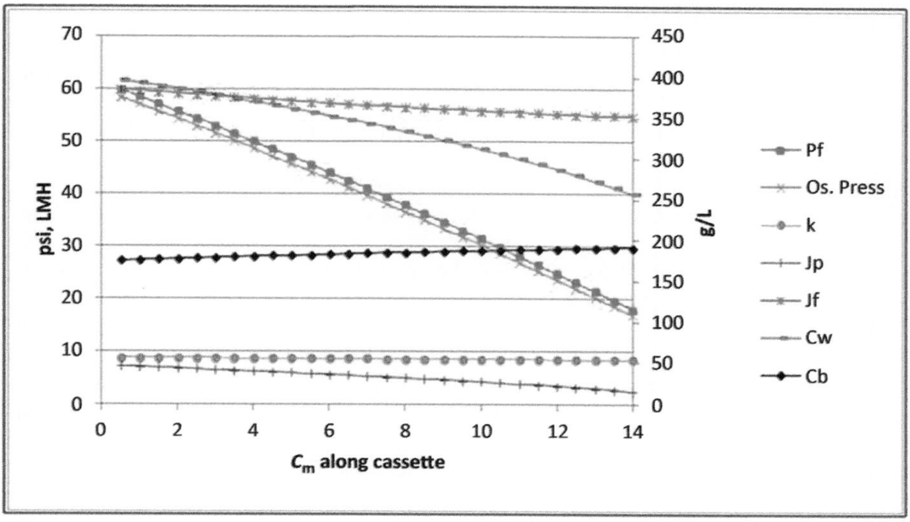

Figure 4.17 Cassette profiles net flux = 5 LMH, feed 175 g/L, 1 LMM, 60 psig. (Courtesy of EMD Millipore Corp.)

flux impact of changes in operating at constant feed flow with variable pressures and concentration vs. constant feed pressure with variable feed flow and concentrations.

4.11.1 Cassette profiles

Consider the spatial variation in properties along the feed channel at the end of a batch UF process with a slow feed flux of 1 LMM, high bulk concentration of 175 g/L and a high feed pressure of 60 psig. For the model predictions shown in Figure 4.17, the TMP is high enough to maintain positive permeate flow and overcome osmotic pressure along the entire cassette path length. The average permeate flux is only 5 LMH (0.083 LMM). The net conversion of 8.3% is low so the feed flux, bulk concentration, and mass transfer coefficients are relatively constant. The feed pressure, osmotic pressure, wall concentration and flux all vary significantly along the feed channel. The main driver is the drop in feed pressure along the channel. The wall concentration and flux respond to this change in feed pressure. The osmotic pressure is never a constraint. Since feed pressure drop determines performance and not osmotic pressure. For the empirical gel model, C_g would be influenced by viscosity, feed flux and channel resistance, not osmotic effects.

References

1. Eckstein EC, Bailey PG, Shapiro AH. Self-diffusion of particles in shear flow and suspension. J Fluid Mech 1977;79:191.
2. Drew DA, Shonberg JA, Belford G. Lateral interial migration of a small sphere in fast laminar flow through a membrane duct. Chem Eng Sci 1991;46:3219.

3. Belford G, Davis RH, Zydney AL. The behavior of suspensions and macromolecular solutions in crossflow microfiltration. J Membr Sci 1994;96:1.
4. Stamatakis K, Tien C. A simple model of cross-flow filtration based on particle adhesion. AIChE J 1992;39:3067.
5. Bird RB, Warren ES, Lightfoot EN. Transport phenomena. 2nd ed. New York, NY: John Wiley; 2002.
6. Dean WM. Analysis of transport phenomenon. New York, NY: Oxford Press; 1998.
7. Sparrow EM, Gregg JL. Mass transfer, flow, and heat transfer about a rotating disk. Trans ASME 1960;294–302.
8. Schwinge J, Wiley DE, Fletcher DF. Simulation of the flowaround spacer filaments between channel walls. 2. Mass-transfer enhancement. Ind Eng Chem Res 2002;41:4879–88.
9. Da Costa AR, Fane AG, Wiley DE. Spacer characterization and pressure drop modeling in spacer filled channels for ultrafiltration. J Membr Sci 1994;87:79.
10. Schlichting H. Boundary layer theory. 6th ed. New York, NY: McGraw-Hill; 1968.
11. von Kármán T. Über laminare und turbulente Reibung. Zeits F Angew Math U Mech 1921;1:233–52.
12. Cochran WG. The flow due to a rotating disc. Proc Cambridge Phil Soc 1934;30: 365–75.
13. Stuart JT. On the effects of uniform suction on the steady flow due to a rotating disk. Quart J Appl Math 1954;7:446–57.
14. Rogers MH, Lance GN. The rotationally symmetric flow of a viscous fluid in the presence of an infinite rotating disk. J Fluid Mech 1960;7:617–31.
15. Levich VI. Physiochemical hydrodynamics. Englewood Cliffs, NJ: Prentice Hall; 1962.
16. Smith KA, Colton CK, Merrill EW, Evans LB. Convective transport in a batch dialyzer: determination of true membrane permeability from a single measurement. Chem Eng Prog Symp Ser 1968;64:45–58.
17. Kaufmann TG, Leonard EF. Mechanism of interfacial mass transfer in membrane transport. AIChE J 1968;14:421–5.
18. Mitchell BD, Deen WM. Effect of concentration on the rejection coefficients of rigid macromolecules in track-etch membranes. J Colloid Interf Sci 1986;113:132–42.
19. Kozinski AA, Lightfoot EN. Protein ultrafiltration: a general example of boundary layer filtration. AIChE J 1972;18:1030–40.
20. Malone DM, Anderson JL. Diffusional boundary-layer resistance for membranes with low porosity. AIChE J 1977;23:177–84.
21. Atkinson B, Brocklebank MP, Card CCH, Smith JM. Measurements of velocity profile in developing liquid flows. AIChE J 1967;13:17–20.
22. Blasius H. Grenzschichten in Flussigkeiten mit kliener Reibung. Z Math U Phys 1908;56:1–37. Engl. Transl. in NACA TM 1256.
23. Howarth L. On the solution of the laminar boundary layer equations. Proc Roy Soc Lond 1938;A164:547–79.
24. Nikuradse J. Laminare Reibungsschichten an der langsangetromten Platte Monograph Zentrale f. wiss. Berlin: Berichtswesen; 1942.
25. Raithby G, Knudsen DC. Hydrodynamic development in a duct with suction and blowing. J Appl Mech 1974;41:896–902.
26. Berman AS. Laminar flow in channels with porous walls. J Appl Phys 1953;24:1232–5.
27. Yuan SW, Finkelstein AB. Laminar pipe flow with injection and suction through a porous wall. Trans ASME 1956;78:719–24.
28. Regirer SA. On the approximate theory of the flow of a viscous incompressible liquid in a tube with permeable walls. Sov Phys Tech Phys 1960;5:602–5.

29. Harnett JP, Koh JCY, McComas ST. A comparison of predicted and measured friction factors for turbulent flow through rectangular ducts. J Heat Transfer 1962;84:82–8.
30. Kozinski AA, Schmidt FP, Lightfoot EN. Velocity profiles in porous-walled ducts. Ind Eng Chem Fund 1970;9:502–5.
31. Raithby G. Laminar heat transfer in the thermal entrance region of circular tubes and two-dimensional rectangular ducts with wall injection and suction. Int J Heat Transfer 1971;14:223–43.
32. Doughty JR, Perkins HC. Thermal and combined entry problems for laminar flow between parallel porous plates. J Heat Transfer 1972;94:233–4.
33. Terrill RM. An exact solution for flow in a porous pipe. Z Angew Math Phys 1982;33: 547–52.
34. Brady JF. Flow development in a porous channel and tube. Phys Fluids 1984;27:1061–7.
35. Karode SK. Laminar flows in channels with porous walls, revisited. J Membr Sci 2001;191:237–41.
36. Tilton N, Martinand D, Serre E, Lueptow RM. Incorporating Darcy's law for pure solvent flow through porous tubes: asymptotic solution and numerical simulations. AIChE J 2012;58:2030–44.
37. Brown GM. Heat or mass transfer in laminar flow in a circular or flat conduit. AIChE J 1960;6:179–83.
38. Graetz L. Uber die Warmeleitfahigkeit von Flussigkeiten. Ann Physik 1883;18(3):79.
39. Hildebrand FB. Advanced calculus for applications. Englewood Cliffs, NJ: Prentice-Hall; 1962.
40. Bender CM, Orszag SA. Advanced mathematical methods for scientists and engineers. New York, NY: McGraw-Hill; 1978.
41. Sellars JR, Tribus M, Klein JS. Heat transfer to laminar flow in a round tube or flat conduit–the graetz problem extended. Trans Am Soc Mech Eng 1956;78:441–448.
42. Leveque AM. Les lois de la transmission de la chaleur par convection. Ann Mines 1928;13:201.
43 Pohlhausen E. Der Warmeaustausch zwichen festen Korpern und Flussigkeiten miot kleiner Reibung und leiner Warmeleitung. ZAMM 1921;1:115.
44. Sherwood TK, Brian PLT, Fischer RE, Dresner L. Salt concentration at phase boundaries in desalination by reverse osmosis. Ind Eng Chem Fund 1965;4(2):113–8.
45. Gill WM, Derzansky LJ, Doshi MR. Convective diffusion in laminar and turbulent hyperfiltration (reverse osmosis) systems. Surface and Colloid Science, (Matejevic E editor), Wiley, NY, 1971;4:261–360.
46. Dresner L. Boundary layer build-up in the demineralization of salt water by reverse osmosis. Oak Ridge Natl. Lab Rept., vol. 3621; May 1964.
47. Opong WS, Zydney AL. Diffusive and convective protein transport through asymmetric membranes. AIChE J 1991;37:1497.
48. Lutz H, Zou Y. High concentration ultrafiltration, 2014, submitted to biotech prog for sublication 2015.
49. Dean WR. The streamline motion of fluid in a curved pipe. Phil Mag 1928;5(7):673–95.

Configurations

5

Herb Lutz
EMD Millipore, Biomanufacturing Sciences Network

5.1 Batch concentration

The most common bioprocessing configuration and application is batch concentration. Fluid is pumped through a module by the feed pump while a retentate valve provides upstream pressure (Figure 5.1).

The conversion (fraction of the feed solution passing through the membrane as permeate) is typically lower than the desired volume reduction. This means that a tank is required to recirculate retentate through the module in multiple passes to achieve the desired final concentration. As the permeate flow is withdrawn from the system, the tank volume drops and the mass retained in the tank becomes more concentrated. The process is completed when the retentate tank concentration reaches the target specification.

The performance of this system can be determined using system balances on the total mass and the component of interest:

Total mass $\frac{dM}{dt} = \rho \frac{dV}{dt} = -\rho q_p = -\rho JA$ for retentate tank mass M, retentate volume V, retentate density ρ, time t, volumetric permeate flow q_p, flux J and area A.

Component mass $\frac{dM_i}{dt} = \frac{d(cV)}{dt} = -c_p q_p = -scJA$ for component mass M_i, component retentate concentration c, component permeate concentration c_p and component sieving coefficient s.

The initial conditions are $V(0) = V_0$, $c(0) = c_0$.

For most components in aqueous systems, one can assume equal densities between the retentate and permeate. In addition, the mass balance assumes that the component concentration is uniform within the retentate loop. The retentate is more concentrated in the line after the module to the feed tank. If this line volume is small and the increased concentration is moderate, a uniform concentration approximation is good. Satisfying the uniform concentration also requires a well-mixed retentate tank with no stratification or bypass.

These coupled non-linear differential equations can be solved by recognizing that $JAdt = dV = d(cV)/sc$ so $\frac{d(cV)}{dV} = sc$ or $c + \frac{dc}{dV} = sc$. Rearranging and integrating gives $\ln(c/c_0) = (s-1)\ln(V/V_0)$, or $c = c_0 X^{1-s}$ where $X = V_0/V$, the volume

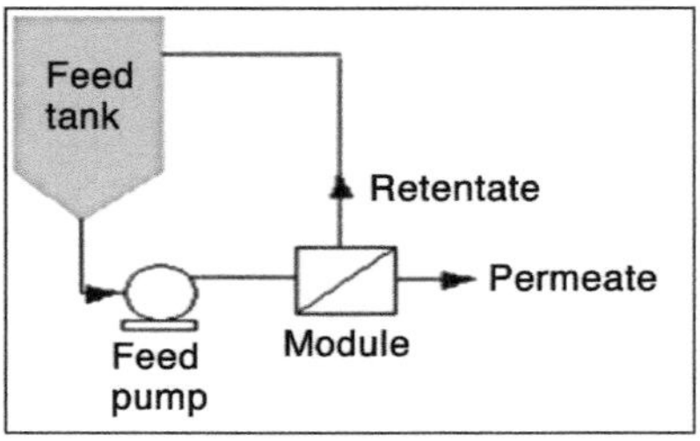

Figure 5.1 Batch concentration.

reduction factor. To achieve a final retentate concentration of c requires a volume reduction factor $X=(c/c_0)^{1/(1-s)}$ and a final volume of $V=V_0(c/c_0)^{1/(s-1)}$. If s = 1 (permeable component), then $c=c_0$ throughout. If s = 0 (retained component), then $c=c_0(V_0/V)=M_0/V$, the retentate component mass is constant and the concentration increase is proportional to the volume reduction (e.g., 1/2 the volume = twice the concentration). The retentate mass is $M=cV=c_0X^{1-s}V=M_0X^{-s}$ and the retentate yield is $Y_R=M/M_0=X^{-s}$. The yield loss has gone into the permeate stream.

Consider the overall mass balance with constant density $\frac{dV}{dt}=-JA$. Integrating this equation requires knowing how the flux varies over the concentration process. For dilute feeds, the flux may be insensitive to solute concentration and constant. In that case $V=V_0-JAt$ and the membrane area required is $A=(V_0-V)/Jt$. The numerator $V_0-V=V_0(1-1/X)$ is the permeate volume. Where the flux varies over the concentration run, consider an average flux $\bar{J}$ where $(V_0-V)/\bar{J}=\int_{V0}^{V}dV/J$. The membrane area required is $A=(1-1/X)V_0/\bar{J}t$. The area scales with the time required, the initial retentate volume, and the average flux (e.g., if the time or average flux is halved or the volume doubled, the area must be doubled).

5.1.1 Gel model

For most applications, the flux is well described by the gel model J = $k\ln(c_g/c)$ with parameter k as the mass transfer coefficient and c_g as the gel point concentration. In this case, one must integrate the non-linear ordinary differential equation $\frac{dV}{dt}=-k\ln\left(c_g/\left[c_0(V_0/V)^{1-s}\right]\right)A$ or $A=\frac{V_0I(X,\alpha)}{kt\ln(c_g/c_0)}$ where $\alpha=\frac{1-s}{\ln(c_g/c_0)}$ and $I(X,\alpha)=\int_1^X\frac{dX}{X^2\{1-\alpha\ln X\}}$. Substituting $t=(1-\alpha\ln X)/\alpha$ and rearranging the integral to $I(t,\alpha)=\frac{-e^{-1/\alpha}}{\alpha}\int_{-1/\alpha}^{t}\frac{e^{-t}dt}{t}$ puts it in a form similar to the exponential integral[1] $E_1(z)=\int_z^\infty\frac{e^{-t}dt}{t}$, but this function is not defined for negative real values of z. No closed form solution of this integral is available and it requires numerical integration. Since the argument of the integral changes slowly, this integration does not require special methods and Euler's method or Simpson's rule work well.

5.1.2 Cheryan approximation

Cheryan[2] proposed the useful average flux approximation using a weighting of the initial and final fluxes over the process run $\bar{J} \approx J_0/3 + 2J_f/3$. This generates the convenient closed form analytical expression $\bar{J} \approx k\ln(c_g/c_0)/3 + 2k\ln(c_g/c_0X^{1-s})/3$ or $\bar{J} \approx J_0\left[1 - \dfrac{2(1-s)\ln X}{3\ln(c_g/c_0)}\right]$. The area then becomes $A = \dfrac{V_0(1-1/X)}{t\bar{J}} \approx \dfrac{V_0 I_{\mathrm{Ch}}(X,\alpha)}{kt\ln(c_g/c_0)}$ where Cheryan integral approximation is $I_{\mathrm{Ch}}(X,\alpha) = \dfrac{1-1/X}{[1-2\alpha\ln X/3]}$. Figure 5.2 shows that the approximation is 5.5% high for $\alpha = 0$ and $X < 2.5$. It is recommended that the Cheryan approximation be used only for rough sizing. More accurate results are obtained from developing a flux model for a particular application and numerically integrating the model.

As the tank volume drops, it is important to make sure that the fluid in the tank stays well mixed to apply the performance equations. If the membrane sees higher concentration due to poor mixing, fluxes will be lower and the time–area requirements will be higher. If the membrane sees a lower concentration due to poor mixing, fluxes will be higher and the time–area requirements will be lower. Poor mixing can also lead to sampling error where the average retentate concentration may not be at the desired specification even though a sample of the retentate indicates it is.

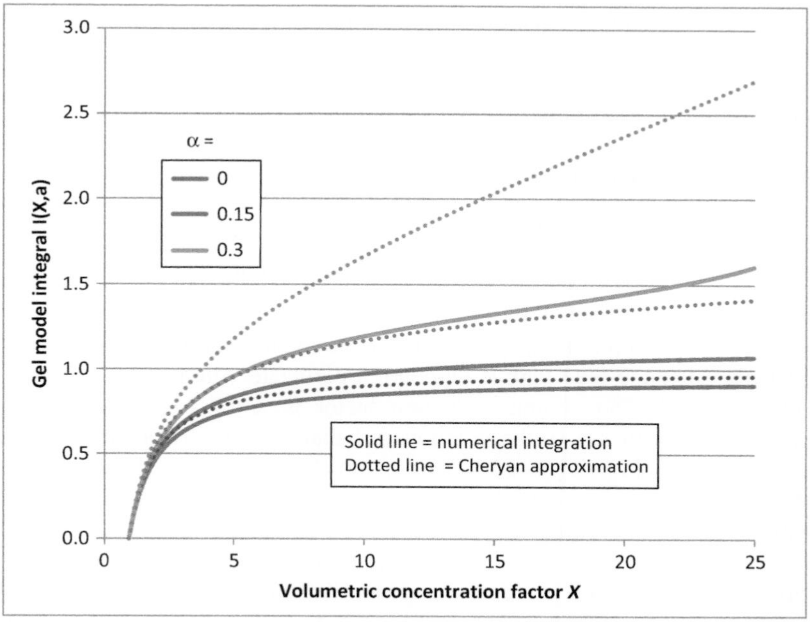

Figure 5.2 Cheryan approximation for batch concentration.

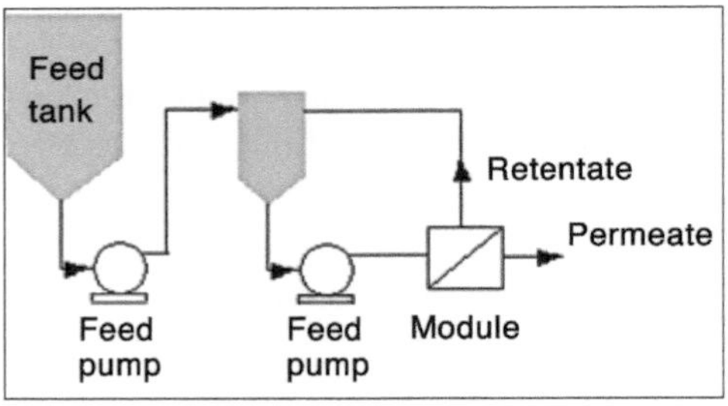

Figure 5.3 Fed-batch concentration.

If the tank level drops below the retentate return line during processing, biopharmaceutical products can foam and denature. These system considerations cause batch concentration volume reduction to be limited[3] to $X \sim 30$. Increased volume reduction can be accommodated by fed-batch concentration.

5.2 Fed-batch concentration

Fed-batch concentration is commonly used to achieve higher volume reduction factors than possible with batch concentration alone. Instead of returning the retentate back to a feed tank containing the entire batch volume, it is returned to a smaller retentate tank (Figure 5.3). Feed is also pumped into this retentate tank at the same rate as permeate is withdrawn from the system so the retentate tank level remains constant during fed-batch concentration. When all the feed has been pumped into the retentate tank, one can concentrate the retentate further in a batch concentration process.

Another way of implementing a fed-batch operation is to put a bypass line in-between the retentate return line and the pump inlet.

The performance of this system can be determined using system balances on the total mass and the component of interest:

Total mass $\frac{dV}{dt} = q_F - q_p = 0$ for volumetric feed flow $q_F = q_p = JA$.

Component mass $V_R \frac{dc}{dt} = c_0 q_F - c_p q_p = (c_0 - sc) JA$ for retentate tank volume V_R.

The initial conditions were $c(0) = c_0$.

The feed batch volume is V_0 with V_R in the retentate tank and the difference in the feed tank. After the feed tank is emptied, a permeate volume of $V_p = V_0 - V_R$ has been generated. The volume reduction factor is $X = V_0 / V_R$. For a fully retained component with $s = 0$, the component balance on the retentate tank yields $cV_R = c_0 V_R + c_0 V_p = c_0 (V_R + V_0(1 - 1/X))$ or $c / c_0 = 1 + r(1 - 1/X)$ where $r = V_0/V_R$ is the tank ratio. For batch concentration with $s = 0$, $c / c_0 = X$. As shown in Figure 5.4, the retentate concentration in fed-batch concentration starts at $X = 1$, $c / c_0 = 1$, and increases, exceeding the retentate concentration using batch concentration. It rises relatively quickly and levels out at $c / c_0 = 1 + r$. When $X = 1 + r$, the feed tank is empty and a conventional batch concentration can begin on the retentate tank along the line $c / c_0 = X$.

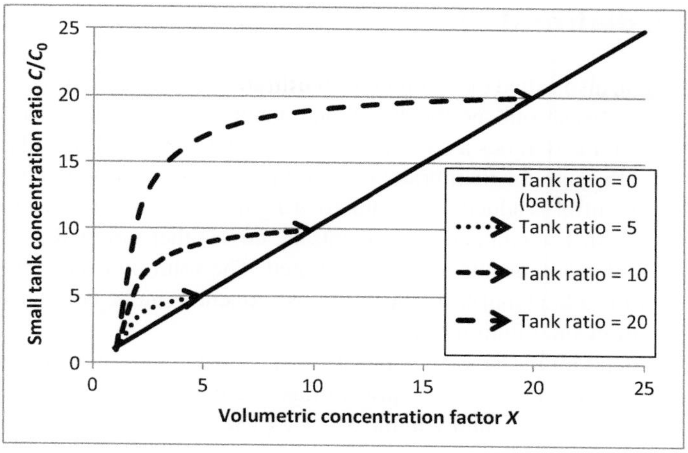

Figure 5.4 Fed-batch concentration.[4]

For a general value of s, the component mass balance integrates to $c = c_0\left[1-(1-s)\exp(-sr\{1-1/X\})\right]$. To achieve a final retentate concentration of c requires a volume reduction factor $X = 1/\left\{1+\ln\left[(1-c/c_0)/(1-s)\right]/sr\right\}$. If $s \to 1$ (permeable component), then $c \to c_0$ throughout. If $s \to 0$ (retained component), then $c = c_0\left[1+r(1-1/X)\right]$. The retentate mass is $M = M_0\left[\dfrac{1+r(1-1/X)}{X}\right]$ and the retentate yield is $Y_R = M/M_0 = \left[1+r(1-1/X)\right]/X$. The yield loss has gone into the permeate stream.

For sizing, one should integrate a flux model $\dfrac{dV}{dt} = -JA$ where flux depends on concentration and incorporating the concentration change with volume derived earlier. Using the Cheryan approximation with the gel model $\bar{J} \approx k\ln(c_g/c_0)/3 + 2k\ln\left(c_g/c_0\left[1+r(1-1/X)\right]\right)/3$ or $\bar{J} \approx J_0\left[1-\dfrac{2\ln(1+r(1-1/X))}{3\ln(c_g/c_0)}\right]$.
As for batch concentration, the Cheryan approximation is recommended for rough sizing, but it is recommended to develop a flux model for a particular application and numerically integrate the model. If the overall process includes both a fed-batch concentration step and a concentration step, one should integrate each separately.

While fed-batch concentration allows one to reach greater final concentrations than batch concentration, Figure 5.4 shows that the membrane will see higher concentrations causing lower fluxes and higher system areas. In addition, higher concentrations may lead to more protein degradation in the system. Application of fed-batch concentration should be considered on a case-by-case basis.

5.3 Batch diafiltration

Batch diafiltration, also called discontinuous diafiltration,[2] is sometimes used at small-scale operation to 'wash out' permeable contaminants from a solution or exchange buffers. First, a fluid undergoes a concentration step where the solvent and permeable components pass through the membrane while retained components are concentrated. Then, a diafiltrate buffer is added to the tank to bring the total volume back to its original level (Figure 5.5). The retained components are now at their original concentration while the permeable components have been diluted. The solution is now closer to the concentration of the added diafiltrate buffer. This sequence can be continued with repeated concentration and dilution steps as desired.

Consider a batch volume V_0 with concentration c_0 undergoing a volume reduction X. After batch concentration, the component has retentate concentration $c = c_0 X^{1-s}$, component mass $M = M_0 X^{-s}$, and volume $V = V_0/X$ (Section 5.1). After dilution back to the original volume V_0, the component concentration is $c = c_0 X^{-s}$, the component mass is $M = M_0 X^{-s}$ and the amount of permeate generated or buffer used is $V_P = V_0 - V_0/X = V_0(1-1/X)$. If this process is repeated n times, the final component concentration becomes $c = c_0 X^{-ns}$, the component mass is $M = M_0 X^{-ns}$ and the volume of permeate generated or buffer used is $V_P = nV_0(1-1/X)$.

To achieve a final retentate concentration of c requires $n = -\dfrac{\ln(c/c_0)}{s \ln X}$ volume reductions and dilutions, each by a factor X. If $s = 0$ (retained component), then $c = c_0$. The component yield in the retentate is $Y_R = M/M_0 = X^{-ns}$. The yield loss has gone into the permeate stream. Sizing the membrane area is equivalent to the batch concentration case described earlier (Figure 5.6).

5.4 Constant volume diafiltration

Constant volume diafiltration is used to 'wash out' permeable contaminants from a solution or exchange buffers. New buffer is added to the retentate tank at the same flow rate as the permeate leaving the system (Figure 5.7). This balancing of flows keeps the volume in the system constant throughout the process. The permeate stream

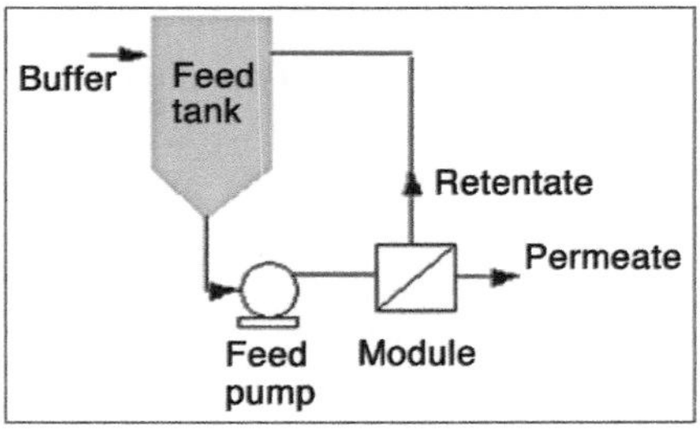

Figure 5.5 Batch diafiltration.

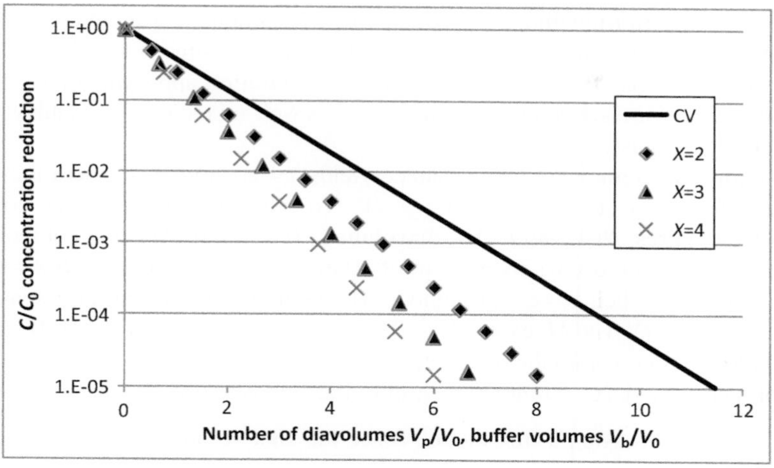

Figure 5.6 Batch and constant volume diafiltration.

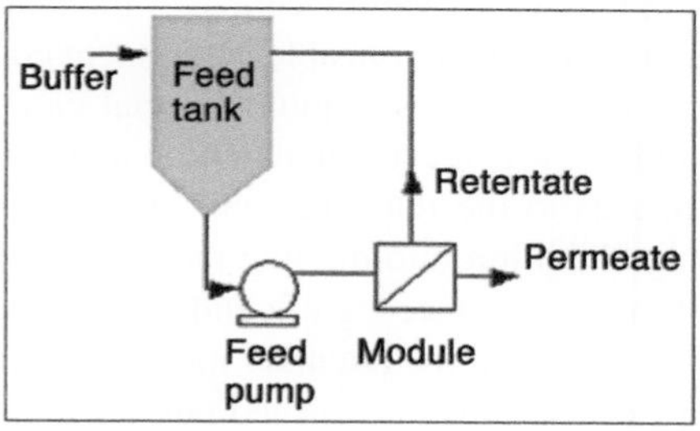

Figure 5.7 Constant volume diafiltration.

contains permeable components in the original feed while the new buffer being added does not. This leads to the net removal of permeable components over the course of the diafiltration process.

The performance of this system can be determined using system balances on the total mass and the component of interest:

Total mass $\frac{dM}{dt} = \rho \frac{dV}{dt} = \rho\left(q_{\mathrm{d}} - q_{\mathrm{p}}\right) = 0$ for diafiltrate flow rate q_{d}

Component mass $\frac{dM_{\mathrm{i}}}{dt} = V\frac{dc}{dt} = -c_{\mathrm{p}}q_{\mathrm{p}} = -scJA$ for component mass M_{i}, component retentate concentration c, component permeate concentration c_{p} and component sieving coefficient s.

The initial conditions are $V(0) = V_0$, $c(0) = c_0$.

For most components in aqueous systems, one can assume equal densities between the retentate and permeate. In addition, the mass balance assumes that the component concentration is uniform within the retentate loop. The retentate is more concentrated in the line after the module to the feed tank. If this line volume is small and the increased concentration is moderate, a uniform concentration approximation is good. Satisfying the uniform concentration also requires a well-mixed retentate tank with no stratification or bypass.

Integrating the component mass balance yields $c = c_0 \exp(-sN)$ where the number of diavolumes is $N = V_{\text{buffer}}/V = V_{\text{permeate}}/V$. To achieve a final retentate concentration of c requires $N = -\ln(c/c_0)/s$ diavolumes. If $s = 1$ (permeable component), then the concentration decays by a factor of 10 every ~2.7 diavolumes. If $s = 0$ (retained component), then $c = c_0$ throughout. The component mass in the retentate is $M = cV = c_0 \exp(-sN)V = M_0 \exp(-sN)$. The retentate yield is $Y_R = M/M_0 = e^{-sN}$. The yield loss has gone into the permeate stream.

Membrane area sizing requires integrating the flux over the diafiltration process $\frac{dV_{\text{perm}}}{dt} = q_{\text{perm}} = -JA$ using a flux model. Since the tank volume is constant over the diafiltration, the retained component concentrations are also constant. The gel model implies that the flux will be constant as well so $At = NV/J$. However, if the different diafiltrate buffer changes the gel point concentration, the mass transfer coefficient, or the fluid viscosity, the flux may vary over the course of the diafiltration.

The mass balance assumes that the permeable component concentration is uniform within the retentate loop and tank. This requires special care with buffer addition, retentate mixing and dead legs. It is most convenient to add diafiltrate buffer into the retentate line before it returns to the retentate tank. Addition directly into the tank requires good tank mixing. Addition into the line feeding the pump will prevent adequate mixing and result in performance significantly worse than model predictions. Stagnant zones or 'dead legs' within the retentate loop will maintain high concentrations while the rest of the system is being washed out. As an example, consider that a 1000 L batch diafiltration with $N = 12$ diavolumes would reduce the concentration in a well-mixed system to 3.1×10^{-7} of its original value. If a 1 mL dead leg at the original concentration mixes with the rest of the 1000 L volume, it will produce a concentration 1×10^{-6} of the original value. As a result, the concentration of permeable components starts deviating from the exponential model for $N > 12$–15 (Figure 5.8).

There are additional factors affecting the sieving coefficient that can make the diafiltration process deviate from the model. Solutes that would be expected to be completely permeable can be partially retained. Sieving coefficients can also change during the course of diafiltration. These are discussed in more detail under later case studies.

5.5 Batch process sizing and optimum diafiltration concentration

A typical ultrafiltration batch process will involve concentrating by a factor X (with retained species concentration from c_0 to c_F) and diafiltering by a factor N. These processing steps can be run in any sequence. That is, one can concentrate by X first and then diafilter at c_F by N second, or one can diafilter at c_0 by N first and then concentrate

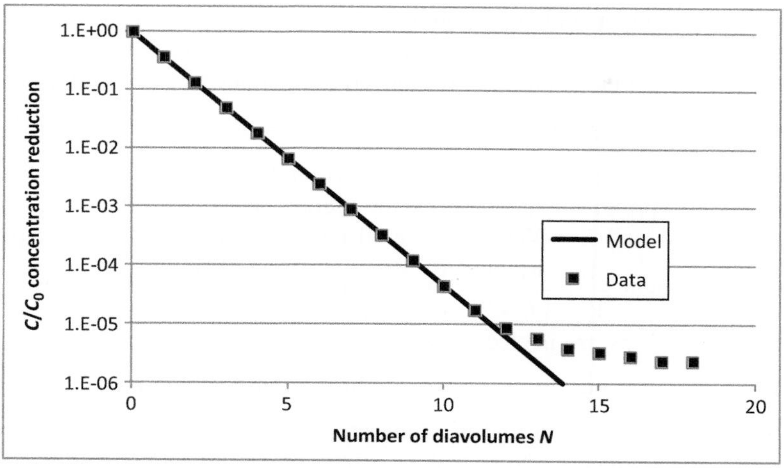

Figure 5.8 Diafiltration concentration model and data.

by X second. At the end of either process, the final retentate concentrations will be the same and meet the process requirements.

Consider a general three-step process involving an initial concentration step or UF1, a diafiltration step or DF and a final concentration step or UF2 as outlined in Table 5.1. We require that the final concentration be c_F and the diafiltration be N. It is convenient to consider the case with complete retention. As one varies the amount of concentration in the initial step from $X_1 = 1$ to c_F/c_0, the concentration at which diafiltration occurs, c_D, varies between c_0 and c_F and the general process moves between the two extremes just described while keeping the final retentate concentration at c_F. The

Table 5.1 UF/DF batch process analysis

Process step	UF1: initial concentration step	DF: diafiltration step	UF2: final concentration step
Description	Concentrate from c_0 to c_D using X_1	Diafilter at c_D using N diavolumes	Concentrate from c_D to c_F using X_2
Initial step retentate volume	V_0	V_0/X_1	V_0/X_1
Final step retentate volume	V_0/X_1	V_0/X_1	$V_0/(X_1X_2)$
Permeate volume	$V_0(1-1/X_1)$	V_0N/X_1	$(V_0/X_1)(1-1/X_2)$
Buffer volume	0	V_0N/X_1	0
Average flux $\bar{J} \approx$	$J_0/3+2J_D/3$	J_D	$J_D/3+2J_F/3$
Time–area required At =	$\dfrac{V_0(1-1/X_1)}{[J_0/3+2J_D/3]}$	$\dfrac{V_0/X_1}{[J_D]}$	$\dfrac{(V_0/X_1)(1-1/X_2)}{[J_D/3+2J_F/3]}$

volumes for each step are obtained from the processing requirements (X_1, N and X_2) along with a mass balance requiring changes in the retentate volume to balance buffer input and permeate output volumes. The time–area requirements for each stage are obtained by integrating the permeate flow $\frac{dV_{\text{perm}}}{dt} = -JA$ as $At = \int_0^{V_{\text{perm}}} dV_{\text{perm}}/J = V_{\text{perm}}/\bar{J}$ for each step. To analyse the impact of varying X_1 on this process, it is convenient to use the Cheryan approximation to describe the average flux in each step.

From Table 5.1 and $X = X_1X_2$, the overall requirements for all three steps can be written as the sum $\frac{At}{V_0} = \frac{3(1-1/X_1)}{[J_0+2J_D]} + \frac{N/X_1}{[J_D]} + \frac{(3/X_1)(1-1/X_2)}{[J_D+2J_F]}$. Following Ng,[5] if the concentration steps are neglected relative to the diafiltration step and we substitute $X_1 = c_D/c_0$, this becomes $\frac{At}{V_0} = \frac{Nc_0}{c_D J_D}$. In general, one can plot flux vs. concentration data as cJ vs. c to find the concentration to maximize cJ and minimize the time–area requirements.[6] Note that this optimum in area–time results from a tradeoff between the permeate volume and the flux. Area–time is determined by the ratio of the permeate volume to the flux. At the optimum the volume reduction is just balanced by the flux decline. Using the gel model to describe the flux, the equation becomes $\frac{At}{V_0} = \frac{Nc_0}{c_D k \ln(c_g/c_D)}$. To minimize, differentiating the expression $\frac{d\{c_D \ln(c_g/c_D)\}}{dc_D} = \ln\left(\frac{c_g}{c_D}\right) + \frac{c_D}{(c_g/c_D)}\left(\frac{-c_g}{c_D^2}\right) = 0$. Solving gives $c_D = c_g/e$ to minimize area–time.

One can also include the concentration steps. Using concentrations instead of volume reduction factors ($X_1 = c_D/c_0$, $X = c_F/c_0$), the gel model to describe the fluxes, normalizing the concentrations by c_g (e.g. $c_0' = c_0/c_g$), and rearranging simplifies to $\frac{kAt}{V_0} = -\frac{3(1-c_0'/c_D')}{[2\ln c_D' + \ln c_0']} - \frac{Nc_0'}{c_D' \ln c_D'} - \frac{3(c_0'/c_D' - c_0'/c_F')}{[\ln c_D' + 2\ln c_F']}$. These terms are plotted as a cumulative sum in Figure 5.9 for the case of $c_0' = 0.04$, $c_F' = 0.6$, $N = 12$. Note that the optimum defined by the diafiltration step alone ($c_D' = 1/e \sim 0.368$) is very shallow so one can run from $c_D' = 0.3$ to 0.45 and get similar requirements. Addition of the concentration steps shifts the optimum very slightly down to $c_D' = 0.335$. The concentration steps seem to balance each other out so the diafiltration step approximation is good.

5.6 Constant mass batch diafiltration[7]

It is convenient and very common to use retentate mass measurements to control diafiltration buffer addition. Retentate tank load cells are less expensive and more reliable than measuring fluid level in the tank or permeate flow rate. However, if the diafiltration buffer density differs significantly from that of the retentate, diafiltration will cause a significant change in the retentate solution density. Buffer addition to

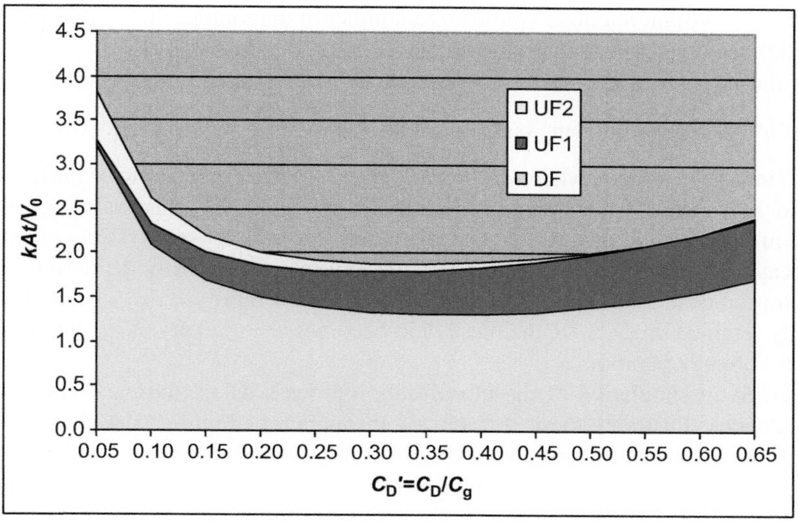

Figure 5.9 Minimum time–area diafiltration point.

keep the retentate mass constant will cause changes in the retentate volume. This alters the standard formulas describing system performance based on constant volume operation and a new constant mass diafiltration model is needed.

Changes in retentate volume can be quantified by considering a mass balance on the retentate:

$\frac{dM_R}{dt} = \frac{d(\rho V)}{dt} = \rho_i q_i - \rho q$, where M_R is the grams of retentate mass, ρ is the g/L density, V is the retentate volume, q is the L/min flow rate and i refers to the DF buffer input.

During diafiltration at constant retentate volume, the flow rates in and out are equal ($q_i = q$) and the retentate mass M_R changes with retentate density as $\frac{dM_R}{dt} = V\frac{d\rho}{dt}$. During diafiltration at constant retentate mass M_R, the flow rates in and out are different (buffer $q_i = q\rho/\rho_i$) and the retentate volume V changes with retentate density as $\frac{dV}{dt} = -\left(\frac{V}{\rho}\right)\frac{d\rho}{dt}$.

Changes in component concentration during diafiltration can be quantified by considering a mass balance on the component:

$\frac{dM}{dt} = \frac{d(cV)}{dt} = -scq$, where M is the g of component mass, c is the g/L component concentration, s is the component sieving coefficient and it is assumed that no component is added in the DF buffer.

For diafiltration at constant retentate volume V, this balance simplifies to $V\frac{dc}{dt} = -scq$ or $d\ln c = -sdN$. Here, $dN = dV_i/V = q_i dt/V$, the ratio of buffer volume added to the constant retentate volume as a number of diavolumes. Integration gives a first-order decay in concentrations with diavolumes as $c = c_0 \exp(-sN)$.

For diafiltration at constant retentate mass M_R, substituting $V = M_R/\rho$ and $q = q_i \rho_i / \rho$ in the component balance yields $M_R \frac{d(c/\rho)}{dt} = -scq_i\rho_i/\rho$. This simplifies to $d\ln(c/\rho) = -sdN'$, where $dN' = dM_i/M_R = q_i\rho_i dt/M_R$, the ratio of buffer mass added to the constant retentate mass as a number of 'diamasses'. Integrating gives the change in concentrations with diamasses as $c = c_0(\rho/\rho_0)\exp(-sN')$. The concentration decays as a result of both diamasses and any retentate density decreases. If the concentration of a component is expressed as the mass ratio $\underline{y} = c_A/c_P$ (*g* of *A* per g of a fully retained product *P*), the mass ratio becomes $y = y_0 \exp(-sN')$, analogous to constant volume operation.

A numerical simulation of the ultrafiltration process using mass and component mass balances illustrates these differences in operation. Figure 5.10 shows simulated constant mass and constant volume operation using a flux model of the form $J = 24\ln(2.8/X)/(1+0.0009G)$. The numerator terms were derived earlier from the concentration data. The denominator term accounts for the effect of excipient concentration *G* on the flux through viscosity changes. Figure 5.10 shows identical flux declines with *X* during the initial concentration step. During constant volume diafiltration, the flux increases as the excipient *G* is washed out and viscosity drops. During constant mass diafiltration, flux increases even more because density drops, volume increases, and *X* decreases. Constant weight diafiltration has higher fluxes for shorter processing times for smaller sizing, an average lower product concentration *P* for less product degradation, and a faster reduction in excipient *G* concentration. However, the retentate volume is larger requiring larger tankage and more buffer volume.

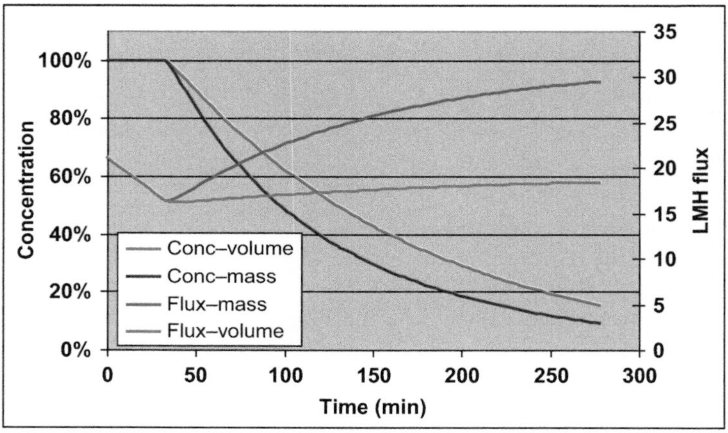

Figure 5.10 Constant mass vs. constant volume diafiltration.

5.7 Single-pass or continuous concentration

Membranes can also be run in a continuous, rather than batch mode. Figure 5.11 shows a single-pass concentration configuration. The retentate exits the module at the target final concentration instead of having to recycle back to the tank as in a batch configuration. This process requires high conversion (i.e., the permeate flow is a large fraction of the inlet feed flow) to reach the target concentration. If the process flux is in the linear region of the flux curve, then increasing the transmembrane pressure will increase the flux and the conversion. Another way to accomplish this is to stage modules in series where the retentate from the first module becomes the feed for a subsequent module. This has the effect of increasing the feed channel length. The permeate flows from each module add to increase the total conversion. If the process flux is in the polarized region of the flux curve, one can also reduce the feed flow rate. The flux is reduced less than the feed flow so conversion increases with lower feed flow rates. The longer feed channel or reduced feed flow both have the effect of increasing residence time in the feed channel to increase conversion. Single pass operation is the standard method of configuring reverse osmosis systems.

The performance of this system can be determined using steady-state system balances on the total mass and the component of interest:

Total mass $m_F = m_R + m_P$ or $q_F = q_R + q_P$ for feed, retentate and permeate mass flows m or volumetric flows q with constant density.
Component mass $c_F q_F = c_R q_R + s c_R q_p$ for component concentrations c and passage s.

This can be written as $c_R = c_F / [1 + (s-1)Y]$ for conversion $Y = q_P / q_F$. For a given feed and target retentate concentration, this gives the required conversion. For a given feed flow q_F, average module flux J and conversion Y, the required membrane area is $A = Y q_F / J$.

The flows and concentrations for the continuous concentration vary along the length of the feed channel instead of with time as in the batch case. Taking differential balances across a feed channel length dz produces:

Total mass $\frac{dq}{dz} = -Ja$ for membrane area/feed channel length a, feed flow q.
Component mass $\frac{d(cq)}{dz} = -csJa$ for feed concentration c.
The initial conditions were $q(0) = q_F$ and $c(0) = c_F$.

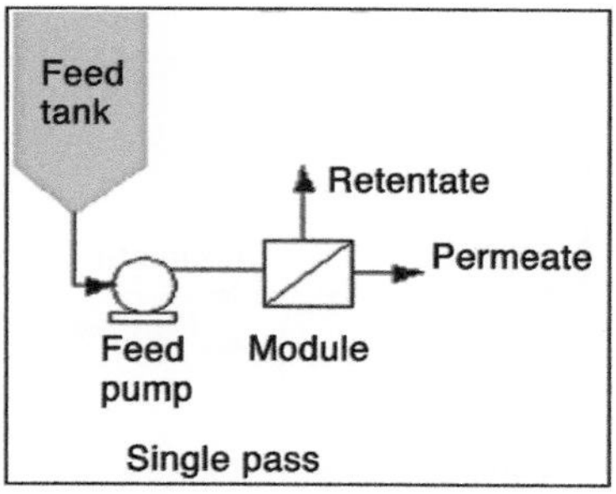

Figure 5.11 Single-pass configuration.

Combining and simplifying, the component balance can be written as $q\frac{dc}{dz} = c(1-s)Ja$. The gel flux model for varying feed channel flows can be written as $J = k_0\left(\frac{q}{q_F}\right)^n \ln(c_g/c)$ where $n \sim 0.5$ for proteins in cassettes. Applying the balances to the most common case where $s = 0$ with conversion $Y = 1 - q/q_F$, the balances reduce to $dY = (k_0 adz/q_F)(1-Y)^n \ln\left[(1-Y)\left(c_g/c_F\right)\right]$ where $k_0 adz/q_F$ represents a dimensionless area. Integration along the feed channel does not have a closed form analytical solution, analogous to the batch case in Section 5.1. One can numerically integrate Y vs. $k_0 adz/q_F$ for a given concentration ratio c_g/c_F, and mass transfer coefficient exponent n.

Following Zeman,[9] the Cheryan approximation $\bar{J} \approx J_F/3 + 2J_R/3$ is applied where J_F is the flux at the inlet of the feed channel and J_R is the flux at the outlet of the channel or retentate. Considering the most common case where $s = 0$, applying the gel flux model yields $\bar{J} \approx k_0 \ln(c_g/c_F)/3 + 2k_0(1-Y)^n \ln(c_g(1-Y)/c_F)/3$ for conversion Y. Recognizing that $Y = \bar{J}aL/q_F$, where L is the length of the feed channel, one has the following implicit relationship between conversion Y, dimensionless area $k_0 aL/q_F$, feed concentration ratio c_g/c_F and mass transfer coefficient exponent n: $Y \approx (k_0 aL/q_F)\ln(c_g/c_F)/3\left\{1+2(1-Y)^n\left[1+\ln(1-Y)/\ln\left(c_g/c_F\right)\right]\right\}$. A comparison[9] between the Cheryan approximation with the numerical integration for $n = 0.5$ and ratio $c_g/c_F = 20$ shows excellent agreement with the error growing to a maximum of $\sim 10\%$ at a dimensionless area of 0.2 and shrinking to insignificance thereafter.

A single stage of a continuous concentration configuration is shown in Figure 5.12. This configuration is also referred to as 'feed and bleed'.[8] Compared to batch concentration, this has a retentate withdrawal. Compared to single-pass concentration, this has an added recycle loop. One can operate this system in batch mode (without retentate withdrawal) until the concentration reaches the target specification. Then the system will run in steady-state with retentate withdrawal. By comparison with single-pass operation, continuous operation increases the flow through the module by the amount of recycle. Fluxes are increased as a result of the higher cross-flow so less membrane area is needed relative to single-pass concentration.

One can link a number of these stages together by having the retentate from the first stage feed a subsequent stage. This breaks up the overall concentration into a series of

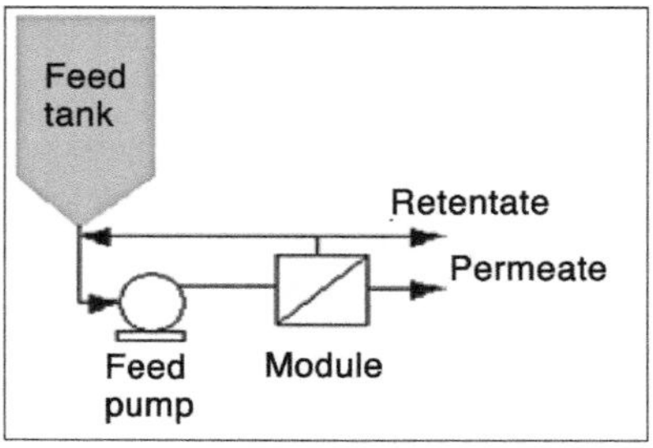

Figure 5.12 Continuous concentration configuration.

staged concentrations. One can optimize the area needed for each stage. This configuration is the standard method of configuring dairy systems.

The performance of this system can be determined using steady-state system balances on the total mass and the component of interest. For a control volume that includes the recycle loop, these balances are identical to those for single-pass operation: $c_{\mathrm{R}} = c_{\mathrm{F}} / [1+(s-1)Y]$ for conversion $Y = q_{\mathrm{P}} / q_{\mathrm{F}}$. The conversion required for a given concentration is $Y = (1 - c_{\mathrm{F}} / c_{\mathrm{R}}) / (1-s)$. For a given feed and target retentate concentration, this gives the required conversion. For a given feed flow q_{F}, average module flux J and conversion Y, the required membrane area is $A = Yq_{\mathrm{F}} / J$.

5.8 Single-pass or continuous diafiltration

Single-pass diafiltration involves a series of concentrations and inline dilutions. Buffer passes into the permeate stream during the concentration step and volume is restored in the inline dilution. This is exactly analogous to the batch diafiltration described in Section 5.5 except that it is taking place sequentially in space along a feed channel length instead of sequentially in time. For n equal stages, the final component concentration becomes $c = c_0 X^{-ns}$ with component mass $M = M_0 X^{-ns}$. To reach a concentration comparable to N diavolumes in constant volume diafiltration $e^{-Ns} = X^{-ns}$, or $X = e^{N/n}$ volume reduction for a given diavolume equivalents and number of stages. As shown in Figure 5.13, one needs at least four stages for each stage to show a reasonable conversion.

For *continuous batch DF*, the area required is $A = \dfrac{NM_0}{C_{\mathrm{d}} J_{\mathrm{d}}^{t}}$ for feed mass M_0, diafiltration concentration C_{d}, flux J_{d} and time t. The optimum area is given by maximizing the term $C_{\mathrm{d}}J_{\mathrm{d}}$. For a gel flux model $J_{\mathrm{d}} = kln\left(\dfrac{C_{\mathrm{g}}}{C_{\mathrm{d}}}\right)$, this optimum is $C_{\mathrm{d}} = C_{\mathrm{g}}/e$.

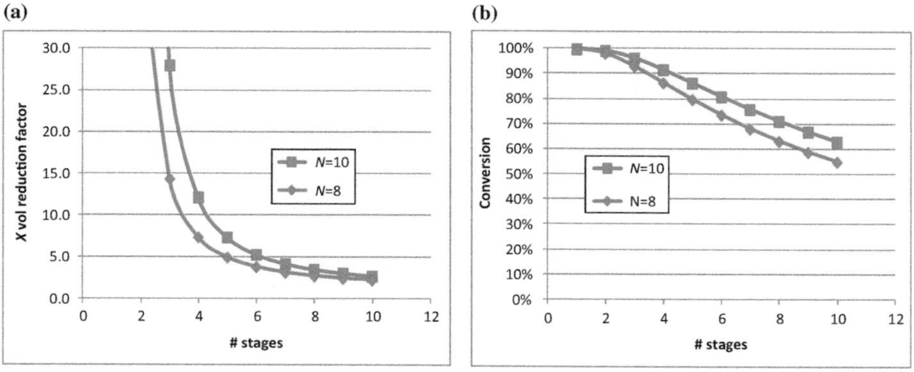

Figure 5.13 Single-pass diafiltration: (a) volume reduction factor required and (b) conversion required.

For *sequential single-pass DF*, the area required per stage is $A = \frac{i\dot{M}_0}{C_d v_d}$ for feed mass flow $\dot{M}_0$, volumetric reduction factor X, i stages and a module diafiltrate feed flux of v_d. The feed flux changes through the length of the module are dramatic due to high (>50%) conversions. In batch DF things are more constant with low (~10–15%) conversions. The optimum area is again given by maximizing the term $C_d v_d$. The average flux is given by the Cheryan approximation $J = J_o/3 + 2J_{ret}/3$. Using a modified gel flux model to account for changing feed channel flux, $\bar{J}_d = \frac{1}{3} k v_d^n \ln\left(\frac{C_g}{C_d}\right) + \frac{2}{3} k v_{ret}^n \ln\left(\frac{C_g}{C_{ret}}\right)$. Using $v_{ret} = v_d/X$, $C_{ret} = C_d X$, $y_d = \ln(C_g/C_d)$, the flux is $\bar{J}_d = \frac{1}{3} k v_d^n [y_d + 2X^{-n}\{y_d - \ln X\}] = v_d/(1-1/X)$. Then, $v_d = \left\{\frac{k[y_d + 2X^{-n}\{y_d - \ln X\}]}{3(1-1/X)}\right\}^{1/(1-n)}$. Maximizing requires $\frac{d}{dy_d} e^{-y_d}[y_d + 2X^{-n}\{y_d - \ln X\}]^{1/(1-n)} = 0$ or $y_d = \frac{1}{1-n} + \frac{\ln X}{1+.5X^n}$ as shown in Figure 5.14 on the left. Increasing X means reducing C_d until about a 10X reduction where things level out.

For *Countercurrent Single-Pass Diafiltration*,[10] diafiltrate buffer is added before the last stage with the last stage permeate used as the diafiltrate to the previous stage. All flows are the same for each stage, protein is fully retained and salt fully passes through the membrane (Figure 5.15).

By mass balances, $F = R + D$. Let $X = F/R$ then $F = RX$ and $D = P = R(X-1)$. Fully retained protein concentrations are the same from stage-to-stage. Within a stage, $C_{pR}R = C_{pF}F$ so $C_{pF} = C_{pR}/X$. Fully passing salt concentrations decrease from stage-to-stage, but are the same within each stage $C_{si} = C_{sRi} = C_{sPn} = C_{sFi}$.

- Stage n: $C_{sDn} = 0$, $C_{sRn-1}R = C_{sFn}F$, or $C_{sn-1}/C_{sn} = X$
- Stage $n-1$: $C_{sRn-2}R + C_{sPn}D = C_{sFn-1}F$, or $C_{sn-2} = C_{sn-1}X - C_{sn}(X-1) = C_{sn}X^2 - C_{sn}(X-1)$,

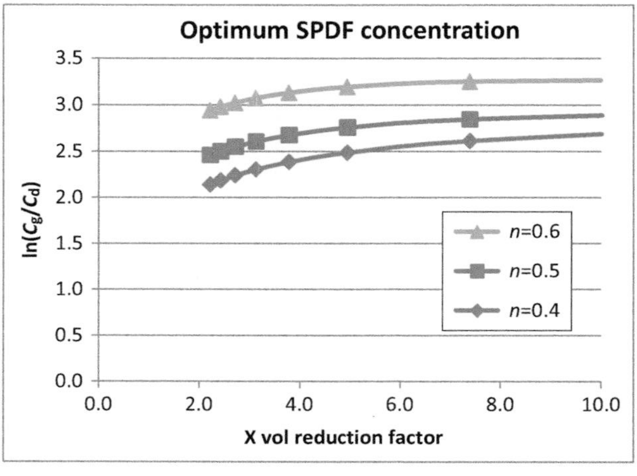

Figure 5.14 Optimum single-pass diafiltration vs. volume reduction factor.

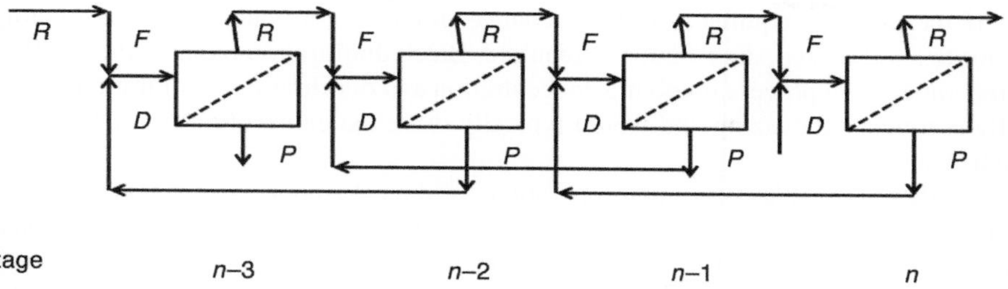

Figure 5.15 Countercurrent single-pass diafiltration.

or $C_{sn-2}/C_{sn} = X^2 - (X - 1)$

- Stage $n - 2$: $C_{sRn-3}R + C_{sPn-1}D = C_{sFn-2}F$, or $C_{sn-3} = C_{sn-2}X - C_{sn-1}(X - 1)$, $C_{sn-3}/C_{sn} = X[X^2 - (X - 1)] - (X - 1)[X] = X^3 - 2X(X - 1)$
- Stage $n - 3$: $C_{sn-4} = C_{sn-3}X - C_{sn-2}(X - 1)$, $C_{sn-4}/C_{sn} = X[X^3 - 2X(X - 1)] - (X - 1)[X^2 - (X - 1)]$
  $= X^4 - 3X^2(X - 1) + (X - 1)^2$
- Stage $n - 4$: $C_{sn-5}/C_{sn} = X[X^4 - 3X^2(X - 1) + (X - 1)^2] - (X - 1)[X^3 - 2X(X - 1)]$
  $= X^5 - 4X^3(X - 1) + 3X(X - 1)^2$

5.9 Selection of system configuration and operating methods

Ultrafiltration applications typically involve a concentration or volume reduction. Batch concentration is generally recommended as it requires the smallest membrane area. However, alternative configurations, such as fed-batch and single-pass, have advantages in some applications. Fed-batch uses a smaller retentate tank volume. This is helpful when using an existing system for a new molecule where the feed volume does not fit in the existing tank. A smaller tank also tends to have a lower minimum working volume to enable a higher final concentration (see Section 7.6). In the extreme, one can completely eliminate the retentate tank and feed the retentate directly into the pump inlet along with the feed from a feed tank. One can use valving to slowly change the retentate flow from feeding back to the retentate tank and instead divert it to a bypass pipe that feeds directly into the feed pump inlet. This bypass loop can be difficult to clean and recover product from. Note that the tank does need to have enough volume to conduct any necessary diafiltration at the target concentration. Fed-batch also requires more pump passes, which can affect protein quality (see Section 7.7). Fed-batch should be used only when it is required.

Single-pass operation also requires more membrane area than batch concentration. However, it significantly reduces the minimum working volume constraint to allow reaching higher concentrations. It also has much less holdup for high product recovery without dilution. It can be a more simple compact system consisting of just a holder and retentate valve. This makes single pass useful for harvest concentration, feed conditioning prior to another step, or final formulation.

If the product requires a buffer exchange, for final formulation or conditioning prior to another step, diafiltration is required. Batch diafiltration requires the smallest area–time. For a process involving concentration and diafiltration, diafiltration generally dominates the sizing and would typically drive the entire ultrafiltration process to be batch.

Single pass enables continuous operation. To the extent one can extend processing times, and reduce changeover time between batches, the manufacturing process can be smaller and less costly, with higher productivity. This does require however, very reliable steps. One can envision single-pass diafiltration being useful in this context.

References

1. Abramowitz M, Stegun IA. Handbook of mathematical functions. New York, NY: Dover; 1965.
2. Cheryan M. Ultrafiltration and microfiltration handbook. Lancaster, PA: Technomic; 1998.
3. Goodrich E. Personal communication 2000.
4. Raghunath B. Personal communication 2005.
5. Ng P, Lundblad R. Optimization of solute separation by diafiltration. Separ Sci 1976;11:499.
6. Beaton NC, Klinkowski PR. Industrial ultrafiltration design and application of diafiltration processes. J Separ Proc Technol 1983;4(2):1–10.
7. Shultz JA. Personal communication.
8. Kulkarni SS, Funk EW, Li NN. Membrane handbook. In: Ho WS, Sirkar KK, editors. New York, NY: Van Nostrand Reinhold; 1992
9. Zeman LJ, Zydney AL. Microfiltration and ultrafiltration. New York, NY: Marcel Dekker; 1996.
10. Westoby M. Personal communication 2014.

Pre-processing operations

6

Herb Lutz
EMD Millipore, Biomanufacturing Sciences Network

Most operating steps besides the processing operation do not require development for each specific protein (Table 6.1). One can use a standard method, developed either by the bioprocessor or as recommended by a vendor. These standard methods are described in accompanying sections. They are verified during a small-scale trial run to ensure they are working properly for the particular application. If these do not function properly, extra development is needed.

6.1 Module installation

New modules must be installed when preparing for a new molecule campaign or when previous modules have reached the end of their operating life. Regulators have not allowed the use of the same modules for multiple drug substances due to concerns over the adequacy of cleaning procedures to reach all the internal module surfaces and prevent batch-to-batch carryover. This may change[1].

Cassettes, spirals, or hollow-fibre modules are inspected for the correct part number and any shipping and handling damage. They may be sprayed with a disinfecting alcohol solution and inserted within an appropriately designed holder. The holder must accommodate a number of modules and direct the flow in and out of each module under pressure without leaks between the flows or external environment while under pressure.

Holder and cassette specifications need to match to ensure consistent sealing. Cassettes are placed vertically in a horizontal row within a holder in 'book shelf' fashion. Some cassettes have built-in gaskets but others require separate gaskets (typically 0.044 in thick silicone) inserted between the cassettes to provide sealing. Gaskets prevent internal bypass (between the feed or retentate distribution ports and the permeate ports) or external leaks (between the ports and the outside of the module). One should be careful to centre the holes in the gaskets to avoid covering the distribution ports and causing high-pressure drops from cassette-to-cassette. For process holders with a centre block (Figure 9.7), installation of an adjacent double thick gasket and a stainless support plate before adding modules is frequently recommended. For single-use assemblies, special end-plate adaptors are also recommended. Rigid end plates at the far end of the holder from the centre block provide the mechanical sealing force to compress the cassettes together. These end plates can be hand-tightened to a torque specification (typically 350–550 in-lbs) or they can be attached to hydraulic cylinders under a pressure specification (typically 1870–2050) psig. Misalignment of the end plates results in a non-uniform compressive force across the surface of the cassette and can lead to leaks. Vendor recommendations should be followed to avoid leaks[2,3].

Ultrafiltration for Bioprocessing

Table 6.1 UF Pre-processing step summary

Step	Process development requirements
Installation	Standard from vendor
Flush	Standard from vendor
Integrity test	Standard from vendor
Equilibration	Standard from vendor

Thermal cycling during cleaning causes expansion and compression of the cassettes. Hydraulic cylinders maintain a uniform compression but repeated manual compression may need to be performed to accommodate these cycles while maintaining seals. In some cases, the cassette urethane has warped under repeated cleaning and required the use of double thick gaskets to accommodate the deformation.

Hollow fibre modules typically have sanitary end caps for installation into piping systems. Large modules are supported with additional clamps that brace the module. Multiple spiral modules are typically inserted within tubular housings. A 'chevron' seal is inserted on the outside of the spirals to ensure there is no bypass flow into the annulus between the spiral and the housing. A 'sealing plug' is used to connect the permeate ports from successive modules and prevent leaks between the pressurized feed/retentate upstream side of the module and the downstream permeate side.

6.2 Flushing

Flushing is performed to (1) wet the modules, (2) remove storage agents from modules, (3) remove cleaning or sanitization agents from modules, (4) displace water in the modules with buffer and to (5) recover product. This section focuses primarily on the first two applications.

Flushing with USP (US Pharmacopoeia) purified water or WFI (water for injection) wets the modules throughout in order to access all the filtration surface area. Some membranes can be stored dry without collapsing. Inadequate wetting leaves trapped air bubbles that can 'blind off' some of the membrane and reduce the flux. Additional air interfaces can also increase protein denaturing by acting as a hydrophobic surface. Increasing retentate back pressure is useful since gas volumes shrink under pressure in accordance with the ideal gas law. This can assist in dislodging air from the module. The fact that gas dissolves more under pressure also helps remove trapped air. System components are also flushed out at the same time as the cassettes.

Modules are typically stored in fluid that (1) keeps the membrane wet so that it does not dry out and collapse (i.e., humectants such as glycerine), and (2) inhibits microbial growth (i.e., sanitant storage solution). Inadequate flushing of components can leave residuals that can contaminate the retained protein product (for storage agents) or denature the protein product (for cleaners and sanitants).

Table 6.2 Cassette hold-up volumes

	Feed channel	Permeate channel	Membrane and support
mL/m^2 hold-up	150	110	170

Flushing should follow vendor recommendations[3]. It is common to start by configuring the permeate line open and the retentate valve open with both lines directed to drain. Then, the feed pump is ramped up to the process cross-flow of typically 5–7 Lpm/m^2. Upon reaching the feed flow target, tighten the retentate valve to provide a little back pressure to get a conversion of about 30%. That is, the permeate flow is about 30% of the feed flow. It is critical to avoid 'dead-ending' the module and make sure that flow continues to exit from both the retentate and the permeate ports. Retentate flow is needed to ensure that air bubbles can be removed from the upstream side of the module. High molecular weight membranes with high permeability are particularly sensitive to having all the feed flow into the permeate stream at low pressures with no flow going to the retentate. This means relatively higher feed flow rates and lower retentate pressures. Permeate flow is needed to ensure the membrane is adequately flushed.

Displacement of air or process solutions from the module require a flush of 20 L/m^2. As shown in Table 6.2, total hold-up volume for a cassette is about 0.43 L/m^2. However, the tight screens hold-up air bubbles by capillary forces and require flushing at process cross-flows of 5–7 Lpm/m^2 over volumes of 20 L/m^2 to drive out most of the air. Allowing 20 L/m^2 of flush volume also ensures that leachables and buffers diffuse out of the 'nooks and crannies' of the module to achieve a quality specification on the flush fluid exiting the retentate or permeate. Even though the product may only be directly exposed to the upstream side of the membrane, current conservative practice is to flush through the membrane and ensure permeate flush quality. Warm flush water (40–50°C) can be more effective in flushing out compounds by increasing their solubility. Large pipe diameters may require increased flow rates to ensure adequate flush through all the pipe segments. Flushing is also a good time to do any system checks. Pumps, valves and sensors should all be active and working properly.

It is recommended that the initial installation of cassettes be followed by a more extensive flushing and cleaning procedure. Cassette compression will intrude the screen into the membrane, reducing part of the membrane effective area. This shows up as a drop in water permeability after the first cleaning. After subsequent cleanings, the permeabilities tend to be stable. In addition, components used for storing and preserving the modules need to be flushed out. For example, initial high concentrations of glycerine (for maintaining membrane moisture) require additional flush to reduce organic levels to purified water specifications. Some preservatives may also adsorb to the membrane and require extended flushing to remove. A cumulative flush of 60 L/m^2 is recommended with 20 L/m^2 as the initial flush, 10–20 L/m^2 used for cleaning, and the remainder used for subsequent flushing.

6.3 Integrity testing

Ultrafiltration system integrity testing is performed by bioprocessors to detect leaks and membrane defects that can compromise product retention by the ultrafiltration system. This helps ensure good yields but is not necessarily a regulatory requirement unless included in a filing as part of the SOP. Pre-processing testing is required to prevent processing losses. It is estimated that 90% of batches include a pre-use integrity test. Examination of a proprietary Millipore data base on module returns with quality issues compared to module sales over an extended time frame indicates that on average, integrity failures occur at a rate of 0.3% per cassette. For systems with n multiple filters, the rate that at least one filter in the housing has a defect p_n is obtained from the single filter rate p_1 using the formula $p_n = 1 - (1 - p_1)^n$. For low rates, this is roughly equivalent to multiplying single filter rates by the number of filters in the housing. For a large assembly of 80 filters of a nominal 1 m^2 each, this implies a failure frequency of 24% of batches. This high failure frequency with associated valuable product losses in case of failure indicates that pre-use testing is very cost-effective. Failures are associated with the initial installation and not with re-use. The failure rate varies with operator experience in installation.

Module integrity and product retention is assured not just by the bioprocessor test but through vendor testing. Vendors use qualified procedures, trained operators and extensive in-process testing to ensure the ultra-filtration module lies within established specifications. This includes different integrity tests on both the membrane and modules. Membrane retention characterization has been described in Chapter 3.

Ultra-filter membrane pore diameters are on the order of 1 nm or 0.001 μm. The air pressure required to overcome water capillary forces and evacuate these pores would have a bubble point on the order of several thousand pounds of pressure according to the Young–Laplace bubble point equation[4,5] (Equation 6.1). For example, sterilizing filters with a pore size of 0.2 μm have a bubble point of about 50 psid. This makes bubble point tests inappropriate as bioprocessor tests.

$$\Delta P = 4k\gamma \cos(\theta) / r \qquad (6.1)$$

where r is the pore diameter, k is a shape correction factor (pore circumference/pore cross sectional area), γ is the air–water surface tension of 72 dynes/cm at 25°C, and θ is the contact angle between the liquid meniscus and the pore wall surface where $\theta = 0$ is fully wet and $\theta = 180°$ is unwet.

The recommended bioprocessor integrity test is based on air–water diffusion. Vendors prescribe test pressures commonly in the 10–50 psid range. This air pressure would evacuate holes in the 1.0–0.2 μm diameter range. This is the range of defect diameters that can be measured. Most all defects that can occur after the module has left the vendor plant due to transportation, handling, installation and processing, are larger than these diameters and readily picked up. A low test pressure of 30 psid is not correlated to a specific marker retention. For example, if regenerated cellulose ultra-filtration membrane is cast on a 0.22 μm substrate and

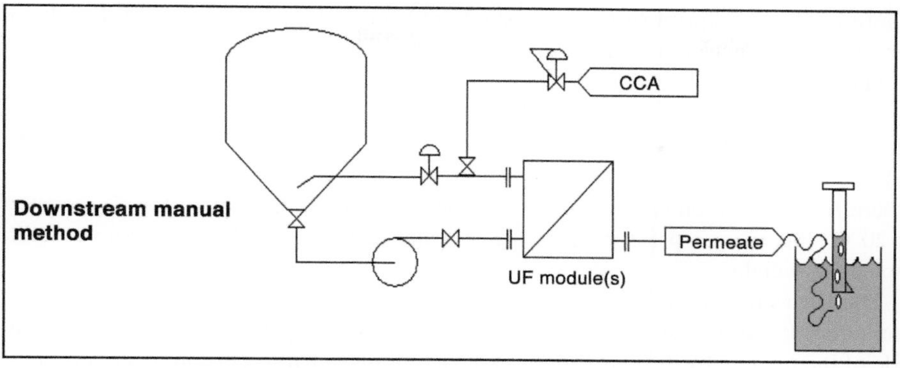

Figure 6.1 Ultrafilter integrity testing.
(Courtesy of John Cyganowski.)

the ultrafilter skin is compromised by exposure to caustic or hypochlorite, the air diffusion may not pick up the destruction of the thin layer of retentive pores. It would however provide assurance that the 0.22 μm substrate is fully wet and without larger defects. Additionally, a diffusion test may not detect seal leaks that could open up at higher operating pressures. The diffusion test should be considered a 'fitness for use' test.

Bioprocessor integrity tests of ultrafiltration systems also use air–water diffusion. An upstream of the membrane air pressure is applied and the diffusive air flow across the wetted membranes is measured. As shown in Figure 6.1, pressurized air is typically introduced into the retentate line with the feed line to the modules valved off, the retentate to the tank line valved off, and the permeate line open. Air flow can be measured in the permeate line using a graduated cylinder and stopwatch as shown in the figure. It is also conveniently measured on the compressed air introduction line using a mass flow meter or pressure decay with an automated integrity tester. This location enables detection of both internal bypass through the membrane and external leaks to the environment.

The ultrafiltration system can be tested as a whole, sections can be tested at a time using sealing gaskets, or modules can be removed and tested individually in a separate holder. Since many of the defects occur as a result of poor alignment and sealing, it is preferred to test the system as a whole. Once a defect has been identified, modules can be re-installed and sections can be tested to locate the defective area. A defect that is large enough to fail the specifications for a system will show a larger airflow effect when testing just a single module. However, the integrity test is sensitive enough to detect system defects that could result in >1% yield losses.

The air flow through the wetted filter is the sum of a diffusion flow through an integral membrane, which serves as background noise, and a high convective flow through defect holes, which is the signal. The background diffusion flow originates from a solution–diffusion mechanism. Pressurized air dissolves in the water contained within the membrane pores on the upstream side of the membrane.

There is also an equilibrium established between low pressure air and dissolved air on the downstream side of the membrane. This equilibrium is governed by Henry's law[6]:

$$c = P\rho / HM, \tag{6.2}$$

where H is the solubility coefficient[7] of 6.64×10^4 atm/mole fraction for air in water at 20°C, P is the gas pressure, ρ is 1 g/cc water density and M is 18 g/gmole water molecular weight.

The high-pressure side has a higher concentration of dissolved air than the low-pressure side. As a result, dissolved air migrates from the upstream to the downstream side of the module by diffusing through the membrane. This diffusion process is governed by Fick's Law[8]:

$$N = -D\varepsilon\Delta c / L, \tag{6.3}$$

where N is the molar flux of the diffusing species, D is the free solution diffusivity, ε is the membrane porosity or void volume, Δc is the concentration difference driving force, and L is the length of the diffusion path or membrane thickness.

Combining the equations along with the ideal gas law yields a formula for the background diffusion rate volumetric gas flow rate in equation (6.4).

$$Q = \frac{\varepsilon DH \Delta P \rho ART}{\mathrm{LMP_a}} \tag{6.4}$$

where Q is the solution–diffusion volumetric gas flow rate, R is the gas constant, T is the absolute temperature, P is the absolute pressure, A is the membrane cross-sectional area, and other symbols as equation (6.3).

This background solution–diffusion flow through integral modules increases linearly with the membrane area and is sensitive to the gas composition, liquid composition, pressure difference, membrane thickness and porosity, and temperature. It can vary from module to module. Once installed, module integrity values can still change due to operating differences in setting the applied pressure, controlling temperature and measuring flow rate. For a 10 kDa Ultracel™ membrane at 23°C, typical module air flows[2] are about 1.2 sccm/psi/m^2 while the maximum air-flow specification is about 1.8 sccm/psi/m^2.

Flow through small diameter evacuated defects with diameters of 0.2–1.0 μm and lengths (membrane thicknesses) of 200 μm can be considered as flow through a small orifice. For a defect area that is a small fraction of the total membrane area, the mass rate of flow of a compressible gas through an orifice can be described by frictionless adiabatic expansion[8]:

$$w_{\mathrm{gas}} = C_{\mathrm{d}}\rho_2 S_0\sqrt{2(P_1 / \rho_1)[\gamma/(\gamma-1)][1-(P_2/P_1)^{(\gamma-1)/\gamma}]} = \rho_2 S_0 v \tag{6.5}$$

where C_d is a dimensionless discharge coefficient[9] ~0.85, S_0 is the defect area, ρ is the gas density, P is the gas absolute pressure, $\gamma = C_p/C_v = 1.4$ for air heat capacity ratio constant, v is the exit velocity, and 1 refers to the upstream conditions and 2 refers to the downstream (atmospheric) conditions.

There is a maximum orifice exit velocity or 'choke' flow given by a Mach number $Ma = v / v_{\text{sound}} \leq 1$ where $v_{\text{sound}} = \sqrt{\gamma RT / M} = 359$ m/s at 23°C. This occurs when the pressure ratio reaches $P_2 / P_1 = \left(2/(\gamma + 1)\right)^{\gamma/(\gamma-1)} = 0.5285$ for air[8], which corresponds to an upstream pressure of >13 psig. Increasing the upstream air pressure above this ratio does not increase the exiting air velocity beyond the speed of sound.

The difference between the air flow specification and the typical area flow at a 30 psig test pressure is the air flow through the largest potential measurable defect or 18 sccm/m^2 of membrane area. The maximum exit flow is the speed of sound $v = 2.15\times10^8$ sccm/m^2 of defect area. Then, $1.8\text{A} = 2.15\times10^8\, S_0$ or $S_0/\text{A} \geq 8.4\times10^{-9}$. If the integrity test can measure the air flow, it can detect a defect that is ≥0.084% of the membrane area.

For a small area defect compared to the membrane area, the mass rate of flow of an incompressible liquid through an orifice can be described by[8]

$$w_{\text{liq}} = C_d S_0 \sqrt{2\rho\Delta P} \tag{6.6}$$

where C_d is a dimensionless discharge coefficient[7] ~0.6, S_0 is the defect area, ρ is the liquid density ~ 1g/cc for water and ΔP is the pressure drop across the orifice.

For a typical process run 15 psid trans-membrane pressure, the defect flow would be 5.18×10^8 ccm/m^2 defect area. A typical average membrane flux is 35 LMH or 583 ccm/m^2 of membrane area. This translates to a fractional liquid flow through the defect, corresponding to the sieving coefficient, of $s = (5.18\times10^8 S_0)/(583\text{A}) = 0.00075$. At a typical processing of $N = 8$ diavolumes and $X = 5$ volumetric reduction factor, this produces a yield loss of $1 - \exp(-s(N + \ln X)) = 0.71\%$. This shows that the ultrafiltration integrity test at typical conditions is sufficient to guarantee <1% yield losses into the permeate stream caused by defects.

6.4 Module equilibration in buffer

The modules are wet with water after integrity testing. Exposure of some protein solutions to purified water can lead to precipitation. This can degrade the quality of the protein product and it can leave deposits on the membrane. These deposits can form an additional resistance layer that reduces the flux. They can also present an extra cleaning load that needs to be removed.

Flushing the system with an equilibration buffer leaves the membrane in a state where protein is less likely to precipitate. Frequently, the diafiltration buffer is used for this pre-flush. The typical flushing conditions of Section 6.3 utilizing 20 L/m^2 of flush volume at 5–7 Lpm/m^2 process feed flow rates and 30% conversion are recommended.

References

1. Mahajan E, Werber J, Kothary K, Larson T. One resin, multiple products: a green approach to purification, In: Kantardjieff A, et al., editors. Developments in biotechnology and bioprocessing, Washington DC: ACS Symposium Series, ACS; 2013. [chapter 6].
2. Lau SY, Pattniak P, Raghunath B. Integrity testing of ultrafiltration systems for biopharmaceutical applications. BioProcess Int October 2012;10(9):52–66.
3. EMD Millipore Corporation, Pellicon 3 cassettes: installation and user guide, Literature # AN1065EN00 Rev. E, 02/2014.
4. Young T. In: Peacock G, editor. Miscellaneous works, 1. London: J. Murray; 1855. p. 418.
5. De Laplace PS. Mechanique Celeste, Supplement to Book 10; 1806.
6. Poling BE, Thomson GH, Friend DG, Rowley RL, Wilding WV. Physical and chemical data. In: Don W Green, Robert H Perry, editors. Perry's chemical engineers' handbook. 8th ed. New York, NY: McGraw-Hill; 2008. pp. 2-130–2-133.
7. International critical tables, vol. 3. New York: McGraw-Hill; 1926. p. 257.
8. Bird BR, Steward WE, Lightfoot EN. Transport phenomena. 2nd ed. New York, NY: John Wiley & Sons; 2002.
9. Shapiro AH. The dynamics and thermodynamics of compressible fluid flow, vol. 1. New York, NY: John Wiley & Sons; 1953.

Processing

7

Herb Lutz
EMD Millipore, Biomanufacturing Sciences Network

7.1 Membrane and module selection

7.1.1 Membrane and module material

Ultrafiltration (UF) membranes and module wetted surfaces must be chemically compatible with the application. This means no impact on the membrane by all fluids employed (i.e., swelling or dissolution) and no impact on the fluid by the membrane (i.e., extractables, adsorption and shedding). Aqueous systems are generally not a problem but solvents can pose compatibility issues (see Appendix and consult with vendors).

As described in Chapter 3, regenerated cellulose (RC) and hydrophilized polyethersulfone (PES) are the two membrane materials commonly used for bioprocessing. Membranes having a composite structure (Figure 3.6) are preferred because of their consistency, higher integrity and higher mechanical robustness. Both RC and PES are acceptable and in wide use on all types of commercial products. RC is sensitive to high pH and hypochlorite while PES fouls requires more and more cleaning. For facilities with bioburden control concerns or plasma products with prion inactivation requirements, the need for higher concentration and stronger cleaning and sanitizing agents leads to a preference for PES. Otherwise, the ease of cleaning and low fouling properties of RC makes it the best choice.

Vendor documentation and quality varies considerably with the industries they serve. Vendors need to be qualified to meet the needs of the application (Tables 7.1 and 7.2).

7.1.2 Membrane retention and nominal molecular weight limit

The nominal molecular weight limit (NMWL) values (Section 2.4) differ by membrane polymer and between manufacturers. There is not an agreed upon method for these ratings. In addition, the shapes of the retention curves vary between polymers due to differences in pore size distributions. Some NMWL values are not available in all polymer chemistries and module formats. This may require some selection trade-offs.

The rule of thumb for the initial selection of an NMWL for globular proteins is that it should be 20–33% of the MW of the solute one is trying to retain. There are a number of factors described later that can significantly change this selection, such as protein conformation, electrical charge, association, polarization, amount of processing and size distribution. For operation in the polarized region of the flux curve, use of a tighter membrane is recommended since there is no significant impact on process flux. One must test the initial selection under the actual process conditions to confirm performance. Some protein solute assays are sensitive to the buffer. As a result, one

Table 7.1 UF processing step summary

Steps	Process development requirements
Processing-UF1: concentrate to Diafiltration conditions	Flux vs. concentration testing with feedstock
Processing-DF: diafilter	Flux vs. concentration testing with final buffer
Processing-UF2: concentrate to final end-point	Flux vs. concentration testing with final buffer and different cross-flows

might see changing concentrations as an artifact due to buffer exchange during diafiltration. Proper assay selection and calibration is needed.

As described in Chapter 3, the membrane retention is based on size, not molecular weight, so the solute *conformation* or shape will matter. The velocity distribution at the surface of the membrane will apply a force on different sides of the solute to align the shape with the flow streamlines. This will align long, thin solutes (e.g., dextran, nucleic acids and PEG) with the long diameter perpendicular to the pores and the small diameter entering the pores.[1] As a result, for a given solute MW, long flexible solutes can pass the membrane while globular solutes do not.

Solutes can also have an effective size that is larger than what their steric size from NMR or crystal structure might imply. *Electrically charged* solutes have an electric field that must change its shape when the solute enters the membrane pore. This requires additional energy and reduces the likelihood of charged solutes from entering pores. In effect, the charge increases the effective size of the solute.[2] One expression developed using BSA[3] is $r_{\text{sff}} = r_s + 0.0137z^2 / r_s\sqrt{I}$ where r_s is the steric size in the absence of charge, I is the ionic strength and z is the net number of charged sites of

Table 7.2 UF processing options

Processing option	Description	Application
Conventional	Typical UF/DF	Most all
High concentration	High final UF2 step	High concentration final formulation
Product aggregation	Minimize pump and polarization damage	Product aggregation propensity
Adsorbed solutes	DF/UF2 washout issues	Final formulation
C_{wall}	Maintain wall concentration	Final formulation
Variable volume DF	Concentrate and diafilter	Final formulation solubility issues
Mass flux	Variable polarization and diafiltration	Sieving varies with flux
Fractionation	Better quality	Purity and yield vary with operation

the same charge on the molecule. Most solutions have an ionic strength above isotonic (150 mM) where charge effects are passivated by the free salts. This makes the charge effect more limited in practice.

Solutes can also *associate* and increase their apparent size. This does not require a covalent or electrostatic bond as in dimerization. However, if solutes tend to temporarily associate, these associated solutes will have a reduced likelihood of entering membrane pores. The high concentrations at the membrane surface due to polarization can increase solute association. If solutes are in a multimeric form (e.g., haemoglobin or insulin), the solution conditions throughout the process must help keep these multimeric forms intact. Otherwise, during some step in the process, the smaller solutes can pass through the membrane and one can see significant yield loss into the permeate.

Polarization at the membrane surface creates higher wall concentrations than in the bulk fluid. As a result, more solutes can enter membrane pores than if they were subject to the lower bulk fluid concentration. This shows that the hydraulics in the feed channel can affect retention by changing the degree of polarization.[4]

Polarization of *large solutes* creates an additional retentive layer on the membrane. For a multi-component mixture, the larger solutes have lower diffusivity, hence less back transport away from the membrane (Figure 4.8). As a result, they will accumulate closer to the membrane surface. This causes a stratification of solutes from large to small as one moves from the membrane surface into free solution. Larger solutes at the membrane surface will occupy space and hinder small solutes from being dragged past them by the permeating solvent. The small solutes will have a reduced likelihood of entering membrane pores in the presence of large solutes. This shows that the feed solution composition can affect membrane retention. One can use a higher NMWL membrane for volume reduction of a multi-component harvest solution while a tighter NMWL membrane is required for final formulation concentration of a purified single component solution.

The membrane must retain solutes through the entire *extent of processing*. Solutes are repeatedly passed through the module in a batch processing system. As described in Chapter 5, the retentate yield after sequential batch concentration and diafiltration steps is $Y = X^{-s}e^{-sN} = e^{-s(\ln X+N)}$ where s is the apparent sieving coefficient for a single pass through the module, assumed to not vary significantly over the course of processing. Figure 7.1 shows how the amount of processing, or $N + \ln X$, along with the module sieving coefficient affects the net loss from the retentate into the permeate. For a typical process with $N + \ln X > 12$, one needs a membrane with a sieving of $\leq 0.05\%$ (retention $\geq 99.95\%$) to achieve process yields of $>99\%$. Some processes (e.g., conjugate vaccine manufacturing) may have toxic reactants that require extensive diafiltration of 70 diavolumes. These processes are particularly sensitive to membrane sieving.

The biologically active product of interest can also exist with a *size distribution*. As examples, one may have different amounts of PEGylation or different conjugate vaccines that are active but with different sizes. The smaller-size products are the most weakly retained and will be lost first into the permeate stream if the membrane employed is open enough.

This description shows that solute retention is governed by the membrane pores, solute shape, charge, association, size distribution, module hydraulics, processing requirements and size distribution. The 'rule of thumb' for a 160 kDa antibody is a

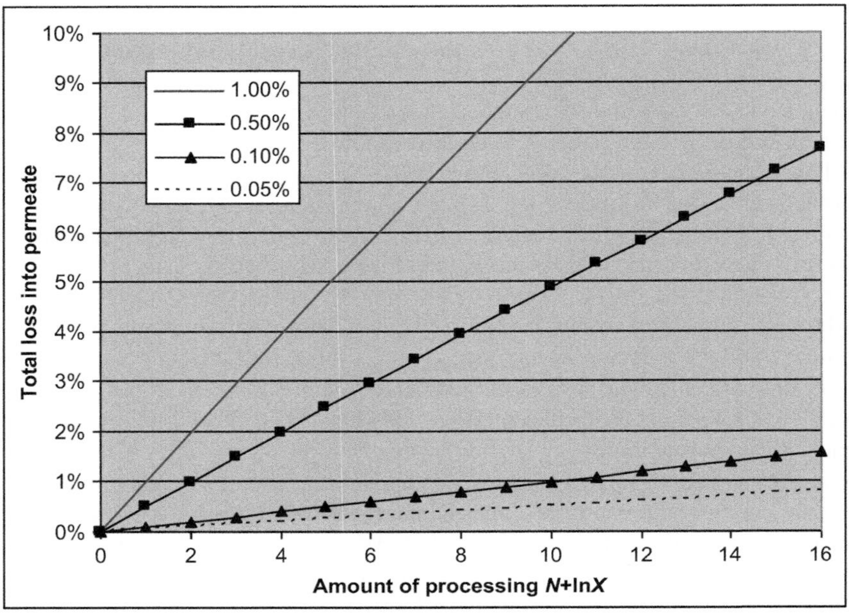

Figure 7.1 Effect of batch processing on retentate yield.

30–50 kDa membrane. Such a membrane should be tested under the expected process conditions to ensure that it retains the product to the level desired. In consideration of some of the effects just described, one may need to shift to a tighter or more open membrane depending on the specific application.

Membranes have a range of specifications. Usually, the membranes are made near the centre of their distribution to increase membrane manufacturing process yields. Process testing with these membranes may show good performance. However, changes in manufacturing location, raw materials and conditions may produce more open membranes that are still within specification, but with a higher sieving. Process changes within the biomanufacturer upstream process (e.g., changed biomolecule product distribution, increased presence of larger solutes, buffer shifts) may also result in higher sieving. It may be useful to use a tighter NMWL membrane to account for future potential process changes to ensure good sieving.

For high MW concentrated solutes where transmembrane pressure (TMP) $\ll \Pi_w$, the membrane permeability is not a significant factor in determining the flux or system area. In this case using a tighter, lower NWML membrane gives one retention assurance with no penalty in flux. For low MW dilute solutes, tighter membranes with higher retention comes at a price of lower fluxes. Usually, any product loss ($300–$3000/g) is valued extremely high compared to the cost of the ultra-filtration system with module reuses ($0.25–0.75/g). Adding an extra membrane area of tighter UF membrane modules is generally a good idea. However, it is useful to verify the economics of each application.

7.1.3 Module selection

As shown in Table 4.1, cassettes, spirals and hollow fibre modules are the widely used formats for bioprocessing use. Of these, cassettes are by far in the widest use because of their mass transfer efficiency and linear scale-up. Cassettes employ tight feed channel screens not used in the other devices that enable high module fluxes at low feed flow rates. The tightest screens generate the best back migration from the membrane surface. For most applications where the polarization and not membrane permeability controls the flux, this thins the polarization boundary layer creating higher fluxes and the need for less membrane area. (Section 8.6 describes an exception to this rule.) The reduced area requirements and smaller pumping requirements for tighter screen cassettes shrink the test system and enable both a low minimum working volume and a high product recovery. These benefits tend to outweigh the fact that cassettes have a higher module cost and a higher feed channel pressure drop than other modules.

7.1.4 Scaling

Scaling generally requires that one keep the same module type (membrane and feed channel design) while increasing the amount of membrane area in parallel. Adding membrane in parallel for scaling up is intended to maintain similar flow fields and concentration distributions along each feed channel. This is straightforward for cassettes or hollow fibres but for spirals it involves also changing the permeate path length (and flow resistance). For applications sensitive to changes in TMP, one can see changes in performance between scales for spirals.

One must be careful about changing the properties along the feed channel. For example, increasing the length of hollow fibres can create significant retentate channel pressure drop while the shell side permeate can remain at a fixed pressure. If the pressure drop is significant enough, the TMP can be reversed and the membrane flux can reverse direction. In that case, adding extra membrane area can significantly reduce the net module flux and even drive it to zero. Where some solute fractionation is also occurring along the feed channel, altering the TMP profile can significantly change the product composition. The small area system module configuration used for process development should mirror that in the large-scale system.

Changing the membrane can change retention, flux and cleanability properties. In this regard, note that a 10 kDa module from one vendor is not the same as a 10 kDa module from another vendor even if the membrane material is the same.

Retention properties of a given membrane module from the same vendor can change from batch to batch. As indicated earlier, for most valuable products one would like consistently high yields for every batch. Some vendors publish specifications on membrane release tests to define the range of variation in retention. However, it is not straightforward to relate the retention of a model solute under one set of conditions (e.g., inert linear dextran, non-fouling, stirred cell, rpm and TMP) to the retention of a particular commercial solute run under a different set of conditions (e.g., charged globular protein, fouling, cassette, cross-flow and TMP). It is generally not possible to get membrane limit samples corresponding to the full range of release specifications because the membranes are more consistently made near a 'centre point'. If one imagines the retention distribution of possible membrane lots made over the lifetime of a particular therapeutic, there is almost no chance that one would actually receive a limit sample membrane. Designing and sizing for such an extreme is not cost effective. However, it is useful to allow some margin and not operate close to the edge where significant yield loss costs are incurred for slight membrane changes that are expected to occur over the drug life and number of batches used to produce the therapeutic drug.

In general, one can test with multiple membrane lots to see if there are significant performance differences above assay and operational variations. It is useful to know where in the distribution these modules may fall. These can also be used to explore a correlation between membrane release test values and customer performance values, that is, protein vs. dextran retention.

7.1.5 Safety factor and sizing

System sizing for a variety of applications has been covered in Chapter 6. For typical MAb final formulation processing, where diafiltration at C_g/e with flux $J = k$ dominates the process, the permeate volume is $NV_D = NMe/C_g = JAt = kAt$. The sizing can then be expressed as $M/At = C_g/Nek \sim$ 200 GMH (g/m²/h) for a typical MAb final formulation.

To allow for process variability and process changes, one should add extra area, time, or both to compensate.[5] In general, ultrafiltration processes are very robust (i.e., consistent and insensitive to variations) so the extra area-time does not need to be much for a given molecule. In the case of ultrafiltration processes, adding extra area increases the system hold-up volume and minimum working volume. This can limit the ability of the system to achieve a final concentration or volume reduction target. One typically uses a 20–30% safety factor or 150–165 GMH for a final formulation MAb sizing. Process time is much more flexible although an extended time processing of over 12 h can raise concerns about bioburden and pumping damage. In cases where extended time processing is costly and working volume is not a constraint, larger safety factors may need to be applied. See Reference 6 for a discussion on safety factors and economic tradeoffs.

7.2 Process operating conditions selection

7.2.1 Hydraulic parameter selection

Matching the module hydraulics between scales is accomplished by matching two flow parameters: LMH feed flux and LMH permeate flux. The retentate LMH flow would then be fixed by a mass balance. One could use the module feed to retentate pressure drop ΔP to represent the feed flow. However, at large scale there is a head height between these pressure ports and there can be some additional drop due to added flow resistances between the pressure gauges. One can calculate or measure the static pressure difference (water filled, no flow) and subtract it from the dynamic readings to compensate for the head height. However, added flow resistances at one scale cause a mismatch in operation between scales. This makes the pressure drop more subject to scaling error than a feed flow measurement.

7.2.2 Feed flux selection

A range of LMH feed fluxes are generally recommended for a particular module type by the vendor. Fibres or open channel cassettes require high LMH feed flux while spirals with spacers requires less and tight screen cassettes the least amount of feed flux (Table 4.1). In general, there is a balance determining reasonable feed fluxes. For operation in the polarized region, higher feed fluxes create higher permeate fluxes requiring less module area. However, higher feed flows also create a higher pressure drop across the module and more pump passes during the course of operation. This can lead to more protein degradation. Pump passes are the total volume pumped divided by the tank volume. As shown in equation (7.1), one can calculate the pump passes through numerical integration or approximate its value by assuming a constant conversion Y = (Permeate flow/feed flow) over the course of processing:

$$n=\int_0^t \frac{q_F dt}{V} \cong \frac{1}{Y}\int_0^{V_P} \frac{dV_P}{V} = \frac{1}{Y}\left[-\int_{V_0}^{V_F} \frac{dV}{V} + N\right] = \left[\ln X + N\right]/Y \tag{7.1}$$

It is useful to keep track of pump passes during scale-down testing to see if protein degradation is occurring and the protein is sensitive to pumping damage. It is difficult to maintain similar pumping across scales but one can maintain a comparable order of magnitude of pump passes during scale-up operation. Note that protein damage will depend not only on pump passes but the magnitude of pressure drops in the retentate loop, the pump type, air–liquid interfaces, temperature, the buffer environment, and the protein being processed.

7.2.3 TMP selection

One could use the average module TMP = $(P_{\text{feed}} + P_{\text{retentate}})/2 - P_{\text{permeate}}$ to control the LMH permeate flux instead of controlling it directly with a permeate pump. Note that one still needs to control the LMH feed flux as well to define hydraulic conditions within the module. Rewriting this in terms of pressure drop as TMP= $P_{\text{feed}} - P_{\text{permeate}} + \Delta P/2$, one needs to measure the permeate pressure to make sure it is at atmospheric pressure. Also, the influence of the pressure drop is halved on module TMP so it is less subject to error than the pressure drop alone. For applications where precise control of hydraulics is required (e.g., solute fractionation), special procedures are used to maintain a constant TMP along the length of the feed channel and flux control is used. For most applications, TMP control is accurate enough to use as a control parameter and scaling parameter. It is the industry standard. Note that due to the added resistances described earlier, feed, retentate and permeate pressures can vary widely

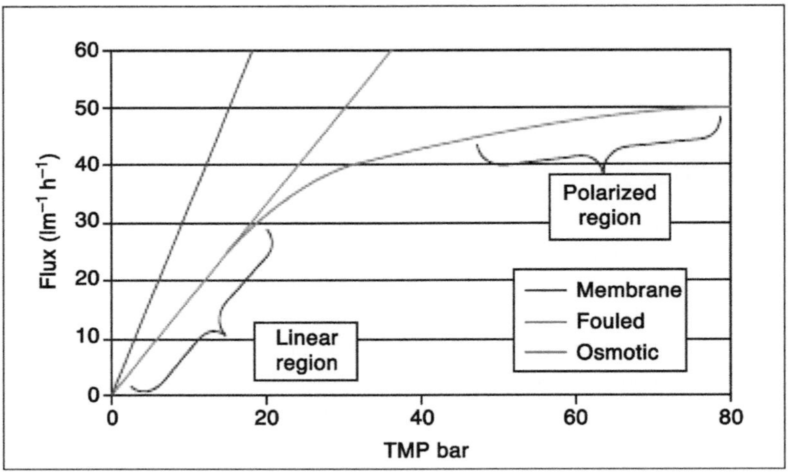

Figure 7.2 Flux curve.

between scales. As a result, it is important to focus on matching TMP values between scales. Other pressures can vary significantly.

The value of TMP selected for operation is based on the flux vs. TMP curve as shown in Figure 7.2. Between the linear region and the polarized region lies the 'knee' of the flux curve where it bends. Below the knee, increases in pressure lead to higher fluxes and reduced system area. Above the knee, increases in pressure provide marginal flux increases at a cost of increasing pumping pressure and potential pumping damage to the protein. The knee represents a point of diminishing return for area reduction based on pressure increases. In addition, the region above the knee represents a point where solute polarization becomes significant, that is, higher solute wall concentrations. For feed solutions that show a tendency to precipitate or aggregate at high concentrations, it is also useful to operate at the knee.[7]

7.2.4 Start-up conditions

Prior to start-up, the system has been flushed with equilibration buffer. It may have been drained or may still have a buffer within the piping system. The feed solution is added gently into the retentate tank. Avoid splashing or any aeration that can denature proteins. For fed-batch operation, only part of the feed batch is added.

Start up the system with the permeate valve open and the retentate valve wide open. Ramp up the feed pump over a period of roughly 1–2 min to a feed flow of 25–30% of the target process cross-flow. Then re-circulate for 5 min or so at a feed flow about 25–30% of the target process cross-flow. Slow ramping or reduced re-circulation minimizes pressure spikes or 'water hammer' effects that can damage membranes. It also displaces air bubbles within the piping system while minimizing breaking them up and creating more air interfaces for protein denaturation. The protein solution is re-circulated to 'condition the membrane'. This step allows any components to adsorb to the membrane. It is also important to check the feed to retentate pressure drop and make sure it is not abnormally high. This would suggest that the feed channel is plugged or that the gaskets may be misaligned and intruding into the feed channel.

Some operators leave the permeate valve closed during start-up ramping and low cross-flow conditioning. When the permeate is closed and there is a pressure drop along the feed channel, one will get Starling flow (Section 4.6) where flow goes into the permeate over the beginning section of the feed channel and flows backward through the membrane, out of the permeate channel, over the end section of the feed channel. The extent of flow has to do with the pressure drop along the module and the membrane permeability.

It is useful to match this start-up procedure between different scales of operation. For clarification applications and fractionation applications, the fouling of the membrane affects the sieving performance. The start-up procedure can significantly influence this performance and it can be worth taking some time to optimize this procedure for these applications.

7.3 Standard batch protein processing

After membrane conditioning, the permeate stream is opened and the retentate valve slowly closed to reach the target TMP. As the permeate starts flowing, the fed-batch feed flow can be initiated. The initial batch or fed-batch concentration proceeds until an end-point is reached. This could be based on retentate volume or tank mass, cumulative permeate volume or mass, or a retentate concentration sensor.

Diafiltration is initiated by starting a diafiltrate pump. The diafiltrate pump flow rate or cumulative diafiltrate buffer volume is controlled to match the permeate flow rate or cumulative permeate volume. It is usually preferred to control the diafiltrate flow to keep the retentate tank liquid level constant. In general, fluxes increase slightly during diafiltration but this may not be true in every case. Diafiltration is ended when the specified number of diavolumes (diafiltrate buffer volume or cumulative permeate volume/retentate volume) have been achieved.

The final volume reduction step generally proceeds to the endpoint just like the previous volume reduction step.

As described in Section 6.2, it is critical to avoid 'dead ending' the feed solution into the membrane by having the retentate closed. This could happen as a transient in the valve sequence. This could permanently foul the membrane.

7.3.1 *Yield analysis*

A mass balance on the system shown in Figure 5.1 is

$$\begin{aligned}\text{Total yield loss} &= \text{in} - \text{recovery out} \\ &= \text{permeate out} + \text{holdup} + \text{adsorption} + \text{activity losses}\end{aligned} \tag{7.2}$$

Table 7.3 **Protein adsorption**[31]

Membrane material	Protein adsorption (g/m^2)
Standard polyethersulfone	0.5
Biomax polyethersulfone	0.2
Regenerated cellulose	0.1

Other sections cover these different terms: permeate losses (Section 7.2), hold-up losses and recovery (Section 8.1) and activity losses (Section 7.6).

Adsorption losses can be obtained using the values in Table 7.3. Although the wetted surfaces include the system components and the membranes, most of the surface is membrane. In addition, stainless surfaces show very low adsorption. For MAb applications sized at 150 g/m^2/h and running for 3 h, the mass loading is 450 g/m^2. The adsorption amounts below are negligible by comparison. Dilute protein batches may require a relatively large system size, say 10 g using 15 m^2 as a result of low molecular weight where the membrane permeability is low and drives sizing. In this case, adsorption losses may be significant and drive membrane selection.

With low adsorbing, tight reliable membranes and good recovery procedures, yields can be consistently >95% and in many cases >98%. Hold-up losses are typically the largest source of yield loss.

7.4 High concentration operation[8]

High dosage and ease of administration drive the trend to higher concentration liquid formulations. Pre-filled syringe administration is limited to about 30 cP liquids or roughly 150–250 g/L for MAbs.[8] The UF processes used to produce these liquid formulations may be constrained and unable to routinely achieve these higher targets. Constraints may include osmotic pressure, cassette feed pressure, minimum working volume, and recovery dilution.

Fluxes drop as concentrations increase following the empirical gel model. IgG solutions at 150–250 g/L have an osmotic pressure of only 2–10 psig (Figure 2.3). This indicates that there is still a driving force for membrane flow at reasonable operating pressures of 30–50 psig.

As the volume is reduced and the protein concentration and viscosity increase during the final UF2 concentration step, the feed to retentate pressure drop across the module will increase and the feed pressure will increase. Eventually, the feed pressure may hit an upper limit, typically governed by a maximum pressure specification on the module of approximately 80 psig.[9] It is driven by the pressure drops across the cassette feed channel, and other piping elements in the retentate line. Each of these pressure drops can be correlated to feed flow, line size, line length and solution viscosity. The cassette feed channel dominates the pressure drop during the final concentration step.

Since the pressure drop across a cassette depends on the fluid viscosity and the feed flow rate, one can alter the control strategy at the end of the UF2 concentration step to allow operation at higher concentrations:

1. Concentrate at diafiltration conditions (TMP and JF). As the feed pressure rises, the retentate valve is slowly opened to maintain TMP. At some point the valve is fully open. At this point, the retentate pressure should be at least 8 psi to overcome osmotic pressure. If not, tighten the retentate valve slightly to maintain a minimum 8 psi retentate pressure.
2. Continue concentrating at J_F and with the retentate valve open (or PR $\geq$ 8 psi). The feed pressure will continue to rise. While roughly 80 psig is a cassette limit, typically a maximum pressure of 50 psig is used to allow for some safety margin.
3. Concentrate at the maximum feed pressure by reducing the feed flow J_F until feed pump flow reaches maximum pump turndown ~1/6 of initial feed flow. Typical operation may start at J_f = 6 lpm/m^2 (LMM) by vendor recommendation and finish at 1 LMM. At this point one can concentrate no further without running up against the maximum feed pressure limit (Figure 7.3).

A cassette feed screen is used to improve back transport of retained solutes away from the membrane surface to increase the mass transfer coefficient k and the process flux *J*. 'Tighter' screens improve transport but also increase the flow resistance and pressure drop. This means that there is a trade-off where opening up the screen allows one to reach higher final concentrations but requires additional membrane area-time for processing. Figures 7.4 and 7.5 show pressure drop and flux data during Bovine IgG (BgG) concentration using cassettes with a conventional 'C screen', a more open 'high viscosity tangential flow filtration (TFF)' screen, and a very open 'V screen'. These figures use a constant feed flux rather than the variable flow strategy described earlier. The high viscosity screen is able to reach higher concentrations at a given pressure drop with only a modest reduction in flux.

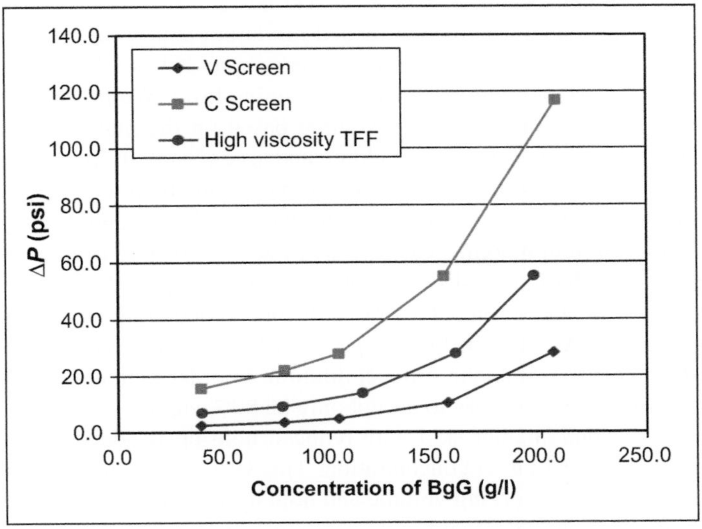

Figure 7.3 Pressure drop vs. BgG concentration at 8 L/m^2/min feed flow. (Courtesy of EMD Millipore.)

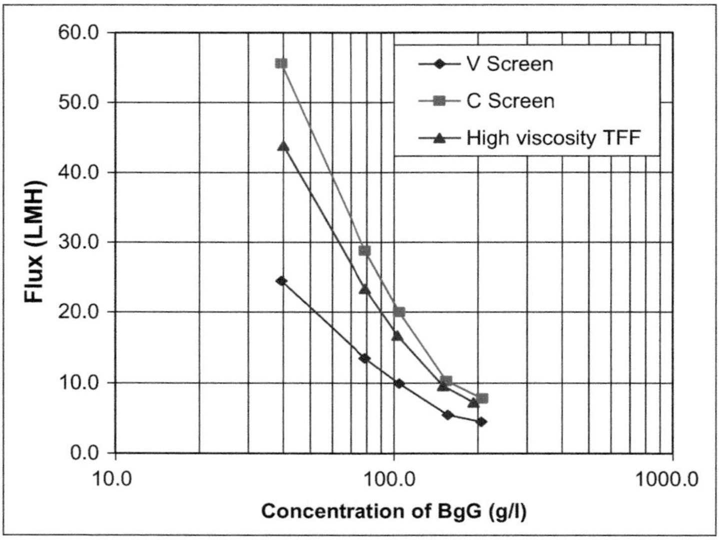

Figure 7.4 Flux vs. BgG concentration at 8 L/m^2/min feed flow. (Courtesy of EMD Millipore.)

One may also be limited in reaching higher concentration by the specific system hardware. During the final concentration step, the liquid level in the retentate tank can fall below a recommended level and air can be entrained into the tank outlet. The conditions at which this occurs depend on the tank design (location and design of retentate return line, tank diameter, vortex breaker design and tank bottom geometry) and the retentate flow rate. The recommended minimum tank level leads to a minimum working volume requirement for the system expressed as L/m^2. One needs to run at a minimum working volume above this foaming level. Well-designed systems can achieve 1 L/m^2 minimum working volume based on the maximum installable area.[2] Systems that are built for ease of welding and access without special attention to hold up can show 3 L/m^2 minimum working volumes. If the final target concentration implies a system volume (i.e., product Kg mass/target g/L concentration-system m^2 area) below the minimum working volume, the system limits one from being able to reach the target. Once built, it is difficult to modify an existing skid to reduce hold up.

Another system limitation on achieving high final concentrations is dilution during product recovery. As described in Chapter 9, achieving good yields of >95% by a buffer flush can cause significant product dilution. In order to maintain a high final concentration, it may be necessary to accept only an 80% yield.[3]

Frequently, a second smaller skid with reduced hold up is added to the process in order to reach the final target concentration. This system with reduced membrane areas allows for a smaller working volume and hold up volume to reach high concentrations without foaming and denaturing the product or taking a big yield loss. This added step can be very costly. A single pass UF system can be particularly useful here due to its compact size and simplicity of operation.

7.5 Protein aggregation and precipitation

Protein products may aggregate or precipitate during an ultrafiltration step. These pose quality issues and can pose yield issues as well. Aggregate concentrations in the final product are of particular concern because they can induce an immune reaction against the product.[4,5] Precipitation occurs when a solubility limit has been reached. Aggregation occurs due to a denaturing reaction.

Protein solubility varies with the buffer system and temperature. During processing, the protein batch concentration and the buffer environment change. For a final formulation application, the protein is in the starting (feed) buffer after the first concentration step. The final product will be in the diafiltration buffer. If the feed has been eluted off a column at high salt conditions, there may be a precipitation issue after the concentration step. It is also possible that during the diafiltration step, as the buffer transitions from the feed to the diafiltrate buffers, one may pass through a region of low solubility, such as the protein isoelectric point. One can use a multi-well plate screening system to incubate different protein concentrations in different buffer solutions and look for turbidity as an indicator of precipitation.[10] Due to polarization in the module, protein concentration is highest at the membrane surface near the module inlet. This region of high concentration is localized and can cause module fouling. Precipitation also varies with time so the screening study is a qualitative indicator of possible precipitation issues during processing.

Strategies to avoid precipitation involve avoiding regions of low solubility. One can diafilter at a bulk concentration lower than the C_g/e optimum (Section 6.5). If the solubility limits are not very constraining, Figure 6.7 shows that modest reductions in this set point do not lead to significant increases in membrane area-time. The 'C_{wall}' control strategy limits average flux through a membrane system to limit polarization (Section 7.9). The variable volume diafiltration strategy changes the path of protein and buffer concentrations during the UF/DF process to avoid regions of low solubility (Section 7.10). If there is a transition through the isoelectric point during diafiltration, it is recommended that the pH be changed quickly through direct addition of buffer concentrate to the retentate tank. This will limit the time that the bulk fluid spends near the isoelectric point and limit precipitation.

For an ultrafiltration step used for harvest concentration, upstream in the manufacturing process, it is not unusual to see precipitation occurring. This tends to be due to cell culture impurities, not product precipitation. It can cause fouling leading to flux decline and cleaning issues. This is particularly true for tight membranes retaining relatively dilute low molecular weight protein products. Here, the process flux may be determined by the process permeability and not protein polarization. The use of additional adsorptive depth filters or flocculation in the harvest step can remove more of these fouling impurities and improve filtration operation.

Denaturing reactions causing aggregation are mainly due to gas–liquid interfaces. The flushing step after integrity testing is designed to eliminate gas from the piping system and membranes. Subsequent steps should be designed to avoid further introduction of gas into the system. There is naturally a gas–liquid interface in a stainless retentate tank but the introduction of collapsible plastic biocontainers enables one to minimize even these interfaces for small to medium retentate tanks. The minimum working volume constraint is set to avoid foaming and aggregation in a stainless retentate tank.

Denaturing has been observed in systems. This has been attributed to shear within an ultra-filtration module although closer investigation does not seem to bear this out (see Section 3.7). Mitigating aggregation based on pumping can involve component selection (pumps, valves, piping), reducing system passes and reducing low pressure regions. As described in Chapter 10, different pumps and valves can cause different fluid stresses and levels of aggregation. Diafiltrate and feed should not be stored under elevated pressures promoting higher concentrations of dissolved gases. Turbidity can be plotted against pump passes to gauge the magnitude of the effect. For a changing retentate volume, one must integrate the number of passes (see equation (7.1)). Aggregation per pass is higher at higher pumping rates, but the number of pump passes are fewer because of the increased flux. In general, flowing slower tends to reduce aggregation.

Temperature can play a role in both precipitation and aggregation. Heat exchangers have been included in systems to remove heat generated by pumps. The diafiltrate contains sufficient thermal mass that a heat exchanger is not generally needed. Only during the final UF2 concentration would a heat exchanger be helpful as the pump heat is going into smaller volumes of fluid. This final step takes place so quickly that generally a heat exchanger is not needed. Care must be taken to insulate the motor drive from the pump head to avoid heat generated in the motor being conducted into the pump head.

7.6 Adsorbed solutes

Solutes that normally would pass through the filter may adsorb to other components in solution. This could be ions or dyes adsorbed to proteins in solution, or excipients adsorbed to adsorption beads in the retentate. Consider an adsorber concentration of C_A g/L with an excipient in equilibrium between its solution concentration of C g/L and bound excipient with an isotherm of $q(C)$ g excipient/g adsorbent. The total mass of excipient in the retentate is $M = CV + C_A q(C)V$ and the excipient mass balance for diafiltration is $\frac{dM}{dt} = V\frac{dC}{dt} + VC_A q'\frac{dC}{dt} = -sCQ_p$ where $q' = dq/dC$, the slope of the isotherm. Rearranging to $d\ln C + C_A q' d\ln C = -sdN$ shows that if q' is a constant, then the excipient sieving coefficient is reduced from s to $s/(1 + C_A q')$ in the presence of adsorption. One can use different isotherm forms (e.g., Freundlich KC^n, Langmuir $KC/[1 + KC]$) and integrate the mass balance to calculate how the diavolumes change with retentate excipient concentration.

One example of adsorbed solutes is aluminium ion binding to human serum albumin (HSA). Aluminium is present in diatomaceous earth used as depth filter but can also release from glass storage vessels and caps. Its concentration in HSA is limited by the British Pharmacopoeia to <200 μg/L to render it suitable for use for renal dialysis and premature infants. Aluminium would normally pass freely through an ultra-filtration membrane with a sieving coefficient of 1. However, in the presence of HSA the sieving coefficient is reduced but constant during diafiltration.

Ethanol from the Cohn precipitation process is typically diafiltered out of HSA as part of its preparation.[11] The aluminium sieving coefficient increases in the presence of NaCl since the positively charged sodium ions can displace the positively charged aluminium ions off the adsorption binding site on the HSA. One can use this behavior to remove aluminium by (1) diafiltering out the ethanol with water, (2) adding salt and diafiltering with a salt solution to reduce the aluminium to acceptable concentrations and (3) diafiltering out the salt with water. Cumulatively, these three steps in sequence take a fair amount of area-time and require roughly 18 diavolumes of buffer. One can optimize this process by recognizing it is possible to wash out more than one excipient at a time. In a more optimized form, one can add salt and wash out the ethanol, aluminium, and salt together with less area-time and diafiltration buffer. Since the salt levels vary over the course of the diafiltration, so does the aluminium sieving. One can generate a differential equation model of this system and run simulations to predict the behavior and design a robust and more economical process. For very high amounts of aluminium, high salt concentrations are required, which may precipitate the HSA. If this occurs, one could diafilter first with NaCl, then with water in two steps.

Excipient adsorption can also lead to buffer shifts during the final UF2 concentration step in final formulation.[12–14] One effect is due to the high protein concentration, which effectively takes up volume and reduces the volume available for other excipients. This would tend to exclude some large excipient like PEG and lead to sieving coefficients above 1. Consider a MAb formulated at a pH around 7.0, below its isolectric point of 8.5. This creates a net positive charge on the IgG that is balanced by negatively charged counter ions in solution. As the charged protein concentrates, OH^- counter ions can pass through the membrane and into the permeate stream. This requires that other negative counter ions, such as Cl^-, be adsorbed or retained to maintain electrical neutrality and avoid a voltage build up across the membrane. As a result, one can end up with a different ion composition than what was present in the diafiltration buffer. This has been termed the Donnan effect. Some excipients can be in higher concentration (e.g., Cl^- in this example) and some lower (e.g., OH^- in this example). Reduced OH^- causes a rise in H^+ concentration to maintain the water equilibrium $K_w = 10^{-14} = [H^+][OH^-]$ and resulting drop in $pH = -\log[H^+]$. Histidine depletion has also been observed. One can adjust the diafiltrate buffer proportionately to reach the target buffer composition after concentration or one can add excipients after concentration to readjust the composition.

7.7 C_{wall} operation[15]

The membrane surface has the highest protein concentration C_{wall} and this concentration controls the osmotic pressure and the flux. Direct control of wall concentration can be a basis for ultrafiltration process operation. The intended benefits are more consistent operation and reduced turbidity and fouling. During a concentration step, the bulk concentration C_b increases as the volume is reduced by the mass balance $C_b = C_{b0}V_0/V$. The wall concentration C_w then increases as well. One can compensate for C_b increases and keep the wall concentration constant by reducing the flux J along with the retentate volume according to the polarization equation for a fully retained solute, $J = k\ln(C_w/C_b) = k\ln(C_w/C_{b0}) + k\ln(V/V_0)$. The flux is controlled either with a permeate flow meter and a retentate valve or a permeate pump.

An average mass transfer coefficient k is obtained from the slope of the flux vs. ln(C) curve $k = (J_2 - J_1)/(\ln C_{b2} - \ln C_{b1})$. This can be improved to compensate for variations in C_w along with bulk concentration. First calculate fouled membrane permeability L_{fm} in buffer, post-exposure, as the slope of the flux vs. TMP curve. Then calculate osmotic pressures as $\Pi_w = J / L_{fm} - TMP$ for each feed concentration. One can then plot J vs. Π_w for different C_b, interpolate to get new fluxes J' at both a given feed concentration C_b and wall concentration or osmotic pressure Π_w. Then $k = (J_2' - J_1')/(\ln C_{b2} - \ln C_{b1})$.

As C_w increases, process time and membrane cost and hold up volumes decrease, while filtrate, adsorption and solubility losses can increase. The high value of yield losses and product quality led the authors to set $C_w = 50$ g/L for their relatively fouling MAb application. This approach to ultrafiltration process control is not widely adopted in the industry. The costs include added complexity of process operation and the use of higher membrane areas. The benefits include a more consistent process time and performance. For problematic proteins sensitive to high concentrations, the following variable volume diafiltration (VVDF) approach may be useful as well.

7.8 Variable volume diafiltration operation[16]

Figure 7.5 shows regions of increased turbidity (lower solubility) as solution protein concentration increases. The usual path for retentate concentrations is shown as a dashed line with (1) constant NaCl concentration at increasing MAb concentrations, and (2) decreasing NaCl concentrations at constant MAb concentration ($=C_g/e$). This path enters the high turbidity region. Consider following the curved path skirting the edge of the high turbidity region where the retentate undergoes both a buffer exchange and a product concentration. Here, a constant diafiltration ratio $\alpha = Q_D/Q_P < 1$ is used where Q_D is the diafiltration buffer flow rate and Q_P is the permeate flow rate.

Differential solvent and component mass balances for the control volume shown in Figure 7.6 are

$$\frac{dV}{dt} = (\alpha - 1)Q_P \text{ and } \frac{dM}{dt} = \frac{d(CV)}{dt} = (-sC)Q_P \tag{7.2}$$

where V is the retentate volume, M is solute mass, densities are assumed constant, C is the retentate concentration and s is the component apparent sieving coefficient ($s = C_{permeate}/C_{retentate}$). Capitalized subscripts refer to the different flows: F for feed, P for permeate and D for diafiltrate.

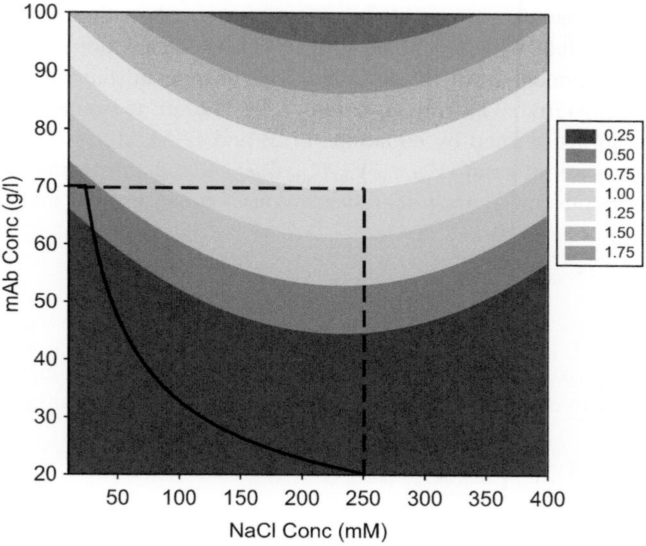

Figure 7.5 Retentate concentration path and solubility limits.
(Courtesy of Biot Prog.)

Using equation (7.2) the solvent balance becomes

$$dV_P = Q_P dt = \frac{dV}{\alpha - 1} \tag{7.3}$$

Simplifying the component balance by expanding the differential $d(CV) = CdV + VdC$ one can write $VdC = (-sC)Q_P dt - C(\alpha - 1)Q_P dt$ using equation (7.2). Substituting dV from the solvent balance (Equation (7.3)) and rearranging using $d\ln V = dV/V$

$$\frac{d\ln V}{\alpha - 1} = \frac{dC}{(\alpha + s - 1)C} = \frac{dCs}{-(\alpha)Cs} = \frac{dCp}{-(\alpha - 1)Cp} dN \tag{7.4}$$

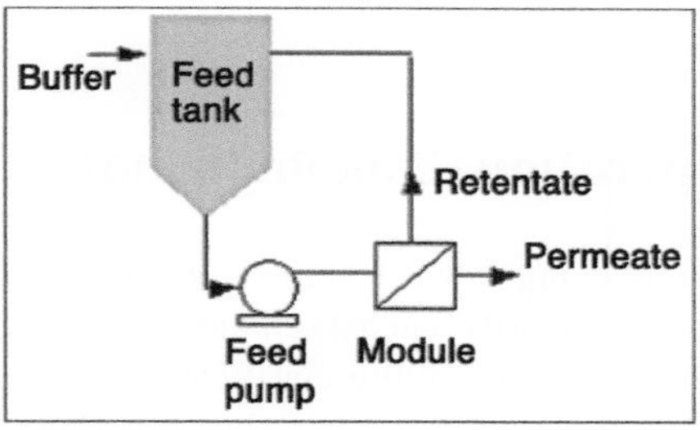

Figure 7.6 Batch diafiltration.

Here, two components are considered: a permeable salt with $s = 1$ (subscript s) and a retentive product with $s = 0$ (subscript p). Lower case nomenclature is used to distinguish component from fluid stream. The change in diavolumes is defined as $dN = dV_P/V = d\ln V/(\alpha - 1)$ from equation (7.2). The set of simultaneous ordinary differential equations defined by equations (7.2) and (7.3) defines the evolution of the system for any initial point (V_{p0}, t_0, V_0, C_{s0}, C_{p0}) and a control strategy defined by parameter α. For constant α, analytical integration of equation (7.4) yields

$$C_p = C_{p0}[C_S / C_{S0}]^n \tag{7.5}$$

where $n = [\alpha - 1]/\alpha$.

Figure 7.5 shows a curved path as the model prediction of how product and salt concentrations evolve over time with a constant value of α during the initial VVDF step.

$$C / C_0 = \exp[-(s - 1 + \alpha)N]. \tag{7.6}$$

This reduces to the familiar form for constant volume diafiltration when $\alpha = 1$ $C = C_0 \exp(-sN)$.

The gel flux model, $Q_P = kA\ln(C_g / C_p)$, is used along a path where no precipitation occurs. The evolution of the system over time can then be calculated by numerically integrating the time and volume terms from equation (7.2) as

$$kA\ln(C_g / C_p)dt = \frac{dV}{\alpha - 1} \tag{7.7}$$

The VVDF processes take more time to reach the diafiltration concentration target than the constant volume diafiltration case. However, since buffer exchange occurs during the VVDF step, the resulting time of constant volume diafiltration is shortened, and the net result is a comparable total processing time for the different diafiltration scenarios. All of the different VVDF paths are preferable to diafiltering at the low initial concentration.

Modelling enables simulation of different control strategies and determination of one that skirts the low solubility region and does not significantly increase process time requirements.

7.9 Mass flux operation (Bala Raghunath and Herb Lutz)

Consider the application of trying to wash some key component out of the retentate. This component could be an impurity, such as ethanol or PEG, or the desired product, such as a vaccine. Mammalian cell culture clarification can be considered a microfiltration application of washing out protein product into the permeate stream. Here, we will stick to ultra-filtration membranes ranging up to 1000 kDa for applications, such as allantoic fluid clarification in egg-based flu process where the product (flu virus) may get concentrated on the feed side of the membrane while the contaminating impurities (ovalbumin, etc.) may get removed into the permeate side.[17]

Ultrafiltration steps are typically designed to maximize solvent flux J and thereby minimize the area-time requirements. For some applications however, the sieving coefficient of a key component that one is trying to wash out of the system can vary with operating conditions. As an example, low fluxes might increase sieving and high fluxes might decrease sieving and increase fouling. Maximizing flux in this case may require excess area. Consider the mass flux G of a key component k that one is trying to remove in units of g/m^2/h. It can be written as

$$G = JC_{\mathrm{pk}} = JC_{\mathrm{k}}s_{\mathrm{k}} \quad (7.8)$$

for flux J, permeate concentration C_{pk}, retentate concentration C_{k} and sieving s_{k}.

Maximizing the mass flux G, rather than J or s_{k} individually, minimizes the area-time requirements (Figure 7.7). G can be dependent on several operating conditions. For example, it can vary with polarization of another, retained component, in the module. At conditions of higher polarization (i.e., high TMP and low cross-flow), the passage and the mass flux can be low while the flux is high. There can be a 'sweet spot' where the operating conditions are only moderately polarizing. This generally falls below the 'knee' of the flux vs. TMP curve. Linear polymers can also extend and change their effective size under cross-flow shear. This can change sieving.[1]

The stability of the mass flux is also a key consideration. If fouling occurs, sieving coefficients typically decline. Very open UF membranes with high NMWL cut-offs are particularly sensitive to fouling.[18] It only takes a small TMP to drive fluid through these highly permeable membranes. As a result the back transport mechanisms for

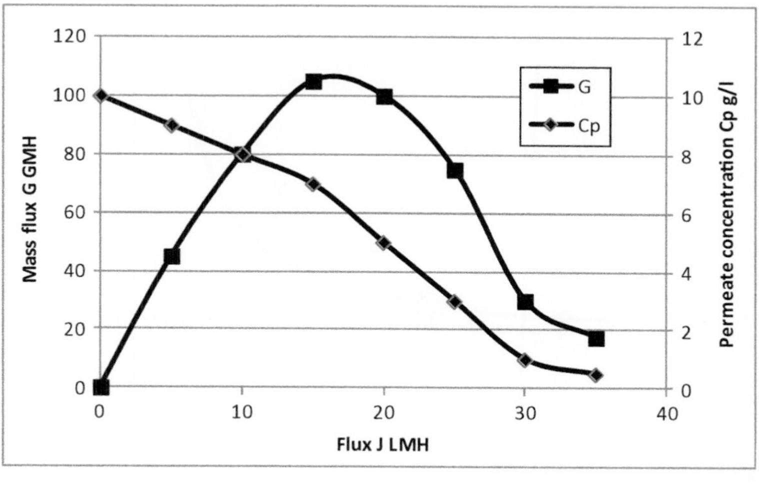

Figure 7.7 Optimizing mass flux.

retained species can become overwhelmed and one can build up high wall concentrations. As shown in Figure 5.1, back transport by molecular diffusion decreases with particle size and other mechanisms do not contribute significantly until the particle size is several micrometers in diameter.

During start-up, the membrane typically adsorbs some fluid components and reduces in permeability. The initial high permeability will draw extra fluid to the surface and retain more fouling species. Variable hydraulic conditions during start-up can result in variable surface concentrations and lead to potentially different sieving coefficients. This is true not only for high NMWL membranes but also for lower NMWL membranes where fractionation is important (see Section 7.14).

For high MW membranes, flat sheet open channel (i.e., without screens or using a suspended screen) and hollow fibre configurations are recommended. These can handle the higher concentrations and sizes of suspended solids and colloids without channel or screen plugging. They can also facilitate ease of cleaning as the screen-free surfaces will be more accessible to cleaning fluids. They will also reduce pressure drop along the feed channel to provide a more uniform TMP for the high permeability membranes.

The direct control of permeate flux (as opposed to controlling TMP in a typical TFF process) can be used to manage and reduce polarization and fouling.[19,20] Selection of an operating flux is guided by the concept of a 'critical flux',[21] beyond which the process becomes unstable (at a given cross-flow rate). Process instability is evidenced in the form of (1) a sudden and consistent rise in TMP (if permeate flux is controlled) or (2) a sudden and consistent decline in flux and (if TMP is controlled) beyond a certain operating point (flux or TMP). Process instability is frequently accompanied by a sharp decline in the sieving or passage of the desired solute (protein) of interest. Figure 7.8 shows the point of critical flux in a TFF micro-filtration process. Variations in membrane permeability will cause flux variations at a constant TMP. This can result

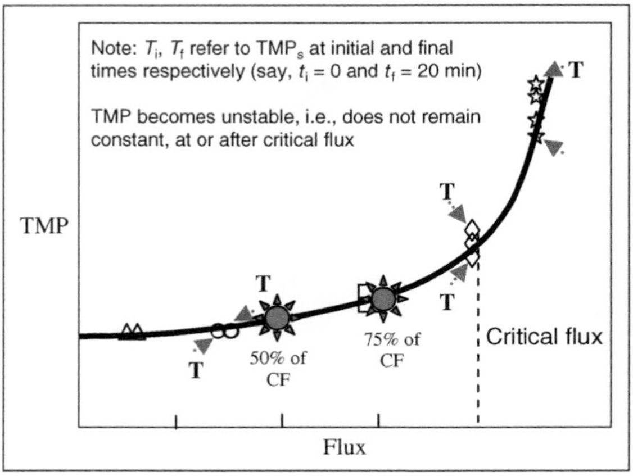

Figure 7.8 Critical flux behavior.
(Courtesy of Bioprocessing Journal.)

in changes in polarization and the resulting mass flux. The permeate pump acts as a valve to create more consistent polarization and mass flux.

Critical flux determination for a given application involves developing TMP vs. flux data at a given cross-flow (in 'total recycle' i.e., permeate back to feed vessel) using an appropriate scale-down module. For lower molecular weight applications, the sieving coefficient is the critical parameter to measure. With the permeate pump off, ramp up the feed pump to the desired cross-flow rate. Slowly ramp the permeate pump to the desired flux level. The starting flux for many applications would be in the 5–10 LMH range. Record inlet, outlet and permeate pressures over a 20–30 min period (i.e., at 5 or 10 min intervals). Monitor stability of TMP and/or sieving coefficient (a rapid increase in TMP indicates membrane fouling – as a rule of thumb, one may consider critical flux to have been reached when $TMP_{final} \div TMP_{initial} > 1.5$–2.0, where initial and final refer to the beginning and end of the 20–30 min time interval). Take samples of feed and permeate at the end of the 20–30 min period and analyze for protein concentration (and sieving). If the TMP and/or sieving are/is stable, then ramp up the flux to the next level (10–15 LMH increments). Repeat the step at increasing flux settings until the critical flux is obtained. (These steps may also be repeated at a different cross-flow.) It is helpful to apply a small amount of pressure on the retentate (by closing the retentate valve) as this will allow for the exploration of a broader range of TMPs. Typical results from the critical flux experiment may be expressed as shown in Figure 7.9.

High MW applications may have the added complexity of cumulative plugging of the TFF membrane by colloids small enough to enter the membrane pores and deposit inside the membrane. For NFF micro-filtration this is typically described as gradual pore plugging.[22] It leads to a maximum L/m^2 throughput, or V_{max} capacity. V_{max} can vary with flux. As flux increases, particles may plug the pore entrances near the surface

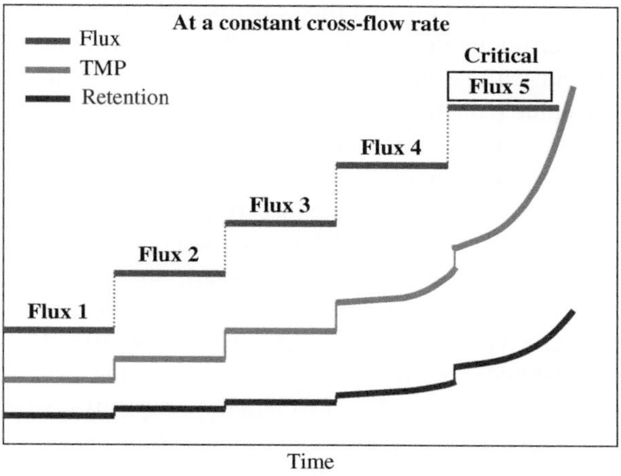

Figure 7.9 Critical flux determination.
(Courtesy of Bioprocessing Journal.)

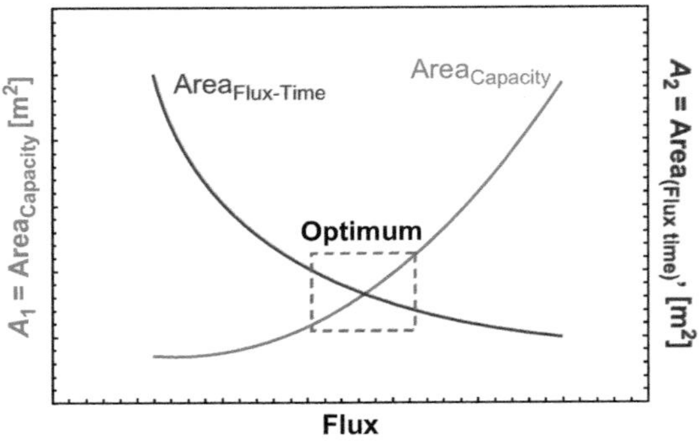

Figure 7.10 Optimizing operation for plugging feed.
(Courtesy of Bioprocessing Journal.)

of the filter and not penetrate further down the pores. This change in particle deposition location can impact the plugged filter flow resistance and membrane capacity.

For V_{max} capacity-limited TFF applications, there may be an optimum flux that minimizes membrane area requirements. Consider sizing to process a batch volume of V_B in batch time t_B using a capacity constraint (Equation (7.9)) and a flux-time constraint (Equation (7.10)):

$$A_1 = V_B / V_{max} \text{ Capacity limited} \quad (7.9)$$

$$A_2 = V_B / (\text{Flux}\, t_B) \text{ Flux - time considerations} \quad (7.10)$$

As flux increases, A_1 increases (as V_{max} decreases with increase in flux) and A_2 decreases; process sizing will be optimum when the areas calculated by both methods are equal to one another, that is, $A_1 = A_2$ as shown in Figure 7.10.

Implementation involves picking two fluxes below critical flux, say 75% and 50% of critical flux and determining the capacity for each using a volume reduction and/or diafiltration experiment with permeate flow control. For cell culture feeds, retentate viscosity can limit volume reduction (e.g., 5–10× for Mammalian cell cultures, 1–2× for yeast). For constant flux the TMP increases with throughput due to membrane plugging. The maximum throughput or capacity is considered to be reached when a predetermined maximum TMP value is attained (typically in the range of 5–10 psi where product passage begins to drop and the process shows diminishing returns).

As an example shown in Figure 7.11, consider clarifying a 1000 l feed in 3 h that has a critical flux (CF) of 45 LMH. At 75% of CF, the test flux of 34 LMH showed a capacity of 40 L/m^2, with A_1 (area based on capacity) = 1000/40 = 25 m^2 and A_2 (area based on flux-time) = 1000/(34 × 3) = 9.9 m^2. At 50% of CF, the test flux of 22.5 LMH showed a capacity of 60 L/m^2, with A_1' (area based on capacity) = 1000/60 = 17 m^2 and A_2 (area based on flux-time) = 1000/(22.5×3) = 14.8 m^2. Extrapolation of these data as shown in Figure 7.11 indicates an optimum flux of 21 LMH with a corresponding optimum membrane area of 15.5 m^2. As seen in Figure 7.11b, operation at 21 LMH with 23.3 m^2 of area, a 1.5× safety factor, would allow for fluctuations in both

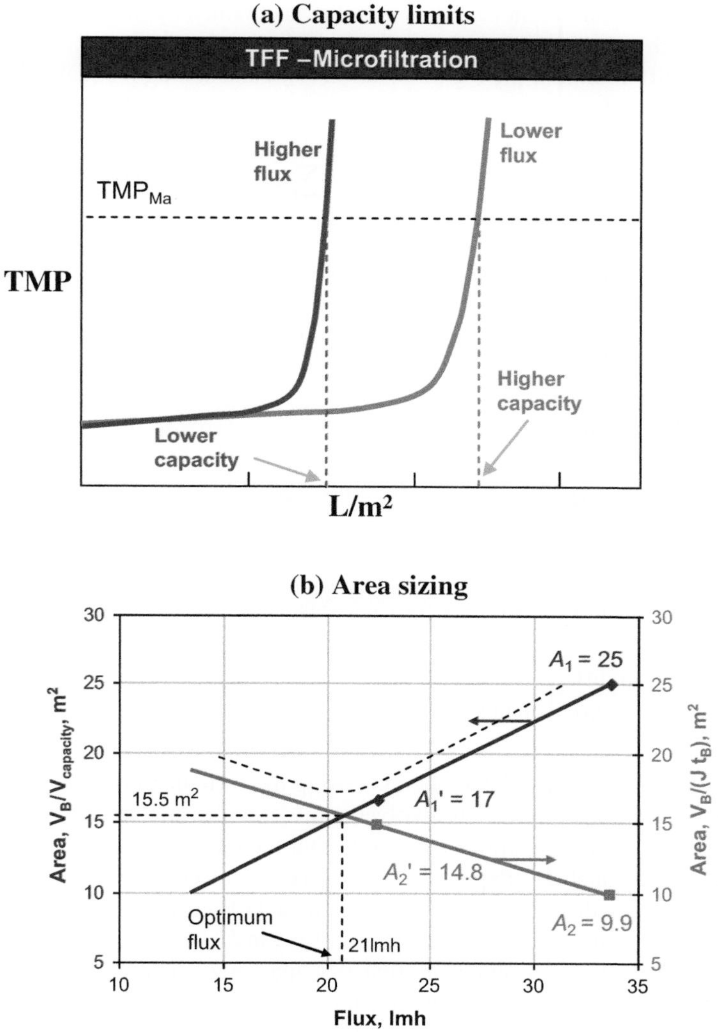

Figure 7.11 Optimizing operation for plugging feed: (a) capacity limits and (b) area sizing. (Courtesy of Bioprocessing Journal.)

batch volume and V_{max} capacity. Decreasing the operating flux while following the blue line capacity constraint increases the area to allow for V_{max} capacity fluctuations.

One can also consider diafiltration strategies to maximize mass flux.[23] Minimum area-time is obtained by maximizing the average mass flux over processing. For processes where the mass flux is stable over time, the mass flux is a state function of the system concentrations only and independent of any particular path or $G(C_i)$ for all components i. In accordance with Bellman's principle of optimality for dynamic programming,[24] the maximum average mass flux is obtained by maximizing the local mass flux at each point in the process.

Taking the differential of the log of equation (7.8) and setting it equal to zero for a local maximum yields

$$d\ln G = \sum_i (\partial \ln J/\partial \ln C_i) d\ln C_i + d\ln C_k + \sum_i (\partial \ln s_k/\partial \ln C_i) d\ln C_i = 0. \qquad (7.11)$$

Mass balances require (equation (8.2)) $\mathrm{d}\ln C_i = -\mathrm{d}\ln V$ so the optimum path is

$$0 = g(C_i) = 1 + \sum_i (\partial \ln J/\partial \ln C_i) + \sum_i (\partial \ln s_k/\partial \ln C_i) \qquad (7.12)$$

Mass balances require

$$\frac{d\ln V}{\alpha - 1} = \frac{dC}{C_D\alpha - (\alpha + s - 1)C} = \frac{dC_S}{-(\alpha)C_S} = \frac{dC_P}{-(\alpha - 1)C_P} dN$$

so the optimum diafiltration strategy for $C_D = 0$ is

$$\alpha(C_k) = 1 + \frac{s_i d\ln C_k - s_k d\ln C_i}{d\ln C_i - d\ln C_k} \qquad (7.13)$$

These equations have also been obtained using the formalism of optimum control theory.[25] This alternative approach involves more complex mathematical manipulation and does not provide the physical insight that the mass flux approach does. Application of these equations requires a mass flux model $G(C_i)$. Consider a system that is dominated by the key component, such as a TFF virus filtration. In this case the optimum path (Equation (7.12)) reverts to $-1 = d\ln J/d\ln C + d\ln s/d\ln C$ and the solution is a single value of concentration that maximizes mass flux. The optimal diafiltration strategy is to maintain this constant concentration using $\alpha = 1 - s$ instead of maintaining a constant volume. One would either concentrate or dilute from the initial point to this optimum concentration. This approach has been termed differential diafiltration.

Consider a two-component system where the objective is to wash out component 2 by maximizing its mass flux. In the conventional case, J is just a function of component 1 and the sieving coefficients are constants. The optimum path (Equation (7.12)) is $0 = 1 + d\ln J/d\ln C_1$ or $0 = d(C_1 J)$, which maximizes the concentration of the retained component times the flux (as shown in Section 6.5). When the flux is a function of both components as a modified gel model, $J = k\ln(C_g / C_1)/[1 + aC_2]$, with sieving coefficients constant at $s_1 = 0$ and $s_2 = 1$, the optimum path (Equation (7.11)) becomes $g(C_1, C_2) = 0 = 1 - 1/\ln(C_g / C_1) - a/[1 + aC_2]$. This simplifies to $C_1 = C_g e^{-[1+aC_2]}$ along the optimum path, which makes the flux constant at J = k. The diafiltration strategy (equation (7.13)) is to stay along this path requires $\alpha(C_2) = 1/[1 + aC_2]$. One uses pure concentration ($\alpha = 0$) or dilution ($\alpha = \infty$) to get to the optimal path from an initial point before proceeding along it. Then, one uses pure concentration ($\alpha = 0$) or dilution ($\alpha = \infty$) to get from the optimal path to the final point.

Consider a whey-lactose case published as an optimum, which is improved upon by the present method.[26] The feed solution containing 3.3 wt% of protein and 5.5% lactose is concentrated and diafiltered using a 50 kDa Romicon hollow fibre module (HF15-43-PM50) to retain protein and pass lactose. The target final retentate concentration is 0.64% lactose. Sequential concentration and diafiltration tests showed constant sieving coefficients of $s_p = 0$ and $s_l = 1$, and a flux correlated to component concentrations of $J = 63.24 - 12.439\ln(C_p) - 7.836\ln(C_l)$ LMH. The optimum path (Equation (7.12)) becomes $0 = 1 - 12.439/J_{opt} - 7.836/J_{opt}$ or $J_{opt} = 202.28$ LMH constant and $C_p = 31.63C_l^{-0.63}$. The diafiltration strategy (Equation (7.13)) to stay along this path requires $\alpha(C_2) = 1/[1 + .63] = 0.613$ constant, a VVDF strategy. Here, one moves from the initial point (3.3, 5.5) to the optimum path at (10.98, 5.5) by concentration ($\alpha = 0$), then along the optimum path from (10.98, 5.5) to (23, 1.7) by VVDF ($\alpha = 0.613$), and finally from the optimum path at (23, 1.7) to the final concentration at (9.04, .64) by dilution ($\alpha = \infty$). Solving for the membrane area-time required for this strategy involves integrating the flux equation to get 0.043 m^2/h per litre of feed. This is about a 10% improvement over the published optimum.

7.10 Fractionation operation

Here we consider both the yield and purity of the retentate and permeate solutions. In the mass flux case, we maximized the rate of removal without explicitly considering the impact on either retentate product yield or permeate product purity. Consider a batch concentration and diafiltration step where the only significant source of yield loss is into the permeate stream. The retentate component mass varies with volume reduction factor X and diavolumes N as $M_i = M_{i0}e^{-s_i(N+\ln X)}$ (Chapter 5). The retentate yield or fraction of initial mass in the retentate is $Y_{Ri} = M_i / M_{i0} = e^{-s_i(N+\ln X)}$. The permeate yield is the remainder $Y_{Pi} = 1 - e^{-s_i(N+\ln X)}$.

The g/L concentration of a product and impurity is C_p and C_i, respectively. The relative impurity concentration is g impurity per gram of product or as ppm (parts per million). Consider an ultrafiltration step taking the feed with a relative impurity ppm of C_{i0}/C_{p0} and generating a product retaining retentate with a relative impurity ppm of C_i/C_p. In keeping with a generalized description of separation processes, one can define a process retentate separation factor $SF_R = (C_i/C_p)/(C_{i0}/C_{p0})$ to characterize the degree of separation from the feed to retentate.[27] Since the retentate impurity concentration can be written as $C_i = C_{i0}e^{\ln X}e^{-s_i(N+\ln X)}$, the retentate separation factor is $SF_R = e^{(s_p - s_i)(N+\ln X)}$. Since $N + \ln X = -\ln Y_{Rp} / s_p$, the retentate separation factor can be written directly in terms of retentate yield as $SF_R = Y_{Rp}^{\psi - 1}$ where $\psi = s_i / s_p$ is the selectivity. Here the product is in the retentate so $s_i \gg s_p$ and $\Psi > 1$. As the

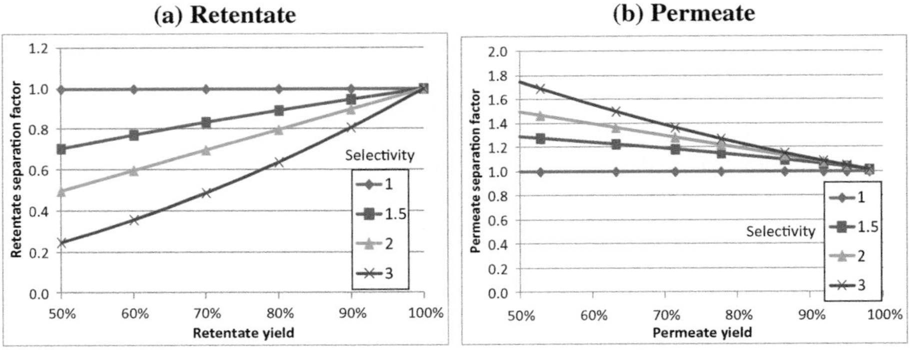

Figure 7.12 Solute fractionation: (a) retentate and (b) permeate.

ultra-filtration process proceeds (Figure 7.12a) from the point where all of the product is in the retentate ($Y_{Rp} = 1$) to a point where some product is lost into the permeate ($Y_{Rp} < 1$), the separation factor decreases from a start of 1.0. For a selectivity of 1.0, the separation factor remains at 1.0 (no relative depletion of impurity) throughout the process as product is lost into the permeate stream at the same rate as impurity. One could also think of a purification factor as 1/separation factor to describe the relative enrichment of the product.[28]

Consider a process where product is recovered in the permeate stream. Then a process permeate separation factor $SF_P = (C_i/C_p)/(C_{i0}/C_{p0})$ characterizes the degree of separation from the feed to the permeate. Since the permeate impurity concentration can be written as $C_i = M_{i0}/V_{perm} = M_{i0}\left[1-e^{-s_i(N+\ln X)}\right]/\left[V_0(1-1/X+N/X)\right]$, the permeate separation factor is $SF_P = \left[1-e^{-s_i(N+\ln X)}\right]/\left[1-e^{-s_p(N+\ln X)}\right]$. Since $N+\ln X = -\ln Y_{Pp}/s_p$, the retentate separation factor can be written directly in terms of permeate yield as $SF_P = \left[1-\left(1-Y_{Pp}\right)^{\psi}\right]/Y_{Pp}$ where $\psi = s_i/s_p$ is the selectivity. Here the product is in the permeate so $s_p \gg s_i$ and $\psi < 1$. As the ultrafiltration process proceeds (Figure 7.12b) from the point where all of the product is in the retentate ($Y_{Pp} = 0$) to a point where the product is recovered in the permeate ($Y_{Rp} < 1$), the separation factor decreases from a value of ψ to one.

This analysis shows that it is possible to fractionate with high purities and yield, regardless of whether the product is recovered from the retentate or permeate, if the selectivity is high enough. The historical rule of thumb is that fractionation is feasible if the impurity and product solute MW differs by a factor of at least 10. Consider the ideal sphere in a cylindrical pore size exclusion model from Section 2.5. It suggests that the highest separation between two different-sized spherical solutes is achieved by using a cylinder with the same diameter as the larger sphere for complete retention ($s_p = 0$), and let the smaller sphere pass through with partial sieving ($s_i > 0$) with $\psi \gg 1$. The existence of a broad, non-uniform pore size distribution, along with other effects, limits selectivities in practice.

One group pursued the development of membrane fractionation and termed it 'high performance tangential flow filtration' or HPTFF.[29] By operating in a non-polarized region, below the 'knee' of the flux curve, the larger solute does not significantly retain the smaller solute by forming a secondary layer on the membrane surface. As noted in Section 7.2, charged solutes 'appear' larger than uncharged ones and are more selectively retained by the membrane.[30] By operating the feed solution at a pH equal to the isoelectric point of one of the solutes and ionic strength of below 100 mM salt, one can enhance selectivity. Polarization conditions can also be controlled by running both the feed and permeate sides of the membrane in cross-flow to maintain uniform TMP along the membrane channel. A permeate pump is also used on the permeate recycle loop to withdraw permeate from the system. These practices have enabled separation of model streams containing the roughly identically sized proteins of haemoglobin (MW = 67 kDa) and BSA (MW = 69 K).[20,30]

One can also consider the use of charged membranes to enhance fractionation.[22,31] This has enabled fractionation of a Fab from HCP with a separation factor of 10 and a yield of 98%. Fractionation processes stretch the limits of membrane separations. They can take time to develop and performance can fluctuate from batch-to-batch depending on the molecule differences and how much one is pushing the separation. This can reduce the purification process cost but one must be prepared for the investment in process development, and processes monitoring and control.

References

1. Latulippe DR, Molek JR, Zydney AL. Importance of biopolymer molecular flexibility in ultrafiltration processes, Ind Eng Chem Res 2009;48(3):2395–2403.
2. Goodrich, EM. Personal communication 2007.
3. Brinkmann A, Souther J, Westoby M. Integrating single-pass tangential flow filtration into biopharmaceutical purification processes. Biogen idec presentation at ACS National Meeting March 2012.
4. Rosenberg AS. Effect of protein aggregates: an immunologic perspective. AAPS J 2006;8(3):E501–7.
5. Carpenter JF, Randolph TW, Jiskoot W, Crommelin DJA, Middaugh CR, Winter G, et al. Overlooking subvisible particles in therapeutic protein products: gaps that may compromise product quality. J Pharm Sci 2009;98(4):1202–5.
6. Lutz H. Rationally defined safety factors for filter sizing. J Membr Sci 2009;341:268–78.
7. Millipore Corporation. Protein concentration and diafiltration by tangential flow filtration, TB032, Rev. C, 2003.
8. Lutz H, Zou IY, High concentration ultrafiltration processing: analysis of pressure constraints and recommended best practices. Planned submission to Biotechnol Prog in 2015.
9. Millipore Corporation. Pellicon 3 Cassettes installation and user guide, Lit.# AN-1065EN00, Rev. C, Jan 2009.
10. Kramarczyk JF, Kelley B, Coffman J. High-throughput screening of chromatographic separations: II. Hydrophobic interaction. Biotechnol Bioeng 2008;100(4):707–20.
11. Jaffrin M, Charrier J. Optimization of ultrafiltration and diafiltration processes for albumin production. J Membr Sci 1994;97:71–81.
12. Bolton GR, Austin WB, Basha J, LaCasse DP, Kelley BD, Acharya H. Effect of protein and solution properties on the Donnan effect during the ultrafiltration of proteins. Biotechnol Prog 2011;27(1):140–52.

13. Miao F, Velayudhan A, DiBella E, Shervin J, Felo M, Teeters M, et al. Theoretical analysis of excipient concentration during the final ultrafiltration/diafiltration step of therapeutic antibody. Biotechnol Prog 2009;25(4):964–72.
14. Stoner MR, Fischer N, Nixon L, Buckel S, Benke M, Austin F, et al. Protein-solute interactions affect the outcome of ultrafiltration/diafiltration operations. J Pharm Sci 2004;93(9):2332–42.
15. van Reis R, Goodrich EM, Yson CL, Frautschy LN, Whiteley R, Zydney AL. Constant C_{wall} ultrafiltration process control. J Membr Sci 1997;130:123–40.
16. Gefroh E, Lutz H. An Alternate diafiltration strategy to mitigate protein precipitation for low solubility proteins. Biotechnol Prog 2013;30(3):646–55.
17. Zeman LJ, Zydney AL. Microfiltation and Ultrafiltration – principles and application. Marcel Dekker Inc. NY; 1996.
18. Persson KM, Capannelli G, Bottino A, Tragardh G. Porosity and protein adsorption of four polymeric microfiltration membranes. J Membr Sci 1993;76:61.
19. Raghunath B, Wang B, Pattnaik P, Janssens J. Best practices for optimization and scale-up of microfiltration TFF processes. Bioprocess J 2012;30–40.
20. van Eijndhoven RH, Saksena S, Zydney AL. Protein fractionation using electrostatic interactions in membrane filtration. Biotechnol Bioeng 1995;48:406.
21. Field RW, Hu D, Howell JA, Gupta BB. Critical flux concept for microfiltration fouling. J Membr Sci 1995;100:259–72.
22. Lebreton B, Brown A, van Reis R. Application of high-performance tangential flow filtration (HPTFF) to the purification of a human pharmaceutical antibody fragment expressed in *Escherichia coli*. Biotechnol Bioeng 2008;100(5):964–74.
23. Lutz H. Membrane filtration with optimized addition of second liquid to maximize flux. U.S. patent 5,597,486, Jan. 28, 1997.
24. Bellman RE, Dryfus SE. Applied dynamic programming. Princeton, NJ: Princeton University Press; 1962.
25. Paulen R, Fikar M, Foley G, Kovacs Z, Czermak P. Optimal feeding strategy of diafiltration buffer in batch membrane processes. J Membr Sci 2012;411:160–72.
26. Rajagopalan N, Cheryan M. Process optimization in ultrafiltration: flux-time considerations in the purification of macromolecules. Chem Eng Comm 1991;106:57–69.
27. King C, Judson. Separation processes. New York, NY: McGraw-Hill; 1971.
28. Saksena S, van Reis R. Optimization diagram for membrane separations. J Membr Sci 1997;129:19–29.
29. van Reis R, Gadam S, Frautschy LN, Orlando S, Goodrich EM, Saksena S, et al. High performance tangential flow filtration. Biotechnol Bioeng 1997;56:71–82.
30. Saksena S, Zydney AL. Effect of solution pH and ionic strength on the separation of albumin from immunoglobulins (IgG) by selective filtration. Biotechnol Bioeng 1994;43:960.
31. van Reis R, Brake JM, Charkoudian J, Burns DB, Zydney AL. High performance tangential flow filtration using charged membranes. J Membr Sci 1999;159:133–42.

Post-processing

8

John Cyganowski, Herb Lutz
EMD Millipore, Biomanufacturing Sciences Network

As shown in Table 8.1, most post-processing steps do not require re-development for each specific protein. One can use a standard method, developed either by the bioprocessor or as recommended by a vendor. These standard methods are described here. They are verified during a small-scale trial run to ensure that they are working properly for the particular application. If these do not function properly, extra development is needed.

8.1 Product recovery

For most ultrafiltration (UF) applications, the product is retained by the membrane and is located on the feed/retentate side of the membrane modules after processing. Product recovery involves removing it from the system and transferring it to subsequent steps. For the final formulation step, this could include excipient addition, sterile filtration, lyophilization, fill-finish and storage.

The objectives for the product recovery step include minimizing impacts on product quality, minimizing recovery loss (maximizing yield), maintaining product concentration within target specifications, and providing a reliable consistent economical process. For a final formulation step, the bulk drug substance (BDS) product does not undergo any further purification steps. Therefore, recovery must not introduce product aggregates.

A few percent recovery loss is significant for products that sell for over a billion dollars per year. For final formulation, the BDS represents nearly the full cost of production. As a result, poor yields in this step can significantly increase overall production costs. In addition, poor yields decrease production capacity and require additional batches to meet demand. Even a modest increase in recovery represents a significant positive impact to both the cost of production and bioavailability to the patient population (Table 8.2). Earlier sections considered losses in the permeate (Section 7.4) and activity losses or precipitation during processing (Section 7.8). Frequently, the greatest source of product loss is unrecovered holdup loss during product recovery.

'Non-recoverable hold-up' is described as material that is not removed from the system during recovery operations. Well-designed systems and recovery procedures limit this to 1–2% of the batch for 98+% overall recovery (Equation 8.1). Poorly designed systems with large minimum working volumes and poor recovery procedures can show 20% loss.[1] Additional product cannot be recovered without diluting the retentate pool below its specified final concentration.

$$\% \text{Recovery} = (V_{\text{pool}} C_{\text{pool}}) / (V_r C_r) \tag{8.1}$$

where r refers to the retentate and pool refers to the recovered bulk drug substance.

The steps involved in recovering the product are shown in Table 8.3. The first two steps are common but there are three different options to recover product from the ultrafiltration modules and piping.

Table 8.1 UF post-processing step summary

Step	Process development requirements
Product recovery	Finalized during engineering runs
Cleaning	Vendor recommendations
• Flush	Verified using scale down system
• Add cleaner	
• Incubate	
• Flush	
• NWP measurement	
Sanitization	Vendor recommendations
• Flush	Verified using scale down system
• Add sanitizer	
Storage	Vendor recommendations
• Flush	Verified using scale down system
• Add storage agent	

Table 8.2 Batch impact of MAb process recovery

- Cell culture harvest every 14 days
- 25 batches/year
- 2% Recovery increase (97–99%)
- Additional 1/2 batch of product/year

Table 8.3 Recovery steps

Step	Purpose	Operation
Depolarization	Decrease product held up at membrane surface	Gentle recirculation with permeate closed
Tank retentate recovery	Recover retentate tank mass	Slow pumping to low drain
Option (1) air blowdown	Displace cassette and piping mass	Low pressure displacement
Option (2) buffer displacement	Displace cassette and piping mass	Low flow displacement
Option (3) buffer recirculation	Displace all mass	Recirculate buffer

8.1.1 Depolarization

During UF operations protein product is transported to the surface of the membrane creating a region of high protein concentration near the membrane wall called the polarization layer (Section 4.5). This can contain about 1 g/m^2 of held up mass (Section 2.9). After UF processing is complete, it is useful to re circulate the feed/retentate for approximately 15 min with the permeate valves closed. This allows the reversibly held polarization layer to back-diffuse into solution. Operating the system with the filtrate valves closed gives rise to filtrate back-pressure so it is generally recommended to recirculate cassettes at a gentle 1 LMM feed flow with the retentate control valve fully open in order to minimize this back-pressure. The membrane device manufacturer should be able to provide a specification for the maximum allowable filtrate back-pressure.

8.1.2 Tank retentate recovery

Retentate in the tank is typically recovered by running the feed pump at low speed into a low-point drain and into a separate recovery vessel. The recovery efficiency depends on the system design and how close one is operating to the minimum working volume. Typical values might be 70–80% recovery, but there is no product dilution. One must ensure a true low point and drain. Systems employing a sterile filtration system at the low point will suffer from holdup loss in the sterile filter assembly. This recovery step can also include gravity draining the fluid held up in piping between the tank and points where either gas or diafiltrate is introduced.

8.1.3 Air blowdown of modules

Low-pressure air or nitrogen at 2–5 psig is introduced to the system at the piping highpoint, typically in the retentate line near the diafiltration addition line. The line to the retentate tank is closed. The gas flows in the reverse direction through the retentate line and displaces viscous product held up in the cassette feed channel screens. Once the gas opens a path to the low-point drain, all the gas will follow this path of least hydraulic resistance. Unrestricted high velocity gas flow will not increase recovery but can lead to product denaturing and aggregation due to entrainment and foaming. Low pressures and careful operator attention to shut off the pressure after low-point breakthrough will minimize this effect. For viscous retentates, only a few additional percent of product is recovered this way but there is no product dilution. For low-viscosity products with well-designed draining, higher recoveries are possible.

8.1.4 Buffer displacement of modules

A diafiltration buffer is slowly introduced to the system at the piping highpoint. The line to the retentate tank is closed. The buffer flows in the reverse direction through the retentate line and displaces viscous product held up in the piping and cassette feed channel screens. The viscosity differences between buffer and product are not as great as with gas and product. As a result, the product concentration at the low-point drain undergoes a sharper transition from product to buffer. The recovery pool can yield 98% recoveries but may also cause 15–30% product dilution. Either experience with the holdup volume of the system or experimental measurement of the product concentration breakthrough at the low-point drain during an engineering run guides the decision as to when to 'cut' the buffer flow to minimize dilution with good yields.

8.1.5 Buffer recirculation

Buffer is added to the emptied system to provide one minimum working volume. This volume is recirculated through the system, with the filtrate line closed, for up to 30 min to allow all the product protein to diffuse out of any 'nooks and crannies' in the system. The recirculation is done at a minimum cross-flow and minimum trans-membrane pressure to reduce polarization and foaming. The system is then recovered with a gravity drain or an air blowdown step. This pool is added to the concentrated recovery pool. This method recovers the entire product available for recovery. Any product remaining in the system is either adsorbed to surfaces or is in holdup volume sectioned off from the flow path. Systems designed with piping complexity (e.g., fed-batch bypass loops, loops around feed pumps to run a very low cassette feed flows and special provisions for sampling) can have such sectioned off holdup. This requires special attention to recover product from all parts of the system. Buffer recirculation can significantly dilute the product and may not allow one to achieve the minimum concentration specification.

8.1.6 Batch recovery case study

The EMD Millipore CUF-1 PETS® (Pre-Engineered TFF System) tangential flow filtration system (Figure 8.1) was employed to assess the losses using the buffer displacement and air blowdown recovery methods.[2] It used a single-level process holder loaded with 20 × 1.14 m^2 Pellicon™ 3 cassettes containing a 10 kDa Ultracel™ membrane. Larger systems are configured by vertically stacking additional holder levels. A 16 L solution of 10 g/L bovine serum albumin was re-circulated at ~4.4 lpm/m^2 in the total recycle mode with the filtrate open for 30 min to create the polarization layer. A solution of 0.9% NaCl was used as the buffer for the displacement method.

The air blowdown recovery gave an 85–90% recovery as shown in Table 8.4. The product concentration at the low-point retentate drain during buffer displacement recovery is shown in Figure 8.2. The recovered volume is normalized by the system holdup volume, estimated at 4 L. If the data followed the plug flow model, one could displace the held up product with 1 holdup volume of buffer and achieve 100% recovery with no product dilution. The dispersion model provides a reasonable description of actual flow.[3]

The fraction of product mass recovered by buffer displacement from the piping segment and modules is calculated from integrating the area under the recovery curve (Figure 8.3). Recoveries approach 100% indicating essentially complete recovery of held up fluid and no significant dead volumes. Deviation of the recovery concentration

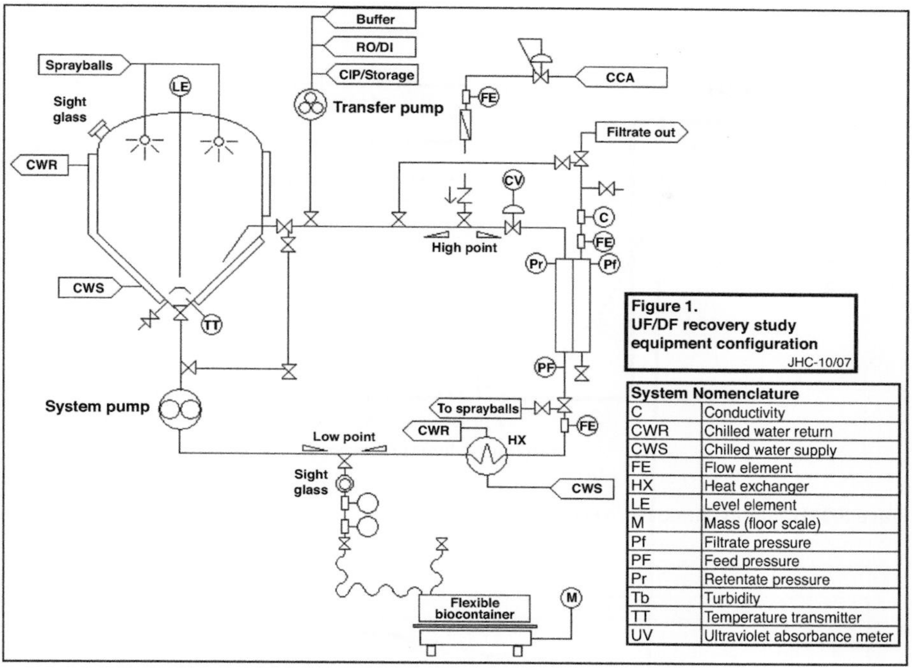

Figure 8.1 PETS system.
(Courtesy EMD Millipore.)

Table 8.4 Recovery method comparison

Recovery method	Air blowdown	Buffer displacement
Description	Introduce compressed air @1–5 psig into the retentate through transfer pump	Introduce DF buffer into the retentate through transfer pump
Advantages	No dilution	Good 98% recovery No denaturing
Disadvantages	Poor 85–90% recovery Product denaturing (foaming)	Product dilution

curve from a step change means that more than one holdup volume of buffer is required to achieve good product recovery. It takes 3–4× holdup volumes to consistently recover 95% of the mass in the holdup volume. For the above test, this means a 4 L of holdup with 12–16 L of buffer. The holdup can be diluted 2–3×. Since 70–75% has already been recovered by the tank drain, the product pool volume in this case is only 24–28 L. Compared to the original 16 L volume, this is a dilution of 1.5–1.75× to recover 98% of the mass from the entire system. An exponential recovery model is also shown.[4]

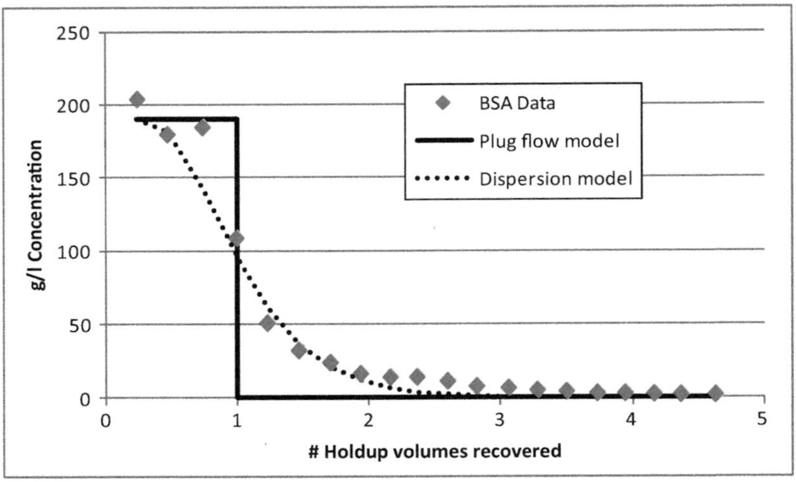

Figure 8.2 Holdup volume recovery curves.

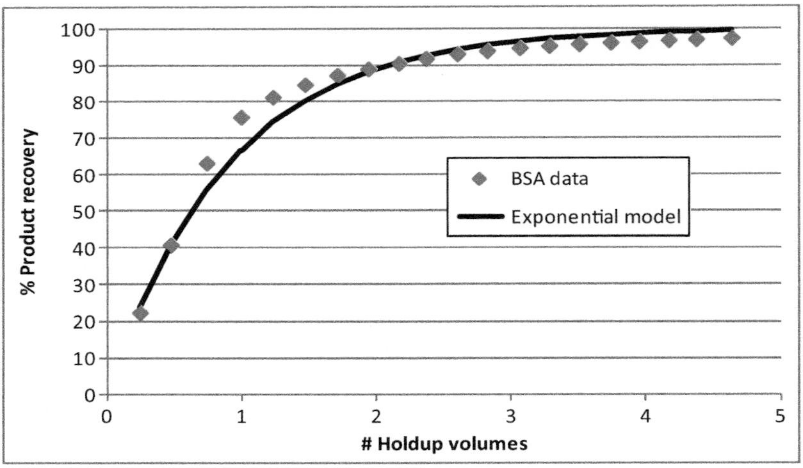

Figure 8.3 Holdup volume recovery yield curves.

8.1.7 Batch recovery method selection and design

Buffer recirculation dilutes the product too much to be practical for most cases so air blowdown and buffer displacement are the two viable options (Table 8.4). In general, the buffer displacement is done without first introducing any air into the retentate lines. Buffer displacement is recommended for general use due to its high recovery with moderate dilution and no foaming. A trade-off can be made between recovery and product dilution.

Table 8.5 System design for recovery

- System drain located at a clearly defined low point in the feed/retentate piping.
- Feed/retentate piping sloped toward the system piping low point for good drainage. Minimum slope should be at least 1/16" of gradation/ft of pipe run.[5]
- Recovery assist fluids (buffer, compressed gas) are introduced to the system at a clearly defined high point in the feed/retentate piping.
- Correct sizing of the process piping (5 ft/s) where possible.
- Orient system components to take advantage of gravity drain where ever possible.
- Minimize the length of piping runs.
- Minimize the number of pipe connections (pulls) on the feed/retentate recirculation loop.

Recovery processes are improved when systems conform to good design practices (Table 8.5). Bench-scale recovery studies are useful for assessing membrane device/process performance. However, they do not accurately predict product recovery at the process scale because of dissimilar component geometries.

8.1.8 Single-pass recovery

Figure 5.9 shows the contributions of the various steps (UF1, DF and UF2) on system time–area requirements. For typical sizing at $c_D = c_g/e$, the final UF2 step represents just a few percent of the overall processing time and is completed in just a few minutes. Consider using a separate, much smaller system for this UF2 step to avoid recovery dilution or minimum working volume system limitations. Operation would involve transferring the diafiltered product to the smaller system along with rinses to ensure high recovery.

It is convenient to use a single-pass system instead of a batch system for this application. A single-pass system would have holdup only within the modules and holder. For a cassette this is about 0.17 L/m^2, much lower than system minimum working volumes of >1 L/m^2. By operating the single-pass system over a longer time period (say >1 h), one can reduce its area and holdup further. Even without any recovery, the holdup can represent $<0.1\%$ of the batch. Conversely, one can use a recovery flush of 2–3 holdup volumes without any significant concentration dilution.

8.2 Cleaning

After the product has been recovered, the system is ready for cleaning. The cleaning step enables module and system reuse for economical operation. However, in several cases, the cleaning costs exceed the module and system costs, and a fresh set of modules, plastic tank liners, and tubing is employed for every run. For example, plant time may be much more valuable than module costs for clinical manufacturing or with small, high value vaccine batches (Table 8.6).

Table 8.6 **Single vs. multiple use**

Mode	Benefits	Where used
Single use	Less time, labor, buffers, no cleaning/reuse validation	Clinical mfg., vaccine mfg.
Reuse	Lower system and module costs	Process development, large commercial mfg.

For modules that are cleaned, it is recommended to use 'cleaning in place' (CIP) utilizing the existing process skid for cost and convenience. For small single-use operations, systems can be disassembled and 'cleaned out of place' (COP) for reuse with new modules. Cleaning objectives include: (1) limiting carryover from one batch-to-the-next to maintain batch identity and limit impurity levels and (2) maintaining consistent run-to-run UF process performance to demonstrate control over the process and meet manufacturing targets. In order to meet these objectives, one must remove all process fluid components (protein, carbohydrates, lipid etc.) that may be adsorbed to the membrane surfaces or inside the pores of the membrane or held up within the system volume. Cleaning is also recommended for new membrane modules before use. This is to flush out extractables, help with bioburden management, and condition the membranes for a baseline measurement of water permeability (see Section 8.6).

ClP flow paths should be engineered to ensure that all the wetted surfaces in the system are exposed to the cleaning solution for the target duration.[5] This requires opening valves at the beginning and end of the cleaning cycle to direct cleaning flow through every component in the system. CIP cycles can be designed so that the system and feed vessel are cleaned as an integrated pair simultaneously. This CIP method can be less complicated from a design and operational standpoint. Sometimes the feed vessel is used for other duties, such as a hold vessel or a transfer vessel. In this case, the feed vessel and some associated system piping are integrated and cleaned with central CIP. The modules are cleaned separately. Some sections of piping may wind up being cleaned twice. This method can also provide acceptable results. However, integration of central CIP to the system piping can impact the hardware design increasing minimum working volume and non-recoverable holdup volume.

A typical cleaning protocol is a series of rinses to drain and recirculations (Table 8.7). First, the system retentate and filtrate (permeate) are rinsed to drain with RO/DI water to remove any remaining process residues (see Section 6.3). In this example the system and tank are cleaned using spray balls to ensure the cleaning fluid reaches all wetted surfaces in the tank, especially the dome. The vessel is first rinsed to drain with a cleaning agent, bypassing the rest of the system. Next, the system piping including the filtrate line is cleaned. It is helpful to open or pulse the valves (open/close/repeat) on the skid that are not on the main flow path (drain valves, divert valves etc.). This assures that all parts of the system are cleaned past the weirs on the valves and can help prevent nagging bioburden issues.

Cleaning agent type, contact time, temperature, and cleaning agent concentration govern the rate (kinetics) of cleaning. The specified cleaning solution (Table 8.8) is

Table 8.7 **System CIP steps**

Step	Control	Volume (L/m^2)	Time (s)	Comments
RO/DI rinse	4.4 LMM*	2.2	–	Rinse single pass, filtrate open to drain
Vessel rinse	100 lpm**	–	300	CIP solution recycle via spray balls
System flush	Level <2 L stop†	2.2	–	CIP solution single pass filtrate open @ 4 LMM
Gravity drain	Time	–	120	All valves open-drains remainder of rinse
System rinse	Level <2 L stop	3.5	600	CIP solution total recycle filtrate open (vessel bypass)
Gravity drain	Time	–	120	All valves open-drains remainder of rinse
System rinse	Level <2 L stop	2.2	1800	CIP solution total recycle filtrate open back to vessel
Gravity drain	Time	–	120	All valves open-drains CIP solution

*Depending on screen type.
**Depending on tank diameter and sprayball design.
†Depending on tank and system volume.

Table 8.8 **Cleaning agents**[6]

Process fluid	Foulants	Primary cleaners	Secondary cleaners
Protein solutions	Protein	NaOH NaOCl	Triton X-100 Henkel P3-11 Tergazyme
Bacterial/yeast whole cell broths	Protein, antifoam, cell debris, lipids, polysaccharides, nucleic acids	NaOH and/or followed by NaOCl	
Bacterial/yeast lysates	Protein, lipopolysaccharides, cell debris, nucleic acids	NaOH followed by H_3PO_4	

introduced into the retentate tank at the specified temperature, typically in the range of 22–50°C. CIP temperatures should not exceed the limits of either the membrane devices or the system components. The feed pump is then slowly ramped up to the target vendor-recommended feed flow (typically the same as processing conditions, around 6 LMM) and the retentate valve tightened to provide a TMP for permeate flow. A conversion of 30% is recommended (permeate flow/feed flow). The cleaning solution is then recirculated through the system for the specified time period, typically 30–60 min. Some cleaning agents, such as chlorine, may get consumed during the cleaning process. When developing a cleaning cycle, it is important to test the retentate fluid over time to verify that the cleaning solution concentrations remain at the target level during the cleaning step. One may need to add extra cleaning agents to make up for material that is consumed.

After the entire system has been exposed to the cleaning solution for the target duration, the cleaning solution is directed to drain through the retentate and permeate streams. This means opening the permeate, directing it and the retentate streams to drain, starting up the feed pump at a low flow and applying a little retentate pressure to induce permeate flow. The system is now ready for flushing as described in Section 6.4.

It is recommended that clean system hold time and dirty system hold time controls be established for UF systems. Post-processing, the system has protein and other residues from the process in contact with the membranes and system internals. Prolonged contact time for these residues on the membranes may result in the growth of bioburden and interactions with the membrane surface making restoration of membrane performance more challenging. Therefore, it is desirable to establish a 'dirty system hold time limit'. The dirty system hold time should be long enough to allow flexibility in manufacturing operations, while setting a time limit to address cleaning the system. A 'clean hold time limit' control ensures that cleanliness is maintained in the process by setting a maximum allowable time period a clean system may be held in idle.

8.2.1 Cleaning agent selection

Selection of cleaning agents (Table 8.8) depends on (1) compatibility with the membrane and system materials of construction (MOC), (2) effectiveness of the cleaning reagents for the specific process residues and (3) ease of cleaning agent removal and validation.

Sodium hydroxide (NaOH) is the most common alkaline cleaning agent at concentrations of 0.1–1.0 M NaOH. It dissolves contaminants, saponifies lipids (hydroxyl ions attack ester bonds in insoluble lipids to form soluble fatty acids and glycerol) and peptizes (protein cleavage into small soluble residues by hydrolysis).[5,7] Sodium hydroxide lacks sequestering (keeping mineral salts in solution), and emulsifying and dispersing action to keep contaminants suspended in solution. In general, NaOH is not significantly consumed during the cleaning process.

$$\begin{aligned}\text{Saponification } & R-C=O-OR'+OH^- \rightarrow R-C=O-OH+{}^-OR' \\ \text{or } & R-C=O-OR'+OH^- \rightarrow R-C=O-O^- +HOR'\end{aligned} \tag{8.2}$$

$$\text{Hydrolysis} \quad R-C=O-NH-R'+H_2O \rightarrow R-C=O-OH+H_2N-R' \tag{8.3}$$

Sodium hypochlorite (NaOCl) is the most common chlorinated alkaline cleaner. It is very effective in peptizing proteins with NaOH (0.1–0.5 M NaOH and 250–300 ppm NaOCl). As shown in Table 8.9, hypochlorite can dissociate and the chlorine can exist

Table 8.9 Chlorine reactions

Reactions	K_{eq}
$Na^+ + OCl^- = NaOCl$	1.0E+00
$OCl^- + H^+ = HOCl$	3.5E+07
$H^+ + OH^- = H_2O$	1.0E−14
$H_2O + Cl_2\ (aq) = H^+ + Cl^- + HOCl$	4.6E−04
$Cl_2\ (aq) = Cl_2\ (g)$	6.1E−02

Adapted from Ref. [8], gmol/L concentrations except water.

in various forms. At elevated pH the chlorine atom is primarily OCl^- ion, which does not corrode the steel. Below pH 4, the chlorine atom exists as free chlorine (Cl^- or Cl_2) in solution and rapidly attacks stainless steel.[9] Chlorine may be consumed during the cleaning process. This requires that chlorine levels be monitored during development to ensure enough chlorine has been added to maintain active levels at the target strength throughout the cleaning cycle.

Acid cleaners (e.g., citric, phosphoric) dissolve mineral precipitates and peptize proteins by hydrolysis. Wetting agents or detergents (e.g., sodium lauryl sulfate, quaternary ammonium compounds, and non-ionic surfactants, such as Triton X-100) lower surface tension to enable liquid penetration and degradation of residues. Emulsifying agents (e.g., sodium tripolyphosphate) dissolve deposits and keep them suspended in solution. Pre-formulated CIP solutions can contain mixtures of different cleaning agents. Assays for these mixtures can be problematic because of the need for a mixture of assays to demonstrate removal. Use complex CIP mixtures when simple CIP agents are ineffective. Consult the membrane manufacturer for compatibility information with specific CIP agents.

Polyethersulfone (PES) membranes are compatible with most cleaners. Regenerated cellulose (RC) is sensitive to high NaOH concentrations and is not compatible with hypochlorite so lower NaOH concentrations are employed. Other common system materials of construction (high-density polyethylene, polyurethanes, platinum cured silicone, EPDM rubber, Teflon and 316L stainless steel) are generally compatible with most cleaners.

8.3 Sanitization

Steel-pipe-based UF systems are not considered to be sterile aseptic operations because UF membranes are not stable or robust under steam-in-place (SIP) conditions. A qualified gamma-irradiated single-use closed system could be considered sterile. A post-cleaning chemical sanitization-in-place strategy is used to maintain bioburden control. Cleaning removes deposits that might shield micro-organisms under them from contact with sanitizing agents. This makes the sanitization-in-place more effective and validateable.

CIP agents themselves may have significant bactericidal capabilities. Used at elevated temperatures and concentrations, these CIP agents are effective even against spore forming organisms. Therefore, cleaning and sanitization may be accomplished in two consecutive steps with the same solution (e.g., CIP with a NaOH and NaOCl mixture), the first to remove deposits, the second to sanitize. Use of separate sanitization agents can provide the broadest spectrum sanitization for micro-organisms (e.g., CIP with NaOH followed by SIP with peracetic acid).

Optional steps to control bioburden could be a rinse cycle with a sanitizing solution, such as peracetic acid. Other practitioners have implemented sanitization procedures for the system hardware. However, as noted earlier, most UF membranes are not compatible with sanitization conditions; therefore, a bypass around the holder is used or the membranes can be removed from the system during SIP operations. A SIP procedure might be used after a series of runs and not after each run.

Hydrogen peroxide is an effective sanitization agent but tends to get consumed very quickly. It is effective against a variety of spores, bacteria and viruses at 500 ppm concentration over an hour of contact.[10] Peracetic acid is more stable and more potent than hydrogen peroxide. It is effective at 100 ppm in inactivating a wide variety of bacteria, yeasts and fungi.[10]

As can be seen in the sanitization agent efficacy tables (Table 8.10), these agents have the ability to kill or inhibit a broad spectrum of micro-organisms. The choice of sanitization agent depends on the agent's efficacy, the compatibility with the membrane materials of construction, and the reagent/facility 'fit'.

8.4 Post CIP measurements

8.4.1 Residuals

Removal of cleaners and sanitizers by flushing is needed to ensure no carryover into the next batch processing. Carryover would degrade the product and cause quality issues as well as fouling the membrane. Simple CIP agents like NaOH, peracetic acid and NaOCl are easily measured by pH, conductivity, or chlorine assay kits (designed for environmental purposes or recreational pool maintenance), respectively. Verifying the removal of more complex CIP preparations (detergents, enzymes, chelating agents, etc.) will require more complex assays or even specific assay development. Consult your CIP chemical vendor for recommendations. Residual total organic carbon (TOC) measurement is a sensitive reliable method to assess organic carbon levels in the flush out streams from the membranes and system. This testing can provide direct evidence of the effectiveness and consistency of the cleaning cycle.

A key consideration for storage agents is the ability to validate the removal of the agent. There is a test-strip-based colorimetric assay for alkyldiethylbenzylammonium chloride (Lysol Brand Sanitizer) and peracetic acid with 1 ppm sensitivity. A simple conductivity meter can be used to demonstrate removal of the other sanitizing agents in the following list (Table 8.10).

Table 8.10 Sanitization agents

Storage agent	Organisms killed[d]	Organisms inhibited	Organisms unaffected
Phosphoric acid, pH 2	*Listeria innocua*[g]	*Bacillus* spp.[b]	Brewers yeasts[h]
	Listeria monocytogenes[g]	*Escherichia coli*[b]	*Obesumbacterium proteus*
	Pediococcus damnosus[i]	*Enterobacter aerogenes*[b]	*Chlamydomonas acidophila*[b]
	Lactobacillus brevis[i]	*Pneumococcus*[b]	*Bacillus subtilis* s. globigil spores[e,f]
	Acetobacter sp.[i]	*Clostridium sporogenes*[b]	*B. subtilis* spores[e,f]
	Enterobacter agglomerans[i]	*Neisseria*[b]	*Bacillus stearothermophilus* spores[e,f]
	Pseudomonas aeruginosa[b]	*Bacillus* spp.[b]	
	Staphylococcus aureus[j,k]	*Erwinia carotovora*[b]	
	Bacteriodes succinogenes[l]	*Lactobacillus acidolphilus*[b]	
		Nitrosobacter spp.[b]	
		Proteus vulgaris[b]	
		Nitrosomonas spp.[b]	
		Synechococcus spp.[b]	
Lysol Brand Sanitizer, 0.1% alkyldimethylbenzylammonium chloride[m,n]	*Salmonella typhosa*		
	S. aureus		
	S. typhosa		
	Staphylococcus epidermidis		
	Pseudomonas cepacia		
	P. aeruginosa		
	Proteus mirabilis		
	Proteus. morganii		
	Serratia marcescens		
	Flavobacterium sp. 82-1-98		
	A. guttatis		
	Alcaligenes faecalis		
	B. subtilis		

(Continued)

Table 8.10 Sanitization agents *(cont.)*

Storage agent	Organisms killed[d]	Organisms inhibited	Organisms unaffected
	Lactobacillus spp.		
	Leuconostoc sp.		
	E. coli		
	Psuedomonas fluorescens		
	Aspergillus niger		
	Chaetosphaeridium globosum		
	Myrothecium verrucaria		
	Trichoderma viride		
0.05 N NaOH, pH 12.7[a]	*E. coli*		Sporulating gram positive organisms
	All gram negative organisms and filterable viruses		Foot-and-mouth virus
	S. typhosa		Hog cholera virus
	Salmonella schottmulleri		*A. niger*
	Salmonella pullorum		*Entamoeba histolytica* cysts
	Salmonella paratyphi		*Bacillus anthracis* spores
	Pasteurella avicida		*Mycobacterium tuberculosis*
	Shigella gallinarum		*Candida albicans*
	Brucella abortus		*B. subtilis* spores
	Fowl pox virus		
0.1 N NaOH, ph 13[a,b,c]	Cocci		Sporulating gram-positive organisms
	Non-sporulating gram-positive organisms		*M. tuberculosis*
	S. aureus		*C. albicans*
	Alal vegetative bacteria		*A. niger*
	Ps. Copacia		*E. histolytica* cysts
	S. marcescens		*B. subtilis* spores
	Pseudomonas maltophilia		*Bacillus antracis* spores
	Alcaligenes sp.		
	Pseudomonas sp.		
	Thiobacillus thiooxidans		

Table 8.10 **Sanitization agents *(cont.)***

Storage agent	Organisms killed[d]	Organisms inhibited	Organisms unaffected
	Lactobacillus acidophilus *Proteus vulgaris* *Enterobacter aerogenes* *Clostridium sporogenes* *P. aeruginosa* *Erwinia carotovora* *Nitrosomonas* spp. *Nitrobacter* spp. *A. faecalis* *Vibrio cholerae*		

Notes:
a. McCulloch EC. Disinfection and sterilization. Lea & Febiger; 1945. pp. 245–262.
b. Atlas RM. Microbiology; fundamentals and applications. Macmillan; 1984. pp. 353–354, 766.
c. Berghof JH, Adner NP, Doversten SY. Control of microbial contamination in chromatographic separation media using sodium hydroxide. XX Congress of International Society of Blood Transfusion, London, UK, 10–15 July, 1988.
d. Killed is defined as >7 LRV in 1–24 h @ 4–20°C.
e. Malle S. Deutsche Milchwirschaft 1984;35(3):100–101; Molkereiung der Milch 1984;37(49):1513–1515.
f. Anon. Lebensmittelltechnik 1983;15(11):587–598.
g. Best M, et al. Appl Environ Microbiol 1990;56 (2):377–380.
h. Simpson WJ, et al. J Inst Brewing 1987;95 (5):347–354.
i. Simpson WJ, et al. J Inst Brewing 1987;93 (4):313–319.
j. Minor TE, et al. J Milk Food Tech 1972.
k. Minor TE, et al. J Milk Food Tech 1971;34 (10):475–477.
l. Russell JB. Appl Environ Microbiol 1987;53 (10):2379–2383.
m. Seymour SB, editor. Disinfection, sterilization and preservation. 4th ed. Philadelphia: Lea & Febiger; 1991. Chapter 13.
n. Makela PM, Korkeala HJ, Sand EK. Effectiveness of commercial germacide products against therapy slide-producing lactic acid bacteria., J Food Protection 1991; 54 (8):632–636.

8.4.2 Normalized water permeability

The normalized water permeability (NWP) test is a recommended method to assess the effectiveness of the CIP process to restore process flux rates. NWP uses water permeability (LMH/psi) as a measure of membrane cleanliness:

$$\text{NWP} = \frac{(\text{litres per hour volumetric permeate flow})(\text{TCF})}{(\text{TMP psid})(\text{membrane area m}^2)} \tag{8.4}$$

TMP = $((P_{\text{feed}} + P_{\text{retentate}})/2 - P_{\text{permeate}})$ as the module average transmembrane pressure. TCF = (viscosity of water at measured temperature)/(viscosity of water @ 25°C) as temperature correction factor (see Appendix).

As described in Section 8.4, the hydraulics of modules are complex and the calculated NWP may be affected by membrane permeability and the selection of feed flow and TMP operating conditions. The calculated NWP value is subject to pressure gauge measurement error so it is useful to run at least at 10 psi TMP to minimize this effect. For high-permeability membranes, low feed flow and high TMP will cause dead-end flow where all the feed goes into the permeate stream. High permeate flows can also cause a pressure drop in the permeate stream so the TMP calculation in equation (8.4) will be thrown off. Air bubbles can build up in the retentate and blind off the membrane to alter the measured NWP. It is critical that the retentate flow is at least 10% of the feed flow to avoid this. This can be achieved through higher feed flows. For lower permeability membranes, there is no need to run at high feed flows as this can accentuate feed channel pressure drop effects.

NWP is measured for a new set of membranes at installation (Section 8.6), after flush and after the first CIP cycle. The first CIP removes membrane preservatives and conditions the membrane prior to the first exposure to process fluid. This measurement becomes the baseline, which is compared to subsequent NWP measurements taken post-CIP. The acceptance criterion for cleaning efficiency is membrane and application specific and may vary between plants. A typical criterion for regenerated cellulose NWP is that if the membrane NWP is ±20% of the baseline NWP (after the initial cleaning), process reproducibility will result.[6] For polyethersulfone membranes and/or fouling feeds, larger variations in NWP of ±40% are seen. Consistency in the NWP measurement rather than an absolute limit is the best indicator of successful cleaning. For example, suppose that for a particular protein, a 40% decrease in NWP is observed upon first use but that subsequent NWP measurements were within ±20% of the reduced NWP. A consistent NWP coupled with consistent process fluxes (process cycle times), consistent yields, acceptable product quality, and passing system integrity tests, would lead one to conclude that the CIP process was efficiently cleaning the membranes and maintaining system performance. NWP is an important indicator of cleaning efficiency, but should be evaluated within the context of process performance.

Application of NWP as a measure of cleaning effectiveness involves tracking NWP from run-to-run. An initial value should be measured on modules that have been initially cleaned but have not seen protein. It has been observed that NWP declines during the first cycle of operation, even without exposure to protein. This is attributed to the cassette screen compressing into the membrane and blinding off some of the membrane area. There is some variability in the measurement of the NWP. For example, it can change with the operating conditions. NWP should be measured under consistent conditions for each run to minimize variability. If the NWP drops by more than 20%, it is possible that the cleaning cycle was not effective and an investigation should be made. If the cleaning agent was incorrectly formulated or lost some of its potency due to age, this could be a cause of the decline. Provision should be made for a re-cleaning option. In addition, if the NWP steadily decreases from run-to-run, this is a sign of inadequate cleaning. It would appear that some component is building up on the membrane and not being removed by the cleaning cycle.

Post-use integrity testing is generally not performed for an ultrafiltration system. Pre-use testing is generally used (Section 6.3). That is because defects are generally introduced with the first time module installation in large systems. After operation, the modules are not disturbed and the integrity is preserved. One exception occurs when modules are subjected to thermal cycles during the cleaning process and system compression is fixed at a given position. Cycling may reduce compression and cause sealing defects. Rechecking compression or the use of hydraulic cups providing a constant load can remedy this. Before extended storage of modules in the holder of 7 days or more, it may be useful to run an integrity test.

8.5 Storage

Drying of UF membranes may cause them to crack or be unwettable and result in a loss of integrity. Therefore, UF membranes must be held or stored in a 'wet' state. The cleaning and sanitization steps may not result in UF system sterility. Low levels of micro-organisms may survive and may proliferate and colonize within the system and membranes over time. These micro-organisms include bacteria (including spore formers), moulds and fungi. If a system is to be idled for longer than its 'clean hold time', then a storage buffer and storage procedure should be specified for the UF system. The objective for the Storage Step is to prevent or impede growth of micro-organisms that may be present. Considerations for the storage step are:

- Capability of storage agent to inhibit (or kill) micro-organism growth.
- Storage expiration time.
- Membrane storage mode (storage-in-place or storage-out of place).
- Storage buffer compatibility with the materials of construction (MOC) of the system and membrane materials of construction.
- Ease of storage agent removal and validation.

Membrane vendor input should be sought when selecting a storage agent. The vendor will likely have supporting documentation and recommendations for storage agents that are compatible with their membranes. Some common storage agents are also sanitization agents:

- 0.05 M NaOH (pH 12.7)
- 0.1 M NaOH (pH 13)
- 0.05 M phosphoric acid (pH 2)
- 0.1% Lysol Brand No Rinse Sanitizer (alkyldiethylbenzylammonium chloride).

Since storage agents may only inhibit growth, some microbial growth may still occur over a long period of time (Table 8.11). A time limit should be established after which the storage solution should be exchanged for fresh solution to keep the bioburden growth low. Some storage agents may slowly affect the retention properties of some membranes (i.e., NaOH/regenerated cellulose), therefore, a maximum total storage time should be imposed for these storage agent – membrane combinations.

Membrane devices are often 'stored-in-place' on the system. This method is typical where the membranes are used in a production process dedicated to a single drug substance. The system hardware and membranes are flooded and stored together with the system in an 'idle' state. The storage reagent is removed from the system and membrane devices prior to the next batch in an operation commonly called the 'critical

Table 8.11 **Storage agents**

Storage agent	Organisms killed	Organisms inhibited	Organisms unaffected
15% NaCl*	Mesophilic gram-negative rods Psychrotrophic bacteria Lactic acid bacteria Spore-forming bacteria		Halobacteria
Ethanol, 40%*	*E. coli* *P. aeruginosa* *S. epidermidis* *S. typhosa* *S. aureus* *Corynebacterium diphtheriae* *S. marcescens* *Streptococcus pyogenes* *Spiroplasma* spp. Yellow fever virus (Asibi stra in) Rabbit fibroma virus Herpes simplex virus Vaccinia virus Influenza virus (Asian) HIV	*M. tuberculosis*	Vaccinia virus Hoof-and-mouth virus E. Equine encephalitis virus *C. albicans* *Cryptococcus neoformans* *Blastomyces dermatidis* *M. audouini* *Coccidioides immitis* *Histoplasma capsulatum* P. tardum Newcastle virus Influenza type A. virus Poliovirus type 1 Coxsackle B-1 Echo type 6 Human rotavirus *Trichophyton gypseum* *C. globosum* *M. verrucaria* *T. viride*

*Seymour SB, editor. Disinfection, sterilization and preservation. 4th ed. Philadelphia: Lea & Febiger; 1991. Chapter 13.

flush'. If the hardware is needed for other purposes or if the system is part of a multi-purpose facility, membrane device storage-in-place is not practical. UF systems in this type of facility have been engineered such that the membrane devices can be isolated in the holder and the entire holder is separated from the system and placed in storage. A second holder, loaded with devices dedicated for the different drug substance, is moved into place and attached to the system. Alternatively, after flooding with storage agent, some practitioners remove membrane devices from the system and submerge them in carboys filled with buffer. Storing the membranes 'out-of-place' risks damage to the devices, by impact or drying if they become unsubmerged. Storage-in-place or in the holder under compression eliminates direct manipulation of the devices but extended compression effects are not well characterized. The trade-off is that storage-in-place on the system ties up the system while separate holders are more capital intensive. Ultimately, 'facility fit' will be an important consideration in the storage mode decision.

References

1. Brinkmann A, Souther J, Westoby M. Integrating single-pass tangential flow filtration into biopharmaceutical purification processes. Presentation at ACS National Mtg., San Diego; Mar. 2012.
2. Arias J, Cyganowski J, Cummings S, Escheverry S, Goodrich E, Ramsey G. Optimization of process scale ultrafiltration/diafiltration systems, Tech Note 31. Millipore Corp. EMD Millipore, Billerica, MA; 2009.
3. Herbert L, Yu Zou I. High concentration ultrafiltration processing: analysis of constraints and recommended best practices., manuscript submitted to Biotechnol Bioeng.
4. Rao S, Gefroh E, Kaltenbrunner O. Recovery modeling of tangential flow systems. Biotechnol Bioeng 2012;109(2):3084–92.
5. Brunkow R, Delucia D, Haft S, Hyde J, Lindsay J, McEntire J, et al. Cleaning and cleaning validation: a biotechnology perspective. PDA Bethesda, MD; 1996.
6. EMD Millipore, Maintenance Guide.
7. Allinger NL, Cava MP, De Jongh DC, Johnson CR, Lebel NA, Stevens CL. Organic Chemistry. New York, NY: Worth Publishers; 1971.
8. Perry's Chemical Engineer's Handbook. 8th ed. New York, NY: McGraw-Hill; 2008 [Figure 28-2].
9. Wagman DD, Evans WH, Parker VB, Schumm RH, Halow I, Bailey SM, et al. The NBS table of chemical thermodynamic properties. J Phys Chem Ref Data 1982;11(2). ACS.
10. Seymour SB. Disinfection, Sterilization and Preservation. 4th ed. Philadelphia, PA: Lea and Febiger; 1991.

Systems

9

John Cyganowski, Herb Lutz, Pietro Perrone
EMD Millipore, Biomanufacturing Sciences Network

9.1 Basic system components and layout

The system piping and instrumentation diagram in Figure 9.1 shows the flowpaths and components needed to perform all the steps in the SOP (standard operating procedure), deliver the desired product and meet any process constraints. At a minimum, a system contains the membrane modules and holder, a recirculation vessel or retentate tank, a feed pump, retentate valve and pressure sensors for the feed and retentate lines. A secondary transfer pump is often provided to support the transfer of fluids (buffers, water and CIP (cleaning-in-place) solution) needed to support the system operation. Sensors are also used to measure tank volumes or cumulative permeate or buffer volumes and control the process. The process flow path may contain a heat exchanger or a vessel jacket to remove pump heat and keep the batch at a constant temperature. Open systems may be acceptable for early clinical phase production but closed systems with hard plumbed lines are needed for validated consistency, short cycle times, and bioburden control manufacturing. Each step of the process is controlled by switching valves to introduce/discontinue the appropriate solutions to the system (air, buffer, diafiltrate, cleaning fluid, etc.).

The flow paths that are active during each step of the SOP should be traced out on the P&ID (piping and instrumentation diagram) to show valve configurations along with the relevant sensors used for monitoring and alarms. This includes the ability to vent, drain, SIP (steaming-in-place), flush the system and respond to system upsets. It is particularly important to ensure that all flow paths are active during the cleaning step, including sample valves. In some cases, the modules may be removed from the system during cleaning. Care should also be taken to make sure that the product is not inadvertently flushed to waste through automation or the use of transfer panels.

Multiproduct facilities process a wide range of batch volumes from clinical trials to marketed products. While fed batch configurations allow flexibility to handle these wide volume ranges, one can encounter issues with cleaning and product degradation. Separate pilot systems are recommended for small batches to avoid these validation issues and speed time to market.

9.2 System design targets

Best practice equipment design considerations span several areas (Table 9.1). Regulatory requirements ensure that processing objectives are consistently met. Economical requirements include lifecycle costs, such as capital, operation, validation,

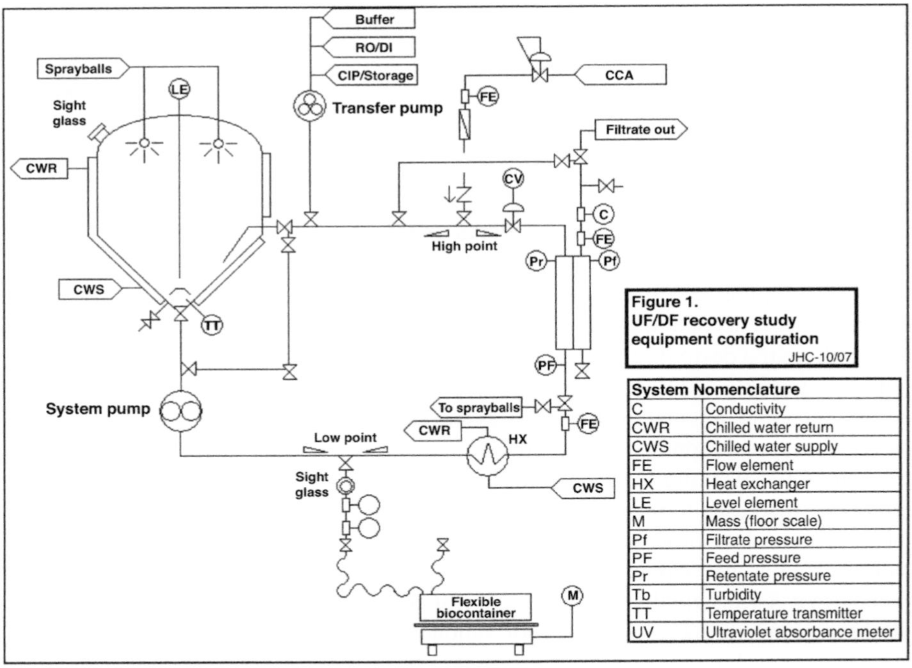

Figure 9.1 Batch ultrafiltration system piping and instrumentation diagram. (Courtesy of EMD Millipore.)

maintenance, cost of replacement (and revalidation if needed), staffing required to support ongoing operation (e.g., programming and calibration) and the scope of the supporting documentation and service from the vendor. Standard, easy-to-use, designs and components are less prone to failure with lower labour costs and maintenance. Components must conform to the standards of the country and operating facility (e.g., metric and voltage) to ensure compatibility and reduce the inventory of spare parts. Additional selection criteria include experience with components and vendors, and scalability. Long lead components, such as specially designed pumps, sensors and tanks, must be designed and ordered early.

Operator safety involves chemical hazards (explosive solvents, biohazards and toxic or corrosive chemicals), physical hazards (high pressures, moving parts, temperature extremes, use of steam and obstacles to operation and maintenance), and electrical hazards (high voltages and currents, and inadequate grounding). Safeguards and alarms should be tested to make sure they work in an emergency. Safety reviews of designs are recommended along with formal hazard and operability analyses (HAZOPs) and control hazard and operability analyses (CHAZOPs).

Integration of skids with each other and the plant may impose additional constraints. Integration requires that skids have compatible mechanical and electrical/control components at their boundaries. Retrofit into an existing facility may impose limitations on skid dimensions for the production floor and for access to the facility.

Table 9.1 System design considerations

Component	Regulatory	Economic	Safety
Wetted surfaces	Non-toxic, cleanable, sanitary, consistent, chemically compatible with all process fluids (no swelling or reacting), non-shedding, non-leaching, non-adsorbing, closed system	Availability, can fabricate	
Wetted volume	Low holdup, drainable, ventable, minimum dead legs, cleanable, flushable		Pressure and temperature rating, sealing
Pipes, valves, heat exchanger		Compact, can fabricate, drainable, available in a variety of formats	Operator protection from moving parts
Vessels	Handles volume range, mixing, avoid foaming	Availability	
Pumps, filter holders	Low protein degradation, consistent, thermal isolation		Operator protection from moving parts
Sensors/ sampling	Reliable, accurate, insensitive to environmental effects, (temperature, pressure changes), calibrate in a closed system	Design in optimum number – not to excess	Electrical shock, closed system
Skid frame	Compatible with cleaners and sanitizers, structure does not pool or trap fluids	Compact	Supports load, ease-of-operation
Display	Capture and store data		Legible
All components	Documentation, easy-to-validate, quality certification	Low cost, reliable, maintainable, conform to plant standards, proven designs, spare parts	Conform to country standards

Best practice ultrafiltration (UF) system design to meet the processing objectives is driven by

- High retentate product recovery to maximize yield.
- A low minimum working volume to permit reaching high concentrations.
- Minimize piping runs and components to yield low parasitic pressure drops.
- Good system mixing to enable efficient diafiltration.
- Bioburden control.
- Ability to clean and reuse the system with minimum carryover.

These considerations are inter-related and should be optimized together to produce an efficient system.

Recovery methods are discussed in Section 8.1 and Table 8.5 lists considerations for recovery. Table 9.2 lists best practice system design principles that will aid in

Table 9.2 System recovery best design practices

- System drain located at a clearly defined low point in the feed/retentate piping.
- Feed/retentate piping sloped towards the system piping low point for good drainage. Minimum slope should be 1/8″ of gradation/ft of pipe run.[1]
- Recovery assist fluids (buffer, compressed gas) are introduced to the system at a clearly defined high point in the feed/retentate piping.
- Correct sizing of the process piping (about 5 ft/s) where ever possible.
- Orient system components to take advantage of gravity drain whereever possible.
- Minimize the length of piping runs.
- Minimize the number of pipe connections (pulls) on the feed/retentate recirculation loop.

recovery operations. A system should be designed with a low-point drain, and the feed/retentate piping should be sloped towards the system low point to facilitate drainage and removal. Otherwise, protein fluid will pool in the system piping and be difficult to recover. Compressed gas (air or nitrogen) employed as a recovery assist fluid is best introduced at a high point in the system piping so as to work with gravity. The pressure should be low (<5 psig) just enough to move the protein fluid while avoiding the creation of foam that may impact the protein product quality. A UF system reflecting these design principles enables flexibility to consider a number different recovery steps and flow paths to maximize recovery.

As the retentate is concentrated and the liquid level in the tank drops, the level can reach a point where air is entrained into the bottom of the tank. The minimum holdup volume in the system needed to avoid foaming and air entrainment is referred to as the minimum working volume. Foaming and air entrainment can lead to product denaturation and aggregation. The minimum holdup volume represents the limit of the system to concentrate the retentate. A well-designed system can have a minimum working volume of 1 and 0.8 L/m^2 has been achieved.[2] However, systems in use with 2–3 L/m^2 minimum working volumes can require a second system to complete the concentration to the target specification. The minimum working volume depends on the module and piping holdup, the tank design, and the retentate flow rate. Optimization of the membrane area requirement during process development means fewer high-flux modules are needed and small short return lines reduce holdup. The location and orientation of the retentate return line into the tank and the use of a vortex breaker can affect the minimum working volume.

High concentrations can result in high viscosities causing elevated cassette feed pressure (Section 8.5). Here, lower flux modules and larger return lines are needed to allow operation to higher concentrations without exceeding pressure limits. To the extent possible, it is useful to minimize components in the return line, such as flow meters and heat exchangers.

Diafiltration buffer exchange is based on efficient mixing within the system (Section 5.4). The presence of a dead leg (Figure 9.2) causes the fluid to remain stagnant and limits efficient solute exchange. Dead legs are spaces or pockets in contact with the product that are difficult to vent, flush, and drain.[3] They arise

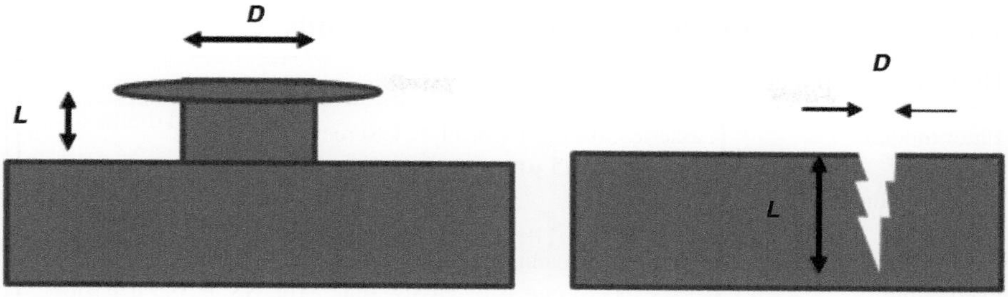

Figure 9.2 Dead leg.

from connecting components to the piping system (e.g., sensors, sampling ports, and rupture disks), within wetted components (e.g., pumps, housings, valves, heat exchangers and tanks), or as surface roughness. In a piping system, a dead leg is a pocket, tee or extension from a primary piping run that exceeds a defined number of pipe diameters from the inner diameter (ID) of the primary pipe. Flow visualization,[4] simulations and testing indicate that the efficiency of cleaning a dead leg is affected by the ratio *L/D*, the average fluid velocity in the pipe, and presence of air pockets in the dead leg.[5] While elimination of dead legs is desirable, current ASME BPE Guidelines currently recommend $L/D < 2$ (based on the internal dimensions of the dead leg).[4]

Solute transport through the dead leg occurs by slow diffusion and not rapid convective flow. The location where diafiltrate is introduced into the system can impact performance. Adding diafiltrate after the retentate tank into the feed pump does not allow for mixing within the tank. Buffer exchange will be poor but fluxes will be high. This also corresponds to a case of bypass through a tank. Adding diafiltrate to the tank but allowing bypass by the retentate line also limits mixing. Buffer exchange will be good but fluxes will be low. As shown in Figure 9.1, it is recommended to add diafiltrate to the retentate return line to facilitate good mixing.

Bioburden control requires internal surfaces that are polished so they do not have dead legs or micro-cracks where micro-organisms can reside without the action of sanitizing agents. Closed systems restrict the ingress of micro-organisms. Pre-sterilized single-use systems also facilitate bioburden control.

Minimizing batch-to-batch carryover also requires polished surfaces in product contact that are readily cleanable and flushable. It is common to use riboflavin as a marker, which fluoresces under UV light. After a cleaning step, the surfaces can be examined to ensure cleaning has been effective.

9.3 Hardware components

The example summary specifications in Table 9.3 are discussed in more detail in the following sections.

Table 9.3 **System component specification example**

Hardware component	Specifications
Filter module	Cassette, 30 kDa, Ultracel™, C screen, 2×2.5 m^2
Holder	316L SS, R_a 0.5 μm, 1 high process
Closure hydraulics	304 SS
Piping	316L SS, R_a 0.5 μm, flow direction labeled, sized for 3–10 ft/s process and cleaning, weld documentation
Feed pump	Rotary lobe/progressing cavity, seal, 316L SS, R_a 0.5 μm, flow and pressure ratings, power requirements
Piping connections	Sanitary
Valves	Weir diaphragm, EPDM, PTFE, Silicone or Viton
Heat exchanger	Shell-and-tube, 316L SS, R_a 0.5 μm, rating, ASME stamp
Tank	Bottleneck design, size, 316L SS, R_a 0.5 μm, well mixed/ agitator, operate over volume range, avoid foaming, cleanable/ spray balls, ASME stamp, no dead legs, sight glass
Air filter	Sterilizing grade, I-line housing 316L SS, R_a 0.5 μm
Sensors	Types, ranges, output forms, reliability, thermocouples in wells, load cell
Sampling	Radial diaphragm valve
Control box	History-vendor and mfr., ISO/cGMP certification, explosion proof
Frame	304 SS, finish, casters, dimension

9.3.1 Piping

System fluid flow and pressure requirements determine the pump size and diameter of the piping/tubing. In turn, the inner diameter of the piping is a major contributor to the working volume of the system.

Pressure drop due to fluid flow through the system may become significant at high linear velocities. High pressure drop may cause cavitation (pressure induced air/fluid interfaces) and misleading pressure sensor reading on the pressure conditions actually at the fluid membrane interfaces. A 'typical' range of linear velocities is 7–12 ft/s. This range provides for hydraulically full piping without large system piping pressure drop. The high CIP flows of 5 ft/s required for wetting internal surfaces are typically more demanding than process flows and form the basis for pump sizing. An economic analysis[6] shows that there is a tradeoff between capital and operating costs for piping systems with optimum velocities in the range of 3–10 ft/s. Velocities >3 ft/s are also recommended for cleaning and flushing as part of the 3A standards. These velocities help ensure that air bubbles in dead legs are flushed out of the piping system so that all the internal surfaces are accessible for cleaning[5] (Table 9.4).

'Piping Pulls' or 'diverts' from the main recirculation line should be carefully engineered so as to avoid dead legs (Figure 9.2). Feed/retentate piping should be sloped towards the system piping low point to facilitate drainage/recovery. Minimum slope should be 1/8″ of gradation per foot of pipe run. 88° ASME BPE elbows will result in approximately 1/2″ of gradation per foot of pipe run.[2] The system piping should

Table 9.4 System pipe diameters, holdups, flow rates maintains reasonable fluid velocity (ft/s)[5]

Pipe size (in.)	Flow range (L/min)	Min. recirc. volume (L)	Undrainable holdup after blowdown (L)
1/2	3–12	1	0.03
3/4	11–36	3	0.1
1	21–71	12	0.4
1 1/2	53–175	25	0.8
2	98–326	45	1.2
3	230–766	90	3

have gravity assist designed into it to facilitate recovery. The piping high point is also used for the introduction of air, diafiltration (DF) buffer, rinse water and CIP fluids. A completely flooded pipe is desirable to avoid air–liquid interfaces that can denature proteins. The presence of air in piping systems can also prevent cleaning or sanitizing fluids from wetting the internal surfaces. Vertical pipe bends that create sections where air or solids can accumulate should be avoided. Liquid retention in undrainable sections represent product loss, growth areas for bioburden, or batch carryover. Venting and draining is aided by a pipe slope of at least 1/8″ drop per foot of pipe length. Fluid velocities of 5 ft/s are required to displace air from dead legs or to completely flood a vertical pipe.[5] Clean steam lines are self-sanitizing and may be plumbed without a vertical slope but need to be sloped to facilitate condensate removal. A low-point drain is required.

Traditionally, systems have been assembled using sanitary tubing made from stainless steel; either 304L for non-wetted surfaces or 316L for wetted surfaces. Single-use flowpath systems have been designed to meet the needs of small-scale clinical manufacturing. Some of these are true single use but most are hybrids (having both single-use flowpaths and reusable membrane devices) due to the cost of UF devices. The engineering of single-use UF systems is rapidly evolving.

9.3.2 Valves

UF systems use both isolation and flow control valves. Isolation valves are simple two-position valves used to open or close a flow path. Sanitary diaphragm valves are most commonly used in the isolation application. These valves can be configured in clusters or 'multi-port' designs to simplify complex process flowpaths. These valves are simple and clean in construction, consisting of a valve body with inlet and outlet ports and a weir. A diaphragm mounted on the valve presses on the weir either with a hand wheel or a pneumatic actuator to isolate the flow. The diaphragm is usually constructed with a Teflon® layer (product wetted) with an elastomeric backing. The Teflon layer of the valve can fatigue and crack after extended service creating a harbouring point for bioburden. Diaphragms are typically replaced on a preventative maintenance schedule.

The retentate valve is a control valve positioned downstream of the membrane holder and the retentate pressure sensor (Figure 9.3). The purpose of this valve is to create pressure on the feed/retentate side of the membrane. This results in the conversion of a portion of the feed flow into filtrate flow (flux). Control valves are typically globe or plug-type valves. Control valves can be obtained in an angled configuration or a linear configuration depending on the system piping design (Figure 9.4).

Proper sizing for the control valve is essential for good system operation. Control valves may be sized using equation (9.1),[6]

$$Q = C_v \times \Delta P / S \tag{9.1}$$

where Q = flow of fluid in gallons per minute, C_v = valve sizing coefficient determined experimentally for each style and size of valve referencing water at standard conditions, ΔP = pressure differential across the valve in psi, S = specific gravity of fluid.

Most control valve manufacturers will have determined the C_v based on water and publish them in their literature.

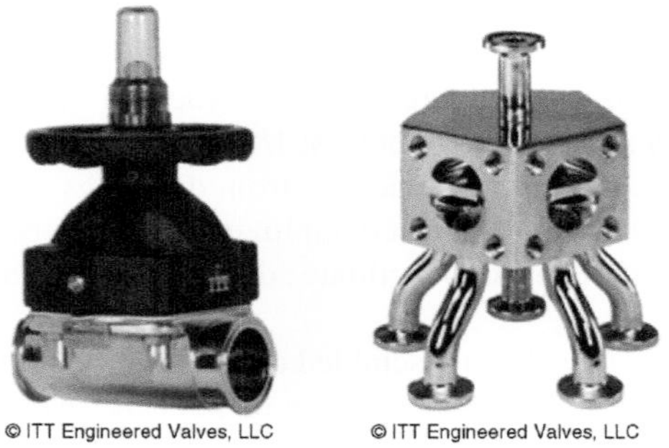

Figure 9.3 Component valves.
(Courtesy of STERIFLOW.)

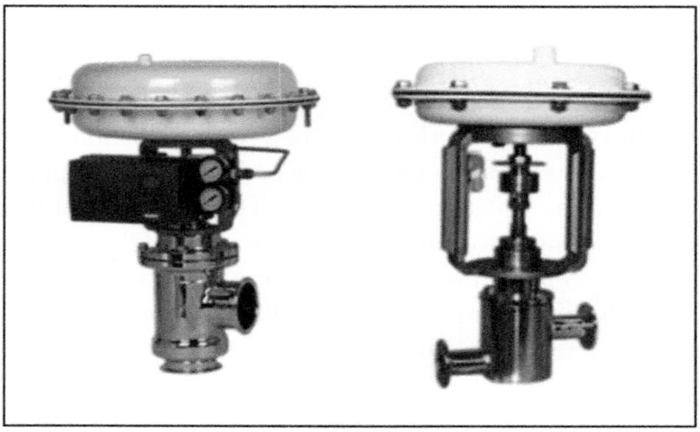

Figure 9.4 Control valves.
(Courtesy of STERIFLOW.)

9.3.3 Pumps

Pumps provide the fluid flow and pressure needed for operation of the UF membrane devices and for the associated operations of flushing, rinsing, CIP and NWP (normalized water permeability). UF systems typically have two pumps associated with them: the recirculation pump and the transfer pump (Figure 9.1). An open UF (100 kDa or greater) may have an additional positive displacement pump in the permeate line to limit flow due to the high permeabilities of these membranes. Considerations for pump selection include flowrate and pressure, sanitary design, non-shedding (no particles generation), compatible materials of construction, single-use service and SIP service.

Pumps are selected to meet the flow, pressure and pulsation requirements of the scale of operation and avoid damaging protein. The recirculation pump in a steel or large scale system is typically a rotary lobe type pump of sanitary design (Figure 9.5). These pumps feature two synchronized counter-rotating impellors called rotary lobes. The lobes rotate in the opposite direction of the flow and trap pockets of fluid between the indents of the rotary lobes and the outer perimeter of the pump case. The pockets are transported to the discharge of the pump and expelled. Many different lobe designs are offered. Bi-wing lobes and tri-wing lobes have been shown to produce minimal protein damage. Adding more lobes will reduce pulsation effects but increase protein degradation. Inlet and outlet connections located in the vertical plane are preferred for easy self-drain. Pumps with connections located in the horizontal plane can retain some fluid inside the casing after the system is drained and possibly harbour bioburden or create corrosion issues. Rotary lobe designs recommended for feed pumps should be run below 500 rpm.

Usually, the recirculation pumps are driven with an A.C. variable speed drive motor/controller. Modern pumps now have decoupled constant rpm cooling fans to facilitate low-speed pump operation. Sanitary centrifugal or peristaltic pumps can be used for buffer or CIP solution transfer and for WFI loop recirculation. Pumps do not in general scale consistently.

Rotary lobe pumps rated for SIP have additional clearance rotors to accommodate thermal expansion. However, the pumping capacities can be significantly reduced by

Figure 9.5 Rotary lobe pumps showing two different lobe design.
(Photos courtesy of Xylem and INOXPA.)

these pumps. Extra care needs to be taken when specifying recirculation pumps for UF systems that are to be SIP'd.

Other types of pumps, such as quaternary diaphragm pumps, have found a place in single-use application and tubing or peristaltic pumps. Pumps are perhaps the least scalable component of a UF system with different types of pumps often being used at different scales.

9.3.4 Heat transfer considerations

It is desirable to carry out UF operations at a controlled constant temperature. Protein viscosities are a strong function of temperature. Variations in temperature can cause flux and pressure drop variations. Protein solutions may have reduced stability at elevated temperatures. The kinetics of bioburden proliferation are more rapid at temperatures that are close to physiologic. UF systems can use a few different methods to control temperature variations. The piping and vessel may be insulated to reduce heat transfer to and from poorly controlled environments. The temperature of the diafiltration buffer can be controlled to help maintain a constant temperature. The retentate vessel may feature a jacket to control heat transfer. An advantage of a vessel jacket is that it does not increase minimum working volume. However, the heat transfer area of the jacket available to the process decreases and the level in the vessel is reduced during concentration. In-line, flow-through heat exchangers can increase the minimum working volume of the system because of their internal volume. The internal volume of the heat exchanger is a function of both the desired flow conditions and the heat transfer area inside the exchanger. The internal volume can be managed by sizing them to a minimum required area to maintain a temperature – that is, removing pump heat as opposed to being able to chill the batch to lower temperatures.

9.3.5 Vessels

The system vessel plays a key role in the UF process step. The vessel design must be carefully evaluated to ensure that the UF operation will be successful. The vessel must be able to support the recirculation of fluid through the system without entraining air at all through the range of fluid volume that it will hold during the process. The vessel is usually equipped with a 'vortex' breaker plate situated at the discharge of the vessel to prevent a vortex from entraining air during recirculation. The smallest volume that can be recirculated without air entrainment is defined as the system 'minimum working volume'. The vessel volume may have to be smaller than the initial batch size (fed-batch mode of operation) if the final concentration specification is greater than $\sim 20\times$. The recirculation vessel size should be no smaller than 1/4–1/5 of the initial batch volume to avoid high concentration/low flux operations. Three types of vessels are most often associated with UF systems: Dish Bottom (ASME), Conical Bottom and Tulip (Figure 9.6).

Regardless of vessel shape, the demands of the process should be considered in the design phase. These parameters are: minimum working volume, agitator location, retentate line return to the vessel, level/volume measurement, CIP sprayer location

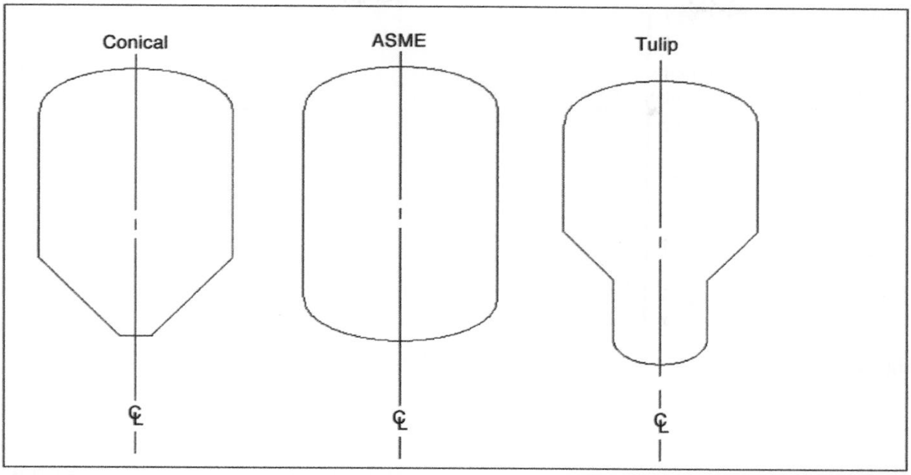

Figure 9.6 Retentate tank or vessel design.

and mixing. Mixing evaluations are often done with computational flow dynamics software. ASME vessels tend to have higher minimum working volumes. Tulip vessels are designed to run at low working volumes due to the dual diameter design. However, retentate return to achieve good mixing at all levels of operation can be challenging. Sometimes separate mixing for the large diameter and the small diameter of the vessel is required for good operation. Conical bottom vessels have found favor as compromise between the ASME and Tulip vessels for low minimum, working volume with simplified mixing operations. Mixer locations vary but are typically found low on the sloped sidewall of the conical-bottomed vessel and offset in the dish of the ASME vessel. Short circuiting, low flux and poor purification progress can result if the retentate line return is not executed correctly. Additionally, the high flow rate of the retentate into the small vessel volumes late in the process can cause foaming and product quality impacts. Retentate line entrance to the vessel should be optimized to reduce the fluid velocity and promote good mixing.

Single-use flexible film vessel liners (Figure 9.7) present their own unique design challenges in addition to those of the steel vessels. Care must be taken during installation of the liners into the support vessel to avoid wrinkles or folds. These irregularities can introduce localized zones of poor mixing resulting in the inefficiency of solute removal during diafiltration.

9.4 Membrane holder

The membrane holder interfaces with the system via standard sanitary connections and then internally directs the fluid into the membrane flow channels. Membrane holders are device specific, but can be classified as either (a) positive containment type or (b) compression containment type. Positive containment membrane devices are contained

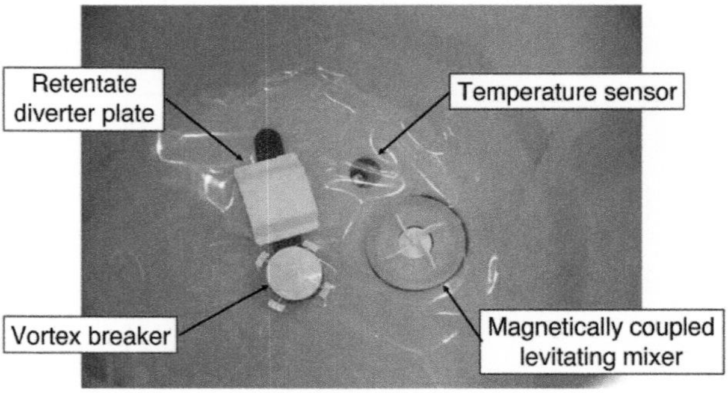

Figure 9.7 Single use UF vessel liner – bottom details.
(Courtesy of EMD Millipore.)

within their own housing and also positively contains the process fluid. The housing may be integral to the membrane device or it may be part of the system. Examples of a positive containment type of holder are spiral wound membrane devices or hollow fibre membrane devices (Figure 9.8).

Compression-type membrane holders are sometimes called 'plate and frame' in that plates and tie rods hold the membrane devices in place under compressive forces. The compressive forces create device to system and device to device seals that contain the process fluid and also support the device flowpaths ensuring proper flow distribution. Cassette modules are an example of compression type containment.

Compression can be imparted to the membranes by tightening fasteners on threaded tie rods to a torque specification or by using hydraulic cylinders for compression. Compression by torque specification can vary depending on the condition of the fasteners, the condition of the threads on the tie rods, and state of lubrication.

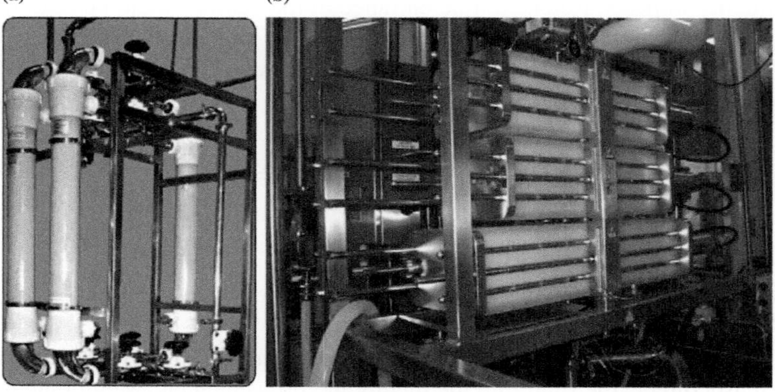

Figure 9.8 UF module holders: (a) hollow fibres and (b) cassettes.
(Courtesy of EMD Millipore.)

Compression by hydraulic cylinders is more easily controlled and provides a more reproducible membrane compression.

System membrane areas are scaled-up by adding holders (and devices) in parallel and or in series. Scale-up must be done carefully keeping in mind pressure and flow requirements for the channels and feed conversion to filtrate. Adding devices/holders in parallel keeps the pressure and flow within the devices similar to the small scale. However, the flow requirement increases linearly with the increase in area. Adding devices/membranes in series, change the flow and pressure profiles within the devices, but conserve on pumping flowrate requirement. Strategies for scale-up and adding membrane area must be well considered to achieve predictable results.

9.5 Process monitoring and control

Operation at a process set-point requires setting crossflow and transmembrane pressure (TMP) for most UF membranes, or crossflow and filtrate flow rate for open UF membranes (≥ 100 kDa). The monitoring, alarm, control, and recording strategy uses sensors to measure permeate flow (water permeability measurement, processing, cleaning), retentate flow, air flow during integrity testing, pressures (feed, retentate, permeate), temperature, tank level or weight, permeate composition (UV absorbance as an indicator of protein content, pH, conductivity as desired), process temperature and coolant temperature. A good design practice is to minimize the number of sensors and identify their optimal placement. These sensor readings may be displayed, logged, used to trigger alarms, and used to trigger a subsequent step in a process. Alarms may be triggered by abnormal pressure excursions (plugged feed channel, valve frozen shut), low tank volumes, high permeate UV absorbance and high temperatures. The duration of each step is determined by time, permeate volume or fluid height in the feed tank. Process data typically are logged for GMP, trending and diagnosis of any unusual excursions.

9.6 Sampling

It is often desirable to withdraw in-process samples from the UF system to assess the actual performance against calculated performance. This may be done during process validation or trouble shooting or as an in-process control (e.g., adjust target protein concentration for bulk drug substance). Sampling locations can include the vessel and the filtrate (permeate) line. Although the UF step is not considered an aseptic operation, aseptic sampling technology can be used for this purpose and minimize the impact of sampling on the bulk drug substance.[7]

9.7 Pressure sensors

Pressure sensors are typically located in the feed, retentate, and filtrate lines of the system and are used to monitor and control a TMP set-point. It is good practice to locate these pressure sensors as close to the feed inlet, retentate, and filtrate outlets of the holder as possible to minimize head height or parasitic pressure drop differences. For a large system with multiple vertical holders, the calculated TMP may need to be adjusted to an 'average holder TMP' for head height differences if one sensor is on the top of the holder compared to another sensor on the bottom.

Pressure sensor types range in complexity from a simple Bourdon-tube mechanical gauge for manual operations to an electronic strain gauge based sensor. The latter type uses a transducer/transmitter unit to change the strain gauge output to a typical 4–20 mA signal for automated process control systems. The range for the pressure sensor depends on its service. It is typical to see a 0–100 psig (6.5 barg) for the feed and retentate sensors. The filtrate pressure is expected to be low for UF systems with membranes $\leq$100 kDa. The range for these sensors may be lower, 30 or 60 psid (2–4 barg). For UF systems with membranes $>$100 kDa, a compound gauge is used to measure pressures above and below gauge pressure.

9.8 Temperature sensors

Temperature monitoring and control ensures consistent operation of the UF/DF step. The viscosity of water and protein solutions changes by ~2% per °C and impacts flux and processing time. Protein stability may be temperature dependent. Typical process ranges are ~6°C for cold plasma fraction processing to 30°C for some high-concentration MAb applications. Clean-in-place operations range from 25 to 50°C depending on the cleaning needs of the system and the tolerance of the UF devices to elevated temperature. Acceptable ranges should be set during the process performance qualification (PPQ).

The RTD (resistance temperature detector) has become the predominant temperature sensing device because of its accuracy (DIN/IEC 60751, Class A $(0.15 \pm 0.002 \times t)$°C) and wide range of capability (−30 to 300°).[8] The RTD can be located near the discharge of the process vessel. If the system is equipped with an in-line heat exchanger, then a second RTD/temperature transmitter may be located in-line at the discharge of the heat exchanger to facilitate measuring the temperature change.

Some instruments, such as Coriolis-effect-based mass flow meters, have integral temperature measurement that can be an output of the instrument in addition to mass flow rate. Manual systems can use RTDs with local manual read-outs or bi-metallic thermometers with local mechanical displays.

9.9 Flow meters and compressed gas supply

Flow meters are commonly used to monitor feed and filtrate flow rates. Sometimes they are located on the retentate (after the UF modules) rather than on the feed (before the UF modules).

Coriolis-effect-based mass flow meters are preferred for the feed/retentate recirculation loop because they measure mass flow rate independently of fluid properties. Coriolis-effect-based mass flow meters for sanitary applications employ a straight flow tube to facilitate cleaning. The drawback of this type of meter is that it is expensive and can be subject to system vibration, which may cause error conditions.

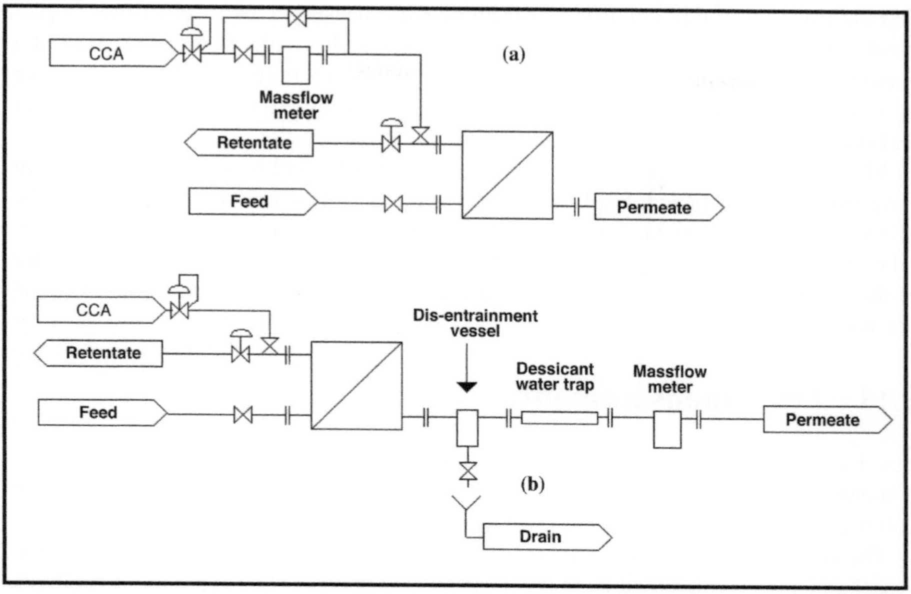

Figure 9.9 UF integrity testing: (a) Upstream implementation and (b) downstream implementation.

Magnetic (Mag) flow meters measure true volumetric flow rate. They are low in cost with a low parasitic pressure drop. However, in order to function, the fluid being measured must have some threshold conductivity so they will not function with water for injection (WFI). This can be overcome by using dilute sodium hydroxide or dilute salt-based buffer solutions to make measurements during preparation and post-use operations. 'Mag' meters are also subject to secondary flow regimes, such as Taylor vortices, and therefore they require some minimum length of straight tubing up and downstream of the sensor.

Integrity test requires the measurement of air diffusion through the membrane. Automated integrity test measurements typically use a differential temperature-type gas mass flow meter. They report airflow in standard cubic centimetres per minute (sccm). These meters can be implemented either on the filtrate side of the membrane or on the feed/retentate side (see Figure 9.9). However, this type of meter is sensitive to liquid and implementation on the filtrate line requires additional consideration.

A good quality self-relieving regulator should be used on the gas supply to accurately set the integrity test pressure.

9.10 Buffer and protein concentrations

Conductivity, pH, and optical density (UV absorbance) are sometimes implemented as online process analytical technology (PAT). Conductivity is often found implemented in the filtrate line. Conductivity can be used to monitor the progress of a diafiltration and the flushout of CIP solutions. Optical density measurement is sometimes implemented into the filtrate line. These sensors work via ultraviolet light absorbance at a wavelength of 280 nm. Most often this is used to detect a failure condition where the protein product leaks into the filtrate side of the system, and to trip an alarm to prevent additional product loss to the filtrate.

Flow-through densitometers have been suggested for online monitors for protein concentration. The principle of operation for these instruments is oscillating a tube of known volume. The dampening of oscillation is a function of the mass fluid in the tube. The mass divided by the volume then is the density of the fluid. If the density of the fluid as a function of protein concentration is known, then concentration of the batch may be directly monitored.

9.11 Level measurement

Level controls are used as set-points to initiate the next phase of operation and to maintain recirculation vessel level during fed-batch operation and during constant volume diafiltration. Methods for level measurement can be contact or non-contact.

The recirculation vessel can be mounted on lead cells to measure the mass in the vessel and a level calculated from the mass measurement. Other non-contact methods include radar (microwave) or ultrasonic transmitter/detectors mounted on top ports of the vessel. These devices can measure the level of the surface of the fluid in the vessel. Level can be measured directly by a magnetostrictive float level detector (Figure 9.10). This technology features a captive magnetic float on a guide. The magnetic field of the float interacts with the magnetic field of a wire inside the guide that is converted to a level signal. This type of level sensor measures the level of the fluid in the vessel directly. The measurement is robust and not influenced by foaming.

Fluid volume can be calculated from a level height measurement. Figure 9.11 shows fluid volume as a function of level height for a conical bottomed vessel. Similar equations exist for bottoms of other shapes including ASME heads (torispherical).[9]

Figure 9.10 Magnetostrictive level bottomed tank.
(Courtesy of Drexelbrook.)

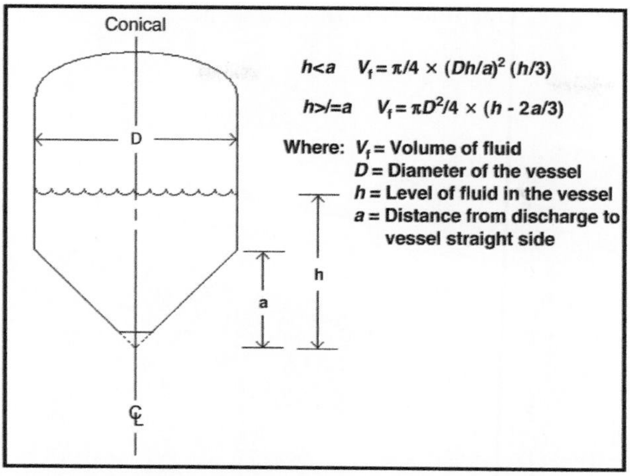

Figure 9.11 Fluid volume for a cone measurement.

9.12 Other sensors

Implementation of various sensors into the UF system presents problems due to space, dead-spots, cleanability and so on. One method of sensor integration has been to create sensor 'blocks' that efficiently and cleanly integrate the sensors into the UF system. Figure 9.12 shows the retentate and filtrate (permeate) pressure gauges integrated into the membrane holder discharge.

9.13 Layout

Skid layout criteria include minimizing holdup volume and dead legs, as well as allowing for flushing, cleaning, venting, draining, mixing, sanitizing/steaming, operating and servicing ergonomics and safety. Holdup volume and the presence of dead legs in the wetted fluid path impacts product recovery, separation efficiency (e.g., poor buffer exchange at high diavolumes), fluid volumes required (cleaning, flushing, processing), system cost and required floor space and the ease of cleaning and sanitizing. Minimization involves reducing line lengths, employing the full 3D space using CADCAM to explore design alternatives and using compact components (e.g., valve assemblies).

Systems must be designed to minimize product degradation by controlling physical and chemical stresses (e.g., hot spots, excessive shear, air interfaces, cavitation and local concentrations). Mixers are used to eliminate concentration gradients in tanks. See Section 9.2 for diafiltrate addition and mixing. In-line mixing (two fluids are pumped into a Tee connection and blended while flowing in a pipe) for buffer preparation from concentrates reduces waiting time, tankage and floor space.

Figure 9.12 Sensor integration in a UF system holder.
(Courtesy of EMD Millipore.)

The 3D layout of process and remote monitoring and control systems should allow operators to setup, operate, and turn around each process without undue strain. In addition, maintenance and service personnel should be able to conduct routine operations (e.g., calibration and gasket replacement) without undue strain. This requires proper orientation of displays and enough space around a processing skid to gain access. Computer 3D models, constructed using CADCAM systems, should be examined for ergonomics during design reviews.

The design, construction, commissioning, and validation of a process skid often involve a team of experts from the biopharmaceutical manufacturer, the skid supplier, and an A&E (architect and engineering) firm. This is facilitated through the use of a process with clear UF step performance requirements, roles of team members and milestones/reviews.

9.14 Cost considerations

This section considers capital and operating costs for ultrafiltration processes. Ultrafiltration system capital cost varies with membrane area, cross-flow rate, flux, level of instrumentation, automation, any added functionality, and by year of manufacture. The CECPI (Chemical Engineering magazine) combines equipment, facility, engineering supervision, and labor costs in an index to convert them to any particular year[2]. A basic manual system, consisting of just a pump, pressure sensors (two or three), valves, and a permeate flow meter, has a very low capital cost. However, this system requires operator monitoring and control throughout its use. This can lead to process deviations and faulty batches.

Adding components to improve control will reduce operator interactions but increase capital cost. The typical system enhancing features include a tank, level sensors, expanded pressure monitoring with more sensors, expanded flow monitoring (feed or retentate) and minimization of system volume (operating and holdup). Once these enhanced features/components are available in the ultrafiltration system, it makes it easy to justify automation in the operations to maximize repeatability of operation and minimize operator interactions. While automation usually requires a strong initial training program for proper implementation, the long-term benefits are gained when running the equipment day in and day out. For example, adding a manual steaming functionality to a system can be done inexpensively. Automating this functionality can increase the cost of this feature several times relative to the manual steaming option.[10]

Automation costs are highly dependent on several key factors: the specific level of automation, the software applied for the automation, and the methods needed to comply with the manufacturing plant requirements. The impact of these factors can result in extremely variable costs. The relative costs of a fully automated system can be double or triple that of a manual system. Once the equipment and the plant operation are defined, it is recommended that the user develop the cost impact of this functionality in conjunction with selected control automation integrators.

As shown in Figure 9.13, capital costs for a biotech ultrafiltration system ranged from \$0.3K to \$0.8K in 2010. Using a 5-year capital cost depreciation makes for annual depreciation costs on the order of \$100K/year. Biotech products typically sell for hundreds of millions of dollars, if not billions of dollars per year. While the capital cost can be significant to a small firm with financial constraints, the ultrafiltration system capital cost is a very minor component of the cost of goods.

Increased membrane area requires proportionately larger flows leading to a bigger pump and larger diameter plumbing components (Figure 9.13). Typically, capital costs are scaled using a capacity term such as membrane area to a 0.6 power.[11]

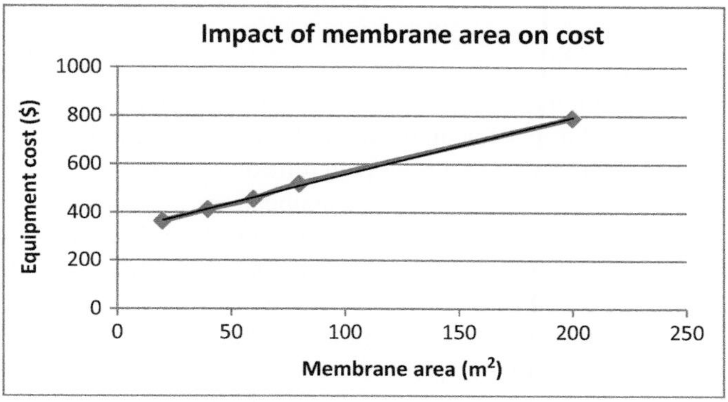

Figure 9.13 System cost vs. membrane area.

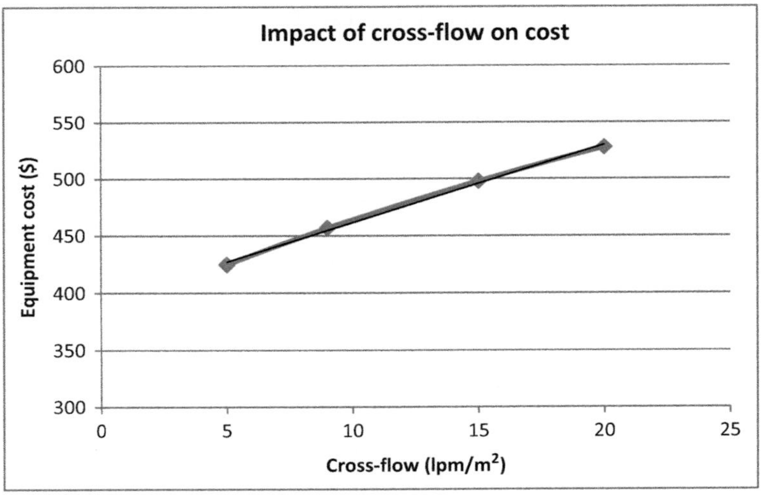

Figure 9.14 System cost vs. cross-flow rate.

The best power law fit to membrane area has an exponent of 0.54 but the linear fit, $ = 2.37A + 319.8, has a better fit with $r^2 = 0.999$.

Increased cross-flow leads to a bigger pump and larger diameter plumbing components (Figure 9.14). Typical process operation is at 5–6 lpm/m^2 but higher flows are commonly used for flushing and cleaning. The best power law fit to cross-flow has an exponent of 0.16 but the linear fit, $ = 393.2 + 6.85Q, has a better fit with $r^2 = 0.997$. In the polarized region of the flux curve, flux increases with cross-flow and the corresponding area requirements decrease. Consider a typical 60 m^2 system run a 5 lpm/m^2 where the area varies with cross-flow as $A = 80(Q/5)^{-0.5}$. Higher feed fluxes can decrease total area while modestly impacting equipment costs (Figure 9.15). Equipment cost goes through a slight minimum but total costs here are driven by membrane area and decrease with higher feed fluxes but at a diminishing rate. A number of constraints limit running at high feed fluxes including feed pressure limits, protein damage limiting pump passes and pump sizes available.

Operating costs vary with number of module reuses, number of modules and module cost, electrical for automation and pump power, compressed air for integrity testing, water for flushing, buffers for diafiltration, buffers for cleaning and sanitizing, maintenance and labour including training with refreshers. Table 9.5 shows some operating cost breakdowns for general industrial applications.

Relative costs differ for an expensive biotech product. For final formulation biotech applications, membrane requirements for a 3 h process are (150 g/m^2/h) (3 h) ~ 0.45 kg/m^2 or \$0.09/g for 50 reuses compared to drug sales prices on the order of \$2000/g. Capital cost depreciation for 50 batches/year, containing about 35 kg/batch, represents \$0.06/g. A yield loss of 1% of a \$100 mil/year drug product represents \$57/g. Diafiltrate buffer consumption for 10 diavolumes at a 50 g/L concentration represents 0.2 L/g of use or \$0.2/g at a \$10/L buffer cost. These

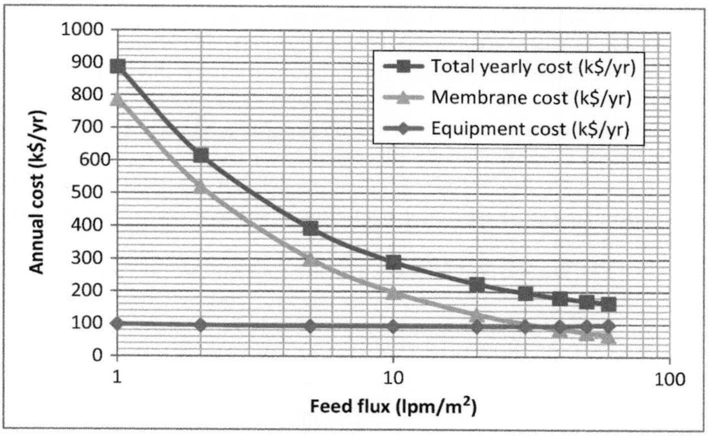

Figure 9.15 Annual membrane and equipment cost vs. feed flux (5 year-depreciation, 1 set membranes/year @$5k/m^2).

Table 9.5 Operating costs

Cost component	Fraction of total (%)[12]	Fraction of total (%)[13]
Capital equipment	38	31
Membrane	27	23
Labor and maintenance	14	11
Energy	10	11
Cleaning	5	24*

* Includes value of protein lost during cleaning.

calculations show that product quality and yield drive the design and operation of a biotech UF process, not capital and membrane costs. For shear sensitive products (e.g., high molecular weight components like vaccines), pump speeds could be reduced further.

References

1. Campbell P. Engineering design standards, North American Customer Engineered Products Plant, EMD Millipore Corporation, November 7, 2006.
2. Chemical Engineering Magazine, Plant Cost Index CEPCI, at Che.com.
3. Bioprocessing Equipment. ASME BPE-2009, Copyright 2009, American Society of Mechanical Engineers, ISBN 13: 978-0-7918-3213-4.
4. Van Dyke M. An album of fluid motion. Stanford, CA: Parabolic Press; 1982.
5. DeLucia D. Fundamentals of CIP design. ASME Bioprocessing Seminars. 1997.

6. Peters M, Timmerhaus KD. Plant design and economics for chemical engineers. 2nd ed. New York: McGraw-Hill; 1968.
7. Gonzalez MM. Stainless steel tubing in the biotechnology industry. Pharmaceut Eng 2001;21(5):48–63.
8. Control Valves Information. <http://www.globalspec.com/learnmore/flow_control_flow_transfer/valves/flow_control_valves> [accessed 26.02.14].
9. Van Reis R, Zydney AL. Bioprocess membrane technology. J Membr Sci 2007;297:16–50.
10. Perrone P. Optimizing your steaming operation: a steam-in-place primer. ISPE Boston Newslett 2008;XVIII(1):10–4.
11. Remer DS, Chai LH. Design cost factors for scaling-up engineering equipment. Chem Eng Prog 1990;86:77–82.
12. Beaton NC, Steadly H. Industrial ultrafiltration. In: Li NN, editor. Recent developments in separation science, 7. Boca Raton, FL: CRC Press; 1982. p. 1–29.
13. Eykamp W. Ultrafiltration. In: Membrane separation systems: a research needs assessment. U.S. Dept. of Energy, OER, OPA; 1990. pp. 7-1–7-35.

Validation

10

Herb Lutz
EMD Millipore, Biomanufacturing Sciences Network

Validation studies are conducted under Good Laboratory Practice with a pre-specified validation master plan defining protocols and end-point specifications, calibrated equipment, and trained personnel. Results are recorded with signed lab notebooks or preapproved test forms and written up in a formatted report.[1]

10.1 Scaling

For scaling, one is interested in the comparable performance between large area, commercial scale, and small area, lab scale, systems. Development of the processing step requires a small area test system that delivers performance, such as flux, yield, impurity removal and cleanability, which is expected to reliably scale-up to large manufacturing area systems. These small area test systems are also useful for scale-down studies to investigate the effect of process changes at large scale. Scaling has been verified by running two different scale systems side-by-side at the same time with identical feed, membrane lots, operating steps, and operating conditions.[2] Regulators require some data to qualify scale-down performance models. Vendors can demonstrate comparable performance on model feed streams.

Linear scaling involves using membrane modules with the same channel length but different channel widths or numbers of channels. This is the most straightforward approach. For ultrafiltration (UF) processes involving solute fractionation, the polarization of solutes within the channel can be critical and it is important to employ linear scaling. However, one can scale between different module types for basic operations, such as concentration and diafiltration using scaling factors with specially set operating conditions. This approach to scaling is not discussed here in any more detail.

Scaling performance should match the average performance seen at the two different scales. Random batch-to-batch variability can be due to a host of factors and does not reflect a systematic difference between the two scales of operation. Scaling performance can be influenced by everything that influences process performance: membrane, module, system design, process operation, pre-processing steps, and feedstock. Specific details of these effects are described in a number of places in this book.

10.2 Quality and process attributes

The UF step is typically designed to produce a retentate with certain quality attributes, including product protein concentration and formulation buffer composition. In addition, the step should minimize the formation of aggregates, feed buffer components, the presence of bioburden or endotoxin, or components carried over from the previous batch, module storage or extractables, cleaning or sanitization. For cases involving membrane fractionation (see Section 7.11), an additional quality attribute of product purity is required. These quality attributes are of interest to regulators since they could potentially affect patients through altering product safety and/or efficacy and purity.

The UF step is also typically designed to meet certain process attributes including yield, costs, and process time. These process attributes do not affect patients and are not of interest to regulators. They are of interest, however, to bioprocessors who need to maintain adequate drug supply and maintain adequate profitability.

10.3 Process parameters

Chapters 7–9 describe the operating steps for an ultrafiltration process. Process parameters are taken as things that can vary from process-to-process. It is convenient to consider them in categories as things that can change in the feed, in the modules and system, and in the operation. Table 10.1 outlines the UF process steps and shows and lists the associated process parameters. These steps are for routine operation after the first batch where a pre-cleaning step is recommended. The potential impact of process parameter changes on quality and process attributes are indicated in Table 10.2.

A failure modes and effects (FMEA) type risk analysis is recommended to rank the parameters by the magnitude of risk they pose in causing the ultrafiltration process to fail to meet its quality and process attributes.[3] The highest risk parameters (associated with severe impacts on product quality or process performance or operator safety, a high likelihood of impact occurrence, and an inadequate parameter control system) would be designated as critical parameters. The omission of ultrafiltration process parameters from a recent MAb case study[4] and a vaccine case study[5] developed by

Table 10.1 UF steps and process parameters

Step	Process parameter(s)
Installation	Module type, area, torque
Flush	L/m^2 retentate and permeate water
Integrity test	Test pressure and air flow spec
Equilibration	L/m^2 retentate and permeate buffer
Processing	Feed composition, feed flow and TMP, concentration end-points and diavolumes, DF buffer composition
Recovery	L/m^2 retentate DF buffer
Cleaning	Water flush as above, cleaner type/temp/concentration/time/feed flow, L/m^2 loading, NWP measurement conditions and specs
Sanitization	Sanitant type/temp/concentration/time/feed flow, L/m^2 loading, environment
Storage	Storage type/concentration, reuse spec

Table 10.2 UF process parameters and attributes

Attribute	Process parameters
Retentate concentration	Feed concentration, final concentration end-point, recovery L/m^2, module type (NMWL)
Impurity concentration	Feed concentration, diafiltrate buffer concentrations, diavolumes, pump passes
Contaminant concentration	Diavolumes, L/m^2 retentate and permeate flush water
Yield	Recovery L/m^2, module type (NMWL), IT value
Processing time	Flux, feed volume, diavolumes, area, concentration end-points
Cost	Reuse spec
Bioburden	System design, cleaning/sanitization, environment

an industry working group indicates that these process parameters, along with the ultrafiltration step, are considered very robust and without critical parameters compared to other steps. The Parenteral Drug Association (PDA) process validation report indicated TMP, diafiltration volumes, and yield as non-key ultrafiltration formulation step parameters one may include in validation studies.[1] For fractionation applications where the UF step is expected to purify the product, process parameters can become more critical.

10.4 Design space (operating set-points and windows)

Outside of the processing step, process parameter values are primarily based on vendor recommendations[6] or plant operations with other molecules or with the specific ultrafiltration skid (e.g., flushing and sanitization). These may be verified for the particular molecule in process development and validation studies (e.g., module re-use studies). These may also be verified on a new manufacturing system (e.g., system cleaning and flushing studies). Processing step parameter values that require testing or specification for each molecule include area, TMP, diafiltrate concentration and concentration end-points. Module type and feed flow rate is set by the vendor recommendation or typical values used within the bioprocessor manufacturing network and verified for each molecule. Chapter 8 describes parameter selection and is consistent with other published Quality by Design (QbD) recommendations.[7] Table 10.3 summarizes process parameter set-point values and the potential impact of these set-points on product quality or process performance attributes.

Figure 10.1 shows a single parameter with various ranges. First, there is the target set-point point where one tries to run each batch. Second, there is the actual or normal range of operation around the set-point. Third, there may be an alarm range where operation outside the range triggers an alarm for investigation. Fourth, there is the validated range where the operation has been shown to produce quality product. Fifth, there may be edge of failure limits, outside of which, the process does not produce

Table 10.3 **UF steps and set-points**

Step	Set-point(s)	Impact
Installation	30 kDa, regenerated cellulose, C Screen, 80 m^2 in parallel	Time, yield, concentration
Flush	L/m^2 retentate and L/m^2 permeate water	Storage contaminants, product degradation
Integrity test	10 psi, <10 sccm/m^2 spec	Yield
Equilibration	L/m^2 buffer	Aggregates, time
UF1	6 LMM feed, 20 psi TMP, volume reduction factor X	Time, aggregates
DF	6 LMM feed, 20 psi TMP, diavolumes N	Time, aggregates, buffer contaminants
UF2	6 LMM feed, 20 psi TMP, Volume reduction factor X	Time, aggregates, final concentration

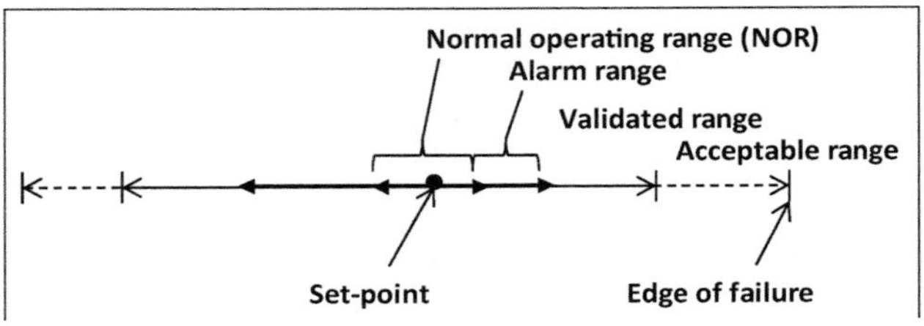

Figure 10.1 Process parameter ranges.[8,9]

acceptable products. An example of this may be a batch that is too small for a particular processing skid so final concentration values cannot be reached. Parameters may show process interactions so a 2D graph may be needed to illustrate the target operating point and associated areas of operation.

In general, UF steps easily accomplish their processing goals. They are very robust (i.e., relatively insensitive to process variations and batch-to-batch variability). This means that operating windows are generally far from an edge of failure and one has considerable freedom to operate without significantly impacting product quality or performance. For example, flush specifications are far from minimum flush requirements to ensure minimal carryover and pressures are far from component maximum specifications. One exception is the case where there is a critical flux (Section 7.10). Exceeding this flux causes the membrane to foul or sieving to decline precipitously. Another exception is operating close to the minimum working volume of a system (Section 9.2). If the protein mass falls below a critical value for that manufacturing system, the target final concentration will not be achieved without falling below the minimum working volume and suffering product quality issues. Finally, if system integrity test values exceed specifications, yields can suffer significantly.

In most cases parameter limits are set by demonstrating control over the manufacturing process and not by exceeding quality or performance goals. For example, one might specify operating with hydraulic parameters of a 5 Lpm/m^2 feed flux and a 10 psid TMP. One should be able to control the feed flux within ±20% and the TMP within ±2 psid. This controllability may be expressed in terms of a Cpk or a six-sigma process capability.[10] For this level of control, one should have a normal operating range of 4–6 Lpm/m^2 feed flux and 8–12 psid TMP. These ranges are controllable and do not impact product quality or performance significantly. Exceeding these ranges does not significantly impact quality or performance either but an alarm could be triggered that process control is not as tight as expected.

10.5 Characterization studies

Characterization studies establish a validated range of operation for the manufacturing plant, within which experimental data show no impact on product quality. This range is filed with regulators in a Biologics License Application. Operation outside this range is possible but there is no characterization data to support a claim of acceptable product quality.

Designing characterization studies involves (1) selecting only the most significant process parameters influencing quality or performance to reduce the number of tests, (2) determining a range for testing and (3) determining a test system. A formal FMAE analysis can be used to rank process parameters by risk. However, a quick review of the parameters in Table 10.3 suggests that the feed flux, TMP and diavolumes are the most significant parameters. Other parameters are considered to have a smaller impact on the process with a huge safety margin. An appropriate range for these parameters is determined by how far manufacturing deviations may extend and where quality might be impacted. Here, we can take 5–20 psid TMP, 2–8 Lpm/m^2 feed flux and 10–12 diavolumes as process limits for a characterization study. This roughly matches a hypothetical case in the PDA validation Tech Report.[1] If a test point shows unacceptable quality, the ranges were set too wide and need to be reduced using a new DOE (design of experiments) test matrix. If a manufacturing plant excursion exceeds these limits without product quality issues, then the limits could be expanded with a new DOE test matrix.

These test points can be explored using a scaled down test system. For this example, the impact of the hydraulic set-points on flux, cleanability, and excipient flushout can be determined. Specific pump-induced damage at manufacturing scale is difficult to model in a scaled down system. Only a general indication of the sensitivity of a particular protein to pumping relative to another protein can be determined. A scaled down excipient flushout curve can indicate the presence of adsorptive effects. Qualification tests on the manufacturing system must be used to determine flushing requirements, product recovery, minimum working volumes, and adequate mixing for excipient clearance.

10.6 Cleaning/reuse studies

Cleaning conditions (cleaner type, concentration, temperature, time) are typically based on vendor recommendations and/or previous plant experience. Scale down module testing is used to demonstrate that process permeability and water permeability are consistent from batch-to-batch for several cycles. Experience indicates that extrapolation of these results to multiple cycles is not risky. Batches are run at large scale using a co-current validation strategy. That is, batches are produced and placed on hold with a possible risk of failure if permeability criteria are not met. They are then released for sale when permeabilities pass specifications. Modules are replaced if: (1) process or water permeabilities are not maintained, and/or (2) storage time limit is exceeded, and/or (3) number of cycles limit is exceeded, and/or (4) module integrity testing fails, and/or (5) module pressure drop exceeds specifications and/or (6) bioburden and/or TOC limits are exceeded.

Equipment cleaning is generally considered separately (see Section 10.8).

Large-scale qualification runs demonstrate cleaning effectiveness. Blank runs involving a buffer feed alone (without protein) may also be included to demonstrate that the retentate product contains no product carried over from the previous batch.

10.7 Extractables

All product contact surfaces need to be considered as potential sources of leaching contaminants into the product. For UF systems, the largest surface is the membrane with contact also from stainless hardware components, films from disposable components, and miscellaneous surfaces from pumps, sensors, O-rings, and valves. Vendors typically furnish information on the concentration and composition of these after a recommended flush or cleaning. A convenient benchmark, generally considered as safe without further testing, is USP[11] water for injection (WFI) or purified water (PW) with a total organic carbon (TOC) spec of $\leq$500 ppb and a 25°C conductivity spec of $<$1.3 μS/cm. A risk assessment[12] of product contamination risk can be made based on these data and operating characteristics of contact time, temperature, solution composition, and so on.

10.8 Equipment and facilities

Equipment validation is commonly broken down into the following:[13,14]

1. Design qualification (DQ) showing that the design and component specifications meet the performance targets.
2. Installation qualification (IQ) showing that the system has been fabricated properly and that the components comply with the manufacturer's installation and design specifications.
3. Operational qualification (OQ) showing that the fabricated system operates according to design requirements and specifications, commonly as factory acceptance testing (FAT).

4. Performance qualification (PQ) showing that the system with feedstock meets the performance targets. Cleaning qualification involves demonstrating consistent equipment surface cleanliness and flush water purity after processing and worst case cleaning. Swabbing of internal surfaces to collect samples for assay is commonly used to verify effective cleaning.[14] UV light will cause riboflavin to fluoresce so solutions applied to internal surfaces can demonstrate effective cleaning coverage and help to identify worst case locations. Showing excipient clearance from the retentate during diafiltration qualifies adequate mixing. Showing consistent protein quality after processing qualifies minimal pumping damage or damage during recovery.

10.9 Monitoring and re-validation

Ultrafiltration process monitoring can include measure of product quality (e.g., aggregates, conductivities, bioburden, and LAL) and process performance (e.g., NWP, fluxes, times, yields, integrity test values, pressure drops, numbers of cycles, and membrane age). Some of this information is used to trigger the change out of modules. Wide deviations from averages or outside of the normal operating range should prompt closer examination to see if there is a trend and whether corrective action is necessary. Sometimes deviations are random and disappear as quickly as they appear. A steady decline in NWP suggests an inadequacy of the cleaning process. Deviations outside the validated range established by characterization studies prompts an investigation. This will place a batch on hold until there is additional validation data generated to show that it is OK and can be released for sale.

Revalidation may be required as part of process changes.[15,16] This is generally performed as part of a comparability protocol demonstrating equivalence.

References

1. Parenteral Drug Association. Technical Report No. 42: process validation of protein manufacturing. PDA J Pharm Sci Technol 2005;59(Sept–Oct).
2. Van Reis R, Goodrich EM, Yson CL, Frautschy LN, Dzengeleski S, Lutz H. Linear scale ultrafiltration. Biotechnol Bioeng 1997;55(5):737–46.
3. International Conference on Harmonization. ICH Topic Q9 Quality Risk Management; 2005.
4. CMC Biotech Working Group. A-Mab: a case study in bioprocess development, Version 2.1, Oct. 2009, available from ISPE and FDA.
5. CMC Vaccines Working Group. A-VAX: Applying Quality by Design to Vaccines, May 2012, available from PDA.
6. EMD Millipore Corporation, Pellicon 3 Cassettes: installation and user guide, Literature #AN1065EN00 Rev. E, 02/2014.
7. Watler PK, Rozembersky J. Application of QbD principles to tangential flow filtration conditions. In: Rathore AS, Mhatre R, editors. Quality by design for biopharmaceuticals. Hoboken, NJ: Wiley; 2009.
8. Chapman KG. The PAR approach to process validation. Pharm Technol 1984;1984:23–36.
9. Seely RJ, Hutchins HV, Luscher MP, Sniff KS, Hassler R. Defining critical variables in well-characterized biotechnology processes. Biopharm Int 1999;April:33–6.

10. Edgar TF, Smith CL, Shinskey FG, Gassman GW, Waite AWH, McAvoy TJ, Seborg DE. Process control. In: Green DW, Perry RH, editors. Perry's chemical engineers' handbook. 8th ed New York, NY: McGraw-Hill; 2008.
11. USP. Section <1231> Water for Pharmaceutical Purposes, United States Pharmacopeial Convention, Rockville, Maryland, USA.
12. Brennan J, VanDeinse H, Seely RJ, Miller D, Boone H, Fernandez J, et al. Evaluation of extractables from product contact surfaces. Biopharm Int 2002;15(12):3–11.
13. Ferenc BM. Qualification and change control. In: Carleton F, Agalloco J, editors. Validation of pharmaceutical processes. 2nd ed New York, NY: Marcel Dekker, Inc; 1999.
14. Bronkow R, Delucia D, Haft S, Hyde J, Lindsay J, McEntire J, Murphy R, Myers J, Nichols K, Terranova B, Voss J, White E. Cleaning and cleaning validation: a biotechnology perspective. Parenteral Drug Association; Bethesda, MD, 1996.
15. European Medicines Agency (EMEA). Annex 15 qualification and validation: change control. EU Guide to Good Manufacturing Practice. EMEA, Brussels, Belgium, July 2001.
16. US FDA. Guidance for industry: changes to an approved application for specified biotechnology and specified synthetic biological products, July 1997.

Troubleshooting

11

Joseph Parrella, Bala Raghunath, Herb Lutz
EMD Millipore, Biomanufacturing Sciences Network

11.1 Troubleshooting methodology

Troubleshooting is often a reactive process, typically in response to process parameters or product attributes deviating from expected or historical values.[1,2] Troubleshooting describes the investigation method used to logically and systematically identify the root cause(s) and determine the necessary corrective actions. An example of a simplified troubleshooting process is summarized in Figure 11.1.

The first step in the troubleshooting process involves identifying the underlying problem or failure mode. The failure mode can be related to an alarm (e.g., high feed pressure alarm), product attribute (e.g., low yield), or a deviation from typical behaviour (i.e., long processing times). Multiple failure modes (e.g., high feed pressure and low permeate flux) may also occur within a single failure. An understanding of what failed, how it failed, when it failed, and what did not fail is essential in identifying likely root causes.

Identifying likely root causes of a particular failure mode(s) is often done by combining a fundamental understanding of ultrafiltration (UF) processes, equipment, and filter devices with personal experiences. A review of historical data trends and change controls related to the process are often helpful in identifying potential root causes. This step typically involves a team with expertise in multiple disciplines. Depending on the nature of the failure, different disciplines may be pulled into the team as needed to provide the required expertise. It is useful to include vendors in this process. Vendors have the breadth and depth of experience in seeing how ultrafiltration processes work in a variety of applications.

The determination of the most likely root cause might include a review of historical data trends, paper exercises, process modelling, and/or scaled-down experimentation. It is most often an iterative process, as new data and information gained through the investigation might eliminate some potential root causes or suggest alternative root causes not initially identified. It might also be necessary to consider the costs, timing, complexity, and risk to process; and product for possible corrective actions when evaluating potential root causes. For some troubleshooting cases, it might be preferred to address a likely root cause requiring simple parameter change (within the validated range) before implementing a corrective action to address a more likely root cause that requires a significant capital investment and/or skid downtime/impact to production schedule. This strategy may also help improve confidence in the more impactful root cause/corrective action by eliminating a likely root cause.

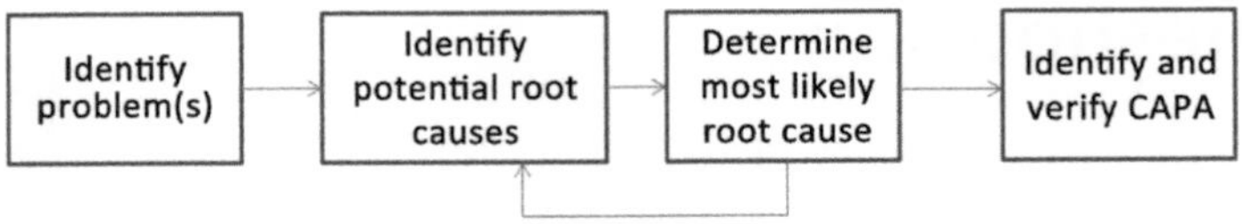

Figure 11.1 Troubleshooting process flow diagram.

Once the likely root cause has been identified, corrective and/or preventative actions (CAPA) are identified and implemented to prevent the failure(s) from re-occurring. Considerations are often given to the capital costs, impact to the validated status, impact to the production schedule and risk to process and product when determining and evaluating corrective actions. Often times, the effectiveness of the corrective actions are verified via scaled-down models to minimize risk prior to large-scale implementation.

11.2 Ultrafiltration troubleshooting

The performance of UF systems and processes can be affected by equipment and system design, membrane and device selection, operating parameters, process design/control and/or product changes. All of these factors must be considered when troubleshooting ultrafiltration processes. Additionally, similar failure modes may manifest themselves in different ways for different control strategies. For example, if the feed channel of a filtration device becomes blocked, a process controlled to a constant pressure drop is expected to see a decrease in feed flow, while a process that is controlled to a constant feed flow rate is expected to see an increase in feed pressure. To develop a more general troubleshooting guide, the failure modes will be described in a normalized form where possible.

Table 11.1 lists the likely failure modes covered in this chapter. For each failure mode, a table reviews potential root causes for a given failure mode. For each potential root cause the table also includes typical observations and suggests ideas to evaluate/verify the potential root causes to determine the most likely root cause. Each table also suggests possible corrective actions for each of the potential root causes. Discussion of the different failure modes is also included.

Table 11.1 Likely failure modes

Title	Table	Comments
Low yield or yield loss	11.2	
Water permeability	11.4	
Process permeability	11.5	High process time
High feed channel resistances	11.6	
Integrity testing	11.7	Likely on first installation w/ gaskets
Excipient clearance	11.8	

11.3 Failure mode: low yield or yield loss

One of the more critical failure modes observed during ultrafiltration processes is low product yield or yield loss (see also Section 7.4 on permeate losses, Section 8.3. on recovery losses). In some cases, this failure mode can be diagnosed before the process is completed if there is a UV meter inline. In other cases, yield loss might be identified after the process is completed when measuring the final concentrations.

The first step in identifying likely root causes is to close the mass balance; to quantify how much product was lost and to determine where the 'lost' product went. The more common failure mode results in product loss into the permeate pool. If there is no UV meter or concentration measurement in process, it may not be possible to determine where the lost product went as the permeate liquid is often discarded to drain. If the location of the lost permeate product cannot be determined, it is generally a fair starting point to assume it was lost to the permeate stream. Table 11.2 reviews possible root causes of yield loss into the permeate stream.

Table 11.2 Potential root causes for low yield via permeate yield losses

Potential root causes	Typical observations	How to evaluate potential root cause	Corrective actions
Membrane too open	• Minor shifts in yield (<15%). • Increased sensitivity of retention to operating conditions.	• Consult with filter vendor to understand impact of membrane manufacturing range. • Review PD data to understand impact of different feed and membrane lots.	• Select tighter membrane. • Modify operating conditions to reduce polarization.
Insufficient internal sealing	• More typical of major shift in yield (>15%). • More common for manual cassette holders. • May coincide with increased permeate liquid flow. • External liquid leak between filter and or hardware. • Possible failure of post-use integrity test.	• High pressure/ramping integrity test for as-found system.* • Inspect nuts and threaded rods for manual holders for galling and damage. Nuts should spin freely. • Inspect tie rod alignment to ensure straight. • Inspect holder floating endplate to ensure flat and not warped. • Inspect silicone gaskets if being re-used for compression set.	• Verify use of stainless steel support plates or manifold adapter plate (for cassettes). • Replace O-rings (spirals). • Increase torque for manual holders (cassettes). • Replace nuts and threaded rods (cassettes). • Use new silicone gaskets for each installation.

(Continued)

Table 11.2 Potential root causes for low yield via permeate yield losses *(cont.)*

Potential root causes	Typical observations	How to evaluate potential root cause	Corrective actions
Membrane damage	• More typical of major shifts in yield (>15%). • Often detected via post-use integrity failure (not always though). • Often detected via changes in water or process permeability. • Possible failure of post-use integrity test. • Inspect membrane for particulates. • Sealing pressure does not help pass integrity test.	• Review membrane compatibility with vendor. • Review operating conditions with vendor, particularly for non-composite membranes. Focus on start up and shut down sequences. • Review SOPs for proper application of chemical agents, particularly cleaning chemicals. • Device autopsy. Typically performed last as it is a destructive test. • Inspect pump and system components to ensure operational and not introducing particles (i.e., passivation complete, pump head not sheared).	• Modify processing sequences or start up/shut down sequences. • Modify SOPs to ensure proper application of chemicals. • Consider alternative membrane and/or device format for improved robustness or compatibility. • Replace shedding components or close system.
Insufficient cleaning	• Retention/yield differences after initial processing (most often between Runs 1 and 2). • More common with more hydrophobic membranes (ex. PES). • Often (but not always) accompanied by reduction in NWP. • Possible failure of post-use integrity test.	• Review vendor recommended cleaning strategies. • Test affected devices (or perform scaled down testing) to evaluate more aggressive cleaning cycle.	• Implement more aggressive cleaning cycle. • Use membrane that is easier to clean (RC instead of PES). • Use tighter membrane.

*A high-pressure/ramping integrity test evaluates diffusive air flow at pressures between the recommended integrity test pressure and the maximum feed pressure observed in the process. Diffusive air flow through the membrane is expected to increase linearly with pressure. A rapid increase in air flow (deviating from linearity) indicates bulk air flow and possible seal bypass. The test should be repeated after implementation of corrective actions to ensure that bulk air flow is not a result of membrane damage.

11.3.1 Incorrect membrane selection (see also Section 7.4)

Rules of thumb are often applied when selecting the appropriate nominal molecular weight limit (NMWL) based on the molecular weight of the product of interest. Retention and yield are verified experimentally during process development. However, retention depends on more than just product size. Retention can be affected by product properties, buffer conditions, and operating conditions. Additionally, process development and pilot-scale testing may only evaluate one or several membrane lots. During development, it is likely the process is not run with the entire expected manufacturing range of the membrane. If the incorrect membrane is selected, it may be possible that product retention may be adversely affected as it is exposed to the entire (but) expected manufacturing range of the filter.

11.3.2 Insufficient internal sealing

Many UF filters require sealing to separate the feed and permeate flow paths. UF cassette-type filters require external compression (applied using hydraulic or manual holders) to provide a feed/permeate sea. UF spiral-type filters typically require O-rings to seal the feed and permeate. If insufficient compression or damaged O-rings are used, the sealing may not be sufficient to prevent product in the feed from bypassing the membrane and entering the permeate channel via a compromised seal. This failure mode is more likely to occur at higher trans-membrane pressures (TMPs), as more force is applied to each of the seals. If the failure point occurs at a transmembrane pressure higher than the recommended integrity test pressure, it is unlikely to be detected by the integrity test. The applied compression is also affected by temperature changes, particularly for cassettes in a manually operated holder where there is no active control of the compressive force.

11.3.3 Membrane damage

Membrane damage can occur due to UF operating conditions. Older, non-composite membranes are particularly susceptible to damage from reverse pressure (i.e., permeate pressure exceeds retentate pressure). These instances most often occur during start up and shutdown sequences. Membrane damage from reverse pressure can typically be detected via integrity testing.

Chemical exposure can also result in membrane damage. Most vendors provide general guidance on chemical compatibility, but it is critical to verify compatibility of a given process and product (see Appendix). Additionally, the proper application of compatible chemicals is critical. For example, when preparing a sodium hydroxide cleaning solution in the recirculation vessel from a concentrated stock solution, it is critical to verify that the solution is properly mixed prior to recirculation. Depending on the standard operating procedure (SOP), tank design, and mixer design, agitation is not always sufficient to ensure adequate mixing. Dense, caustic concentrate may chemically burn the membrane (at >1 N) when the recirculation pump is turned on. Short-term high-concentration caustic exposures have been shown to irreversibly damage the tight retentive pores of regenerated cellulose membranes. This failure mode is not always detected via low-pressure air integrity tests.

11.3.4 Insufficient cleaning

Insufficient cleaning has been shown to impact product retention and yield. It has been hypothesized that insufficient cleaning alters the membrane pore size distribution. Fouling plugs the smaller pores so a greater portion of flow passes through large, less-retentive pores.

Since flow is proportional to pore diameter to the fourth power, slight changes in the population or characteristics of smaller pores may not be detected via permeability measurements. This failure mode has been more common for membrane types that are often more difficult to clean (PES membranes).

Table 11.3 reviews possible root causes when yield loss is not related to permeate losses. These potential root causes should be considered when no product is detected in permeate samples.

Table 11.3 Potential root causes for low yield not related to permeate losses

Potential root causes	Typical observations	How to evaluate potential root cause	Corrective actions
System hold up losses	• No product passage into the permeate. • Typically good recovery / yield during PD. Large scale process yields not consistent with PD.	• Evaluate modified recovery strategies (ex. increase buffer flushing. Sample recovery flushes separately to determine flush out. • System walk down to verify proper slope and evaluate potential dead legs.	• Modify recovery strategy or recovery flow paths. • Modify system design to improve drainability. • SPTFF for over concentration if concentration targets limit buffer flush.
Absorption to membrane/device/ system components	• Relevant for very dilute applications. • Relevant for applications with low g/m^2 loadings. • More common for PES membranes. • Yields consistent with process development (assuming similar g/L and g/m^2).	• Paper exercise to evaluate impact of potential membrane binding based on published values. • Evaluate impact of changing g/m^2 loading. • Compare process performance with process development history. • Analytical testing to evaluate protein bound to membrane surface (destructive test).	• Increase g/m^2 (decrease membrane area). • Evaluate less absorptive membrane (RC instead of PES).

Table 11.3 **Potential root causes for low yield not related to permeate losses *(cont.)***

Potential root causes	Typical observations	How to evaluate potential root cause	Corrective actions
Product quality losses	• Aggregate levels out of specification. • Protein activity out of specification. • Product 'foaming' during process or product recovery. • In rare cases, aggregate formation correspond to increase in feed channel resistance. • Solubility issues may occur during concentration steps.	• Analytical testing to quantify product quality changes (compare to PD) and characterize formation during process. • Compare to process development results to identify skid effects. • Review system for possible causes of product quality issues: air/liquid interfaces (pump microcavitation, air entrainment at mixer during over concentration, J-tube/dip tube design, DF buffer degassing), shear induced damage (pumping damage). • Excessive polarization during operation.	• Eliminate or minimize potential air water interfaces (apply overlay pressure to recirculation tank to minimize degassing and microcavitation, consider alternative mixer designs if mixer entraining air, consider alternative recovery strategies if foaming observed during recovery). • Reduce polarization. • Couple with SPTFF to minimize pump passes if product is shear sensitive.

11.3.5 System hold up losses (see also Sections 7.7 and 9.4)

Yield losses might arise when the concentrated feed pool cannot be efficiently removed from the system. Yield losses might result from a poor product recovery strategy and/or a poor system design. Product recovery steps, most typically buffer flushes or air blow-downs, are required to remove product remaining in the system. Product recovery steps are often system specific, as the recovery strategies used for process development might not apply to the larger (fixed) stainless systems. Product recovery is often more challenging for highly viscous (i.e., high g/L concentrations) products.

Good system design is critical to ensuring good recovery. Sanitary system design ensures proper sloping and drainability. Zero dead leg valves are often used in the feed–retentate recirculation loop to eliminate dead legs and maximize product recovery.

11.3.6 Absorption to membrane/device/system components (see also Section 7.6)

Various biological products have been shown to bind to the surface of ultrafiltration membranes (primarily through hydrophobic interactions). The binding capacities are typically very low (1–35 $\mu g/cm^2$ for RC membranes and 10–125 $\mu g/cm^2$ for PES membranes, compared to typical protein loadings of >100 g/m^2 or $>10{,}000$ $\mu g/cm^2$). Because of the low binding capacities, yield losses due to membrane losses are typically more common for very dilute processes or very low g/m^2 membrane loadings.

11.3.7 Product quality/solubility (see also Section 7.11)

Product quality issues (aggregation/activity) have often been attributed to the air/protein interfaces. These interfaces might result from the retentate returning to the recirculation tank, micro-cavitation at the pump, or air entrainment/foaming at the mixer at low recirculation tank volumes or during product recovery. Troubleshooting often starts by comparing product quality profiles to process development runs to identify skid effects and/or working to eliminate or minimize potential air water interfaces. In some rarer cases, pump shear has been shown to impact product quality. Buffer conditions can also impact solubility.

11.4 Failure mode: changes in water permeability

Another failure mode observed during ultrafiltration processes is changes in water permeability beyond typical or historical values. This might be observed during the initial (pre-use) water permeability testing (changes relative to process development or pilot-scale testing) or during post-use water permeability testing (changes relative to pre-use values).

Table 11.4 reviews possible root causes for changes in water permeability.

11.4.1 Head pressure affects and friction losses

Larger systems often have the feed, retentate, and permeate pressure gauges placed farther from the filter devices than the smaller systems. Larger systems add additional filtration area vertically (to allow for good drainability), so the relative difference in heights between pressure gauges can have a measureable impact on pressures and TMPs. Height differences of 5–10 ft between the feed and retentate/permeate gauges

Table 11.4 **Potential root causes for changes in water permeability**

Potential root causes	Typical observations	How to evaluate potential root cause	Corrective actions
Head pressure effects and friction losses	• Initial (pre-use) NWP values lower than pilot and PD testing.	• Paper study to quantify impact of head pressure differences on feed pressure, retentate pressure, and permeate pressure (and ultimately TMP). • Paper study to quantify friction affects/piping pressure loss effects.	• Adjust TMP calculations to account for known differences in head pressures.
Insufficient cleaning / membrane fouling	• Reduction in NWP (relative to pre-use values) after processing with each batch. • Trends typically consistent with PD and pilot scale. • Might correspond to an atypical batch.	• Repeat cleaning step with more aggressive cleaning procedure (higher temps, higher cleaning agent concentrations).	• Implement more aggressive cleaning procedure. • Evaluate membrane that is easier to clean.
Membrane damage	• Membrane permeability increases or decreases depending on chemical/ cleaning agent relative to pre-use values. • Typically consistent with PD and pilot scale testing.	• Review membrane compatibility with vendor. • Review SOPs for proper application of chemical agents, particularly cleaning chemicals. • Device autopsy. Typically performed last as it is a destructive test.	• Modify SOPs to ensure proper application of chemicals. • Consider alternative membrane and/ or device format for improved robustness or compatibility. • Consider alternative chemicals.

are commonly observed. The vertical stack of cassettes might also span 5 ft. Most control systems use gauge readings to control the TMP. When comparing water permeability to smaller systems, the differences in pressure and TMP must be accounted for to facilitate accurate comparisons.

Additionally, friction losses resulting from liquid flow in the piping system must also be considered. Before accounting for head and friction pressure effects, larger systems show lower water permeabilities when compared to smaller systems.

11.4.2 Insufficient cleaning/membrane fouling

Membrane fouling might occur rapidly after a single batch or gradually over repeated batches. Membrane fouling of a more gradual nature can be difficult to detect during process development as extended re-use is often validated concurrently during commercial manufacturing (due to feed and cost limitations).

11.4.2.1 Membrane damage

Membrane damage via chemical exposure can result in either an increase or decrease in permeability, depending on the chemical and exposure time. Refer to Table 11.1 for additional details.

11.5 Failure mode: changes in process permeability

Changes in process permeability typically result in processes running shorter or longer than expected. Longer runs might impact the overall production schedule or downstream unit operation schedule. Unexplained changes in process time might indicate a lack of control or understanding of the process.

Table 11.5 reviews possible root causes for changes in process permeability.

11.5.1 Changes in water permeability

For processes with dilute products and/or tighter UF membranes ($<$10 kDa membrane), changes in water permeability are expected to directly correlate to product flux/permeability because permeability is the more significant flow resistance. These types of processes are more sensitive to small changes in water permeability. For processes with more concentrated products or more open membranes, process permeability is less sensitive to changes in membrane water permeability because the osmotic pressure (polarization) is the more significant to flow resistance.

11.5.2 Poor mixing

Poor mixing can result in stratification in the recirculation tank. More concentrated product pools in the lower fraction of the recirculation tank due to density differences. The more concentrated product is recirculated through the cassettes,

Table 11.5 **Potential root causes for changes in process permeability**

Potential root causes	Typical observations	How to evaluate potential root cause	Corrective actions
Changes in membrane water permeability	Refer to Table 11.4		
poor mixing	• Pre- and post-use water permeabilities similar. • Lower process fluxes often correspond to higher feed channel resistances. Higher process fluxes often correspond to low lower feed channel resistances. • Water flow vs. pressure drop trends consistent with vendor and historical data but product flows vs. pressure drop trends differ from historical data. • Typically related to the skid and skid operating conditions. (i.e., likely to occur during first run/every run for a given skid/process). • Non-ideal excipient profiles. • More common for tulip tank design or tanks/bags with passive agitation. • More common for highly viscous solutions.	• Process modelling to characterize impact of poor mixing. • Compare to PD performance and/or small scale experimentation. • Compare pre-use and post-use feed channel resistances with water (typically from NWP steps).	• Add active agitation or modify existing mixer design.
Device compression out of range (for cassettes)	• Overcompression might result in higher process fluxes. Under compression might result in lower process fluxes. Due to mass transfer changes. • Often coincides with changes in feed channel resistance. • Polarized processes more sensitive to device compression. • More common for processes with temperature changes and/or those that use manual holders.	• Compare water flow vs. pressure drop trends to vendor and historical data. • Inspect existing installation and re-torque for manual holders. • Re-install devices into holder at correct torque/compression and evaluate performance.	• Verify usage of proper torque/compression.

(Continued)

Table 11.5 **Potential root causes for changes in process permeability *(cont.)***

Potential root causes	Typical observations	How to evaluate potential root cause	Corrective actions
Feed changes	• Pre- and post-use water permeabilities similar. • A typical behaviour from previous unit operations. • Polarized processes more sensitive to feed changes.	• Review performance trends from previous unit operations. • Analytical testing on feed composition.	• Failure may result from performance of other unit operations.

resulting in a reduction in permeate flow because of the unexpected increase in product concentration. The reduction in permeate flow is related to an increase in polarization so this failure mode is generally reversible. If the system is properly mixed after observing this failure mode, the process flux typically returns to its expected value.

11.5.3 Device compression out of range

This root cause is applicable for cassette devices requiring compression to engage internal seals. At higher than specified compressions, membranes can occlude into the feed channel. This increases feed channel velocity (due to the smaller cross-section) and can result in an increase in permeate flux via improved mass transfer. Alternatively, compressions lower than the specified ranges can have the opposite effects and allow the feed channel to slightly expand. This reduces feed channel velocity and can result in a decrease in permeate flux via reduced mass transfer.

Device compressions out of range are more often observed with manual holders and processes that experience significant temperature changes. The torque recommendations are dependent on the friction between the nuts, spacers, holder and threaded rod. If these components are galled, damaged, or even lubricated, the compressive force applied at the recommended torque can vary greatly. Temperature changes can affect the applied compressive force at the given torque as the devices and holders expand and contract with temperature changes.

11.5.4 Feed changes

Changes in feed composition are generally more difficult to identify as the root cause lies with another unit operation upstream. The change might result from a difference in bioreactor or chromatography column performance. Analytical testing can often help diagnose differences in the feed or product composition.

11.6 Failure mode: high feed channel resistances

High feed channel resistance may manifest in different ways for different control strategies. Processes at constant flow will observe higher than expected feed pressures, while processes at constant pressure drop will observe a lower than expected feed flow rate. High feed pressures might prevent a UF process from completing due to burst disc limits or process alarms, while low feed flow rates might result in a process that runs much longer than anticipated.

Table 11.6 reviews possible root causes for changes in feed channel resistance

11.6.1 Poor mixing

Poor mixing can result in stratification in the recirculation tank. More concentrated product pools in the lower fraction of the recirculation tank due to density differences.

Table 11.6 **Potential root causes for changes in feed channel resistance**

Potential root causes	Typical observations	How to evaluate potential root cause	Corrective actions
Poor mixing	• Pre- and post-use water permeabilities similar. • Reversible phenomenon. Post-use feed channel resistances (measured with water) typically similar to pre-use feed channel resistances. • Higher feed channel resistances often coincide with lower process flux. • Typically related to the skid and skid operating conditions (i.e., likely to occur during first run/ every run for a given skid/process). • Non-ideal excipient profiles. • More common for tulip tank design or tanks/ bags with passive agitation. • More common for highly viscous solutions.	• Process modelling to characterize impact of poor mixing. • Compare to PD performance and/ or small scale experimentation. • Compare pre-use and post-use feed channel resistances with water (typically from NWP steps).	• Add active agitation or modify existing mixer design.

(Continued)

Table 11.6 Potential root causes for changes in feed channel resistance *(cont.)*

Potential root causes	Typical observations	How to evaluate potential root cause	Corrective actions
Device compression out of range (for cassettes)	• Over compression might result in higher feed channel resistances. Under compression might result in lower feed channel resistances. • For polarized processes, might coincide with changes in process permeate flux. • More common for processes with temperature changes and/or those that use manual holders.	• Compare water flow vs. pressure drop trends to vendor and historical data. • Inspect existing installation and re-torque for manual holders. • Re-install devices into holder at correct torque/ compression and evaluate performance.	• Verify usage of proper torque/ compression. • Verify proper installation for filter being used, i.e., silicone gasket type, manifold adapter plate.
Feed channel fouling	• More typical of products containing particulate matter. • Often irreversible. Differences likely to be observed with pre-use and post-use feed channel resistance (with water).	• Compare pre-use and post-use feed channel resistances (with water). • Test devices offline and 'reverse' feed-retentate flow to displace material. • Device autopsy.	• Add prefilters or guard filters to feeds and buffers, as applicable, to liquids entering the recirculation tank. • Verify sanitary connection gaskets are not worn and shedding, replace if required.

The more concentrated product is recirculated through the cassettes, resulting in an increased feed channel resistance due to the increased viscosity. If the system is properly mixed after observing this failure mode, the process flux typically returns to its expected value.

11.6.2 Device compression out of range

This root cause is applicable for cassette devices requiring compression to engage internal seals. At higher than specified compressions, membranes can occlude into the feed channel. This decreases the feed channel cross-sections, resulting in an increase in feed channel resistance. Alternatively, compressions lower than the specified ranges can have the opposite effects and allow the feed channel to slightly expand. This increases the feed channel cross-section and can result in a decrease in feed channel resistance.

As discussed in Section 11.3, this failure mode is more common with manual holders and processes experiences significant temperature changes.

11.6.3 Feed channel fouling

The accumulation of particulate matter or even aggregates has been shown to obstruct or partially obstruct the feed channel of filter devices resulting in an increase in feed channel resistance. This is often more common during initial commissioning as external material has more opportunity to enter the system during construction. Traditional cleaning procedures are often not effective at displacing or breaking up these particulates or aggregates. As a result, the increase in feed channel resistance is typically observed after the cleaning during the post-use water permeability test.

Reversing flow through the feed channel (flow from retentate to feed) can often be helpful in displacing any accumulated material. Most systems are not configured to flow in this direction so this testing is typically performed offline.

11.7 Failure mode: post-use integrity test failures

Integrity test failures typically coincide with instances of low yield or yield loss through the permeate stream. The possible root causes discussed in Table 11.1 are relevant to most integrity failures. Table 11.7 can be used as a guide to troubleshooting

Table 11.7 Potential root causes for integrity test failures

Potential root causes	Typical observations	How to evaluate potential root cause	Corrective actions
Insufficient internal sealing/device compression	Refer to Table 11.1		
Membrane damage	Refer to Table 11.1		
Insufficient cleaning	Refer to Table 11.1		
Membrane not properly wet	• Integrity failure might not correspond to yield losses. • Membranes with hydrophobic properties more susceptible.	• Repeat wetting/ flushing steps and repeat integrity test. • Evaluate IPA flushing step before water flushing and repeat integrity.	• Modify flushing procedures. • Add additional flushing step IPA.
Different wetting fluid used	• Integrity failure does not correspond to yield losses. • Integrity test not performed in water but with specifications based on water.	• Repeat integrity test with water.	• Validate alternative diffusion set point or perform integrity test with water.

integrity failures for possible root causes where integrity failures are mentioned as possible observation. This table focuses on the UF system but it is worth noting that false failure can occur with an incorrectly conducted integrity test (e.g., wrong pressure, poor flow measurement, temperature/stabilization effects and using the wrong test). Where no air flow is measured in the permeate stream, it is often an indication that the permeate flow path is not clear preventing accurate measurements.

11.8 Failure mode: excipient clearance

Non-ideal removal of initial buffer components or excipient offsets in the final product pool can affect product efficacy and/or stability. Table 11.8 reviews possible root causes for non-ideal buffer clearance exchange.

11.8.1 Poor mixing

The fraction of unmixed pool volume that can prevent ideal removal of the initial buffer is very small (~0.1%) leaving initial buffer excipients in the final bulk pool. This unmixed fraction can be caused by dead legs in the recirculation loop as well as unmixed regions in the tank. The process may appear to trend similar to ideal performance based on permeate grab samples. The problem is likely detected after the final pool is recovered from the system when the unmixed fractions combine with the remainder of the final pool.

The relative volume fraction that can cause problems with buffer clearance is far less than what might be needed to observe some of the hydraulic problems referenced earlier in this chapter. It is very possible to detect higher than ideal initial buffer excipients in the bulk pool but not observe any other hydraulic problems related to poor mixing.

11.8.2 High concentration effects

High concentration effects can result in an enrichment and depletion of excipients resulting from protein-excipient charge effects (Donnan effect), volume exclusion effects, and preferential hydration effects. Protein charge works to enrich opposite-charged excipients and deplete like-charge excipients. Volume exclusion effects reduce the concentration of excipients regardless of charge. These effects are typically a function of protein concentration and are more noticeable at higher protein concentrations (>50 g/L for monoclonal antibodies).

11.9 Failure mode: poor product quality (aggregation/activity)

Refer to the product quality losses section in Table 11.8. Table 11.8 focuses on cases where product quality issues affect yield. In some cases, product quality may be impacted without affecting yield. The possible root causes and corrective actions listed in Table 11.8 are still applicable for those cases where poor product quality does not affect yield.

Table 11.8 Potential root causes for poor excipient clearance

Potential root causes	Typical observations	How to evaluate potential root cause	Corrective actions
Poor mixing	• Clearance of initial buffer solutes lower than predicted. • Diafiltration buffer. • Permeate grab samples might be consistent with predicted values. Differences detected in final pool. • More common for higher viscosity solutions. • More common for systems with passive agitation. • Typically related to skid and skid operating conditions (i.e., performance may vary for different systems).	• Process modelling to characterize impact of unmixed fraction. • Sample permeate during process to profile buffer removal. • Compare buffer clearance data for PD performance.	• Modify mixer or tank design. • Replace passive agitation with active agitation.
High concentration effects (product charge effect, preferential hydration, volume exclusion)	• Effects can result in enrichment and/ or depletion of diafiltration buffer excipients. • More common for higher concentration (>50 g/L) processes. • Clearance of initial buffer components generally unaffected. • Permeate grab samples show offset between feed and permeate (and ideal values). • Generally unaffected by scale. PD results expected to be similar.	• Modelling to estimate and understand high concentration effects. • Sample permeate to characterize buffer removal. • Compare to process development data.	• Modify recovery flush buffer to adjust final bulk. • Adjust diafiltration buffer to account for offsets so that final bulk is at target. • Spike in excipients at end of process (for case of depleted excipients) to meet targets.

11.10 Troubleshooting case study

This case study is intended to serve as an example for how to work through a problem using some of the tables and other information in this chapter.

11.10.1 Problem description

A biotechnology company uses 30 kDa polyethersulfone membranes for the final UF/DF step of their monoclonal antibody process. Process development was performed using mini cassettes (0.1 m^2). The first clinical run was performed using maxi cassettes (5 m^2) and performed as expected based on process development. The second clinical run reused the same maxi cassettes and showed lower yields. Key processing values are summarized in Table 11.9. A review of processing trends indicated no significant differences in other processing values (pressures and flow rates) were noticed between process development and clinical runs.

The post-use integrity tests for both clinical runs passed. The devices were cleaned with 0.5 N NaOH at 40 °C for 60 min after each run. A permeate UV meter was not used on the system and it was not known if the yield losses were a result of permeate losses.

11.10.2 Troubleshooting approach

11.10.2.1 Identify problem

The reduction in yield was the primary problem. However, a reduction in water permeability was also observed after the first clinical run. In some instances, an initial reduction in water permeability is observed after the first run, after which permeabilities remain stable. Without re-use data beyond two runs, it was unclear if the reduction in water permeability was also a problem. The initial investigation focused primarily on low yields with additional consideration given to root causes that might affect both yield and permeability.

11.10.2.2 Identify potential root causes

Tables 11.2 and 11.3 were reviewed to assess likely root causes and rule out unlikely root causes. The initial assessment is described in Table 11.10.

Table 11.9 Process summary for troubleshooting case study

	Process development	First clinical run	Second clinical run
Pre-use-NWP (LMH/psi)	21	19	15
Process time (h)	4.2	4.1	4.2
Yield (%)	98	98	85

Table 11.10 Initial assessment of potential root causes for low yield

Potential root causes	Initial assessment	Assessment justification	Next steps
Incorrect membrane selection	Unlikely	Acceptable yields observed during process development and first clinical run.	N/A
Insufficient internal sealing	Possible	Operating pressures exceed integrity pressure. Combined with temperature shifts and use of old holder. Poor sealing may not be detected by 10 psi integrity test.	Perform ramping integrity test on stalled devices.
Membrane damage*	Unlikely	Similar chemicals and procedures used in PD.	N/A
Insufficient cleaning*	Possible	Cleaning PES membranes challenging with only 0.5 N NaOH.	Process development cleaning evaluation.
Membrane absorption	Unlikely	Different yields between clinical runs. Absorption issues expected to be similar for similar mass loadings.	N/A
Product quality losses*	Unlikely	Different yields between clinical runs. Quality issues expected to be similar for a given system, scale and process.	N/A

*Identified as potential root causes for reduced water permeability as well.

11.10.2.3 Determine most likely root cause

The initial assessment of root causes identified two possible root causes: insufficient cassette sealing and inadequate cleaning. Insufficient cassette sealing can typically be verified by performing a high pressure integrity test. This test typically takes 1–2 h and must be performed while the cassettes are still installed. If the cassettes are removed from the system after detecting the failure, the test is not as valuable. Insufficient cleaning is typically evaluated via process development testing to minimize the amount of protein required. Process development testing may also provide flexibility to evaluate alternative cleaning conditions or chemicals without consideration to validation impact.

Initial testing evaluated cassette sealing because the cassettes needed to be removed as soon as possible. Additionally, time and resources were required to generate material to perform the cleaning evaluation. A ramping integrity test was performed on the as-installed cassettes and the results are shown in Figure 11.2

The ramping integrity did not indicate bulk air flow to the permeate at higher pressures. For comparison, a similar system with insufficient sealing showed an increase in permeate flow from 22 mL/min at 20 psi to 800 mL/min at 30 psi.

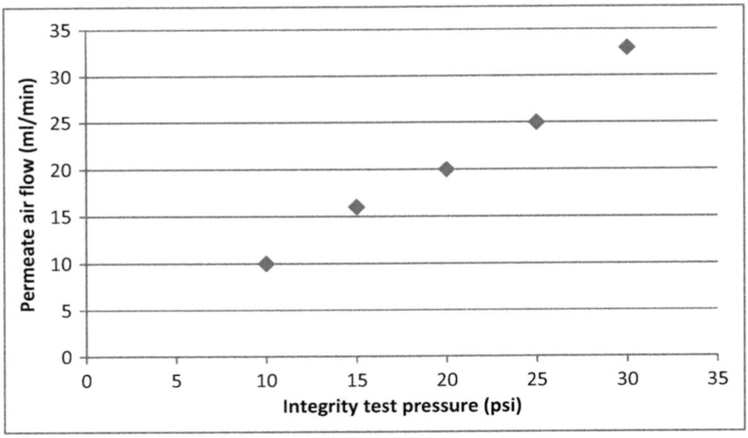

Figure 11.2 Permeate air flow during ramping integrity test.

The effectiveness of the cleaning cycle and impact of more aggressive cleaning cycles was then evaluated. A scaled-down process simulation was performed with mini cassettes to evaluate the standard cleaning process (0.5 N NaOH at 40°C) followed by a more aggressive cleaning cycle (0.5 N NaOH with 500 ppm bleach). Permeate concentrations were trended during the concentration and diafiltration step and shown in Figure 11.3.

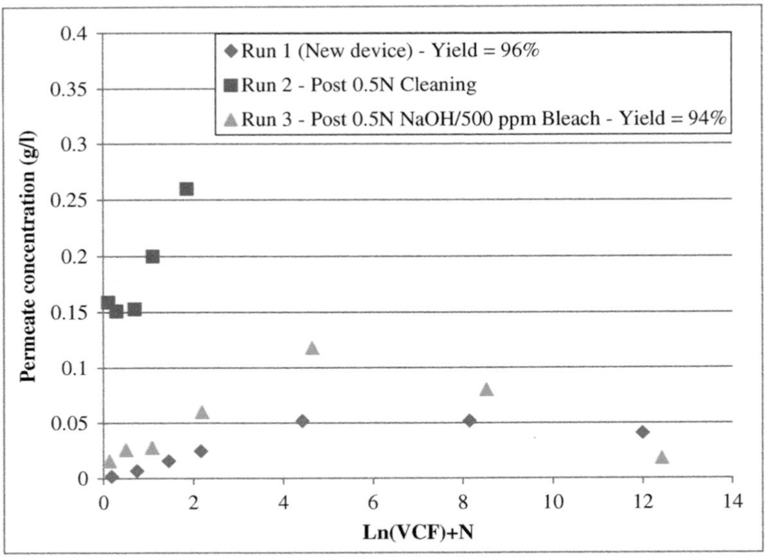

Figure 11.3 Impact of cleaning cycles on protein retention.

The scaled-down cleaning simulations indicated that the initial 0.5 N NaOH cleaning was not effective. Product retention decreased with reprocessing resulting in lower yields. Run 2 was stopped to minimize product lost due to the higher concentrations measured in the permeate. Product retention and yields were improved with a more aggressive cleaning cycle.

Additional review of the permeate concentration data with consideration of the root causes assessed in Table 11.10 also suggested that the incorrect membrane may be used for this process. Permeate yield losses of 4–6% are generally considered high. Processes with this baseline retention are typically more sensitive to the membrane manufacturing range and cleaning processes. Additional evaluation of a 10 kDa PES membrane showed improved yield robustness with minimal (<5%) impact on process time.

References

1. International Conference on Harmonization. Topic Q9, Quality Risk Management; 2005.
2. Kepner CH, Tregoe BB. The new rational manager. Princeton, NJ: Princeton Research Press; 1981.

Conclusion

12

Herb Lutz
EMD Millipore, Biomanufacturing Sciences Network

12.1 Historical overview

The first large commercial application of ultrafiltration (UF) was paint recycling, followed by dairy whey recovery in the mid-1970s.[1] Paint recycling enabled the use of electropainting of automobiles to prevent rust. Dairy whey recovery solved a waste disposal problem by increasing yields for cheese production. Early firms involved in commercializing ultrafiltration technology included Abcor and AMICON. These were spun off from the Massachusetts Institute of Technology Chemical Engineering Department by Professors Raymond Baddour and Alan Michaels respectively. Fundamental work on membrane transport and module design was funded by the US Office of Saline Water Research for Reverse Osmosis desalination applications.

Juice clarification emerged as a useful application for ultrafiltration along with waste treatment through volume reduction. Biotechnology was just getting started in the late 1970s and many solutes of interest were in the size range for retention by ultrafiltration membranes. The biotechnology industry has displaced classical pharmaceuticals as the fastest growing sector for producing new therapeutics. Antibody proteins have emerged as the dominant platform.

In all these applications, ultrafiltration fills a need that other technologies cannot fill. Applications can be described as clarification of a permeate product (e.g., water purification), concentration of a retentate product (e.g., paint, dairy, and pharmaceuticals) or solute purification (e.g., virus/protein separation and buffer exchange). UF applications are enabled by low-temperature and low-cost operation, particularly with membrane reuse.

UF processes can vary significantly between applications due to differing requirements and relative costs. The industrial markets are less concerned with yield than with operating costs, particularly pumping costs. The heavily regulated biotechnology market is more concerned with consistency and performance documentation. This has led to different suppliers serving these different customer needs with different membranes, devices, levels of documentation, quality control and support. Millipore spun-off a separate firm, Mykrolis, to address the microelectronics industry separately because the high-performance product needs common biotechnology and microelectronics industries were not enough to overcome the different industry dynamics and increased stock valuations were obtained.

12.2 Recent developments and current research

Biotechnology has become the biggest ultrafiltration application. It is expected to continue to grow out into the future at declining rates. Antibodies and their variants are expected to continue to drive revenue. Despite much promise in being able to cure rather than treat diseases, gene therapies using nucleic acid therapeutics have not proved safe and effective.

Ultrafiltration retains solutes of nanometer size so the active area of nanotechnology research may commercialize nanometer-sized products employing ultrafiltration processes. It is too early to tell.

Academic research has focused on novel membrane substrates, such as ceramics, carbon nanotubes or sheets and nanofibres. Coatings designed to prevent fouling and extend life in challenging applications have proved useful for reverse osmosis membranes. Investigation into membrane fractionation has so far failed to deliver on its initial promise.

12.3 Trends and likely future developments

The adoption of new technologies by the biotech industry tends to be slow. Process developers are focused on developing and validating robust processes for new molecules. Many of them are interested in exploring new technologies and there is a diversity to allow variation between groups in adoption and a role for champions or lead adopters. However, patient requirements to produce a consistent product that is always available drives risk reduction and the use of historically proven technologies. Adoption becomes driven by clear unmet needs.[2] As the industry matures, this becomes more widely applicable across the industry. The concern for plugging in MF TFF harvest technology drove an interest in centrifuge adoption. This took 6 years to reach pilot manufacturing and another 6 years to reach commercial manufacturing.[3]

Membrane casting has become more and more controlled over time to enable casting membranes with reduced lot-to-lot variability and tighter specifications. This has enabled a proliferation of different molecular weight cutoffs that can be tailored for different applications. Permeability tends to increase with more open pores. For most applications, the process flux is controlled by the retained solute and not membrane permeability. This means that one generally does not pay a penalty in flux for selecting a tighter membrane unless the solute is very dilute or has a low molecular weight. For the more open membranes, the fluxes through the membrane tend to be too high causing non-uniformity along the feed channel. It may be useful to actually reduce the permeability in these membranes.

A tight pore size distribution is useful for fractionation. This would enable more of the smaller molecular weight impurities to be removed in a tangential flow filtration (TFF) step. For an ideal spherical solute in a cylindrical pore, the selectivity between two solutes can be extremely large when the membrane pore size is selected to be comparable to the largest solute. In practice, pore size distributions reduce selectivities. It is expected that there will be increased control over pore size distribution using better equipment, advances in casting technology or new type of materials for membrane formation.

Ease-of-use drives improvements in modules. Less fouling membranes make cleaning easier. More consistent sealing modules reduce integrity failures. Pre-sterilized modules could potentially reduce flushing requirements.

Single-pass TFF (SPTFF) is gaining interest in niche biotech applications. It is expected that SPTFF modules could be more integrated with other steps so that these operate as one assembly with improved performance, such as reduced cost or faster flow. SPTFF also enables continuous manufacturing. This is currently a topic of active interest and it is expected that some robust continuous processes will get commercialized. As larger molecular weight therapeutics such as vaccines are commercialized, low-shear SPTFF may also be implemented to reduce product degradation.

Single-use assemblies are popular and growing in use at clinical and pilot scale to reduce time and facilitate closed system processing. It is expected that this trend will continue with integrated UF modules in batch or SPTFF operating modes.

12.4 Summary

Ultrafiltration is a relatively young unit operation, only about 40 years old. Much of the technology has been developed as evidenced by the content of this technical monograph. Nevertheless, there appears to be opportunities for new developments and it is expected that improved products and performance will continue into the future.

References

1. Zeman LJ, Zydney AL. Microfiltration and ultrafiltration. New York, NY: Marcel Dekker; 1996.
2. Kelley B. Industrialization of mAb production technology, the bioprocessing industry at a crossroads. MAbs 2009;1(5):443–52.
3. Wolk B. Design considerations for a multi-use stack centrifuge. Presentation at Recovery of Biologicals XI Conference, Banff, Canada; 2003.

Appendix A

A.1 Chemical compatibility

Chemical compatibility for cassette filters

Chemical	PES	RC	PVDF
Acetic acid (0.05%)	+	+	+
Acetic acid (5.0%)	+	+	+
Acetic acid (25%)	+	+	+
Acetic anhydride (100%)	−	−	−
Acetone	−	−	−
Acetonitrile	−	−	−
Alconox™ (100%)	+	+	+
Aliphatic esters	−	−	−
Amines	+	+	−
Ammonium chloride (1.0%)	+	+	+
Ammonium fluoride (2.0%)	+	+	+
Ammonium hydroxide (5.0%)	+	S	−
Aromatic hydrocarbons	−	−	−
Aromatic esters	−	−	−
Benzene (100%)	−	−	−
Butanol (1.0%)	+	+	+
Butyl acetate (100%)	−	−	−
Butyl cellosolve (10%)	−	−	−
Calcium chloride (5.0%)	+	+	+
Chloroform (0.8%)	S	+	+
Citric acid (1.0%)	+	+	+
Cyclic ketones	−	−	−
Cyclohexanone	−	−	−
Decon-90 (1.0%)	+	+	
Dichlorobenzene (100%)	−	−	−
Diethanolamine (5.0%)	+	+	−
Dimethylacetimide (DMAC)	−	−	−
N,N dimethylformamide (1.0%)	−	−	−
Dimethyl sulfoxide	−	−	−
Disodium salt of EDTA (10.0%)	+	+	
Ethanol (25.0%)	+	+	
Ethers	−	+	−
2-Ethoxyemanol (1.0%)	+	+	+
Ethyl acetate (100%)	−	−	−
Ethylene dichloride (100%)	−	−	−
Formaldehyde (1.0%)	+	+	+
Formic acid (0.1%)	+	+	+
Formic acid (5.0%)	+	+	+

+, Acceptable; −, not recommended; S, may shorten life with extended use.

(Continued)

Chemical compatibility for cassette filters

Chemical	PES	RC	PVDF
Furfural	−	−	−
Gluteraldehyde (0.5%)	+	+	+
Glycerine (25.0%)	+	+	+
Guanidine HCl (6 M)	+	+	+
Halogenated hydrocarbons	−	−	−
HCl (0.01 N)	+	+	+
HCl (0.5 N)	+	+	+
Hydrogen peroxide (1.0%)	S	−	+
Isopar®	−	+	+
Isopropyl acetate (1.0%)	−	+	−
Isopropyl alcohol (5.0%)	+	+	+
Isopropyl alcohol (25%)	S	+	+
Kerosene		−	
Ketones (other)	−	−	−
Lactic acid (5.0%)	+	+	+
Mercaptoethanol (0.1 M)	+	+	+
Mercaptoethanol (1.0%)	S	+	+
Methyl alcohol (5.0%)	+	+	+
Methyl alcohol (25%)	+	+	+
Methylene chloride (1.0%)	−	−	−
MEK (1.0%)	−	+	−
Methyl isobutyl ketone (1.0%)	−	+	−
M-Pyrol (1.0%)	−	−	−
Nitric acid 1 N	S	−	+
Nitrobenzene (100%)	−	−	−
Nitroethane/Nitromethane	−	−	−
Oleic acid (5.0%)	+	+	+
Oxalic acid (1.0%)	+	+	+
Peracetic acid 500 ppm	+	+	+
Phosphate buffer (pH 8.2; 1.0 M)	+	+	+
Phosphoric acid (<10%)	+	+	+
Polar aromatics	−	−	−
Sodium azide (1.0%)	S	S	+
Sodium chloride (5.0%)	+	+	+
Sodium deoxycholate (5.0%)	+	+	+
Na dodecyl sulfate (0.001 M)	+	+	+
Na dodecyl sulfate (0.01 M)	+	+	+
NaOH (0.01 N)	+	+	−
NaOH (0.1 N)	+		−
NaOH (0.5 N)	S	−	−
Sodium hypochlorite	S	S	S
Sodium nitrite (1.0%)	+	+	+
Sodium thermerosal (0.1%)	+	+	+
Styrene monomer (100%)	−	−	−
Sulfamic acid (5.0%)	+	+	+
Sulphuric acid (1.0%)	+	+	+

+, Acceptable; −, not recommended; S, may shorten life with extended use.

Chemical compatibility for cassette filters

Chemical	PES	RC	PVDF
Terg-a-zyme® (1.0%)	+	+	+
THF (5.0%)	–	–	–
Toluene (1.0%)	–	–	–
Tris buffer (pH 8.2; 1.0 M)	+	+	+
Triton® X-100 (0.0002 M)	+	+	+
Triton® X-l00 (0.1%)	+	+	+
Urea 6M	+	+	+

+, Acceptable; –, not recommended; S, may shorten life with extended use.

A.2 Temperature correction factor

$$\text{NWP} = \frac{(\text{litres per hour volumetric permeate flow})(\text{TCF})}{(\text{TMP psid})(\text{membrane area m}^2)}$$

$\text{TMP} = ((P_{\text{feed}} + P_{\text{retentate}})/2 - P_{\text{permeate}})$ as the module average transmembrane pressure. TCF = (viscosity of water at measured temperature)/(viscosity of water @ 25°C) as temperature correction factor (see Appendix).

These units yield LMH/bar [L/m^2·h·bar]. Calculate:

R = permeate flow rate in L/h; P_{in} = feed pressure in bar; P_{out} = retentate pressure in bar; T = water temperatures in °C; P_{p} = permeate pressure (if non-zero) in bar; A = total filter area in m$_2$; F = temperature correction factor from Table A.1.

A.3 Nomenclature

A = m^2 area
F = feed (as subscript) or temperature correction factor (Section A.2)
J = LMH flux in permeate
J_{F} = LMH feed flux
NWP = LMH/psi normalized water permeability
P = permeate (as subscript)
Q = Lpm volumetric flowrate
R = retentate (as subscript)

A.4 Conversion factors and constants

Area: 1 m^2 = 10.76 ft^2, 1 in^2 = 6.45 cm^2
Concentration: 1 wt% = 10 g/L = 10,000 ppm
Flux (velocity): 1 LMH = 0.59 gal/ft^2/day = 2.78×10^{-5} cm/s = 1.67×10^{-3} cm/min
Length: 1 m = 3.28 ft = 10^2 cm = 10^3 mm = 10^6 μm = 10^9 nm = 10^{10} Å

Table A.1 Viscosity temperature correction factor calculated from CRC handbook data[1]

T (°F)	T (°C)	F	T (°F)	T (°C)	F	T (°F)	T (°C)	F
125.6	52	0.595	96.8	36	0.793	68.0	20	1.125
123.8	51	0.605	95.0	35	0.808	66.2	13	1.152
122.0	50	0.615	93.2	34	0.825	64.4	18	1.181
120.2	49	0.625	91.4	33	0.842	62.6	17	1.212
118.4	48	0.636	89.6	32	0.859	60.8	16	1.243
116.6	47	0.647	87.8	31	0.877	59.0	15	1.276
114.8	46	0.658	86.0	30	0.896	57.2	14	1.310
113.0	45	0.670	84.2	29	0.915	55.4	13	1.346
111.2	44	0.682	82.4	28	0.935	53.6	12	1.383
109.4	43	0.694	80.6	27	0.956	51.8	11	1.422
107.6	42	0.707	78.8	26	0.978	50.0	10	1.463
105.8	41	0.720	77.0	25	1.000	48.2	9	1.506
104.0	40	0.734	75.2	24	1.023	46.4	8	1.551
102.2	39	0.748	73.4	23	1.047	44.6	7	1.598
100.4	38	0.762	71.6	22	1.072	42.8	6	1.648
98.6	37	0.777	69.8	21	1.098	41.0	5	1.699

NWP temperature correction factor (FT) is based on water fluidity relative to 25°C (77°F) fluidity value $F = (\mu_{T\,°C}/\mu_{25°C})$ or $(\mu_{T\,°F}/\mu_{77°F})$.
Source: Courtesy of EMD Millipore Corp.

Pressure: 1 bar = 14.5 psi = 10^5 Pa = 10^5 N/m^2 = 10^6 g/cm/s^2 = 0.9869 atm = 7.501 mmHg = 33.48 ft H_2O
Temperature: °C = (°F-32)·5/9, °F = °C·9/5 + 32, °K = °C + 273
Viscosity: 1 poise = 100 cp= 1 g/cm-s = 10^{-3} Pa-s
Volume: 1 L = 10^3 mL = 10^3 cm^3 = 10^{-3} m^3 = 0.03534 ft^3 = 61 in^3 = 0.2642 gal
Volumetric flow: 1 Lpm = 0.2642 gal/min

Constants

$\pi \approx 3.1416$, $e \approx 2.7183$
Gas constant R = 1.98721 cal/gmol-°K = 82.06 cm^3-atm/gmol-°K

Reference

1. CRC Handbook of Chemistry and Physics. In: Lide DR, editor, 73rd ed., CRC Press, Boca Raton, FL; 1992, 6-166.

Index

CPI Antony Rowe
Eastbourne, UK
March 11, 2019

Differential Geometry and Relativity Theory

An Introduction

Richard L. Faber

Department of Mathematics
Boston College
Chestnut Hill, Massachusetts

MARCEL DEKKER, INC. New York and Basel

Library of Congress Cataloging in Publication Data

Faber, Richard L.
Differential geometry and relativity theory.

(Monographs and textbooks in prue and applied mathematics : v. 75)
Bibliography: p.
Includes index.
1. Relativity (Physics) 2. Geometry, Differential.
3. Geometry, Riemannian. I. Title. II. Series.
QC173.55.F33 1983 530.1'1'0151636 82-19938
ISBN 0-8247-1749-X

MARCEL DEKKER, INC.
270 Madison Avenue, New York, New York 10016

Current printing (last digit):
10 9 8 7 6 5 4 3 2

PRINTED IN THE UNITED STATES OF AMERICA

To Susan and Lynn

Preface

This text is an introduction to differential and Riemannian geometry and to the rudiments of special and general relativity. It is the author's hope that, for perhaps the first time, the general theory of relativity is herein made accessible to mathematics majors whose background includes only multivariable calculus and linear algebra.

Chapter I is in itself a mini-course in differential geometry. Topics include curves in the plane and 3-space, the first and second fundamental forms of surfaces in 3-space, curvature and geodesics, the *Theorema Egregium* of Gauss, and abstract surfaces and manifolds.

Chapter II presents the physical foundations of special relativity and its geometric interpretation as the geometry of flat spacetime. Here the emphasis is on coordinates, the Lorentz metric, and spacetime diagrams.

Chapter III begins with an intuitive and non-technical account of the key ideas of Einstein's general theory of relativity. Following this is a more mathematical treatment using the tools developed in Chapter I. The requirement that general relativity should give close agreement with classical Newtonian physics is used to explain how Einstein arrived at his field equations. The Schwarzschild solution is derived, and the orbital equations are deduced, both for classical physics and general relativity. Additional consequences of general relativity, such as the "bending" of light, the gravitational redshift and perihelial precession, are also discussed.

Although all of the details are worked out, the reader is required only to have mastered the mathematical machinery set out in the first chapter for two-dimensional surfaces. Indeed the calculation of relativistic orbits is seen as a higher dimensional version of the calculation of geodesics in the plane in terms of polar coordinates.

The point of view of the treatment is that relativity is a geometric theory of space and time, in which gravity is not a force, as Newton conceived it, but rather a manifestation of the curvature of (4-dimensional) spacetime. It is the distribution of matter throughout the universe that determines this curvature, which in turn defines the geodesics along which both matter and light are constrained to move.

Abundant exercises appear at the end of nearly all sections. These are indispensable for the student's progress through the book. Most of the exercises are designed to give the student experience in applying the concepts presented within the text; others call for verification of basic facts used in the text; some give additional details or alternative explanations of the more technical matters; and some introduce optional new material.

For students of average preparation and ability, there is probably enough material here to fill three quarters or even an entire year, especially if a majority of the exercises are to be assigned and discussed in class. On the other hand, with some trimming, I have been able to teach most of the text to better prepared classes in one semester. I have also found Chapter I useful for independent reading projects.

Acknowledgments

First and foremost, I would like to thank my beautiful wife Susan, who typed much of the manuscript, for abundant patience, understanding and love. I wish also to thank the following mathematicians, physicists, and historians of science for their helpful comments and suggestions: Jay P. Fillmore, Thomas Hawkins, Meyer Jordan, John Kenelly, Bennett Kivel, Edwin Taylor, Thomas Tucker, and R. L. Wilder.

I am indebted also to my students for many text improvements which resulted from their questions and criticisms.

Contents

I

Surfaces and the Concept of Curvature

A single curve, drawn in the manner of the curve of prices of cotton, describes all that the ear can possibly hear as the result of the most complicated musical performance . . . That to my mind is a wonderful proof of the potency of mathematics.
—Lord Kelvin

INTRODUCTION

A pivotal concept to be developed in this book is that of *curvature*, the eventual goal being to present Einstein's theory of relativity as a geometric theory of physical space and time, in which gravitation is a manifestation of the curvature of spacetime. The concept of curvature is more complicated in higher dimensions, and so we shall begin with a discussion of the curvature of one- and two-dimensional objects: curves and surfaces in Euclidean 3-space, E^3. Vector notation and standard facts from vector analysis are reviewed in Appendix A. You should look at these briefly before going on.

1. CURVES

A (*smooth*) *curve* in E^3 is given by a vector-valued function of a real variable or *parameter* t defined on some interval [a,b] :

$$\boldsymbol{\alpha}(t) = (x(t),y(t),z(t))$$

For convenience, we assume the coordinate functions x, y, and z have continuous second derivatives. Let $s = s(t)$ denote arc length along the curve from the initial point $\boldsymbol{\alpha}(a)$ to the point $\boldsymbol{\alpha}(t)$. Thus $s(a) = 0$ and $s(b) =$ the total length of the curve.

The *derivative vector* of the curve $\boldsymbol{\alpha}$ is the vector

$$\begin{aligned}\boldsymbol{\alpha}'(t) &= \lim_{\Delta t \to 0} \frac{1}{\Delta t} [\boldsymbol{\alpha}(t + \Delta t) - \boldsymbol{\alpha}(t)] \\ &= \lim_{\Delta t \to 0} \left[\frac{x(t+\Delta t) - x(t)}{\Delta t}, \frac{y(t+\Delta t) - y(t)}{\Delta t}, \frac{z(t+\Delta t) - z(t)}{\Delta t}\right] \\ &= (x'(t), y'(t), z'(t))\end{aligned}$$

$\boldsymbol{\alpha}'(t)$, if drawn as an arrow emanating from the point $\boldsymbol{\alpha}(t)$, is tangent to the curve and points in the direction of increasing t. To see this, notice that in Figure I-1, the vector $\Delta\boldsymbol{\alpha} = \boldsymbol{\alpha}(t + \Delta t) - \boldsymbol{\alpha}(t)$ is directed along the chord from $\boldsymbol{\alpha}(t)$ to $\boldsymbol{\alpha}(t + \Delta t)$. Multiplying $\Delta\boldsymbol{\alpha}$ by the scalar $1/\Delta t$ (for $\Delta t \neq 0$) will result in a vector pointing in the same direction if $\Delta t > 0$, or in the opposite direction if $\Delta t < 0$. In either case, $\Delta\boldsymbol{\alpha}/\Delta t$ points in the direction of increasing t, and hence increasing s. As $\Delta t \to 0$, $\boldsymbol{\alpha}(t + \Delta t)$ approaches $\boldsymbol{\alpha}(t)$, and so the direction of the chord (and of $\Delta\boldsymbol{\alpha}/\Delta t$) approaches that of the tangent line to the curve at $\boldsymbol{\alpha}(t)$.

Furthermore, for Δt small, the length of $\Delta\boldsymbol{\alpha}$, i.e., the straight line distance from $\boldsymbol{\alpha}(t)$ to $\boldsymbol{\alpha}(t + \Delta t)$, is a good approximation to the arc length $|\Delta s|$ between these points. If we interpret t as time, and think of the curve as the path of a moving point, then $\|(\boldsymbol{\alpha}(t + \Delta t) - \boldsymbol{\alpha}(t))/\Delta t\|$ approximates the average speed $\Delta s/\Delta t$ of the moving point

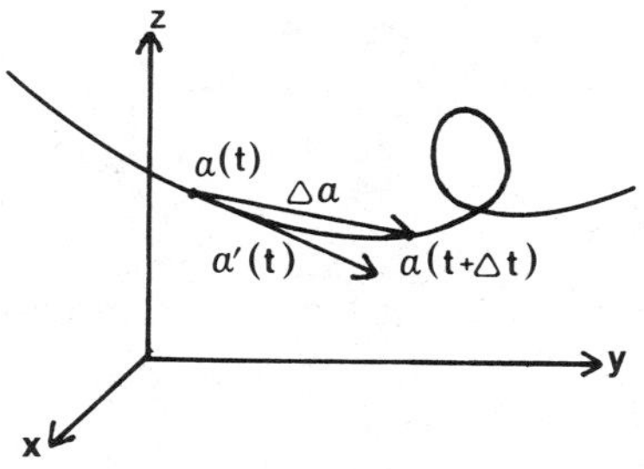

Figure I - 1.

in traversing the arc from $\boldsymbol{\alpha}(t)$ to $\boldsymbol{\alpha}(t + \Delta t)$. In the limit as $\Delta t \to 0$, $\|\boldsymbol{\alpha}'(t)\|$ is the instantaneous *speed*, ds/dt, at time t. In this context, $\boldsymbol{\alpha}'(t)$ is often called the *velocity vector.* Since $ds/dt = \|\boldsymbol{\alpha}'(t)\|$, we have

$$s(t) = \int_a^t \|\boldsymbol{\alpha}'(u)\| du$$

and the length L of the entire curve $\boldsymbol{\alpha}(t)$, $a \leqslant t \leqslant b$, is obtained by integrating the speed over the interval [a,b].

All of the above applies as well to plane curves, i.e., to curves in E^2, except that there all vectors have two components: $\boldsymbol{\alpha}(t) = (x(t), y(t))$.

Example 1. Let

$$\boldsymbol{\alpha}(t) = (r \cos t, r \sin t),\ 0 \leqslant t \leqslant 2\pi,$$

where r is a positive constant. Since $x^2 + y^2 = r^2$, the image of α is a circle of radius r centered at the origin of E^2. The velocity vector,

$$\boldsymbol{\alpha}'(t) = (-r \sin t, r \cos t),$$

is perpendicular to $\boldsymbol{\alpha}(t)$—because the dot product $\boldsymbol{\alpha}(t) \cdot \boldsymbol{\alpha}'(t)$ is zero—and this expresses the familiar fact that a tangent line to a circle is perpendicular to the radius drawn to the point of tangency (see Fig. I-2). [In the previous sentence, picture $\boldsymbol{\alpha}(t)$ as the arrow from the

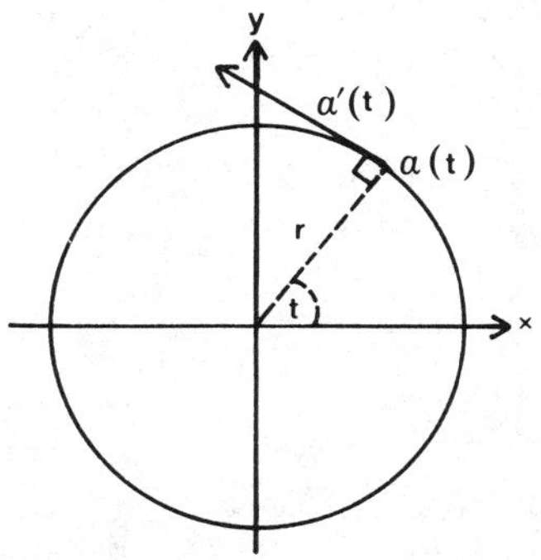

Figure I - 2.

origin to the point $\boldsymbol{\alpha}(t)$.] The speed is $\|\boldsymbol{\alpha}'(t)\| = r$, and so α represents motion in a circle at constant speed [t can be interpreted either as time or as the angle between the vectors $\boldsymbol{\alpha}(0)$ and $\boldsymbol{\alpha}(t)$, in radians]. The length of $\boldsymbol{\alpha}$ is

$$L = \int_0^{2\pi} \|\boldsymbol{\alpha}'(t)\|\, dt = \int_0^{2\pi} r\, dt = 2\pi r$$

the circumference of a circle of radius r.

Example 2. The graph of a smooth function f(x), $a \leqslant x \leqslant b$, may be parametrized as the curve

$$\boldsymbol{\alpha}(t) = (t, f(t)), \quad a \leqslant t \leqslant b$$

The velocity vector is $\boldsymbol{\alpha}'(t) = (1, f'(t))$, and so the length of this curve is

$$L = \int_a^b [1 + f'(t)^2]^{1/2}\, dt$$

The derivative of the velocity vector,

$$\boldsymbol{\alpha}''(t) = (x''(t), y''(t), z''(t))$$

is called the *acceleration vector*. The direction and magnitude of $\boldsymbol{\alpha}''(t)$ are not always easy to tell at a glance, but often we can gain some intuition by imagining the force that would be required to make a material particle of mass m traverse the curve $\boldsymbol{\alpha}$. According to Newton's Second Law of Motion, the force $\mathbf{F}(t)$, acting on a particle when it is at the point $\boldsymbol{\alpha}(t)$, satisfies $\mathbf{F}(t) = m\boldsymbol{\alpha}''(t)$. Thus $\boldsymbol{\alpha}''(t)$ has direction and magnitude proportional to the force needed to keep the particle "on the track."

In Example 1, the acceleration $\boldsymbol{\alpha}''(t) = (-r \cos t, -r \sin t) = -\boldsymbol{\alpha}(t)$ is directed toward the center and has constant magnitude. In other words, an object will be maintained in uniform circular motion if acted upon by a center-directed force of constant magnitude. In Exercise 3, you will verify that a ball in flight along a parabolic path undergoes acceleration that is directed vertically downward and has constant magnitude, since the force of gravity (the ball's weight) has those properties.

In general, many different curves can have the same image. For example,

$$\begin{aligned} \alpha(t) &= (t, t^2), && 0 \leqslant t \leqslant 2 \\ \beta(t) &= (2t, 4t^2), && 0 \leqslant t \leqslant 1 \\ \gamma(t) &= (t^{1/3}, t^{2/3}), && 0 \leqslant t \leqslant 8 \end{aligned}$$

all have the same portion of the parabola $y = x^2$ as image curve. They are, as we say, different parametrizations of the parabola. Often, it is convenient to choose, as a standard parameter, the arc length s, along the curve, measured from the initial point. Under this choice, $t = s$, $ds/dt = 1$, and so $\boldsymbol{\alpha}'(s)$ is a *unit tangent vector*, which will be denoted $\mathbf{T}(s)$:

$$\mathbf{T}(s) = \boldsymbol{\alpha}'(s)$$

[We then sometimes speak of $\boldsymbol{\alpha}(s)$ as a *unit speed curve.*]

For example, in the case of the circle $\boldsymbol{\beta}(t) = (r \cos t, r \sin t)$, $0 \leqslant t \leqslant 2\pi$, $ds/dt = \|\boldsymbol{\beta}'(t)\| = r$, and so $s = rt$. We may therefore replace t by s/r and reparametrize the circle in terms of arc length:

$$\boldsymbol{\alpha}(s) = \boldsymbol{\beta}(s/r) = \left(r \cos \frac{s}{r}, r \sin \frac{s}{r}\right), \quad 0 \leqslant s \leqslant 2\pi r$$

Returning now to the general case, let $\boldsymbol{\alpha}(s)$, $a \leqslant s \leqslant b$, be any smooth curve parametrized in terms of arc length. At each point $\boldsymbol{\alpha}(s)$ of the curve, $\mathbf{T}(s)$ is the unit tangent vector pointing in the direction of increasing s. Since the vector-valued function $\mathbf{T}(s)$ has constant length, only its direction changes, and the rate of this change (or turning) is a measure of the curvature of $\boldsymbol{\alpha}$.

Specifically, $\mathbf{T}(s) \cdot \mathbf{T}(s) = 1$ implies $\mathbf{T}(s) \cdot \mathbf{T}'(s) = 0$ (just differentiate by the product rule), so that the acceleration vector $\boldsymbol{\alpha}''(s) = \mathbf{T}'(s)$ (also often called the *curvature vector*) is orthogonal to $\mathbf{T}$. The faster $\mathbf{T}$ turns, the larger the components of $\mathbf{T}'$, and the larger $\|\mathbf{T}'\|$.

Definition I-1

The *curvature* of $\boldsymbol{\alpha}$ at $\boldsymbol{\alpha}(s)$, denoted k(s), is the length of $\mathbf{T}'(s)$:

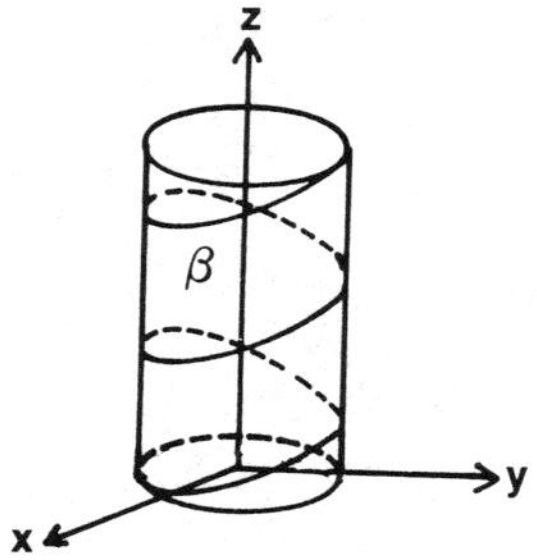

Figure I - 3.

$$k(s) = \|\mathbf{T}'(s)\| = \|\boldsymbol{\alpha}''(s)\|$$

If $\mathbf{T}'(s) \neq 0$, the unit vector in the direction of $\mathbf{T}'(s)$ is called the *principal normal vector*, denoted $\mathbf{N}(s)$:

$$\mathbf{N}(s) = \mathbf{T}'(s)/\|\mathbf{T}'(s)\| = \mathbf{T}'(s)/k(s)$$

Thus,

$$\mathbf{T}'(s) = k(s)\mathbf{N}(s)$$

Example 3. A circular *helix* (Fig. I-3) is given by the vector equation

$$\boldsymbol{\beta}(t) = (a \cos t, a \sin t, bt)$$

with a, b constant. (If $b \neq 0$, this curve resembles a coil spring contained in a cylinder of radius a.) Since $\boldsymbol{\beta}'(t) = (-a \sin t, a \cos t, b)$, $ds/dt = \|\boldsymbol{\beta}'(t)\| = (a^2 + b^2)^{1/2}$, and so $s = t(a^2 + b^2)^{1/2}$ plus a constant, which we may take as zero if we assume $s = 0$ when $t = 0$. We can therefore reparametrize the helix in terms of arc length as

$$\boldsymbol{\alpha}(s) = \left(a \cos \frac{s}{\sqrt{(a^2+b^2)}}, a \sin \frac{s}{\sqrt{(a^2+b^2)}}, \frac{bs}{\sqrt{(a^2+b^2)}}\right)$$

Then

$$\mathbf{T}(s) = \boldsymbol{\alpha}'(s) = \frac{1}{\sqrt{(a^2+b^2)}}\left(-a \sin \frac{s}{\sqrt{(a^2+b^2)}},\ a \cos \frac{s}{\sqrt{(a^2+b^2)}},\ b\right)$$

which is a unit vector, and

$$\mathbf{T}'(s) = k(s)\mathbf{N}(s) = \frac{1}{a^2+b^2}\left(-a \cos \frac{s}{\sqrt{(a^2+b^2)}},\ -a \sin \frac{s}{\sqrt{(a^2+b^2)}},\ 0\right)$$

Thus

$$k(s) = a/(a^2 + b^2)$$

and

$$\mathbf{N}(s) = \left(-\cos \frac{s}{\sqrt{(a^2+b^2)}},\ -\sin \frac{s}{\sqrt{(a^2+b^2)}},\ 0\right)$$

The helix, then, has constant curvature. In the special case $a > 0$ and $b = 0$, the curve is a circle in the xy-plane and $k = 1/a$, the reciprocal of the radius. Thus, the smaller the radius, the greater the circle's curvature, as intuition suggests (see Fig. I-4). If $a = 0$ and $b \neq 0$, the curve is a straight line, $k(s) = 0$, and $\mathbf{N}(s)$ is undefined.

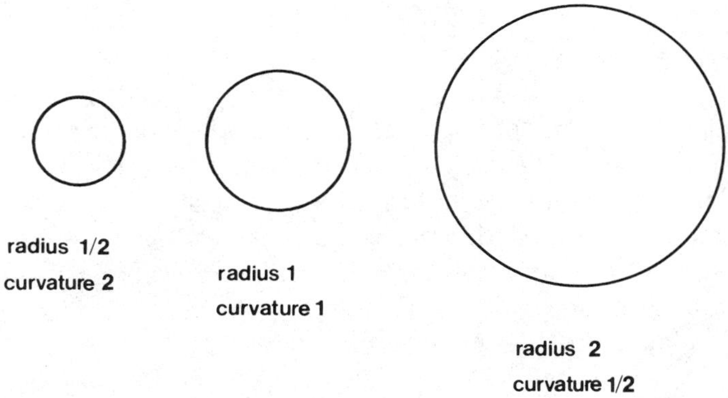

Figure I - 4.

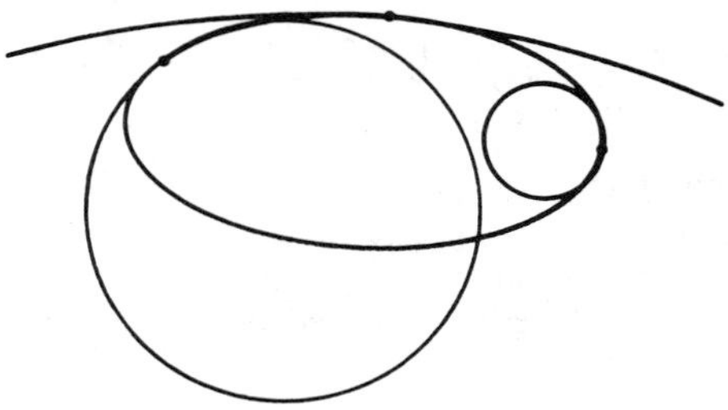

Figure I - 5.

For an arbitrary curve $\boldsymbol{\alpha}$, the curvature k(s) is in general not constant but varies from point to point. However, where $k(s) \neq 0$, the curve is closely approximated by the circle of radius $1/k(s)$ which is tangent to $\boldsymbol{\alpha}$ at $\boldsymbol{\alpha}(s)$ and which lies in the plane of $\mathbf{T}(s)$ and $\mathbf{N}(s)$. This circle, called the *osculating* circle*, has the same tangent vector, curvature, and principal normal vector at $\boldsymbol{\alpha}(s)$ as does the curve $\boldsymbol{\alpha}$ itself. It is the circle which best approximates the curve at the point of tangency. Figure I-5 shows the osculating circle at several points of an ellipse. You will be able to compute their centers and radii in the exercises.

The center of the osculating circle, called the *center of curvature* of $\boldsymbol{\alpha}$ at $\boldsymbol{\alpha}(s)$, is obtained by going a distance $1/k(s)$ away from $\boldsymbol{\alpha}(s)$ in the direction of $\mathbf{N}(s)$. Thus, the center of curvature at $\boldsymbol{\alpha}(s)$ is the point

$$c(s) = \boldsymbol{\alpha}(s) + [1/k(s)]\,\mathbf{N}(s)$$

The plane of the osculating circle, i.e., the plane determined by $\mathbf{T}(s)$ and $\mathbf{N}(s)$, is called the *osculating plane*.

To gain a complete description of a space curve's gyrations, we need also a knowledge of how the osculating plane turns or tilts as we move along the curve. This information is furnished by the magnitude of $\mathbf{B}'$, where $\mathbf{B} = \mathbf{T} \times \mathbf{N}$ is the unit normal to this plane, the so-called *binormal vector*† (see Fig. I-6).

---

*From the latin *osculari,* to kiss. In geometry, the word signifies a close touching, as of curves tangent to one another.

†To save space, we shall often forego affixing (s) to the name of a vector. It is always to be understood that **N**, **T**, **B**, etc., are vector-valued *functions* of s.

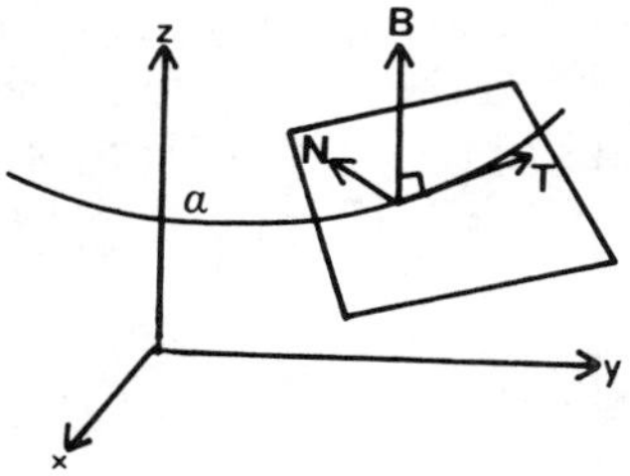

Figure I - 6.

$\mathbf{B}' = (\mathbf{T} \times \mathbf{N})' = \mathbf{T}' \times \mathbf{N} + \mathbf{T} \times \mathbf{N}' = \mathbf{T} \times \mathbf{N}'$, since $\mathbf{T}' = k\mathbf{N}$. Since $\mathbf{N}$ (like $\mathbf{B}'$) is orthogonal to both $\mathbf{T}$ and $\mathbf{N}'$, $\mathbf{B}'$ is a multiple of $\mathbf{N}$. The *torsion* of $\boldsymbol{\alpha}$ is the function $\tau = \tau(s)$ defined by the equation

$$\mathbf{B}' = -\tau \mathbf{N}$$

The function τ measures the rate of turning of the osculating plane, and hence to what extent $\boldsymbol{\alpha}$ departs from being a plane curve. For a curve which is planar, the osculating plane is the plane in which the curve lies, $\mathbf{B}$ has constant components, and so $\tau = 0$.

In the example of the helix, computation of the cross product gives

$$\mathbf{B} = \mathbf{T} \times \mathbf{N} = \frac{1}{\sqrt{(a^2+b^2)}}\left(b \sin \frac{s}{\sqrt{(a^2+b^2)}}, -b \cos \frac{s}{\sqrt{(a^2+b^2)}}, a\right)$$

Since

$$\mathbf{B}' = \frac{b}{a^2+b^2}\left(\cos \frac{s}{\sqrt{(a^2+b^2)}}, \sin \frac{s}{\sqrt{(a^2+b^2)}}, 0\right) = -\tau \mathbf{N}$$

we obtain $\tau = b/(a^2 + b^2)$, a constant.

The fundamental theorem of the differential geometry of curves in E^3 states that a curve is completely determined by its curvature and torsion functions. Specifically, suppose $k(s)$ and $\tau(s)$ are given functions defined on some interval I, with k positive and continuously differentiable and τ continuous. Then there exists a curve $\boldsymbol{\alpha}(s)$ defined on I for which s is arc length and k and τ are, respectively, the curvature and torsion. Moreover, any two such curves are congruent.

Consequently, if we in addition specify the values of $\boldsymbol{\alpha}$, **T**, **N**, and **B** at any one point s of I, then the curve is uniquely determined. (For proofs, consult Laugwitz [31, pp. 15-17], Stoker [45, pp. 65-67], Struik [46, pp. 29-31], O'Neill [37, pp. 117-118].)

Exercises I-1

1. Compute the length of one turn of the helix

$$\boldsymbol{\beta}(t) = (a \cos t, a \sin t, bt).$$

2. The parabola $y = x^2/2$ may be parameterized as the curve $\boldsymbol{\alpha}(t) = (t, t^2/2)$. Sketch this curve, as well as its velocity and acceleration vectors at $t = -1, 0, 1$, and 2.

3. A ball is thrown with initial speed v_0 and angle of elevation θ relative to the horizontal. Let coordinate axes be set up as in Figure I-7. While in flight, the ball follows the curve

$$\boldsymbol{\alpha}(t) = (v_0 t \cos\theta, v_0 t \sin\theta - 16t^2)$$

(a) Compute the velocity vector and verify that $\| \boldsymbol{\alpha}(0) \| = v_0$.
(b) Compute the acceleration vector. What is its direction and magnitude?

4. A particle of matter moves along the curve

$$\boldsymbol{\alpha}(t) = (a \cosh t, a \sinh t)$$

where a is a positive constant. (Since $x^2 - y^2 = a^2$ and $x > 0$, the path is one branch of a hyperbola.) Show that the force on the particle is directed away from the origin (i.e., is proportional to the

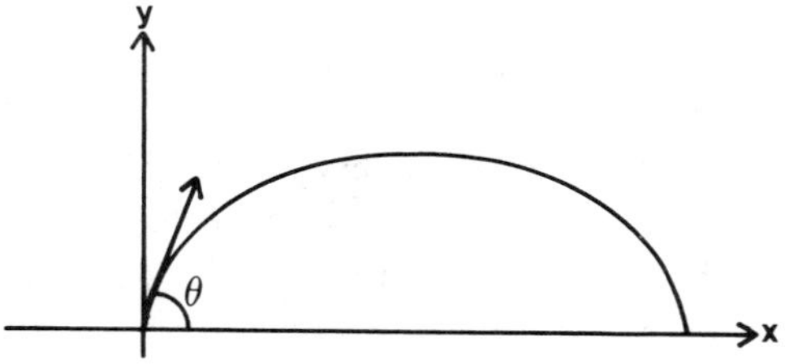

Figure I - 7.

vector $\boldsymbol{\alpha}$ itself, with positive proportionality factor), and that its magnitude is proportional to the particle's distance from the origin. (See Appendix B for the definitions and properties of the hyperbolic functions *cosh* and *sinh*.)

5. In many cases, it is difficult or impossible to reparametrize a curve in terms of arc length. For the parabola $\boldsymbol{\alpha}(t) = (t, t^2/2)$, $0 \leqslant t \leqslant 1$, $ds/dt = \|\boldsymbol{\alpha}'(t)\| = (1 + t^2)^{1/2}$, and so $s = \int_0^t (1 + u^2)^{1/2}\,du = \frac{t}{2}(1 + t^2)^{1/2} + \frac{1}{2}\ln\left[t + (1 + t^2)^{1/2}\right]$ (from a table of integrals). We are unable in this case to solve for t as a function of s and reexpress $\boldsymbol{\alpha}$ in terms of s: $\boldsymbol{\alpha}(t(s))$. To compute the curvature, we must therefore proceed another way, using the chain rule to compute derivatives with respect to s.

$$\mathbf{T} = \frac{d\alpha}{ds} = \frac{d\alpha/dt}{ds/dt} = \frac{1}{(1 + t^2)^{1/2}}(1,t) = \left(\frac{1}{(1 + t^2)^{1/2}}, \frac{t}{(1 + t^2)^{1/2}}\right)$$

$$\frac{d\mathbf{T}}{ds} = \frac{1}{ds/dt}\,\frac{d\mathbf{T}}{dt} = \frac{1}{(1 + t^2)^{1/2}}\left(\frac{-t}{(1 + t^2)^{3/2}}, \frac{1}{(1 + t^2)^{3/2}}\right)$$

$$= \frac{1}{(1 + t^2)^2}(-t, 1)$$

$$k(t) = \left\|\frac{d\mathbf{T}}{ds}\right\| = 1/(1 + t^2)^{3/2}$$

(a) Find the equation of the curve's osculating circle at $\boldsymbol{\alpha}(0)$. (b) Find the equation of the curve's osculating circle at $\boldsymbol{\alpha}(2)$. [Hint: its center lies on the normal line to the curve, at a distance $1/k(2)$ from $\boldsymbol{\alpha}(2)$.]

6. (a) By mimicking the derivation in Exercise 5, show that the plane curve $\boldsymbol{\alpha}(t) = (x(t), y(t))$ has curvature

$$k(t) = \left|\frac{x'(t)y''(t) - x''(t)y'(t)}{(x'(t)^2 + y'(t)^2)^{3/2}}\right|$$

at $\boldsymbol{\alpha}(t)$. (b) As a special case, show that the graph of $y = f(x)$ has curvature

$$k(x) = \left| \frac{f''(x)}{(1 + f'(x)^2)^{3/2}} \right|$$

at (x, f(x)). (c) Using the latter formula, compute the curvature of the semicircle $y = (a^2 - x^2)^{1/2}$.

7. Let $\boldsymbol{\alpha}(t) = (a \cos t, b \sin t)$, $0 \leqslant t \leqslant 2\pi$. Since $x^2/a^2 + y^2/b^2 = 1$, the image of $\boldsymbol{\alpha}$ is an ellipse. (a) Compute its curvature k(t) by the formula of Exercise 6(a), at t = 0 and t = $\pi/2$. (b) Sketch the ellipse $x^2/4 + y^2 = 1$ and its osculating circles at the points (2,0) and (0,1).

8. Let $\boldsymbol{\alpha}(t) = (x(t), y(t))$ be a plane curve, and let θ be the angle between the positive x direction and $\boldsymbol{\alpha}'(t)$, measured counterclockwise from the positive x direction (Fig. I-8). Then

$$\theta = \tan^{-1} \frac{dy}{dx} = \tan^{-1} \left(\frac{dy/dt}{dx/dt} \right).$$

Show that $k(t) = \left| \frac{d\theta}{ds} \right|$, where s = arc length along $\boldsymbol{\alpha}$. [Hint: $d\theta/ds = \frac{d\theta/dt}{ds/dt}$; cf Exercise 6(a).]

9. Let $\boldsymbol{\alpha}(t)$ be a smooth curve in E^3, where t is an arbitrary parameter. Let $v(t) = ds/dt$, the speed at parameter value t. Then

$$\boldsymbol{\alpha}'(t) = \frac{d\boldsymbol{\alpha}}{ds} \frac{ds}{dt} = v\mathbf{T}$$

and

$$\mathbf{T}'(t) = \frac{d\mathbf{T}}{ds} \frac{ds}{dt} = kv\mathbf{N}$$

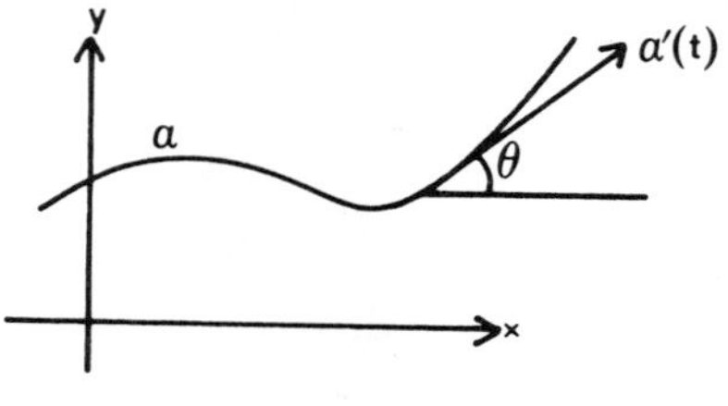

Figure I - 8.

(see Definition I-1). Show that

$$k = \|\boldsymbol{\alpha}' \times \boldsymbol{\alpha}''\| / \|\boldsymbol{\alpha}'\|^3$$

(primes signify differentiation with respect to t here).

10. Recompute the curvature of the helix

$$\boldsymbol{\beta}(t) = (a \cos t, a \sin t, bt)$$

of Example 3 by means of the formula derived in Exercise 9.

11. Compute the curvature k(t) for

(a) $\boldsymbol{\alpha}(t) = (3t - t^3, 3t^2, 3t + t^3)$
(b) $\boldsymbol{\alpha}(t) = (\cos t, \sin t, \cos t)$
(c) $\boldsymbol{\alpha}(t) = (\cos t, \sin t, e^t)$

12. Find the center of curvature, **c**(s), of the helix of Example 3,

$$\boldsymbol{\alpha}(s) = \left(a \cos \frac{s}{\sqrt{(a^2+b^2)}}, a \sin \frac{s}{\sqrt{(a^2+b^2)}}, \frac{bs}{\sqrt{(a^2+b^2)}}\right)$$

and verify that **c**(s) is itself a helix (s is arc length).

13. Let α be the unit speed curve

$$\boldsymbol{\alpha}(s) = \left(\frac{3}{10} \cos 2s, \frac{2}{5} \cos 2s, \frac{1}{2} \sin 2s - 1\right)$$

(a) Find **T**(s), **N**(s), k(s), **B**(s), and τ(s). (b) It can be shown (see below) that a curve with zero torsion at every point lies in a plane. Assuming this fact, prove that $\boldsymbol{\alpha}$ is a circle and find its radius and center.

[If $\tau \equiv 0$, then **B** has constant components. We can then show that $\boldsymbol{\alpha}$ lies in the plane through $\boldsymbol{\alpha}(0)$ with normal vector **B** (Fig. I-9). Since this plane consists of all points **P** satisfying $(\mathbf{P} - \boldsymbol{\alpha}(0)) \cdot \mathbf{B} = 0$, it suffices to prove that $(\boldsymbol{\alpha}(s) - \boldsymbol{\alpha}(0)) \cdot \mathbf{B} = 0$. Consider the function $f(s) = (\boldsymbol{\alpha}(s) - \boldsymbol{\alpha}(0)) \cdot \mathbf{B}$. This function has derivative $f'(s) = \boldsymbol{\alpha}'(s) \cdot \mathbf{B} = \mathbf{T}(s) \cdot \mathbf{B} = 0$ and so is constant, with value $f(0) = 0$.]

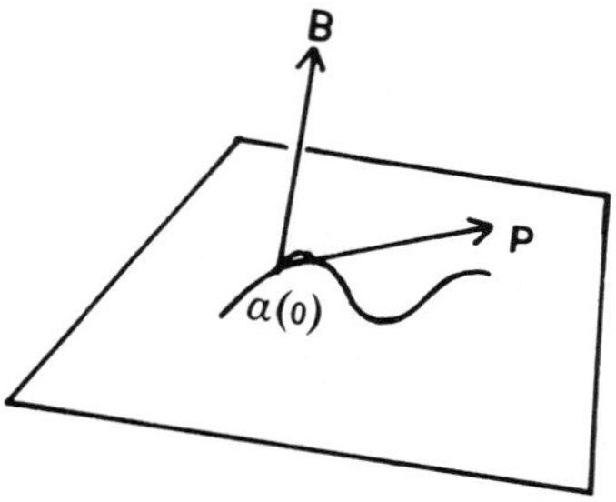

Figure I - 9.

14. Let $\boldsymbol{\alpha} = \boldsymbol{\alpha}(s)$ be a unit speed curve in E^3. The formulas

$$\begin{aligned} \mathbf{T}' &= \quad k\mathbf{N} \\ \mathbf{N}' &= -k\mathbf{T} \qquad + \tau\mathbf{B} \\ \mathbf{B}' &= \quad -\tau\mathbf{N} \end{aligned}$$

are called the *formulas of Frenet*, or the *Serret-Frenet* formulas, after F. Frenet, who first obtained them in his 1847 dissertation, and J. A. Serret, who first published them in 1851. The first and third formulas are essentially the definitions of curvature and torsion. Prove the second. (Hint: Any vector can be written as a linear combination of **T**, **N**, and **B**.)

15. (a) Prove that if all tangent lines to a curve $\boldsymbol{\alpha}$ pass through a fixed point **P**, then $\boldsymbol{\alpha}$ is a straight line. [Hint: we can write $\mathbf{P} = \boldsymbol{\alpha}(s) + \lambda(s)\mathbf{T}(s)$, for some function λ. Differentiate and use the Frenet formulas.]

(b) Assume $\boldsymbol{\alpha}$ has non-zero curvature at each point. Prove that if all osculating planes of $\boldsymbol{\alpha}$ pass through a fixed point **P**, then $\boldsymbol{\alpha}$ lies in a plane. (Hint: We can write $\mathbf{P} = \boldsymbol{\alpha} + \lambda\mathbf{T} + \mu\mathbf{N}$.)

2. GAUSS CURVATURE (INFORMAL TREATMENT)

Let us now consider how we might describe the curvature of a 2-dimensional object: a surface. Let **U** be a unit vector normal to a smooth surface $M \subset E^3$ at an arbitrary point **P** of M. (We begin informally, leaving the details, including a precise definition of the word "surface," for later.) If **v** is any unit vector tangent to M at **P**, then the plane through **P** determined by **v** and **U** intersects the

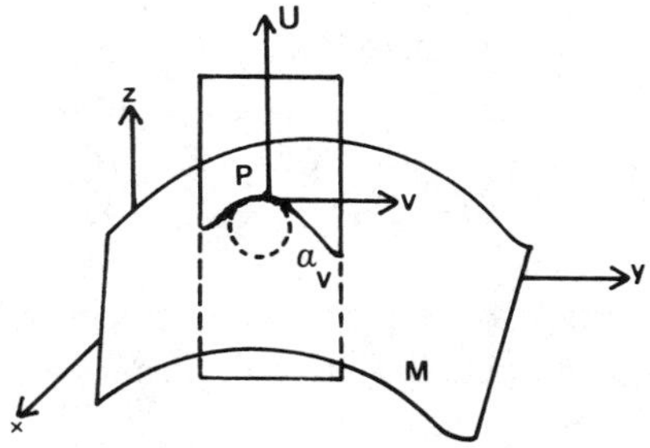

Figure I - 10.

surface in a curve, which we will denote $\boldsymbol{\alpha}_{\mathbf{v}}$, to indicate its dependence on **v** (**P** and **U** are fixed). $\boldsymbol{\alpha}_{\mathbf{v}}$ is called the *normal section* of M at P *in the* **v** *direction* (Fig. I-10).

For example, if M is a plane in E^3 (Fig. 1-11a), then normal sections of M at **P** are lines through **P**. If M is a sphere (Fig. I-11b), the normal sections at **P** are the great circles through **P**.

Example 4. Figure I-11c shows a right circular cylinder M of radius R and an arbitrary point **P** on M. **U** is the inward-pointing unit normal vector at P. $\mathbf{v}_2$ is a unit tangent vector in the direction of the cylinder's element passing through **P**, and $\mathbf{v}_1$ is a unit tangent vector perpendicular to $\mathbf{v}_2$. Then $\boldsymbol{\alpha}_{\mathbf{v}_1}$ is a circle of radius R, and $\boldsymbol{\alpha}_{\mathbf{v}_2}$ is a straight line. (Actually, the plane of **U** and $\mathbf{v}_2$ cuts M in a

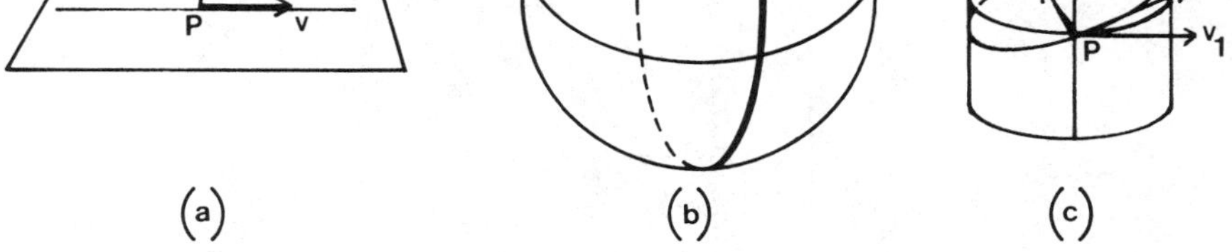

(a) (b) (c)

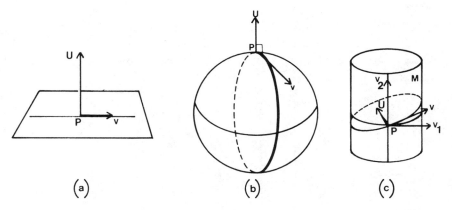

Figure I - 11.

pair of parallel straight lines, but we are interested here only in the component of the intersection that contains **P**.) If **v** is any other unit vector tangent to M at **P**, then $\boldsymbol{\alpha}_{\mathbf{v}}$ is an ellipse, and it is not too difficult to see that **P** is one end point of this ellipse's minor axis.

In general, each normal section $\boldsymbol{\alpha}_{\mathbf{v}}$ is (if $k \neq 0$ at **P**) approximated near **P** by its osculating circle at **P**, whose radius we shall call $R(\mathbf{v})$. The curvature of this osculating circle, $1/R(\mathbf{v})$, taken with an appropriate sign, is called the *normal curvature* of M at **P** *in the* **v** *direction* and is denoted $k_n(\mathbf{v})$:

$$k_n(\mathbf{v}) = \pm \frac{1}{R(\mathbf{v})}$$

The sign is taken to be positive or negative, respectively, according as the principal normal vector of $\boldsymbol{\alpha}_{\mathbf{v}}$ has direction the same as, or opposite to, that of **U**, the surface normal. If $\boldsymbol{\alpha}_{\mathbf{v}}$ has zero curvature at **P**, then we leave $R(\mathbf{v})$ undefined—or call it infinite—and take $k_n(\mathbf{v}) = 0$. A more formal definition of $k_n(\mathbf{v})$ will be given in Section 5.

In Example 4 (the cylinder), the arbitrary unit tangent vector **v** is a linear combination of the orthogonal unit vectors $\mathbf{v}_1$ and $\mathbf{v}_2$ and in fact may be written

$$\mathbf{v} = (\cos\theta)\,\mathbf{v}_1 + (\sin\theta)\,\mathbf{v}_2$$

where θ is the angle between $\mathbf{v}_1$ and **v**, measured counterclockwise from $\mathbf{v}_1$ (Fig. I-11c).

Figure I-12 depicts the slice of the cylinder made by a plane which contains the cylinder's axis (the vertical dotted line) and which is parallel to the plane through **P** that is determined by $\mathbf{v}_1$

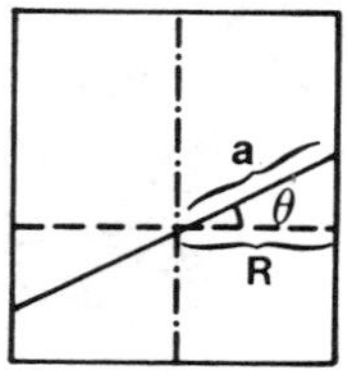

Figure I - 12.

and $\mathbf{v}_2$. The sloping line of length 2a in Figure I-12 is the major axis of $\boldsymbol{\alpha}_\mathbf{v}$. This ellipse has minor semi-axis of length R and major semi-axis of length $a = R/|\cos\theta|$, and therefore is congruent to the ellipse with equation

$$\frac{x^2}{R^2/\cos^2\theta} + \frac{y^2}{R^2} = 1$$

By Exercise 7 of Section 1 on the curvature of ellipses, we have

$$k_n(\mathbf{v}) = +(\cos^2\theta)/R \tag{1}$$

(The sign is positive because **U** and the principal normal vector of $\boldsymbol{\alpha}_\mathbf{v}$ at **P** have the same direction; if we had chosen for **U** the outward-pointing normal to M, then the sign would have been negative.)

Of course on any surface, different normal sections at **P** in general have different normal curvatures. However, as it turns out, there always are two orthogonal directions—called *principal directions* and specified by unit tangent vectors $\mathbf{v}_1$ and $\mathbf{v}_2$ at **P**—for which $k_n(\mathbf{v})$ assumes maximum and minimum values, k_1 and k_2, respectively. The product of these two *principal curvatures* is called the *Gauss curvature* K (of M at **P**):

$$K(\mathbf{P}) = k_1 k_2$$

Note that although the signs of k_1 and k_2 depend on the choice between two possible unit normal vectors, $\pm$ **U**, the sign of their product does not. The details will be given in Section 6.

In our cylinder example (Fig. I-11c) $\mathbf{v}_1$ and $\mathbf{v}_2$ correspond respectively to $\theta = 0$, and $\theta = \pi/2$. Since $k_n(\mathbf{v}) = (\cos^2\theta)/R$, we have

$$k_n(\mathbf{v}_2) = 0 \leqslant k_n(\mathbf{v}) \leqslant 1/R = k_n(\mathbf{v}_1)$$

Accordingly, $k_1 = 1/R$, $k_2 = 0$, $\mathbf{v}_1$ and $\mathbf{v}_2$ give the principal directions at **P**, and the cylinder has Gauss curvature 0 at **P**, and in fact at every point.

Although it may at first seem strange to say that a cylinder has zero curvature, this simply reflects the well known fact that a paper cylinder can be slit along any element and unrolled into a flat sheet without stretching or tearing, and without affecting the lengths of any curves drawn on the cylinder. In other words the geometry of a cylinder is locally (i.e., in a sufficiently small neighborhood of any point) indistinguishable from that of the plane, as far as surface measurements of distances, angles, areas, etc., are concerned. We say that the cylinder and the plane are *locally isometric*. (Globally, however, a plane and a cylinder are quite dissimilar: a bug crawling along a circular cross section of a cylinder—$\boldsymbol{\alpha}_{v_1}$ in Figure I-11c for example—even though it is "following its nose," so to speak, and is turning neither toward the left or right, will eventually return to its starting point. This would not happen on a plane.) Likewise, a right circular cone has zero Gauss curvature.

At any point on a sphere of radius R, every normal section is a great circle. (Here any two orthogonal directions may be taken as the principal directions.) Therefore $k_n(\mathbf{v}) = \pm\, 1/R$ (the sign depending on the choice of normal vector), and so $K = 1/R^2$.

Example 5. An example of a surface with negative Gauss curvature is the hyperbolic paraboloid

$$z = \frac{1}{2}(y^2 - x^2)$$

(Fig. I-13).* It can be shown that the principal directions at the origin are along the x- and y-axes. Since the normal section in one of these directions is concave downward, while that in the other is concave upward, the two principal curvatures have opposite signs,

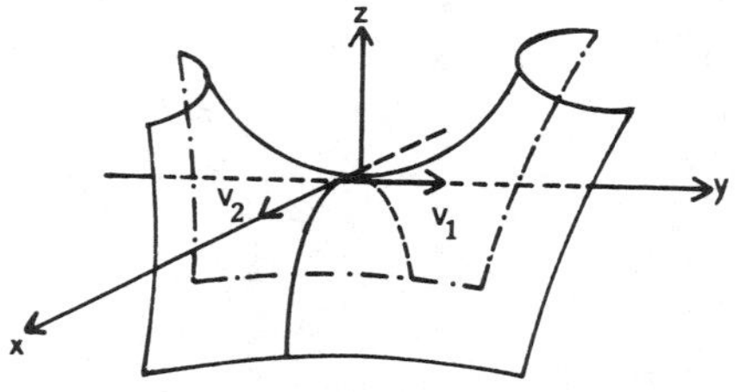

Figure I - 13.

*The coefficient 1/2 is included for convenience: it will simplify calculations in later examples and exercises.

and so K is negative at the origin. In fact, K is negative everywhere on this surface, which is "saddle-shaped" throughout. (However, in contrast to the preceding examples, K is not constant on this surface.)

Example 6. See Figure I-14a. The surface swept out by revolving circle **ABCD** about line ℓ is called a *torus* and resembles the surface of a doughnut (Fig. I-14b). The "outside" portion of the torus, generated by points of arc $\widehat{ABC}$ (with end points excluded), consists of points where the surface bends away from its tangent plane in all directions, and so $K > 0$. The "inside" portion, generated by the points of the open arc $\widehat{ADC}$, consists of points where the surface is saddle-shaped, and so $K < 0$. At points of the two circles swept out by **A** and **C**, $K = 0$. (This may not be obvious, but will be verified later when we compute the Gauss curvature of the torus.)

Now the development up to this point has been made essential use of the embedding of the surface M in 3-dimensional space. Our definitions of normal and Gauss curvature involved normal sections and the normal vector, which "points out" of the surface. If M were populated by 2-dimensional creatures who lacked the ability to leave the surface and so had no knowledge of the "outside world" of a third dimension, then our exposition would be completely unintelligible to them. Like the characters in Edwin A. Abbott's *Flatland* [1], their knowledge would be obtainable only through measurements made on the surface itself. (We would have as much difficulty trying to comprehend a fourth spatial dimension orthogonal to our own 3-dimensional world of existence.)

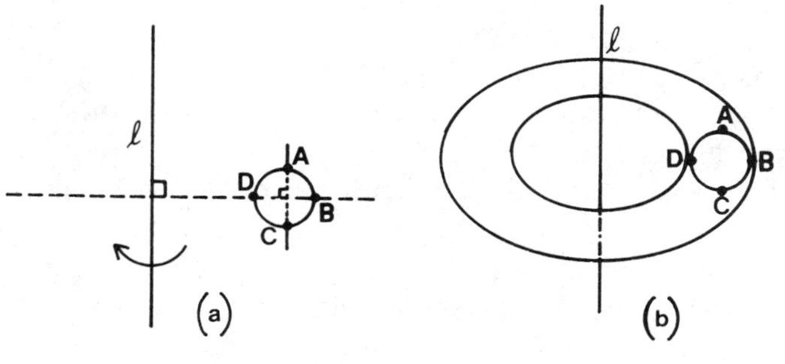

Figure I - 14.

In short, our description of curvature has been *extrinsic* (involving concepts or measurements external to M), rather than *intrinsic* (involving only measurements on the surface). This is in no way unnatural, since we seldom envision a curved object without considering at the same time the space in which its curvature takes place. For example, when we think of a spherical surface, we naturally picture in our mind's eye the 3-dimensional ball of which it is the boundary, as well as a portion of the space exterior to the ball.

However, if we are to give meaning to the concept of curvature in higher dimensions–in particular, the curvature of physical space itself–we must eventually abandon the extrinsic point of view. It would be most inconvenient to have to conjure up some containing 4-dimensional space in order to define what is meant by the curvature of 3-dimensional physical space; and even more so to require a space of five dimensions to define the curvature of the 4-dimensional universe of space and time in Einstein's general theory of relativity (Chapter III).

Fortunately, Gauss had shown, in what was his crowning achievement in surface theory, his *Theorema Egregium* (or "extraordinary theorem"), that the curvature of a surface can actually be redefined (and computed) in terms of measurements made within the surface, and so is an intrinsic quantity, comprehensible to "Flatlanders." The surrounding space is not needed! For example even if the earth had a permanent cloud cover, preventing us from looking down upon it, we could still determine that its surface is roughly spherical and not a plane by making sufficiently careful ground measurements.

It is important to appreciate the significance of Gauss's result, since it proved to be a major turning point in the development of geometry. Imagine a surface made out of paper or aluminum foil, or some other unstretchable material. If we deform or "bend" this surface without stretching, shrinking, tearing, or creasing, then the arc length of any curve drawn on it will remain unchanged. Any 2-dimensional creatures inhabiting the surface would be unaware of the deformation, since all their surface measurements (e.g., lengths, angles, areas) would be unaffected. Intrinsic geometry consists of those properties or concepts which are left invariant by this type of distance-preserving deformation. Gauss's *Theorema Egregium* is

Figure I - 15.

thus the statement that Gauss curvature is *invariant under bending*. For example, the surfaces in Figure I-15 all have zero curvature because they are obtainable by bending a portion of a flat plane. (More generally, K is invariant under any smooth one-to-one mapping of a surface M onto another surface M′ such that the length of any curve in M equals the length of its image in M′. Such a mapping is called an *isometry*. Isometric surfaces have equal Gauss curvature at corresponding points.)

As another example, there can be no distance-preserving transformation of a portion of a sphere onto a portion of a plane, since the two surfaces have unequal Gauss curvatures. (An orange peel cannot be pressed flat without tearing or stretching.)

Gauss's result is quite surprising in view of the way curvature has been defined and the way we intuitively visualize curvature. It says that even though the principal curvatures vary with bending and depend upon how the surface is embedded in E^3, their product nevertheless does not.

As might be expected, a reformulation of curvature that is totally intrinsic will allow us to generalize to spaces of higher dimensions (called *manifolds*) and so speak of "curved space." This is exactly what Riemann did in extending the work on surface theory of his mentor Gauss. A proof of Gauss's *Theorema Egregium* will be given later in this chapter. Next we proceed to some of the mathematical details of the geometry of surfaces in E^3.

Exercises I-2

1. Which of the following surfaces have positive Gauss curvature at every point?

a. a circular cone
b. a paraboliod of revolution
c. a hyperboloid of one sheet
d. a hyperboloid of two sheets

2. According to Exercise 6b of Section 1, the parabola $y = x^2/2$ has curvature 1 at the origin. Use this to compute the Gauss curvature of the hyperbolic paraboloid $z = (1/2)(y^2 - x^2)$ at the origin (cf. Example 5).

3. Find the Gauss curvature of the ellipsoid

$$\frac{x^2}{a^2} + \frac{y^2}{b^2} + \frac{z^2}{c^2} = 1$$

at the end points of its three axes, i.e., at the points $\pm(a, 0, 0)$, $\pm(0, b, 0)$, and $\pm(0, 0, c)$. (Use Exercise 7 of Section 1.)

4. In Figure I-14, assume that the radius of circle **ABCD** is r, and that the distance from the center of circle **ABCD** to line ℓ is R. Find the Gauss curvature of the torus (a) at any point of the circle generated by point **B**; (b) at any point of the circle generated by point **D**.

3. SURFACES IN E^3

Just as a curve is given by a vector-valued function of one real variable, a surface M may be described (at least locally) as the image of a vector-valued function of two variables, $\mathbf{X}: D \to E^3$, defined on an open subset D of $\mathbb{R}^2$:

$$\mathbf{X} = \mathbf{X}(u,v) = (x(u,v), y(u,v), z(u,v)) \qquad (2)$$

For convenience, we shall assume that the coordinate functions have continuous partial derivatives up to at least third order on D, and that **X** is *regular*, i.e., the vectors

$$\mathbf{X}_1(u,v) = \partial\mathbf{X}/\partial u = (\partial x/\partial u, \partial y/\partial u, \partial z/\partial u)$$

$$\mathbf{X}_2(u,v) = \partial\mathbf{X}/\partial v = (\partial x/\partial v, \partial y/\partial v, \partial z/\partial v)$$

are linearly independent for each (u,v) in D. Equivalently, the cross product $\mathbf{X}_1 \times \mathbf{X}_2$ is non-zero throughout D.*

Example 7. Let

$$\mathbf{X}(u,v) = (R \cos u \cos v, R \sin u \cos v, R \sin v),$$

where R is a positive constant. (D will be specified below.) Since $\|\mathbf{X}\| = (\mathbf{X} \cdot \mathbf{X})^{1/2} = R$, the image of $\mathbf{X}$ lies on the sphere of radius R centered at the origin of E^3. If we view this as a model for the earth's surface, with the point (0, 0, R) representing the north pole, then u and v correspond to longitude and latitude (Fig. I-16). Differentiating, we obtain

$$\mathbf{X}_1 = (-R \sin u \cos v, R \cos u \cos v, 0)$$

$$\mathbf{X}_2 = (-R \cos u \sin v, -R \sin u \sin v, R \cos v)$$

and so, computing the cross product, we have

$$\mathbf{X}_1 \times \mathbf{X}_2 = R^2 (\cos u \cos^2 v, \sin u \cos^2 v, \sin v \cos v)$$

Since $\|\mathbf{X}_1 \times \mathbf{X}_2\| = R^2 |\cos v|$, $\mathbf{X}$ is regular except where $\cos v = 0$, which holds only at the north and south poles. Therefore we can choose for the domain of $\mathbf{X}$ a region of the uv-plane such as $D = \{(u,v) | -\pi < u < \pi, -\pi/2 < v < \pi/2\}$. The north and south poles

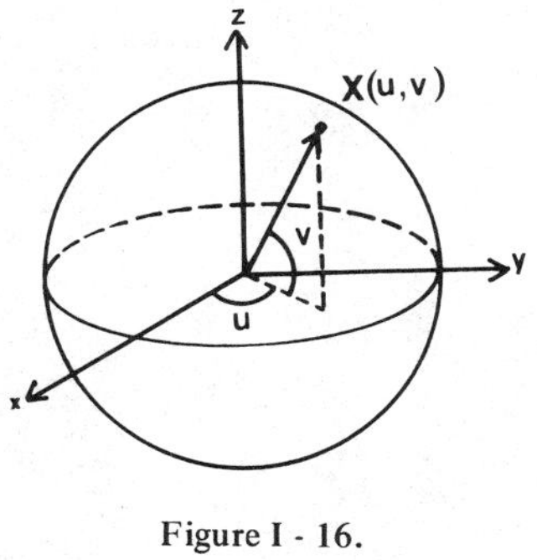

Figure I - 16.

*A more general definition of surface will be given in Section 9.

are then excluded from the image of **X**. At these points, the geographic coordinate system breaks down: there is no unique longitude for either pole. Unfortunately, with D as above, the meridian at 180° longitude (the "international date line") is also excluded, but on the other hand, **X** is one-to-one and has a continuous inverse $\mathbf{X}^{-1} : \mathbf{X}(D) \to R^2$. Each point of **X**(D) is assigned unique longitude and latitude, and these quantities vary continuously as one moves over **X**(D) (see Fig. I-16).

In general, regularity assures us that every point of D has a neighborhood Ω on which **X** is one-to-one with a continuous inverse function $\mathbf{X}(\Omega) \to \Omega$. Thus, to every point of $\mathbf{X}(\Omega)$ there corresponds a unique pair (u,v), and we may think of this correspondence as a system of curvilinear coordinates, u,v, on $\mathbf{X}(\Omega) \subset M$ (Fig. I-17).

You may perhaps be more familiar with the representation of a surface as the graph of a smooth function $z = f(x,y)$, where (x,y) varies over a region D of the xy-plane (cf. Example 5). Such a graph (Fig. I-18) may be expressed in parametric form as $\mathbf{X}(u,v) = (u,v, f(u,v))$, for (u,v) in D. Since $\mathbf{X}_1 = (1, 0, \partial f/\partial u)$ and $\mathbf{X}_2 = (0, 1, \partial f/\partial v)$ are independent, **X** fulfills our definition of a surface if f is sufficiently smooth.

Example 8. The hemisphere

$$x^2 + y^2 + z^2 = R^2, \quad z > 0$$

being the graph of $z = (R^2 - x^2 - y^2)^{1/2}$, may be parametrized as

$$\mathbf{X}(u,v) = (u, v, (R^2 - u^2 - v^2)^{1/2})$$

with $D = \{(u,v) | u^2 + v^2 < R^2\}$. The inverse map, $\mathbf{X}^{-1}$, is simply the orthogonal projection from the hemisphere onto the xy-plane (Fig. I-19).

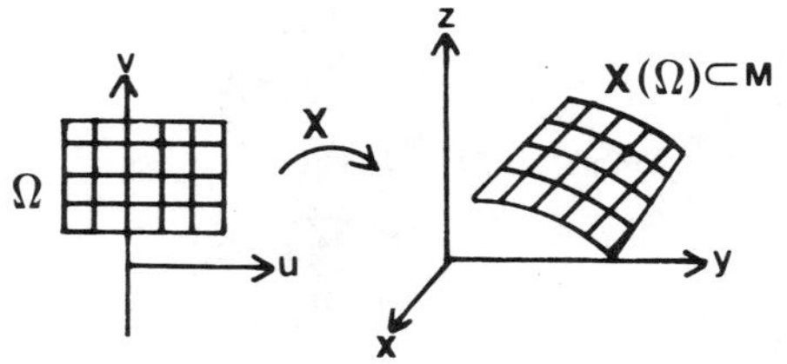

Figure I - 17.

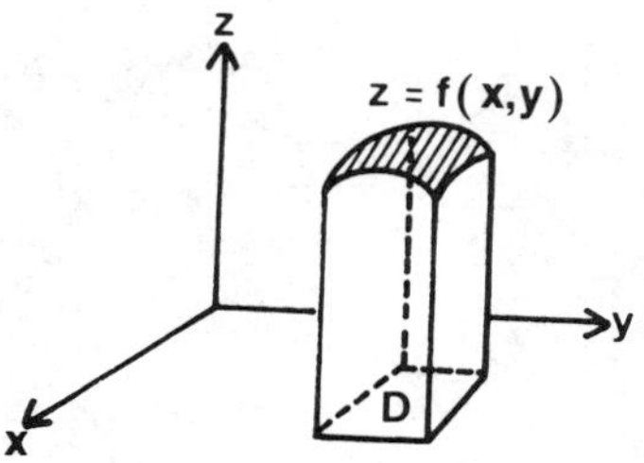

Figure I - 18.

If $u = u(t)$, $v = v(t)$ (t ranging over some interval) is a smooth curve in D, then the image of this curve under **X** is a smooth curve $\boldsymbol{\alpha}$ in M:

$$\boldsymbol{\alpha}(t) = \mathbf{X}(u(t), v(t))$$

(Fig. I-20). The velocity vector $\boldsymbol{\alpha}'(t)$ is given, in accordance with the chain rule, by

$$\boldsymbol{\alpha}'(t) = \frac{\partial \mathbf{X}}{\partial u}\frac{du}{dt} + \frac{\partial \mathbf{X}}{\partial v}\frac{dv}{dt}$$

or, in abbreviated notation, by

$$\boldsymbol{\alpha}'(t) = u'\mathbf{X}_1 + v'\mathbf{X}_2 \qquad (3)$$

[It is to be understood here and elsewhere that u' and v' are evaluated at t, and $\mathbf{X}_1$ and $\mathbf{X}_2$ are evaluated at $(u(t), v(t))$.] $\boldsymbol{\alpha}'(t)$ is tangent to the curve and so is tangent to M itself.

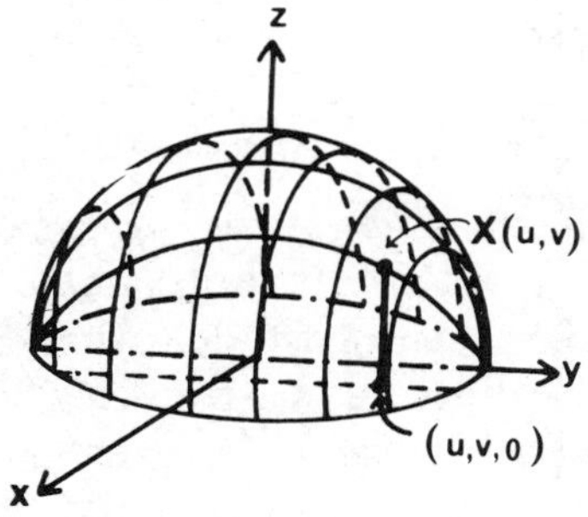

Figure I - 19.

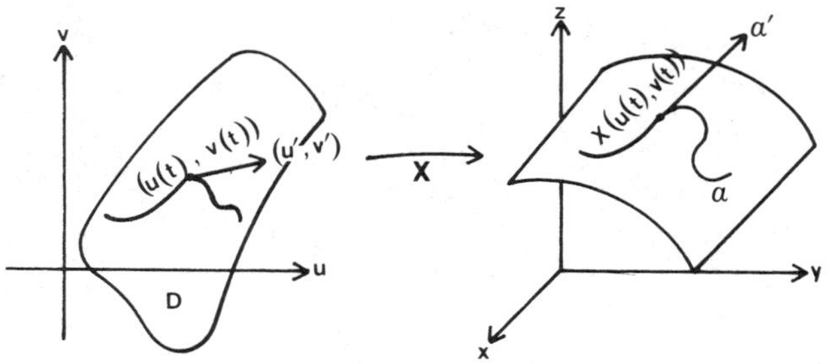

Figure I - 20.

Definition I-2

A vector **v** is called a *tangent vector* to M at **P** if there exists a curve on M which passes through **P** and has velocity vector **v** at **P**. The set of all tangent vectors to M at **P** is called the *tangent plane* of M at **P**, and is denoted $T_P M$.

By Eq. (3), any tangent vector at $\mathbf{P} = \mathbf{X}(u_0, v_0)$ is a linear combination of $\mathbf{X}_1(u_0, v_0)$ and $\mathbf{X}_2(u_0, v_0)$. Conversely, any linear combination $\mathbf{v} = (a\mathbf{X}_1 + b\mathbf{X}_2)(u_0, v_0)$, with real coefficients a, b, is the velocity vector of some curve on M; e.g., $\boldsymbol{\alpha}(t) = \mathbf{X}(u_0 + at, v_0 + bt)$ has $\boldsymbol{\alpha}'(0) = \mathbf{v}$ (as you should check). The regularity of **X** implies that $T_P M$ is a 2-dimensional vector space at each point **P** (with $\mathbf{X}_1$ and $\mathbf{X}_2$ as a basis).

$\mathbf{X}_1(u_0, v_0)$ and $\mathbf{X}_2(u_0, v_0)$ are themselves the velocity vectors of the curves

$$u \to \mathbf{X}(u, v_0) \quad (v = v_0, \text{ fixed}),$$

$$v \to \mathbf{X}(u_0, v) \quad (u = u_0, \text{ fixed}).$$

The first of these is called a *u-parameter curve*, since it is obtained by fixing v and allowing only u to vary. In Example 7, the u-parameter curves are the circles of constant latitude (often called "parallels"), and at every point, $\mathbf{X}_1$ is directed eastward. The second type is called a v-*parameter curve*. In Example 7, the v-parameter curves are the "meridians" (great semicircles of constant longitude), and at every point, $\mathbf{X}_2$ points northward.

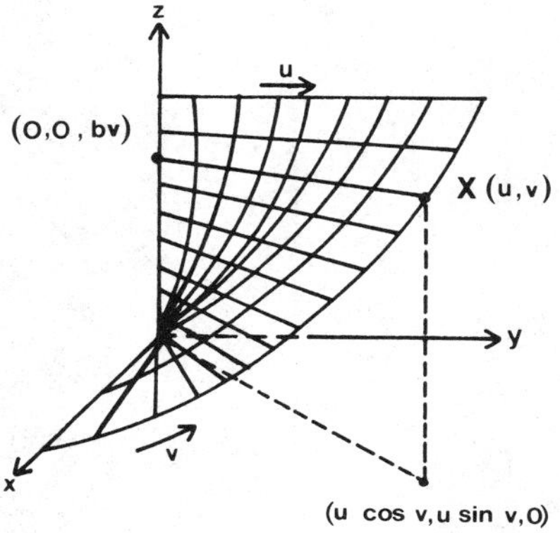

Figure I - 21.

In Example 8, the u- and v-parameter curves are the intersections of the hemisphere with planes parallel respectively to the xz-plane and to the yz-plane (Fig. I-19).

Example 9. The surface

$$\mathbf{X}(u,v) = (u \cos v,\, u \sin v,\, bv)$$

where b is a non-zero constant, is called a *helicoid*. The v-parameter curves (u fixed) are helices (compare Example 3), and the u-parameter curves (v fixed) are lines passing through the z-axis and parallel to the xy-plane (Fig. I-21). If we restrict $u > 0$, then the surface resembles a spiral ramp such as is found in many parking garages.

Exercises I-3

1. If a smooth curve of the form $\boldsymbol{\alpha}(u) = (f(u), 0, g(u))$ in the xz-plane is revolved about the z-axis, the resulting *surface of revolution* is given by

$$\mathbf{X}(u,v) = (f(u) \cos v,\ f(u) \sin v,\, g(u))$$

$\boldsymbol{\alpha}$ is called the profile curve. (a) Show that $\mathbf{X}$ is regular provided $f(u) \neq 0$ and $\boldsymbol{\alpha}'(u) \neq \mathbf{O}$ for all u. (b) Describe the u- and v-parameter curves and show they intersect orthogonally ($\mathbf{X}_1 \cdot \mathbf{X}_2 = 0$). (The

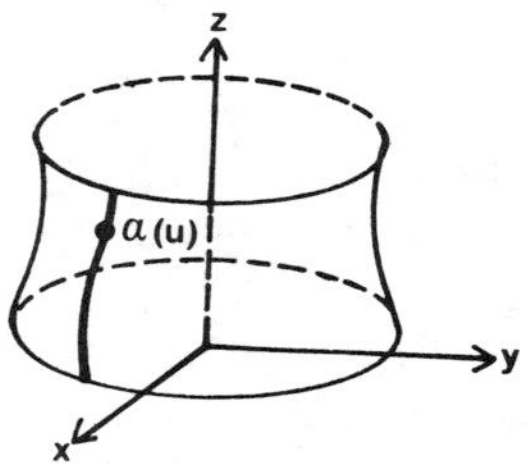

Figure I - 22.

u-parameter curves are called *meridians*, and the v-parameter curves are called *parallels*, by analogy with the sphere of Example 7, which can be put in the form above by interchanging u and v.)

2. For each of the following surfaces of revolution, sketch the profile curve (v = 0) in the xz-plane, and then sketch the surface. In each case, prove that **X** is regular and give an equation for the surface in the form g(x,y,z) = 0. (Assume a and b are positive constants.)

(a) $\mathbf{X}(u,v) = (u \cos v, u \sin v, au), \qquad u \neq 0$

(b) $\mathbf{X}(u,v) = (u \cos v, u \sin v, au^2), \quad u > 0$

(c) $\mathbf{X}(u,v) = (a \cos u \cos v, a \cos u \sin v, b \sin u),$
$-\pi/2 < u < \pi/2$

(d) $\mathbf{X}(u,v) = (a \cosh u \cos v, a \cosh u \sin v, b \sinh u)$

(e) $\mathbf{X}(u,v) = (a \sinh u \cos v, a \sinh u \sin v, b \cosh u),$
$u \neq 0$

(f) $\mathbf{X}(u,v) = \left(a \cosh \dfrac{u}{a} \cos v, a \cosh \dfrac{u}{a} \sin v, u\right)$

3. A *ruled surface* is a surface with the property that through each of its points passes a line (called a *ruling*) wholely contained in the surface. Such a surface may be thought of as swept out or generated by a moving line. If $\boldsymbol{\alpha} = \boldsymbol{\alpha}(u)$ is a curve on the surface that cuts across all the rulings, and if $\boldsymbol{\beta}(u)$ is a non-zero vector in the direction of the ruling through $\boldsymbol{\alpha}(u)$, then the surface has the parametrization

(i) $\mathbf{X}(u,v) = \boldsymbol{\alpha}(u) + v\boldsymbol{\beta}(u)$

The v-parameter curves are rulings. The parameters can be interchanged of course, so that a surface of the form

(ii) $\mathbf{X}(u,v) = \boldsymbol{\alpha}(v) + u\boldsymbol{\beta}(v)$

is also a ruled surface, with u-parameter curves as rulings. (a) Show that the helicoid of Example 9 is a ruled surface. (b) Show that

$$\mathbf{X}(u,v) = (u - v, u + v, 2uv)$$

is a regular parametrization of the hyperbolic paraboloid of Example 5, and that this surface is doubly ruled, i.e., expressible as a ruled surface in two different ways, with different rulings.

4. If $\boldsymbol{\alpha} = \boldsymbol{\alpha}(u)$ is a smooth curve in E^3 and $\boldsymbol{\beta}$ is a non-zero vector with constant components, then the ruled surface $\mathbf{X}(u,v) = \boldsymbol{\alpha}(u) + v\boldsymbol{\beta}$ is called a *cylinder*. (The special case of a right circular cylinder results when $\boldsymbol{\alpha}$ is a circle and $\boldsymbol{\beta}$ is perpendicular to the plane of $\boldsymbol{\alpha}$.) (a) Show that $\mathbf{X}$ is regular except where $\boldsymbol{\alpha}' \times \boldsymbol{\beta} = 0$. Describe the u- and v-parameter curves. (b) Sketch the cylinder for which

$$\boldsymbol{\alpha}(u) = (u, u^2, 0), \quad \boldsymbol{\beta} = (0, -1, 2)$$

5. A *cone* is a ruled surface of the form $\mathbf{X}(u,v) = \mathbf{P} + v\boldsymbol{\beta}(u)$. The point $\mathbf{P}$ is called the *vertex* of the cone. Show that $\mathbf{X}$ is regular except where $v = 0$ (the vertex) or $\boldsymbol{\beta} \times \boldsymbol{\beta}' = \mathbf{O}$. (The special case of a right circular cone results when the vectors $\boldsymbol{\beta}(u)$ make a constant acute angle θ with a fixed line through $\mathbf{P}$. This line is then the cone's *axis* and θ is called the *semi-vertex angle* of the cone.)

6. Let $\boldsymbol{\alpha}(u) = (\cos u, \sin u, 0)$, a unit circle in the xy-plane. Through each point $\boldsymbol{\alpha}(u)$ of this circle pass the line with direction vector $\boldsymbol{\beta}(u) = (-\sin u, \cos u, 1)$. This line is perpendicular to the circle's radius and makes an angle of $\pi/4$ radians with the xy-plane. (a) Write out the coordinate functions of the resulting ruled surface $\mathbf{X}(u,v) = \boldsymbol{\alpha}(u) + v\boldsymbol{\beta}(u)$, and show that $\mathbf{X}$ is regular. (b) Show that the image of $\mathbf{X}$ is the one-sheeted hyperboloid of revolution $x^2 + y^2 - z^2 = 1$. (c) Show that if $\boldsymbol{\beta}(u)$ is replaced by $(-\sin u, \cos u, -1)$ the same image surface results. Thus this hyperboloid is doubly ruled.

7. Let $\boldsymbol{\alpha}(u) = (\cos u, \sin u, 0)$. Through each point of $\boldsymbol{\alpha}(u)$, pass a unit line segment with midpoint $\boldsymbol{\alpha}(u)$ and direction vector

$$\boldsymbol{\beta}(u) = \left(\sin \frac{u}{2}\right) \boldsymbol{\alpha}(u) + \left(\cos \frac{u}{2}\right) (0, 0, 1)$$

The resulting surface,

$$\mathbf{X}(u,v) = \boldsymbol{\alpha}(u) + v\boldsymbol{\beta}(u), \quad -\frac{1}{2} \leqslant v \leqslant \frac{1}{2}$$

is called a *Möbius strip*.

a. Write out the coordinate functions of $\mathbf{X}(u,v)$.
b. Sketch the rulings for $u = 0, \pi/2, \pi$, and $3\pi/2$.
c. Connect the end points of these rulings by drawing the u-parameter curve $\mathbf{X}(u, 1/2)$, $0 \leqslant u \leqslant 4\pi$. [Note that the u-parameter curves are closed and have period 4π, except for $\boldsymbol{\alpha}(u) = \mathbf{X}(u,0)$, which has period 2π.]

8. Let $\mathbf{P}, \mathbf{A}, \mathbf{B} \in E^3$, and let

$$\mathbf{X}(u,v) = \mathbf{P} + u\mathbf{A} + v\mathbf{B}$$

(a) Prove that $\mathbf{X}$ is regular if and only if $\mathbf{A} \times \mathbf{B} \neq \mathbf{O}$. (b) Let $\mathbf{C} = \mathbf{A} \times \mathbf{B}$ and assume $\mathbf{C} \neq \mathbf{O}$. Show that any point of the form $\mathbf{X} = \mathbf{P} + u\mathbf{A} + v\mathbf{B}$ satisfies

$$\mathbf{C} \cdot (\mathbf{X} - \mathbf{P}) = 0$$

and conversely, that any point $\mathbf{X}$ satisfying the latter equation must be of the form $\mathbf{P} + u\mathbf{A} + v\mathbf{B}$. (The set of all such points comprise the plane through $\mathbf{P}$ with normal vector $\mathbf{C}$.)

4. THE FIRST FUNDAMENTAL FORM

If $\boldsymbol{\alpha}(t) = \mathbf{X}(u(t), v(t))$, $a \leqslant t \leqslant b$, is a curve on a surface and if $s = s(t)$ is the arc length along $\boldsymbol{\alpha}$ from $\boldsymbol{\alpha}(a)$ to $\boldsymbol{\alpha}(t)$, then as we saw in Section 1, the total length of this curve is obtained by integrating $ds/dt = \|\boldsymbol{\alpha}'(t)\|$ over the interval from a to b:

$$L = s(b) = \int_a^b \|\boldsymbol{\alpha}'(t)\| \, dt$$

Since $\boldsymbol{\alpha}'(t) = u'\mathbf{X}_1 + v'\mathbf{X}_2$ [see Eq. (3)], we have

$$\left(\frac{ds}{dt}\right)^2 = \|\boldsymbol{\alpha}'\|^2 = \boldsymbol{\alpha}' \cdot \boldsymbol{\alpha}'$$

$$= (u'\mathbf{X}_1 + v'\mathbf{X}_2) \cdot (u'\mathbf{X}_1 + v'\mathbf{X}_2)$$

$$= u'^2 (\mathbf{X}_1 \cdot \mathbf{X}_1) + 2u'v' (\mathbf{X}_1 \cdot \mathbf{X}_2) + v'^2 (\mathbf{X}_2 \cdot \mathbf{X}_2)$$

Following Gauss, let us put

$$E = \mathbf{X}_1 \cdot \mathbf{X}_1, \quad F = \mathbf{X}_1 \cdot \mathbf{X}_2, \quad G = \mathbf{X}_2 \cdot \mathbf{X}_2 \tag{4}$$

We then have

$$\left(\frac{ds}{dt}\right)^2 = E\left(\frac{du}{dt}\right)^2 + 2F\left(\frac{du}{dt}\frac{dv}{dt}\right) + G\left(\frac{dv}{dt}\right)^2 \tag{5}$$

Once the coefficients E, F, and G have been calculated from Eq. (4), we can find the length of any curve on the surface by substituting the derivatives du/dt and dv/dt into Eq. (5) and integrating the square root:

$$L = \int_a^b \left[E\left(\frac{du}{dt}\right)^2 + 2F\left(\frac{du}{dt}\frac{dv}{dt}\right) + G\left(\frac{dv}{dt}\right)^2\right]^{1/2} dt$$

We often abbreviate this last equation by writing

$$L = \int_{\alpha} ds = \int_{\alpha} (E\,du^2 + 2F\,du\,dv + G\,dv^2)^{1/2}$$

or by writing

$$ds^2 = E\,du^2 + 2F\,du\,dv + G\,dv^2 \tag{6}$$

The expression on the right in Eq. (5), or abbreviated in differential notation in Eq. (6), is called the *first fundamental form* or *metric form* of the surface. As we shall see, the metric form completely determines the intrinsic geometry of the surface, including its curvature.

Example 10. Let $\mathbf{X}(u,v) = (u,v,0)$. This surface is the ordinary Euclidean plane embedded in E^3 as the plane $z = 0$, with $x = u$ and $y = v$ as *Cartesian coordinates.* $\mathbf{X}_1 = (1,0,0)$ and $\mathbf{X}_2 = (0,1,)$. Consequently, $E = G = 1$ and $F = 0$, so $ds^2 = du^2 + dv^2 = dx^2 + dy^2$, the Pythagorean formula for the differential of arc length. In other words, the length of a curve in the xy-plane is given by

$$L = \int_a^b \left[\left(\frac{dx}{dt}\right)^2 + \left(\frac{dy}{dt}\right)^2\right]^{1/2} dt$$

If the curve can be represented as the graph of a smooth function, $y = f(x)$, then we may take $x = t$, $y = f(t)$, and the formula for L reduces to

$$L = \int_a^b [1 + f'(x)^2]^{1/2}\, dx$$

found previously in Example 2.

Example 11. Points in the Euclidean plane also may be assigned *polar coordinates*, r, θ. These are related to Cartesian coordinates, x, y, by the formulas

$$x = r\cos\theta, \quad y = r\sin\theta.$$

Let us therefore consider the function

$$\mathbf{X}(r,\theta) = (r\cos\theta, r\sin\theta, 0)$$

for $r \neq 0$. Then $\mathbf{X}_1 = (\cos\theta, \sin\theta, 0)$, $\mathbf{X}_2 = (-r\sin\theta, r\cos\theta, 0)$, and so $E = 1$, $F = 0$, and $G = r^2$. The first fundamental form becomes

$$ds^2 = dr^2 + r^2 d\theta^2$$

which you may recall from calculus as the formula for the differential of arc length in polar coordinates. This formula means that if a curve in E^2 is given in polar coordinates by $r = r(t)$, $\theta = \theta(t)$, $a \leqslant t \leqslant b$, then its length is

$$L = \int_a^b \left[\left(\frac{dr}{dt}\right)^2 + r(t)^2 \left(\frac{d\theta}{dt}\right)^2\right]^{1/2} dt$$

If the curve is given instead by an equation of form $r = f(\theta)$, $\alpha \leqslant \theta \leqslant \beta$, then choosing θ as parameter results in the formula

$$L = \int_{\alpha}^{\beta} [f'(\theta)^2 + f(\theta)^2]^{1/2}\, d\theta \tag{7}$$

As a further example, you are urged to verify that the metric form of the sphere (Example 7) is

$$ds^2 = R^2 \cos^2 v\, du^2 + R^2\, dv^2$$

If $\mathbf{v} = a\mathbf{X}_1 + b\mathbf{X}_2$ and $\mathbf{w} = c\mathbf{X}_1 + d\mathbf{X}_2$ (a, b, c, and d real numbers) are any two tangent vectors at a point of M, then $\mathbf{v} \cdot \mathbf{w} = (a\mathbf{X}_1 + b\mathbf{X}_2) \cdot (c\mathbf{X}_1 + d\mathbf{X}_2) = Eac + F(ad + bc) + Gbd$, which can be written as the matrix product

$$\mathbf{v} \cdot \mathbf{w} = (a, b) \begin{pmatrix} E & F \\ F & G \end{pmatrix} \begin{pmatrix} c \\ d \end{pmatrix} \tag{8}$$

The two-by-two matrix on the right hand side is the matrix of the first fundamental form. According to Eq. (8), this matrix determines dot products of tangent vectors (and hence lengths and angles).

Since the cross product, $\mathbf{X}_1 \times \mathbf{X}_2$, is orthogonal to both $\mathbf{X}_1$ and $\mathbf{X}_2$, it follows that at each point $\mathbf{X}(u,v)$ the *unit normal vector,*

$$\mathbf{U} = \frac{\mathbf{X}_1 \times \mathbf{X}_2}{\|\mathbf{X}_1 \times \mathbf{X}_2\|}$$

is perpendicular to the tangent plane at $\mathbf{X}(u,v)$. An important identity (of Lagrange) is

$$\|\mathbf{X}_1 \times \mathbf{X}_2\|^2 = (\mathbf{X}_1 \cdot \mathbf{X}_1)(\mathbf{X}_2 \cdot \mathbf{X}_2) - (\mathbf{X}_1 \cdot \mathbf{X}_2)^2 = EG - F^2 \tag{9}$$

which is the determinant of the matrix of the metric form. Equation (9) can be obtained from the identities

$$\|\mathbf{X}_1 \times \mathbf{X}_2\| = \|\mathbf{X}_1\|\, \|\mathbf{X}_2\| \sin\theta, \quad \mathbf{X}_1 \cdot \mathbf{X}_2 = \|\mathbf{X}_1\|\, \|\mathbf{X}_2\| \cos\theta$$

or by a straightforward calculation in coordinates. By regularity, $EG - F^2 \neq 0$.

Example 12. Let

$$\mathbf{X}(u,v) = ((R + r\cos u)\cos v, (R + r\cos u)\sin v, r\sin u),$$

where R and r are constants satisfying $R > r > 0$. This is the torus of revolution obtained by revolving the circle $x = R + r\cos u$, $y = 0$, $z = r\sin u$ in the xz-plane [centered at (R,0,0)] about the z-axis (cf. Example 6). The geometric significance of u and v (as angles) is indicated in Figure I-23.

$$\mathbf{X}_1 = (-r\sin u\cos v, -r\sin u\sin v, r\cos u)$$

$$\mathbf{X}_2 = (-(R + r\cos u)\sin v, (R + r\cos u)\cos v, 0)$$

$$E = r^2, \quad F = 0, \quad G = (R + r\cos u)^2$$

Accordingly, $\|\mathbf{X}_1 \times \mathbf{X}_2\|^2 = EG - F^2 = r^2(R + r\cos u)^2 \neq 0$, since $R > r > 0$. Any open subset of R^2 may serve as the domain of $\mathbf{X}$.

As you can check, the u- and v-parameter curves are circles, $\mathbf{X}_1$ and $\mathbf{X}_2$ are orthogonal at every point (because $F = 0$, but you can see it geometrically from the figure as well), and $\mathbf{U} = (\mathbf{X}_1 \times \mathbf{X}_2)/\|\mathbf{X}_1 \times \mathbf{X}_2\|$ is directed into the interior of the torus.

At this point, in order to simplify the writing of formulas, we shall make some notational changes. First of all, we relabel the coefficients of the metric form by setting

$$g_{11} = E, \quad g_{12} = g_{21} = F, \quad g_{22} = G$$

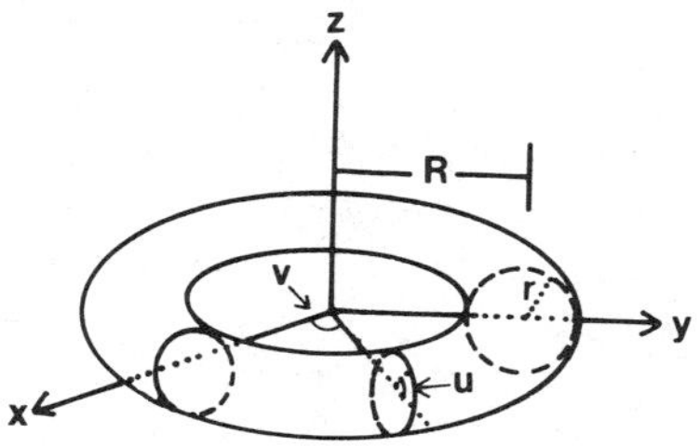

Figure I - 23.

We then have [see Eq. (4)]

$$g_{ij} = \mathbf{X}_i \cdot \mathbf{X}_j$$

and the matrix of the geometric form is

$$\begin{pmatrix} g_{11} & g_{12} \\ g_{21} & g_{22} \end{pmatrix} = \begin{pmatrix} E & F \\ F & G \end{pmatrix}$$

Remember, the entries g_{ij} are functions of u and v. Let $g = \det(g_{ij}) = EG - F^2$. Then Eq. (9) becomes

$$\|\mathbf{X}_1 \times \mathbf{X}_2\|^2 = g \qquad (10)$$

It follows that $\mathbf{X}_1$ and $\mathbf{X}_2$ are independent at (u,v) if and only if the matrix (g_{ij}) is non-singular there.

Next the parameters u and v will usually be written u^1 and u^2. (The superscripts are indices, not exponents.) In the new rotation, the first fundamental form is given by

$$ds^2 = g_{11}(du^1)^2 + 2g_{12}\,du^1\,du^2 + g_{22}(du^2)^2$$

$$= \sum_{i,j} g_{ij}\,du^i du^j$$

where (here and elsewhere in this chapter) i and j are each summed from 1 to 2. (In passing to the final sum, we used the symmetry of the matrix (g_{ij}) to replace $2g_{12}\,du^1\,du^2$ by $g_{12}\,du^1\,du^2 + g_{21}\,du^2\,du^1$.) In Chapter III, we shall deal with similar sums in which the indices are summed from 1 to 4. An advantage of the present notation is that it immediately extends to higher dimensions.

The components of a tangent vector $\mathbf{v} = a\mathbf{X}_1 + b\mathbf{X}_2$ (at some pont of M) will be written v^1 and v^2, so that

$$v = v^1\mathbf{X}_1 + v^2\mathbf{X}_2 = \sum_i v^i\mathbf{X}_i$$

(i is just an index of summation or "dummy variable," and any other letter could be used in its place). If $\mathbf{v} = \Sigma_i\, v^i\mathbf{X}_i$ and $\mathbf{w} = \Sigma_j\, w^j\mathbf{X}_j$ are two tangent vectors at the same point of M, then

$$\mathbf{v}\cdot\mathbf{w} = \sum_{i,j} (v^i\mathbf{X}_i)\cdot(w^j\mathbf{X}_j) = \sum_{i,j} v^i w^j (\mathbf{X}_i\cdot\mathbf{X}_j)$$

or

$$\mathbf{v}\cdot\mathbf{w} = \sum_{i,j} g_{ij} v^i w^j \tag{11}$$

Thus (g_{ij}) is simply the matrix for the ordinary dot product of tangent vectors to M relative to the basis vectors $\mathbf{X}_1$ and $\mathbf{X}_2$. Tangent vectors $\mathbf{v}$ and $\mathbf{w}$ are orthogonal if and only if $\Sigma_{i,j}\, g_{ij} v^i w^j = 0$.

The velocity vector of a curve on M, $\alpha(t) = \mathbf{X}(u^1(t), u^2(t))$, can now be written

$$\alpha'(t) = u^{1\prime}(t)\mathbf{X}_1 + u^{2\prime}(t)\mathbf{X}_2 = \sum_i u^{i\prime}\mathbf{X}_i$$

where, as usual, the $\mathbf{X}_i$ are evaluated at $(u^1(t), u^2(t))$.

We shall denote by g^{ij} the components of the matrix inverse of (g_{ij}), i.e., the matrix (g^{ij}) satisfies the condition

$$\begin{pmatrix} g_{11} & g_{12} \\ g_{21} & g_{22} \end{pmatrix} \begin{pmatrix} g^{11} & g^{12} \\ g^{21} & g^{22} \end{pmatrix} = \begin{pmatrix} 1 & 0 \\ 0 & 1 \end{pmatrix}$$

or, in abbreviated form,

$$\sum_j g_{ij} g^{jk} = \delta_i^k \tag{12}$$

where the symbol δ_i^k is defined by

$$\delta_i^k = \begin{cases} 1 & \text{if } i = k \\ 0 & \text{if } i \neq k \end{cases}$$

As you can check by substituting into Eq. (12),

$$g^{11} = \frac{g_{22}}{g}, \quad g^{12} = g^{21} = \frac{-g_{12}}{g}, \quad g^{22} = \frac{g_{11}}{g}$$

where $g = \det(g_{ij})$ (formerly written $EG - F^2$).

Knowledge of the first fundamental form not only enables us to compute distances and angles, but areas as well. Roughly, here is

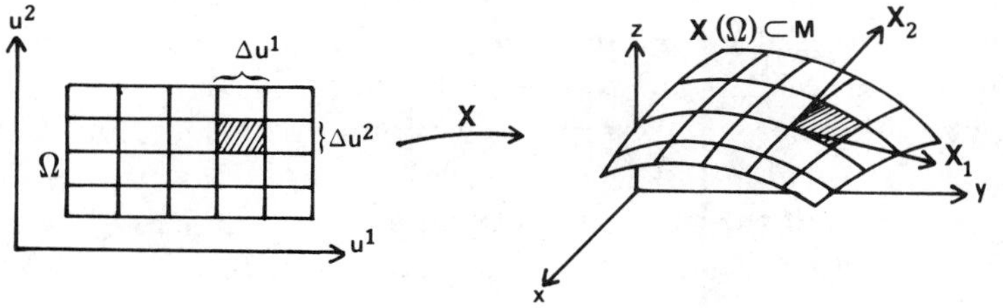

Figure I - 24.

why. Suppose $\mathbf{X}: D \to E^3$ is a surface in E^3 and $\Omega \subset D$ is a region on which $\mathbf{X}$ is one-to-one. To find the area of $\mathbf{X}(\Omega)$, we partition Ω into small rectangular subregions by grid lines parallel to the u^1- and u^2-axes. In Figure I-24, we see that a typical subregion with sides Δu^1 and Δu^2 is mapped into an approximately parallelogram-shaped piece of surface, whose boundary is comprised of arcs of u^1- and u^2-parameter curves.

Now since the distance traveled along any small parametrized curve segment is approximately the product of the speed and the parameter increment ("distance = rate × time"), adjacent sides of this "parallelogram" have the approximate lengths $\|\mathbf{X}_1\|\Delta u^1$ and $\|\mathbf{X}_2\|\Delta u^2$, where $\mathbf{X}_1$ and $\mathbf{X}_2$ may be evaluated at any point of the subregion, since the values of these vector functions will not vary appreciably over a very small subregion. If θ is the angle between these sides, i.e., the angle between $\mathbf{X}_1$ and $\mathbf{X}_2$, then the area ΔA of the "parallelogram" is

$$\Delta A \approx \|\mathbf{X}_1\| \, \|\mathbf{X}_2\| \sin\theta \, \Delta u^1 \Delta u^2$$

$$= \|\mathbf{X}_1 \times \mathbf{X}_2\| \, \Delta u^1 \Delta u^2 = \sqrt{g}\, \Delta u^1 \Delta u^2$$

by Eq. (10). Summing over all the subregions of Ω and passing to the limit of finer and finer partitions, we obtain

$$A = \iint_\Omega \|\mathbf{X}_1 \times \mathbf{X}_2\| \, du^1 du^2 = \iint_\Omega \sqrt{g}\, du^1 du^2 \tag{13}$$

as the area of $\mathbf{X}(\Omega)$ (provided the integral exists and is finite).

This formula is valid also for any smooth function $\mathbf{X}:\Omega \to E^3$ when Ω is a closed subset of the $u^1 u^2$-plane and the restriction of $\mathbf{X}$ to the interior of Ω is one-to-one and regular.

For the sphere of Example 7, we found that $\sqrt{g} = \|\mathbf{X}_1 \times \mathbf{X}_2\| = R^2 \cos v$. The region $\Omega = \{(u,v) | -\pi/2 \leqslant v \leqslant \pi/2, -\pi \leqslant u \leqslant \pi\}$ satisfies the conditions of the previous paragraph and is mapped by $\mathbf{X}$ onto the entire sphere. Therefore the area of a sphere of radius R is

$$A = \int_{-\pi/2}^{\pi/2} \int_{-\pi}^{\pi} R^2 \cos v \, du \, dv = 2\pi R^2 \int_{-\pi/2}^{\pi/2} \cos v \, dv = 4\pi R^2$$

Exercises I-4

1. Derive the formula $ds^2 = dr^2 + r^2 d\theta^2$ for the differential of arc length in polar coordinates by substituting

$$x = r \cos \theta, \quad y = r \sin \theta$$

into the formula $ds^2 = dx^2 + dy^2$ for the differential of arc length in Cartesian coordinates.

2. Use Eq. (7) to find the lengths of the following curves, given in terms of polar coordinates in E^2.

(a) $r = e^\theta, \; 0 \leqslant \theta \leqslant \ln 4$

(b) $r = a(1 + \cos \theta), \; -\pi \leqslant \theta \leqslant \pi$ (a is a positive constant)

(c) $r = \cos^3 \dfrac{\theta}{3}, \quad 0 \leqslant \theta \leqslant 3\pi$

3. For each of the following surfaces, compute the matrix (g_{ij}), its determinant g, the inverse matrix (g^{ij}), and the unit normal vector $\mathbf{U}$:

a. the sphere of Example 7
b. the hemisphere of Example 8
c. the helicoid of Example 9
d. the torus of Example 12
e. the catenoid of Exercise 2f, Section 3

(Retain a copy of your answers for later reference.)

4. Compute the surface area of a torus (Example 12).

5. Compute the metric form and the unit normal vector **U** for the general surface of revolution (Exercise 1, Section 3).

$$\mathbf{X}(u,v) = (f(u)\cos v,\, f(u)\sin v,\, g(u))$$

6. (a) Show that the area A of the surface of revolution

$$\mathbf{X}(u,v) = (f(u)\cos v,\, f(u)\sin v,\, g(u))$$
$$a \leqslant u \leqslant b, \quad 0 \leqslant v \leqslant 2\pi$$

is given by

$$A = 2\pi \int_a^b f(u)[f'(u)^2 + g'(u)^2]^{1/2}\, du$$

(b) Show that the area of the surface obtained by revolving the graph $y = f(x)$, $a \leqslant x \leqslant b$, about the x-axis is given by

$$A = 2\pi \int_a^b f(x)[1 + f'(x)^2]^{1/2}\, dx$$

7. Find the surface area of a right circular cone of base radius r and height h. [This is the surface generated by revolving the line segment $\alpha(u) = (ru, 0, hu)$, $0 \leqslant u \leqslant 1$, about the z-axis.]

8. (a) Show that the surface $z = f(x,y)$, $(x,y) \in D$, has area

$$A = \iint_D \left[1 + \left(\frac{\partial f}{\partial x}\right)^2 + \left(\frac{\partial f}{\partial y}\right)^2\right]^{1/2} dx\, dy$$

(b) Use this formula to find the area of the paraboloid $z = (x^2 + y^2)/2$, $0 \leqslant x^2 + y^2 \leqslant 3$. (Transform to a double integral in polar coordinates.) (c) Compute the surface area of the same paraboloid by representing it as a surface of revolution and using Exercise 6.

9. The image of the vector function

$$\mathbf{X}(u,v) = (a\cos u\cos v,\, a\cos u\sin v,\, b\sin u)$$
$$-\pi/2 \leqslant u \leqslant \pi/2, \quad 0 \leqslant v \leqslant 2\pi$$

is the *ellipsoid of revolution* obtained by revolving the ellipse $x^2/a^2 + z^2/b^2 = 1$ about the z-axis. The ellipsoid is called *oblate* if $a > b$, and *prolate* if $a < b$. Find its surface area in either case. [Hint: If $a > b$, make the substitution $w = (a^2 - b^2)^{1/2}\sin u$; if $a < b$, use $w = (b^2 - a^2)^{1/2}\sin u$. Consult a table of integrals.]

10. If the curve represented in polar coordinates as $r = f(\theta)$, $\alpha \leqslant \theta \leqslant \beta$, is revolved about the polar axis (the line $\theta = 0$), the resulting surface of revolution is

$$\mathbf{X}(\theta, v) = (f(\theta)\cos\theta,\ f(\theta)\sin\theta\ \cos v,\ f(\theta)\sin\theta\ \sin v)$$

(a) Show that the area of this surface is

$$A = 2\pi \int_{\alpha}^{\beta} f(\theta)\sin\theta\ [f(\theta)^2 + f'(\theta)^2]^{1/2}\ d\theta$$

(b) The half *cardioid*

$$r = a(1 + \cos\theta), \quad 0 \leqslant \theta \leqslant \pi$$

(a constant) is revolved about the polar axis. Find the area of the resulting surface.

11. Convince a fellow student (or your instructor) that the area formulas of exercises 6a, 6b, and 10 are all particular cases of the general formula

$$A = 2\pi \int \ell\ ds$$

where ds is the element of arc along the profile curve and ℓ is the perpendicular distance from this element to the axis of revolution.

12. Heuristically, one can think of the differential of arc, ds, as the distance between two infinitesimally close points, (x,y,z) and $(x + dx, y + dy, z + dz)$. If these two points happen to lie on the surface $X(u,v)$, then

$$dx = \frac{\partial x}{\partial u}\ du + \frac{\partial x}{\partial v}\ dv$$

$$dy = \frac{\partial y}{\partial u}\ du + \frac{\partial y}{\partial v}\ dv$$

$$dz = \frac{\partial z}{\partial u}\ du + \frac{\partial z}{\partial v}\ dv$$

Substitute these expressions into the formula $ds^2 = dx^2 + dy^2 + dz^2$ and, using Eq. (4), show that Eq. (6) results.

13. The right circular cylinder $x^2 + y^2 = R^2$ may be parametrized as

$$\mathbf{X}(u,v) = \left(R\cos\frac{u}{R},\ R\sin\frac{u}{R},\ v\right)$$

Compute the metric form. (If we endow the uv-plane with the Euclidean metric $ds^2 = du^2 + dv^2$, then the result of this exercise shows that any curve in the uv-plane and its image under **X** on the cylinder have the same length. A smooth mapping, such as **X**, which preserves lengths of curves is called a *local isometry*. An *isometry* is a local isometry that is one-to-one and onto. Recall our comments about the cylinder in Section 2.)

14. Let M be the plane with polar coordinate. (a) Show that the area of a region Ω of M is given by

$$A(\Omega) = \iint_\Omega r\,dr\,d\theta$$

(b) Find the area enclosed by the cardioid

$$r = a(1 + \cos\theta), \quad 0 \leqslant \theta \leqslant 2\pi$$

5. THE SECOND FUNDAMENTAL FORM

The Einstein Summation Convention. It will be convenient to introduce one more notational convention, due to Albert Einstein. From now on, the summation symbol, Σ, will be omitted in any sum in which each index of summation appears in each term both as a subscript and as a superscript.

For example, we shall write

$$v^i\mathbf{X}_i \quad \text{instead of} \quad \sum_i v^i\mathbf{X}_i$$

$$g_{ij}v^iw^j \quad \text{instead of} \quad \sum_{i,j} g_{ij}v^iw^j$$

(All indices are summed from 1 to 2 in this chapter.) We may of course replace any index letter (wherever it occurs in a sum) by any other letter (that does not occur elsewhere in the sum), so that, for instance, $g_{ik}v^iw^k$, $g_{mn}v^mw^n$, and $g_{rs}v^rw^s$ are the same sum. If

(a_{ij}) is any symmetric matrix ($a_{ij} = a_{ji}$, for all i and j), then $a_{ij}v^i w^j = a_{ji}v^i w^j = a_{ij}v^j w^i$, the last equality resulting from interchanging i and j.

In an expression where an index appears only as a subscript or only as a superscript (e.g., g_{ii}), we do not sum over that index, but we may sum over other indices (as in $g_{ij}u^{j'}$). For a final example, suppose $A = (a^i_j)$ and $B = (b^i_j)$ are two n-by-n matrices. If superscripts are row indices and subscripts column indices, then $a^i_j b^j_k$ is the ith row - kth column entry of the matrix product AB. (In this chapter, n = 2, but the summation convention applies equally to higher-dimensional context, where all indices vary from 1 to n.) See also Exercise 1 of this section.

The Gauss Formulas. Let $\boldsymbol{\alpha}(s) = \mathbf{X}(u^1(s), u^2(s))$ be a curve on M, and assume s is arc length. Then the unit tangent vector to $\boldsymbol{\alpha}$ is

$$\mathbf{T} = \boldsymbol{\alpha}' = u^{i'}\mathbf{X}_i$$

(the summation convention). In Section 1, we defined the curvature of a curve in E^3 to be the length of the acceleration or curvature vector, $\boldsymbol{\alpha}'' = \mathbf{T}'$, a vector normal to the curve. In this section, we are dealing with a curve that lies on a surface M in E^3. In order to relate the curvature of $\boldsymbol{\alpha}$ to the geometry of M, we decompose $\boldsymbol{\alpha}''$ into components tangent and normal to the surface:

$$\boldsymbol{\alpha}'' = \boldsymbol{\alpha}''_{\tan} + \boldsymbol{\alpha}''_{\text{nor}} \tag{14}$$

An example of this decomposition is illustrated in Figure I-25. The next few paragraphs are concerned with computing the two components in Eq. (14).

Now since $\boldsymbol{\alpha}' = u^{i'}\mathbf{X}_i$ – where $\mathbf{X}_i$ means $\mathbf{X}_i(u^1(s),\ u^2(s))$, a function of s – the product rule gives

$$\boldsymbol{\alpha}'' = u^{i''}\mathbf{X}_i + u^{i'}\frac{d}{ds}\mathbf{X}_i \tag{15}$$

In order to decompose $\boldsymbol{\alpha}''$ as in Eq. (14), we must first decompose the vector $d\mathbf{X}_i/ds$ into tangent and normal components. Recall (Section 3) that

$$\mathbf{X}_i = \frac{\partial \mathbf{X}}{\partial u^i} = \left(\frac{\partial x}{\partial u^i}, \frac{\partial y}{\partial u^i}, \frac{\partial z}{\partial u^i}\right), \qquad i = 1,2$$

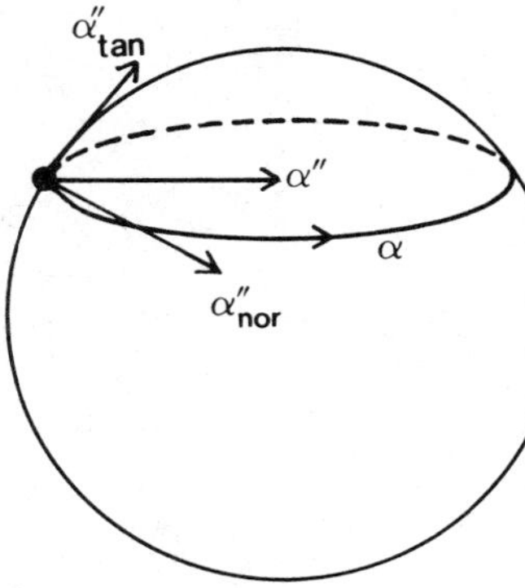

Figure I - 25.

Similarly, define

$$\mathbf{X}_{ij} = \frac{\partial^2 \mathbf{X}}{\partial u^j \partial u^i} = \frac{\partial \mathbf{X}_i}{\partial u^j} = \left(\frac{\partial^2 x}{\partial u^j \partial u^i}, \frac{\partial^2 y}{\partial u^j \partial u^i}, \frac{\partial^2 z}{\partial u^j \partial u^i} \right), \; i,j = 1,2$$

(the order of differentiation is immaterial). Applying the chain rule to $\mathbf{X}_i = \mathbf{X}_i(u^1(s), u^2(s))$, we now have

$$\frac{d}{ds} \mathbf{X}_i = \frac{\partial \mathbf{X}_i}{\partial u^1} u^{1\prime} + \frac{\partial \mathbf{X}_i}{\partial u^2} u^{2\prime} = \frac{\partial \mathbf{X}_i}{\partial u^j} u^{j\prime} = u^{j\prime} \mathbf{X}_{ij}$$

and so, substituting this into Eq. (15) (and changing the dummy variable in the first term), we can write

$$\alpha'' = u^{r\prime\prime} \mathbf{X}_r + u^{i\prime} u^{j\prime} \mathbf{X}_{ij} \tag{16}$$

(In the second term on the right, both i and j are summed.)

Next, we break $\mathbf{X}_{ij}$ into components. The tangent component will be a linear combination of $\mathbf{X}_1$ and $\mathbf{X}_2$, and the normal component, a multiple of the unit normal vector, $\mathbf{U}$. Accordingly, define the functions Γ^r_{ij} and L_{ij} by

$$\boxed{\mathbf{X}_{ij} = \Gamma^r_{ij} \mathbf{X}_r + L_{ij} \mathbf{U}} \tag{17}$$

for i,j = 1,2. These are known as the *formulas of Gauss*. For each i, j, and r, Γ^r_{ij} and L_{ij} are of course functions of u^1 and u^2. Finally, substituting Eq. (17) into Eq. (16), we have

$$\boldsymbol{\alpha}'' = (u^{r\prime\prime} + \Gamma^r_{ij}\, u^{i\prime} u^{j\prime})\,\mathbf{X}_r + (L_{ij}\, u^{i\prime} u^{j\prime})\,\mathbf{U} \tag{18}$$

Thus,

$$\boldsymbol{\alpha}''_{tan} = (u^{r\prime\prime} + \Gamma^r_{ij}\, u^{i\prime} u^{j\prime})\,\mathbf{X}_r \tag{19a}$$

$$\boldsymbol{\alpha}''_{nor} = (L_{ij}\, u^{i\prime} u^{j\prime})\,\mathbf{U} \tag{19b}$$

If Eq. (18) were written out in full, it would look like this:

$$\begin{aligned} \boldsymbol{\alpha}'' = {} & [u^{1\prime\prime} + \Gamma^1_{11}(u^{1\prime})^2 + 2\Gamma^1_{12}u^{1\prime}u^{2\prime} + \Gamma^1_{22}(u^{2\prime})^2]\,\mathbf{X}_1 \\ & + [u^{2\prime\prime} + \Gamma^2_{11}(u^{1\prime})^2 + 2\Gamma^2_{12}u^{1\prime}u^{2\prime} + \Gamma^2_{22}(u^{2\prime})^2]\,\mathbf{X}_2 \\ & + [L_{11}(u^{1\prime})^2 + 2L_{12}u^{1\prime}u^{2\prime} + L_{22}(u^{2\prime})^2]\,\mathbf{U} \end{aligned}$$

You can plainly see the advantages of index notation and the summation convention.

The Second Fundamental Form. The coefficient of **U** in Eq. (19b), $L_{ij}\, u^{i\prime} u^{j\prime}$, is called the *second fundamental form* of the surface. Whereas the first fundamental form determines the intrinsic geometry of M, the second fundamental form reflects the extrinsic geometry, i.e., the way M is imbedded in the ambient 3-dimensional Euclidean space, and how it curves relative to that space.

The coefficients of the second fundamental form are easily computed from the relations

$$\boxed{L_{ij} = \mathbf{X}_{ij} \cdot \mathbf{U}, \quad i,j = 1,2} \tag{20}$$

which follow immediately from the Gauss formulas, Eq. (17). Note that Eq. (20) implies that $L_{ij} = L_{ji}$, i.e., the matrix (L_{ij}) is symmetric.

Example 13. In Example 7 (the sphere parametrized) by geographic coordinates), we found that

$$\mathbf{X}_1 = (-R \sin u \cos v, R \cos u \cos v, 0)$$

$$\mathbf{X}_2 = (-R \cos u \sin v, -R \sin u \sin v, R \cos v)$$

$$\mathbf{U} = (\cos u \cos v, \sin u \cos v, \sin v)$$

Differentiating, we obtain

$$\mathbf{X}_{11} = (-R \cos u \cos v, -R \sin u \cos v, 0)$$

$$\mathbf{X}_{12} = (R \sin u \sin v, -R \cos u \sin v, 0)$$

$$\mathbf{X}_{22} = (-R \cos u \cos v, -R \sin u \cos v, -R \sin v)$$

Therefore,

$$L_{11} = \mathbf{X}_{11} \cdot \mathbf{U} = -R \cos^2 v$$

$$L_{21} = L_{12} = \mathbf{X}_{12} \cdot \mathbf{U} = 0$$

$$L_{22} = X_{22} \cdot \mathbf{U} = -R$$

Example 14. Let

$$\mathbf{X}(u,v) = (u,v,f(u,v)), \quad (u,v) \in D$$

Then $M = \mathbf{X}(D)$ is the graph of f (see Fig. I-18). Using subscripts to indicate partial differentiation, we have

$$\mathbf{X}_1 = (1,0,f_u), \quad \mathbf{X}_2 = (0,1,f_v)$$

and so

$$(g_{ij}) = \begin{pmatrix} 1 + f_u^2 & f_u f_v \\ f_u f_v & 1 + f_v^2 \end{pmatrix}$$

and the determinant of (g_{ij}) is

$$g = 1 + f_u^2 + f_v^2$$

The unit normal, $\mathbf{U} = (\mathbf{X}_1 \times \mathbf{X}_2)/\|\mathbf{X}_1 \times \mathbf{X}_2\|$, is

$$U = \frac{1}{\sqrt{g}}(-f_u, -f_v, 1)$$

The second derivatives of $\mathbf{X}$ are

$$\mathbf{X}_{11} = (0,0,f_{uu}), \quad \mathbf{X}_{12} = (0,0,f_{uv}), \quad \mathbf{X}_{22} = (0,0,f_{vv})$$

or, for short, $\mathbf{X}_{ij} = (0,0,f_{ij})$. Consequently,

$$L_{ij} = f_{ij}/\sqrt{g}, \quad i,j, = 1,2$$

Computation of the Γ^r_{ij} will be deferred until Section 7, where we shall derive formulas for those functions. Next, we shall focus on the geometric significance of the second fundamental form.

Normal Curvature

Definition I-3

Let $\mathbf{v} = v^i\mathbf{X}_i$ be a unit vector tangent to M at $\mathbf{P}$. The *normal curvature* of M at $\mathbf{P}$ *in the* $\mathbf{v}$ *direction*, denoted $k_n(\mathbf{v})$, is defined by

$$k_n(\mathbf{v}) = L_{ij}v^iv^j \tag{21}$$

If $\boldsymbol{\alpha}$ is any curve on M satisfying $\boldsymbol{\alpha}(s_0) = \mathbf{P}$, $\boldsymbol{\alpha}'(s_0) = \mathbf{v}$, then from $\boldsymbol{\alpha}'(s_0) = u^{i\prime}(s_0)\mathbf{X}_i(u^1(s_0),u^2(s_0))$ we conclude that $v^i = u^{i\prime}(s_0)$. From Eq. (16) then, it follows that

$$k_n(\mathbf{v}) = L_{ij}u^{i\prime}u^{j\prime} = \boldsymbol{\alpha}'' \cdot \mathbf{U} \tag{22}$$

[It is to be understood that $u^{i\prime}$ and $\boldsymbol{\alpha}''$ are evaluated at s_0, and L_{ij} and $\mathbf{U}$ at $(u^1(s_0),u^2(s_0))$.] Clearly, all curves passing through $\mathbf{P}$ with the same unit tangent vector $\mathbf{v}$ give the same value for $\boldsymbol{\alpha}'' \cdot \mathbf{U}$.

If $\boldsymbol{\alpha}$ happens to be the normal section at $\mathbf{P}$ in the $\mathbf{v}$ direction, then (at $\mathbf{P}$) $\boldsymbol{\alpha}''$ will be parallel to $\pm\mathbf{U}$, and so $k_n(\mathbf{v}) = \boldsymbol{\alpha}'' \cdot \mathbf{U} = \pm\|\boldsymbol{\alpha}''\|$, where the sign is positive or negative according as the acceleration vector has direction the same as, or opposite to, that of $\mathbf{U}$. Our present definition of normal curvature therefore coincides with the one given informally in Section 2.

Moreover, we can extend the concept of normal curvature to vectors **v** of arbitrary non-zero length simply by replacing **v** in the right side of Eq. (21) by $\mathbf{v}/\|\mathbf{v}\|$. Thus, the normal curvature at **P** in the **v** direction is given by

$$k_n(\mathbf{v}) = \frac{L_{ij}v^iv^j}{\|\mathbf{v}\|^2} = \frac{L_{ij}v^iv^j}{g_{mn}v^mv^n}$$

for any non-zero tangent vector $\mathbf{v} = v^iX^i$ at **P**. Since $k_n(\mathbf{v}) = k_n(c\mathbf{v})$ for any non-zero constant c, $k_n(\mathbf{v})$ depends only on the direction of **v**, i.e., on the one-dimensional subspace of $T_{\mathbf{P}}(M)$ spanned by **v**.

Our next task will be to prove that there exist two orthogonal directions (the *principal directions* at P) in which k_n assumes its maximum and minimum values. We shall then derive an expression for the Gauss curvature, and show eventually (in Section 8) that it is an intrinsic quantity, dependent only on the metric form. The geometric significance of the Γ^r_{ij} will be revealed later.

Exercises I-5

1. (Assume the Einstein summation convention is in force.) Suppose (a_{ij}) is a real symmetric $n \times n$ matrix: $a_{ij} = a_{ji}$. A function of the form

$$f(x^1, x^2, \ldots, x^n) = a_{ij}x^ix^j$$

is called a *quadratic form*. Examples (in which n = 2) are the first and second fundamental forms of a surface at any one of its points. Show that

$$\partial f/\partial x^r = 2a_{rj}x^j, \quad r = 1, 2.$$

[Hint: to see what is going on, take the case n = 2 or 3 and write out all terms. Then argue for the general case, where f may be written $f = \Sigma_{i=1}^n a_{ii}(x^i)^2 + 2\,\Sigma_{i<j}\,a_{ij}x^ix^j$.]

2. Compute the second fundamental form of the surface of revolution

$$\mathbf{X}(u,v) = (f(u)\cos v, f(u)\sin v, g(u)).$$

3. Compute the second fundamental form of (a) the helicoid

$$\mathbf{X}(u,v) = (u \cos v, u \sin v, bv), \quad b \text{ constant}$$

(b) the catenoid

$$\mathbf{X}(u,v) = \left(a \cosh \frac{u}{a} \cos v, a \cosh \frac{u}{a} \sin v, u\right), \quad a \text{ constant}$$

4. Let M be the hyperbolic paraboloid $z = (y^2 - x^2)/2$, and let $\mathbf{P}$ be the origin of E^3. Then $T_{\mathbf{P}}M$ is the xy-plane. Show that the normal curvature of M at $\mathbf{P}$ in the direction of a unit vector $\mathbf{v} = (\cos\theta, \sin\theta, 0)$ is

$$k_n(\mathbf{v}) = -\cos^2\theta + \sin^2\theta = -\cos 2\theta$$

What values of θ give the principal directions?

5. Find the normal curvature of the surface $z = f(x,y)$ at an arbitrary point, in the direction of a unit tangent vector (a, b, c) at that point.

6. Let $\boldsymbol{\alpha}$ be a curve on a surface M. Show that the normal curvature of M at a point of $\boldsymbol{\alpha}$, in the direction of $\boldsymbol{\alpha}'$, is given by

$$k_n(\boldsymbol{\alpha}') = k \cos\phi$$

where $k = \|\boldsymbol{\alpha}''\|$ is the curvature of $\boldsymbol{\alpha}$ (as a curve in E^3) and ϕ is the angle between the surface normal $\mathbf{U}$ and the principal normal vector of $\boldsymbol{\alpha}$. (This result is known as *Meusnier's Theorem.*)

7. The ellipse $x^2/a^2 + y^2/b^2 = 1$ may be obtained as the intersection of a right circular cylinder of radius b with a suitably inclined plane. Use the result of Exercise 6 to compute the curvature k of the ellipse at the ends of its major and minor axes.

8. Let $\mathbf{X}(u^1,u^2) = (x(u^1,u^2), y(u^1,u^2), z(u^1,u^2))$ be a surface. Show that

$$L_{ij} = \frac{\det\begin{bmatrix} x_{ij} & y_{ij} & z_{ij} \\ x_1 & y_1 & z_1 \\ x_2 & y_2 & z_2 \end{bmatrix}}{\sqrt{g}}$$

where, on the right-hand side, subscripts indicate partial differentiation, e.g., $y_2 = \partial y/\partial u^2$, $z_{ij} = \partial^2 z/\partial u^j \partial u^i$.

6. THE GAUSS CURVATURE IN DETAIL

Let us now find the maximum and minimum values of the normal curvature $k_n(\mathbf{v})$ at a given point P of M. This means that we wish to extremize $L_{ij}v^iv^j$ subject to the constraint $g_{ij}v^iv^j = 1$. Since we are dealing with a continuous function ($\mathbf{v} \to L_{ij}v^iv^j$) defined on a closed bounded set (the circle $\|\mathbf{v}\| = 1$ in the plane $T_\mathbf{P}M$), we are assured, by a standard theorem about continuous functions, that k_n assumes both a maximum value k_1 and a minimum value k_2 on this set ($k_1 \geqslant k_2$, with equality holding only when $k_n(\mathbf{v})$ is constant at **P** independently of **v**).

Definition I-4

Let k_1 and k_2 be, respectively, the maximum and minimum values of the normal curvature $k_n(\mathbf{v})$ at a point **P** of the surface M. Then k_1 and k_2 are called the *principle curvatures* of M at **P**, and the corresponding directions are called *principal directions*. The product, $K = K(\mathbf{P}) = k_1 k_2$, is called the *Gauss curvature* of M at **P**.

The usual determination of k_1 and k_2 proceeds by the method of Lagrange multipliers, but since you may be unfamiliar with this method, we shall give a different derivation. We may drop the restriction $g_{ij}v^iv^j = 1$ and extremize the function

$$k = k_n(\mathbf{v}) = \frac{L_{ij}v^iv^j}{g_{mn}v^m v^n}$$

where **v** ranges over all non-zero vectors of $T_\mathbf{P}M$. If $k_n(\mathbf{v})$ is an extreme value of k, i.e., a principle curvature (with $\mathbf{v} = v^i\mathbf{X}_i$ a vector in the associated principal direction), then $\partial k/\partial v^1 = \partial k/\partial v^2 = 0$ at **v**. By the quotient rule (and Exercise 1 of Section 5),

$$\frac{\partial k}{\partial v^r} = \frac{2g_{mn}v^m v^n L_{rj}v^j - 2L_{ij}v^iv^j g_{rn}v^n}{(g_{mn}v^m v^n)^2}, \quad r = 1,2$$

Replacing $L_{ij}v^iv^j$ by $kg_{mn}v^mv^n$ in the numerator and canceling $g_{mn}v^mv^n$ from top and bottom, we have

$$\frac{\partial k}{\partial v^r} = \frac{2(L_{rj} - kg_{rj})v^j}{g_{mn}v^mv^n}, \quad r = 1,2$$

Accordingly, at an extreme value, we must have

$$(L_{ij} - kg_{ij})v^j = 0, \quad i = 1,2 \tag{24}$$

The only way a non-zero vector **v** can satisfy this pair of linear equations is for $\det(L_{ij} - kg_{ij})$ to be zero. Upon expanding this determinant, we obtain the quadratic equation (in the unknown k)

$$k^2 g - k(g_{11}L_{22} + g_{22}L_{11} - 2g_{12}L_{12}) + L = 0$$

or

$$k^2 - k\frac{(g_{11}L_{22} + g_{22}L_{11} - 2g_{12}L_{12})}{g} + \frac{L}{g} = 0$$

where $L = \det(L_{ij})$, $g = \det(g_{ij})$. Since k_1 and k_2 are the roots of this equation, the equation must factor into $(k - k_1)(k - k_2) = 0$, which tells us that the constant term, L/g, equals k_1k_2. This gives us the following formula for the Gauss curvature in terms of the determinants of the first and second fundamental forms.

Theorem I-5

The Gauss curvature at any point P of M is given by $K(P) = L/g$.

We can now compute the Gauss curvature of any of the surfaces we have studied thus far. For instance, the sphere with geographic coordinates has

$$(g_{ij}) = \begin{pmatrix} R^2\cos^2 v & 0 \\ 0 & R^2 \end{pmatrix}, \quad (L_{ij}) = \begin{pmatrix} -R\cos^2 v & 0 \\ 0 & -R \end{pmatrix}$$

from Example 13. Accordingly, $L = R^2\cos^2 v$, $g = R^4\cos^2 v$, and $K = L/g = 1/R^2$, in agreement with our earlier determination in Section 2.

Example 15. The torus,

$$\mathbf{X}(u,v) = ((R + r\cos u)\cos v, (R + r\cos u)\sin v, r\sin u),$$

(refer to Example 12) has

$$\mathbf{U} = (\mathbf{X}_1 \times \mathbf{X}_2)/\|\mathbf{X}_1 \times \mathbf{X}_2\| = -(\cos u\cos v, \cos u\sin v, \sin u)$$

$$\mathbf{X}_{11} = -(r\cos u\cos v, r\cos u\sin v, r\sin u)$$

$$\mathbf{X}_{12} = (r\sin u\sin v, -r\sin u\cos v, 0)$$

$$\mathbf{X}_{22} = -((R + r\cos u)\cos v, (R + r\cos u)\sin v, 0)$$

and so

$$L_{11} = r, \quad L_{12} = 0, \quad L_{22} = (R + r\cos u)\cos u$$

Since $g = r^2(R + \cos u)^2$, we have

$$K = \frac{\cos u}{r(R + r\cos u)}$$

You can now verify that the sign of K varies as described in Example 6. (See also Exercise 4 of Section 2.)

Example 16. From the results of Example 14, you can show that the curvature of the graph of a function $f(x,y)$ is

$$K = \frac{f_{xx}f_{yy} - f_{xy}^2}{(1 + f_x^2 + f_y^2)^2}$$

Let us now prove that the two principal directions are orthogonal. Let $\mathbf{v} = v^i X_i$ and $\mathbf{w} = w^i X_i$ be non-zero vectors in the two principal directions at $\mathbf{P}$, so that, by Eq. (24),

$$(L_{ij} - k_1 g_{ij})v^j = 0$$

$$(L_{ij} - k_2 g_{ij})w^j = 0$$

for i = 1,2. Interchanging i and j in the first equation and taking advantage of the symmetry of the two fundamental forms, we may rewrite this equation as

$$L_{ij}v^i = k_1 g_{ij}v^i, \quad j = 1,2 \tag{25}$$

If we multiply the second equation by v^i, sum over i, and substitute from Eq. (25), we obtain

$$(k_1 - k_2)g_{ij}v^i w^j = 0.$$

Therefore (since $g_{ij}v^i w^j = \mathbf{v} \cdot \mathbf{w}$), if $k_1 \neq k_2$, the two principal directions are orthogonal. If $k_1 = k_2$, then $k_n(\mathbf{v})$ is constant, and we arbitrarily select any two orthogonal directions and call them principal directions.

We conclude this section with another interpretation of Gauss curvature which may agree more with your intuitive notion of what the word curvature means. In the course of presenting this, we shall derive an important equation, (28), which will be useful later.

Suppose $\mathbf{P} = \mathbf{X}(u_0^1, u_0^2)$, and let Ω be a neighborhood of (u_0^1, u_0^2) on which $\mathbf{X}$ is one-to-one with continuous inverse $\mathbf{X}(\Omega) \to \Omega$. To each point $\mathbf{X}(u^1, u^2)$ of $\mathbf{X}(\Omega)$, we may associate the normal vector $\mathbf{U}(u^1, u^2)$ parallel-translated to the origin of E^3. Since $\|\mathbf{U}\| = 1$, the translated vector may be viewed as a point of the unit sphere, S^2. We therefore have a mapping from $\mathbf{X}(\Omega)$ to S^2, called the *sphere mapping* or *Gauss mapping*. Its image, $\mathbf{U}(\Omega)$, is sometimes called the *spherical normal image* of $\mathbf{X}(\Omega)$. Intuitively, the "more curved" M is in the vicinity of $\mathbf{P}$, the more the translated normal vectors will spread apart (Fig. I-26), and the greater will be the ratio of the area

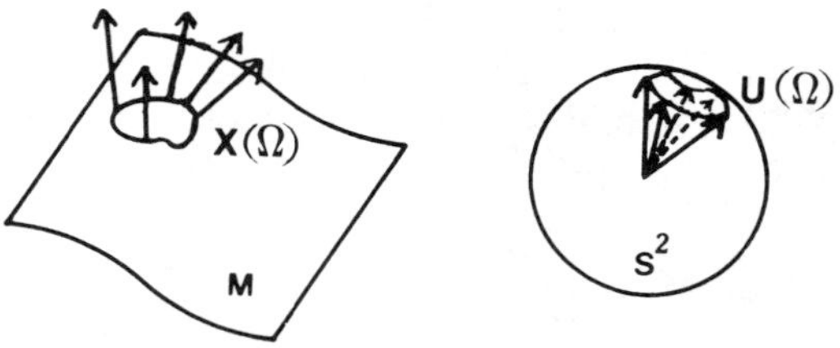

Figure I - 26.

of $\mathbf{U}(\Omega)$ to the area of $\mathbf{X}(\Omega)$. We would therefore expect this ratio to be a measure of the curvature of M near **P**. In fact we shall prove that the limit of this ratio, as the domain Ω shrinks down to the single point (u_0^1, u_0^2), equals $K(\mathbf{P})$, provided the area of $\mathbf{U}(\Omega)$ is taken with an appropriate sign as described below. [We could, for example, approach the limit by taking Ω to be a disk of radius r centered at (u_0^1, u_0^2) and lettering $r \to 0$.]

In terms of the local coordinates u^1 and u^2, the sphere mapping is given by $(u^1, u^2) \to \mathbf{U}(u^1, u^2)$. In this form, we may view **U** as a parametrization of (a part of) S^2, except that the vectors $\mathbf{U}_1 = \partial \mathbf{U}/\partial u^1$ and $\mathbf{U}_2 = \partial \mathbf{U}/\partial u^2$ (the velocity vectors of the u^1- and u^2-parameter curves) are not necessarily independent at all points. (These vectors play the same roles for S^2 as do $\mathbf{X}_1$ and $\mathbf{X}_2$ for M.)

Since **U** (being a radius vector) is normal to the sphere, the tangent plane of S^2 at $\mathbf{U}(u^1, u^2)$, namely $T_{\mathbf{U}} S^2$, is parallel to the tangent plane of M at $\mathbf{X}(u^1, u^2)$, namely $T_{\mathbf{X}} M$. Consequently, if $\mathbf{U}_1 \times \mathbf{U}_2 \neq \mathbf{O}$, then $(\mathbf{U}_1 \times \mathbf{U}_2)/\|\mathbf{U}_1 \times \mathbf{U}_2\|$ and **U** are both unit normal vectors to S^2 at the point **U**, and so can differ at most in sign: $\mathbf{U} = \pm (\mathbf{U}_1 \times \mathbf{U}_2)/\|\mathbf{U}_1 \times \mathbf{U}_2\|$, or

$$\mathbf{U} \cdot \mathbf{U}_1 \times \mathbf{U}_2 = \pm \|\mathbf{U}_1 \times \mathbf{U}_2\|.$$

If $(\mathbf{U}_1 \times \mathbf{U}_2)(u_0^1, u_0^2) \neq \mathbf{O}$, then by regularity and continuity, **U** will be one-to-one on Ω and the sign will be fixed throughout Ω, if Ω is sufficiently small. We then make the area of $\mathbf{U}(\Omega)$ a "signed area" by affixing this sign to it. Therefore, by Eq. (13) and the definition of **U** in Section 4,

$$\text{Area } \mathbf{U}(\Omega) = \pm \iint_{\Omega} \|\mathbf{U}_1 \times \mathbf{U}_2\| \, du^1 \, du^2 = \iint_{\Omega} \mathbf{U} \cdot \mathbf{U}_1 \times \mathbf{U}_2 \, du^1 \, du^2$$

$$\text{Area } \mathbf{X}(\Omega) = \iint_{\Omega} \|\mathbf{X}_1 \times \mathbf{X}_2\| \, du^1 \, du^2 = \iint_{\Omega} \mathbf{U} \cdot \mathbf{X}_1 \times \mathbf{X}_2 \, du^1 \, du^2$$

If $(\mathbf{U}_1 \times \mathbf{U}_2)(u_0^1, u_0^2) = \mathbf{O}$, we shall still refer to the integral of $\mathbf{U} \cdot \mathbf{U}_1 \times \mathbf{U}_2$ as the signed area of $\mathbf{U}(\Omega)$, although in this case it might be more accurate to call it a "net area" since **U** need not be one-to-one and $\mathbf{U} \cdot \mathbf{U}_1 \times \mathbf{U}_2$ may change sign on Ω.

The claim is that

$$K(P) = \lim_{\Omega \to (u_0^1, u_0^2)} \frac{\text{Area } \mathbf{U}(\Omega)}{\text{Area } \mathbf{X}(\Omega)} \tag{26}$$

For Ω sufficiently small, the integrands in the double integrals above, being continuous functions, will not vary appreciably from their values at **P**, and so the two areas are approximately (integrals of constant functions)

$$\text{Area } \mathbf{U}(\Omega) \approx (\mathbf{U} \cdot \mathbf{U}_1 \times \mathbf{U}_2)(\mathbf{P})\, A(\Omega)$$

$$\text{Area } \mathbf{X}(\Omega) \approx (\mathbf{U} \cdot \mathbf{X}_1 \times \mathbf{X}_2)(\mathbf{P})\, A(\Omega)$$

where $A(\Omega)$ is the area of the domain Ω in the $u^1 u^2$-plane. The ratio is therefore approximately

$$\frac{\text{Area } \mathbf{U}(\Omega)}{\text{Area } \mathbf{X}(\Omega)} \approx \frac{\mathbf{U} \cdot \mathbf{U}_1 \times \mathbf{U}_2}{\mathbf{U} \cdot \mathbf{X}_1 \times \mathbf{X}_2}(\mathbf{P})$$

and the limit equals the right side of this approximation (exactly). Equation (26) then results from the following:

Lemma I-6

$$\mathbf{U}_1 \times \mathbf{U}_2 = K(\mathbf{X}_1 \times \mathbf{X}_2)$$

Proof. Define the functions $L_j^i = L_j^i(u^1, u^2)$ by

$$L_j^i = L_{jk}\, g^{ki}, \quad i,j = 1,2 \tag{27}$$

(Sum over k from 1 to 2.) Note that $L_j^i g_{im} = L_{jk} g^{ki} g_{im} = L_{jk} \delta_m^k = L_{jm}$, by Eq. (12). Since $\mathbf{U} \cdot \mathbf{U} = 1$, then $\mathbf{U} \cdot \mathbf{U}_j = 0$ (by differentiation), and so $\mathbf{U}_j$ is a tangent vector to M. As such, it may be expressed as a linear combination of $\mathbf{X}_1$ and $\mathbf{X}_2$:

$$\mathbf{U}_j = a_j^r\, \mathbf{X}_r, \quad j = 1,2$$

where the coefficients a_j^r are functions to be determined. Furthermore, differentiation of the equation $\mathbf{U}\cdot\mathbf{X}_k = 0$ with respect to u^j yields $\mathbf{U}_j\cdot\mathbf{X}_k = -\mathbf{U}\cdot\mathbf{X}_{jk} = -L_{jk}$ [from Eq. (20)]. Therefore

$$-L_{jk} = \mathbf{U}_j\cdot\mathbf{X}_k = a_j^r\,\mathbf{X}_r\cdot\mathbf{X}_k = a_j^r\,g_{rk}, \quad j,k = 1,2$$

To solve these equations for the coefficients a_j^r, multiply both sides by g^{ki} and sum over k. This produces

$$-g^{ki}L_{jk} = a_j^r\,g_{rk}\,g^{ki} = a_j^r\,\delta_r^i = a_j^i$$

or, by Eq. (27), $a_j^i = -L_j^i$. Therefore, we have established the relations

$$\boxed{\mathbf{U}_j = -L_j^i\,\mathbf{X}_i, \quad j = 1,2} \tag{28}$$

(These are known as the *equations of Weingarten.*) From Eq. (28) it follows that

$$\begin{aligned}\mathbf{U}_1 \times \mathbf{U}_2 &= (-L_1^i\,\mathbf{X}_i) \times (-L_2^k\,\mathbf{X}_k)\\ &= (L_1^1 L_2^2 - L_1^2 L_2^1)(\mathbf{X}_1 \times \mathbf{X}_2)\\ &= \det(L_j^i)(\mathbf{X}_1 \times \mathbf{X}_2)\end{aligned}$$

However since $L_j^i = L_{jk}\,g^{ki}$, and (g^{ki}) is the inverse matrix of (g_{ki}), we have $\det(L_j^i) = \det(L_{jk})\det(g^{ki}) = L/g = K$, and the proof is complete.

Exercises I-6

1. Use the results of Example 16 to compute the Gauss curvature of

   a. the hyperbolic paraboloid, $z = (y^2 - x^2)/2$
   b. the hemisphere of radius R, $z = (R^2 - x^2 - y^2)^{1/2}$

2. (a) Show that the surface of revolution

$$\mathbf{X}(u,v) = (f(u)\cos v,\ f(u)\sin v,\ g(u))$$

has Gauss curvature

$$K = \frac{g' \det \begin{pmatrix} f' & f'' \\ g' & g'' \end{pmatrix}}{f(f'^2 + g'^2)^2}$$

(b) Show that if the profile curve has unit speed (i.e., $f'^2 + g'^2 = 1$), then $K = -f''/f$.

3. Compute the Gauss curvature of an ellipsoid of revolution (Exercise 2c of Section 3).

4. Compute the Gauss curvature of the helicoid and the catenoid (Exercise 3 of Section 5).

5. (a) Prove that if $L_{ij} = 0$ at every point, then **U** has constant components. (Hint: it suffices to show that, for $i = 1, 2$, $\mathbf{U}_i$ is orthogonal to each of the vectors $\mathbf{U}$, $\mathbf{X}_1$, $\mathbf{X}_2$, since these form a basis of R^3.) (b) Prove that if **U** has constant components, then the surface is part of a plane. (Hint: a plane with normal vector **U** has equation $\mathbf{U} \cdot \mathbf{X} =$ constant.)

6. (a) Show that if (at a certain point) $g_{12} = L_{12} = 0$, then the principal curvatures are L_{11}/g_{11}, L_{22}/g_{22}, and the associated principal directions are $\mathbf{X}_1$, $\mathbf{X}_2$, respectively. (b) Conversely, show that if $\mathbf{X}_1$ and $\mathbf{X}_2$ are principal, then $g_{12} = L_{12} = 0$. [Hint: use Eq. (24) for both parts.]

7. Show that at each point of a surface of revolution, the two principal directions are along the meridian and parallel through that point. (Use Exercise 6.)

8. Let $\mathbf{X} = \mathbf{X}(u,v) \in D$, be a parametrization of a surface M. The (signed) area of the spherical normal image of M, $\iint_D \mathbf{U} \cdot \mathbf{U}_1 \times \mathbf{U}_2 \, du \, dv$, is called the *total curvature* of M (assuming the integral, which may be improper, exists). (a) Show that the total curvature of M is

$$\iint_D K \sqrt{g} \, du \, dv$$

(b) Show that if the Gauss mapping is one-to-one on M and if K does not change sign on M, then the total curvature is plus or minus the (absolute) area of its spherical normal image.

9. For each of the following surfaces, describe the spherical normal image of the entire surface and compute the total curvature:

a. an ellipsoid
b. an elliptic paraboloid, $z = x^2/a^2 + y^2/b^2$
c. a hyperbolic paraboloid, $z = x^2/a^2 - y^2/b^2$
d. a torus (Example 12)
e. a catenoid (Exercise 2f of Section 3)

10. For $\mathbf{v} = v^i \mathbf{X}_i$ and $\mathbf{w} = w^j \mathbf{X}_j$ in $T_P M$, define

$$L(\mathbf{v},\mathbf{w}) = L_{ij} v^i w^j$$

Then if $\mathbf{v}$ is a unit tangent vector at **P**, $k_n(\mathbf{v}) = L(\mathbf{v},\mathbf{v})$. Suppose $\mathbf{v}_1$ and $\mathbf{v}_2$ are unit tangent vectors in the two principal directions at **P**, with associated principal curvatures k_1 and k_2, respectively. Any unit tangent vector $\mathbf{v}$ at **P** can be written

$$\mathbf{v} = (\cos\theta)\mathbf{v}_1 + (\sin\theta)\mathbf{v}_2$$

where θ is the angle between $\mathbf{v}_1$ and $\mathbf{v}$. The normal curvature k_n then becomes a function of θ,

$$k_n = L((\cos\theta)\mathbf{v}_1 + (\sin\theta)\mathbf{v}_2, (\cos\theta)\mathbf{v}_1 + (\sin\theta)\mathbf{v}_2)$$

$$= (\cos^2\theta)k_1 + 2\sin\theta\cos\theta\, L(\mathbf{v}_1,\mathbf{v}_2) + (\sin^2\theta)k_2$$

Use the maximality of k_n at $\theta = 0$ to prove *Euler's formula*,

$$k_n(\mathbf{v}) = k_1\cos^2\theta + k_2\sin^2\theta$$

(the principal curvatures determine the normal curvatures in all directions). Equation (1) in Section 2 and the formula in Exercise 4 of Section 5 are particular cases of Euler's formula.

7. GEODESICS

In general, a curve $\alpha(s)$ on a surface M has curvature for two reasons. First, if the surface itself is curved, and bends relative to the surrounding 3-dimensional space that contains it, then in general $\boldsymbol{\alpha}$ will be forced to bend along with M. This type of curvature is essentially the *normal curvature* discussed previously, and is exemplified by a great circle on a sphere. A great circle is the straightest possible curve on a sphere: a bug crawling along the sphere's surface, turning neither toward the left nor to the right, will follow a great circle. Yet a great circle has curvature because a sphere is curved.

Second, whether M is curved or not, $\boldsymbol{\alpha}$ may curve within or relative to M. This type of curvature is called *geodesic curvature* and will be defined precisely below. For example, a circle of radius r on a flat plane has curvature 1/r even though the plane has zero curvature. In this case, the curvature of $\boldsymbol{\alpha}$ is, so to speak, its own doing and does not come from surface curvature.

Assume the parameter s is arc length. Then $\boldsymbol{\alpha}' \cdot \boldsymbol{\alpha}' = 1$ and so $\boldsymbol{\alpha}' \cdot \boldsymbol{\alpha}'' = 0$. As we saw in Section 5, the acceleration or curvature vector, $\boldsymbol{\alpha}''(s)$, may be decomposed into components tangent and normal to M:

$$\boldsymbol{\alpha}'' = \boldsymbol{\alpha}''_{\tan} + \boldsymbol{\alpha}''_{\mathrm{nor}}$$

where

$$\begin{aligned} \boldsymbol{\alpha}''_{\tan} &= (u^{r\,\prime\prime} + \Gamma^r_{ij} u^{i\,\prime} u^{j\,\prime})\mathbf{X}_r \\ \boldsymbol{\alpha}''_{\mathrm{nor}} &= (L_{ij} u^{i\,\prime} u^{j\,\prime})\mathbf{U} \end{aligned} \tag{29}$$

The normal component, whose length, with appropriate sign, is the normal curvature of $\boldsymbol{\alpha}$, reflects the curvature of $\boldsymbol{\alpha}$ due to the bending of M in 3-space. The tangent component reflects the curvature of $\boldsymbol{\alpha}$ relative to the surface.

Since $\boldsymbol{\alpha}'_{\tan} \cdot \mathbf{U} = 0$ and $\boldsymbol{\alpha}''_{\tan} \cdot \boldsymbol{\alpha}' = (\boldsymbol{\alpha}''_{\tan} + \boldsymbol{\alpha}''_{\mathrm{nor}}) \cdot \boldsymbol{\alpha}' = \boldsymbol{\alpha}'' \cdot \boldsymbol{\alpha}' = 0$, $\boldsymbol{\alpha}''_{\tan}$ is orthogonal to both $\mathbf{U}$ and $\boldsymbol{\alpha}'$. It is therefore proportional to the unit vector $\mathbf{w} = \mathbf{U} \times \boldsymbol{\alpha}'$ (a vector tangent to M), the *geodesic normal vector*. The proportionality factor is called the *geodesic curvature*.

Definition I-7

Let $\boldsymbol{\alpha} = \boldsymbol{\alpha}(s)$ be a curve on M, where s is arc length. The geodesic curvature of $\boldsymbol{\alpha}$ at $\boldsymbol{\alpha}(s)$ is the function $k_g = k_g(s)$ defined by the equation

$$\boldsymbol{\alpha}''_{\tan} = k_g \mathbf{w} = k_g \mathbf{U} \times \boldsymbol{\alpha}' \tag{30}$$

Dotting Eq. (30) with **w**, we have $k_g = \boldsymbol{\alpha}''_{tan} \cdot \mathbf{w} = \boldsymbol{\alpha}'' \cdot \mathbf{w} = \boldsymbol{\alpha}'' \cdot \mathbf{U} \times \boldsymbol{\alpha}'$, or, by cyclic permutation of the vectors in a triple scalar product,

$$k_g = \mathbf{U} \cdot \boldsymbol{\alpha}' \times \boldsymbol{\alpha}'' \tag{31}$$

Definition I-8

Let $\boldsymbol{\alpha} = \boldsymbol{\alpha}(s)$ be a curve on M, where s is arc length. Then $\boldsymbol{\alpha}$ is called a *geodesic* if $\boldsymbol{\alpha}''_{tan} = \mathbf{O}$ (or, equivalently, if $\boldsymbol{\alpha}'' = \boldsymbol{\alpha}''_{nor}$) at every point of $\boldsymbol{\alpha}$.

From Eqs. (29), (30), and (31), we see that $\boldsymbol{\alpha}$ is a geodesic if and only if either of the following equivalent conditions hold:

$$u^{r''} + \Gamma^r_{ij} u^{i'} u^{j'} = 0, \quad r = 1,2 \tag{32a}$$

$$\mathbf{U} \cdot \boldsymbol{\alpha}' \times \boldsymbol{\alpha}'' = 0 \tag{32b}$$

Examples are given in the exercises.

According to Theorem I-9, geodesics are the surface analogues of straight lines, in the sense that if a curve realizes the shortest distance on M between its end points, then it is part of a geodesic. Before we can prove this, we need certain identities expressing the functions Γ^r_{ij} [defined by Eq. (17) and called *Christoffel symbols of the second kind*] in terms of the g_{ij} and their first derivatives. These identities will be basic for all that follows. [E. B. Christoffel (1829–1901) first introduced his symbols in 1869 in a paper on differential forms. The modern notation Γ^r_{ij} supersedes Christoffel's original notation, $\left\{ {ij \atop r} \right\}$ (still found in some earlier texts), which is incompatible with the summation convention.]

We define the functions $\Gamma_{ijk}(u^1, u^2)$ by

$$\Gamma_{ijk} = \Gamma^r_{ij} g_{rk}, \quad i,j,k = 1,2 \tag{33}$$

(These are known at the *Christoffel symbols of the first kind*.) Since $\Gamma^r_{ij} = \Gamma^r_{ji}$ [from Eq. (17)], $\Gamma_{ijk} = \Gamma_{jik}$. Note also $\Gamma^m_{ij} = \Gamma_{ijk} g^{km}$, by Eq. (12). Upon dotting both sides of Eq. (17) with $\mathbf{X}_k$ we have

$$\mathbf{X}_{ij} \cdot \mathbf{X}_k = \Gamma^r_{ij} \mathbf{X}_r \cdot \mathbf{X}_k = \Gamma^r_{ij} g_{rk} = \Gamma_{ijk} \tag{34}$$

Accordingly,

$$\frac{\partial g_{ik}}{\partial u^j} = \frac{\partial}{\partial u^j}(\mathbf{X}_i \cdot \mathbf{X}_k) = \mathbf{X}_{ij} \cdot \mathbf{X}_k + \mathbf{X}_{kj} \cdot \mathbf{X}_i$$

or,

$$\frac{\partial g_{ik}}{\partial u^j} = \Gamma_{ijk} + \Gamma_{kji} \tag{35a}$$

Permuting the indices yields

$$\frac{\partial g_{ji}}{\partial u^k} = \Gamma_{jki} + \Gamma_{ikj} \tag{35b}$$

$$\frac{\partial g_{kj}}{\partial u^i} = \Gamma_{kij} + \Gamma_{jik} \tag{35c}$$

Since Γ_{ijk} is symmetric in its first two indices, subtraction of Eq. (35b) from the sum of Eqs. (35a) and (35c) gives

$$\Gamma_{ijk} = \frac{1}{2}\left(\frac{\partial g_{ik}}{\partial u^j} + \frac{\partial g_{jk}}{\partial u^i} - \frac{\partial g_{ij}}{\partial u^k}\right) \tag{36}$$

Multiplying by g^{kr} and summing over k, we obtain finally

$$\Gamma^r_{ij} = \frac{1}{2} g^{kr}\left(\frac{\partial g_{ik}}{\partial u^j} + \frac{\partial g_{jk}}{\partial u^i} - \frac{\partial g_{ij}}{\partial u^k}\right) \tag{37}$$

Thus, the Christoffel symbols, although originally defined extrinsically, in terms of the decomposition of $\mathbf{X}_{ij}$ into tangent and normal components, depend only on the first fundamental form and so belong to the intrinsic geometry of M. From Eq. (32a), we see therefore that the concept of geodesic is intrinsic.

Before proceeding to the next theorem, it might be well to run through some computational examples. It turns out that for *orthogonal coordinates*, i.e., for $g_{12} = g_{21} = 0$, the necessary calculations are somewhat easier. In this case, $g^{12} = g^{21} = 0$, $g^{11} = 1/g_{11}$, $g^{22} = 1/g_{22}$, and Eq. (37) takes the form

$$\Gamma^r_{ij} = \frac{1}{2g_{rr}}\left(\frac{\partial g_{ir}}{\partial u^j} + \frac{\partial g_{jr}}{\partial u^i} - \frac{\partial g_{ij}}{\partial u^r}\right) \quad \text{(no sum)}, \tag{38}$$

where (by our conventions) we do not sum over an index that appears only as a subscript.* As you can easily verify, we now have the following cases (for orthogonal coordinates).

Case 1. For $j = r$,

$$\Gamma^r_{ir} = \frac{1}{2g_{rr}}\frac{\partial g_{rr}}{\partial u^i} = \frac{1}{2}\frac{\partial}{\partial u^i}(\ln g_{rr}) \quad \text{(no sum)} \tag{39a}$$

Case 2. For $i = j \neq r$,

$$\Gamma^r_{ii} = \frac{1}{2g_{rr}}\left(-\frac{\partial g_{ii}}{\partial u^r}\right) \quad \text{(no sum)*} \tag{39b}$$

Because $\Gamma^r_{ij} = \Gamma^r_{ji}$, and because each index can assume a value of either 1 or 2 only, every combination of index values is included in one of these two cases. In Chapter III, when we take up general relativity, we will be applying the geodesic equations in four dimensions, where each index may take on a value from 1 to 4. In that context, there is a third case, not present in dimension two.

Case 3 (in dimension > 2). For i, j, r all distinct,

$$\Gamma^r_{ij} = 0 \tag{39c}$$

In the present 2-dimensional case, Eqs. (39a) and (39b) may be rewritten (in terms of the coefficients of the metric form $ds^2 = E\,du^2 + G\,dv^2$) as

*A denominator superscript, as in $\partial/\partial u^r$, counts as a subscript, for purposes of the summation convention. Hence no sum over r in Eq. (38) or (39b).

$$
\begin{aligned}
\Gamma^1_{11} &= \frac{E_u}{2E}, & \Gamma^2_{22} &= \frac{G_v}{2G} \\
\Gamma^1_{12} &= \Gamma^1_{21} = \frac{E_v}{2E}, & \Gamma^2_{21} &= \Gamma^2_{12} = \frac{G_u}{2G} \\
\Gamma^1_{22} &= \frac{-G_u}{2E}, & \Gamma^2_{11} &= \frac{-E_v}{2G}
\end{aligned} \tag{40}
$$

where, as usual, u or v as a subscript denotes partial differentiation.

Example 17. For the plane with Cartesian coordinates

$$ds^2 = du^2 + dv^2$$

and since E and G are constants, all the Christoffel symbols vanish. Therefore the geodesics [see Eq. (32a)] are given by $u'' = v'' = 0$, and so u and v are linear functions of s:

$$u = as + b, \qquad v = cs + d$$

Consequently, the geodesics of the Euclidean plane are the straight lines.

Example 18. For the sphere with geographic coordinates,

$$ds^2 = R^2 \cos^2 v \, du^2 + R^2 \, dv^2$$

The only non-zero Christoffel symbols are found to be

$$\Gamma^1_{12} = \Gamma^1_{21} = \frac{E_v}{2E} = -\tan v$$

$$\Gamma^2_{11} = -\frac{E_v}{2G} = \sin v \cos v$$

The geodesics will be computed from Eq. (32a) in Exercise 14. They are of course the great circles.

Example 19. For the Euclidean plane with polar coordinates,

$$ds^2 = dr^2 + r^2 \, d\theta^2$$

Replacing u by r and v by θ in Eq. (40), we find as the only non-zero Christoffel symbols,

$$\Gamma^2_{12} = \Gamma^2_{21} = 1/r, \quad \Gamma^1_{22} = -r$$

The equations for a geodesic, Eq. (32a) reduce to

$$\frac{d^2 r}{ds^2} - r\left(\frac{d\theta}{ds}\right)^2 = 0 \tag{41a}$$

$$\frac{d^2\theta}{ds^2} + \frac{2}{r}\frac{dr}{ds}\frac{d\theta}{ds} = 0 \tag{41b}$$

Let us again verify that the geodesics are straight lines. The calculations will be similar to some we shall perform in Chapter III.

If $d\theta/ds$ is identically zero, then θ is constant and our geodesic is a straight line through the origin. Otherwise, if we divide Eq. (41b) by $\theta' = d\theta/ds$, we have

$$\frac{1}{\theta'}\frac{d\theta'}{ds} + \frac{2}{r}\frac{dr}{ds} = 0$$

which can be integrated with respect to s to give

$$\ln|\theta'| + \ln r^2 = c$$

a constant. Exponentiating, we have $|r^2\theta'| = e^c$, or

$$r^2\frac{d\theta}{ds} = h \tag{42}$$

with h a non-zero constant.

From the first fundamental form and Eq. (42), we get

$$1 = \left(\frac{dr}{ds}\right)^2 + r^2\left(\frac{d\theta}{ds}\right)^2 = \left(\frac{dr}{ds}\right)^2 + \frac{h^2}{r^2} \tag{43}$$

[This is actually an integral of Eq. (41a), since Eq. (41a) can be obtained from Eq. (43) by differentiation with respect to s and division by 2 dr/ds.] Solving for dr/ds, we have

$$\frac{dr}{ds} = \pm \frac{1}{r} (r^2 - h^2)^{1/2}$$

and dividing this into the equation $d\theta/ds = h/r^2$, we obtain (using the chain rule)

$$\frac{d\theta}{dr} = \pm \frac{h}{r(r^2 - h^2)^{1/2}} = \pm \frac{d}{dr}\left(\cos^{-1} \frac{h}{r}\right)$$

(The integration of the middle term can be carried out by the trig substitution r = h sec t, or found in a table of integrals.)

Accordingly, $\theta - \theta_0 = \pm \cos^{-1}(h/r)$, where θ_0 is a constant of integration, or

$$\frac{h}{r} = \cos(\theta - \theta_0) \tag{44}$$

This is the equation of a straight line in polar coordinates. The constant h is the perpendicular distance from the origin to the line, and θ_0 is the angle between the ray $\theta = 0$ and the perpendicular dropped from the origin to the line. [In Fig. I-27, **P**, with polar

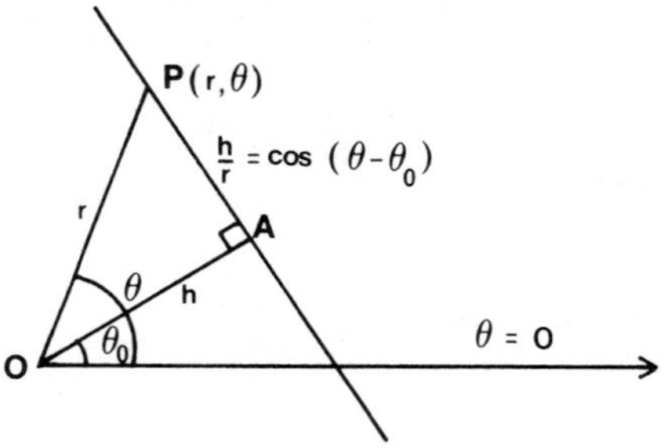

Figure I - 27.

coordinates (r,θ), is a variable point on the line $\overleftrightarrow{\mathbf{PA}}$, and in right $\Delta\mathbf{POA}$, $h/r = \cos(\angle\mathbf{POA}) = \cos(\theta - \theta_0)$.]

Conversely, it is not hard to show [using the left hand equality of Eq. (43)], that Eq. (44) is a solution of Eq. (41) (see Exercise 11). Rays through the origin are included in the general solution Eq. (44) if we allow h to be zero. Thus the geodesics of the plane are all the straight lines.

A few additional examples are given in the exercises. In general, however, the geodesic equations (32a) are often quite difficult or impossible to solve in terms of elementary functions. In Chapter III, we shall compute some geodesics in 4-dimensional spacetime, namely the orbit of a planet and the path of a light ray. The calculations will somewhat resemble those of Example 19.

Theorem I-9

Let $\boldsymbol{\alpha} = \boldsymbol{\alpha}(s)$, $a \leqslant s \leqslant b$, be a curve on the surface M, where s is arc length. If $\boldsymbol{\alpha}$ is the shortest possible curve on M connecting its two end points, then $\boldsymbol{\alpha}$ is a geodesic.

Proof. $\boldsymbol{\alpha}$ is given by $\boldsymbol{\alpha}(s) = \mathbf{X}(u^1(s), u^2(s))$. By assumption, any other curve connecting $\boldsymbol{\alpha}(a)$ to $\boldsymbol{\alpha}(b)$ has a length greater than or equal to that of $\boldsymbol{\alpha}$. In particular, consider the family of curves $\boldsymbol{\alpha}_\epsilon(s)$ (indexed by the real variable ϵ) obtained by replacing $u^i(s)$ by

$$U^i(s, \epsilon) = u^i(s) + \epsilon v^i(s)$$

for $i = 1,2$, $a \leqslant s \leqslant b$, where the v^i are smooth functions satisfying $v^i(a) = v^i(b) = 0$, $i = 1,2$, but otherwise chosen arbitrarily (subject to the requirement (U^1, U^2) in domain of $\mathbf{X}$).

For each ϵ, the curve $\boldsymbol{\alpha}_\epsilon(s) = \mathbf{X}(U^1(s,\epsilon), U^2(s,\epsilon))$ connects $\boldsymbol{\alpha}(a)$ to $\boldsymbol{\alpha}(b)$ and may be viewed, for ϵ small, as a slight variation or distortion of $\boldsymbol{\alpha}$ (Fig. I-28). (Although s is arc length along $\boldsymbol{\alpha}$, it is not in

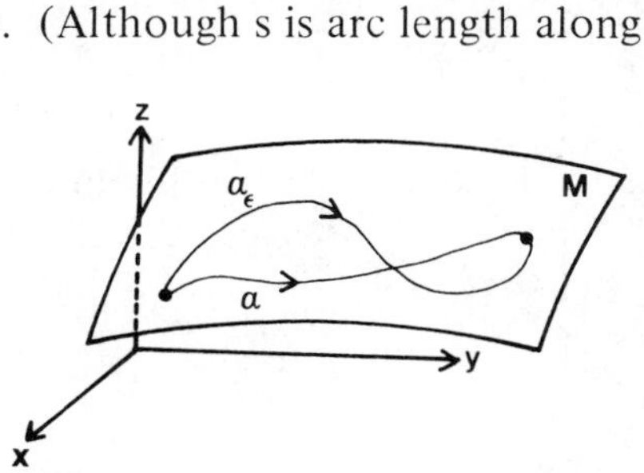

Figure I - 28.

general arc length along $\boldsymbol{\alpha}_\epsilon$ for $\epsilon \neq 0$.) Let $L(\epsilon)$ denote the length of $\boldsymbol{\alpha}_\epsilon$:

$$L(\epsilon) = \int_a^b \lambda(s,\epsilon)\, ds$$

where

$$\lambda(s,\epsilon) = \left[g_{ij}(U^1,U^2) \frac{\partial U^i}{\partial s} \frac{\partial U^j}{\partial s} \right]^{1/2}$$

The minimality of $L(0)$ implies that

$$L'(0) = \int_a^b \frac{\partial \lambda}{\partial \epsilon}(s,0)\, ds = 0$$

and this will be used to show that $\boldsymbol{\alpha}$ must be a geodesic.* Now, using $\partial U^k/\partial\epsilon = v^k$ and the result of Exercise 1, Section 5, we have

$$\frac{\partial \lambda}{\partial \epsilon}(s,\epsilon) = \frac{1}{2\lambda} \left[\frac{\partial g_{ij}}{\partial U^k} v^k \frac{\partial U^i}{\partial s} \frac{\partial U^j}{\partial s} + 2g_{ij} \frac{\partial U^i}{\partial s} \frac{\partial^2 U^j}{\partial s \partial \epsilon} \right]$$

Substituting $\epsilon = 0$ and utilizing $\partial U^j/\partial\epsilon = v^j$ and $\lambda(s,0) = 1$ (s is arc length on $\boldsymbol{\alpha} = \boldsymbol{\alpha}_0$), we obtain

$$L'(0) = \frac{1}{2} \int_a^b \left(\frac{\partial g_{ij}}{\partial u^k} u^{i\prime} u^{j\prime} v^k + 2g_{ik} u^{i\prime} v^{k\prime} \right) ds = 0$$

Because $v^k(a) = v^k(b) = 0$, integration of the second term by parts converts the integral into

$$L'(0) = \frac{1}{2} \int_a^b \left(\frac{\partial g_{ij}}{\partial u^k} u^{i\prime} u^{j\prime} - 2 \frac{\partial (g_{ik} u^{i\prime})}{\partial s} \right) v^k\, ds = 0$$

$L'(0)$ must vanish for all possible smooth functions v^k satisfying $v^k(a) = v^k(b) = 0$, for $k = 1,2$. Because the choice of the v^k is so arbitrary, this implies that the expression in parentheses must vanish identically for all s:

---

*Differentiation under the integral sign is justified by the continuity of $\lambda(s,\epsilon)$ and $\partial\lambda/\partial\epsilon(s,\epsilon)$.

$$\frac{1}{2}\frac{\partial g_{ij}}{\partial u^k} u^{i\prime} u^{j\prime} - \frac{\partial (g_{ik} u^{i\prime})}{\partial s} = 0, \quad k = 1,2$$

We shall now deduce Eq. (32a) from the latter.

We differentiate $g_{ik} u^{i\prime}$ by the product and chain rules, and then employ the identities Eq. (35):

$$0 = \frac{1}{2}\frac{\partial g_{ij}}{\partial u^k} u^{i\prime} u^{j\prime} - \frac{\partial g_{ik}}{\partial u^j} u^{i\prime} u^{j\prime} - g_{mk} u^{m\prime\prime}$$

$$= \left[\frac{1}{2}(\Gamma_{ikj} + \Gamma_{jki}) - \Gamma_{kji} - \Gamma_{ijk}\right] u^{i\prime} u^{j\prime} - g_{mk} u^{m\prime\prime}$$

Since $\Gamma_{ikj} u^{i\prime} u^{j\prime} = \Gamma_{jki} u^{i\prime} u^{j\prime}$ (interchange dummy indices) and $\Gamma_{kji} = \Gamma_{jki}$, the first three terms in brackets drop out, and we are left with

$$g_{mk} u^{m\prime\prime} + \Gamma_{ijk} u^{i\prime} u^{j\prime} = 0$$

Multiplying by g^{kr} and summing over k, we obtain

$$u^{r\prime\prime} + \Gamma^r_{ij} u^{i\prime} u^{j\prime} = 0, \quad r = 1,2$$

Since s is arc length along α, α is a geodesic. The proof is complete.

In connection with Theorem I-9, two important points should be made. First, the converse is false. In general, a geodesic does not necessarily give the minimum distance on the surface between its end-points. [$L'(0)$ would be zero for a maximum length as well as for a minimum, or for any critical point of $L(\epsilon)$.] For example, if **P** and **Q** are two points of a right circular cylinder that lie on the same element of the cylinder, then there are infinitely many geodesics joining **P** to **Q**, only one of which (the element itself) gives the shortest possible path (see Exercise 7). Similarly, on a sphere, a great circle arc longer than a semicircle is clearly not the shortest path on the sphere between its end-points.

Second, given two arbitrary points **P** and **Q** on a surface M, a geodesic joining **P** to **Q** may not exist. For example, if M is the Cartesian plane with the origin removed, then no geodesic in M connects the points (1,1) and (−1,−1).

On the other hand, points that are sufficiently close can be connected by a length-minimizing geodesic. Specifically, it can be shown that any point **P** of M has a neighborhood Ω with the property that any point **Q** of Ω can be joined to P by a geodesic which lies in Ω and is shorter than any other curve in M that joins **P** to **Q**. Moreover, Ω can be chosen "geodesically convex," i.e., so that any pair of its points can be joined by an arc of shortest length that lies inside Ω. We shall not stop to prove these facts. We will, however, need the following result.

Theorem I-10

Given a point **P** of M and a unit tangent vector **v** at **P**, there exists a unique geodesic $\boldsymbol{\alpha}(s)$ such that $\boldsymbol{\alpha}(0) = \mathbf{P}$ and $\boldsymbol{\alpha}'(0) = \mathbf{v}$.

Proof. Let $\mathbf{P} = \mathbf{X}(u_0^1, u_0^2)$ and $\mathbf{v} = v^i \mathbf{X}_i(u_0^1, u_0^2)$. According to the definition of geodesic and Eq. (32a) we must demonstrate the existence of unique functions $u^r(s)$ satisfying

$$u^{r''} + \Gamma^r_{ij} u^{i'} u^{j'} = 0$$

$$u^r(0) = u_0^r, \qquad u^{r'}(0) = v^r$$

for r = 1,2, and then show that s is arc length along the curve $\alpha(s) = \mathbf{X}(u^1(s), u^2(s))$ determined by these equations.

That such functions exist and are unique in a neighborhood of (u_0^1, u_0^2) is a consequence of a standard theorem in the theory of differential equations. To show that s is arc length, it will suffice to show that the function

$$f(s) = g_{ij} u^{i'} u^{j'}$$

is constant, since its value at s = 0 is unity (the length of **v**). Differentiating and making use of the identities Eqs. (33) and (35), we have

$$\begin{aligned} f'(s) &= \frac{\partial g_{ij}}{\partial u^k} u^{i'} u^{j'} u^{k'} + g_{ij} u^{i''} u^{j'} + g_{ij} u^{i'} u^{j''} \\ &= (g_{jr}\Gamma^r_{ik} + g_{ir}\Gamma^r_{jk}) u^{i'} u^{j'} u^{k'} + g_{rj} u^{r''} u^{j'} + g_{ir} u^{i'} u^{r''} \\ &= g_{ir} u^{i'} (u^{r''} + \Gamma^r_{jk} u^{j'} u^{k'}) + g_{rj} u^{j'} (u^{r''} + \Gamma^r_{ik} u^{i'} u^{k'}) \\ &= 0 \end{aligned}$$

because the expressions in parentheses vanish, and so $f(s) \equiv 1$.

Exercises I-7

1. Show that the geodesic curvature, $k_g = \mathbf{U} \cdot \boldsymbol{\alpha}' \times \boldsymbol{\alpha}''$, is given by

$$k_g = k\mathbf{U} \cdot \mathbf{B} = k \cos \theta$$

where k is the curvature of $\boldsymbol{\alpha}$ (as a curve in E^3), $\mathbf{B}$ is the binormal vector of $\boldsymbol{\alpha}$, and θ is the angle between the osculating plane of $\boldsymbol{\alpha}$ and the tangent plane of the surface.

2. (a) Use Exercise 1 to find the geodesic curvature of a circle of latitude on a sphere (a u-parameter curve in Example 7). (b) Prove that the geodesics of a sphere are its great circles. (Use Exercise 15b of Section 1.)

3. Prove that a curve on a surface in E^3 is a geodesic of that surface if and only if $k = |k_n|$ at every point of the curve.

4. Show that on a surface of revolution (a) every meridian is a geodesic, (b) but a parallel is a geodesic if and only if it intersects the profile curve $\boldsymbol{\alpha}$ in a point where $\boldsymbol{\alpha}'$ is parallel to the axis of revolution. [Hint: for (a) assume the meridian is parametrized by arc length.]

5. Definition I-8 and Eqs. (32) give necessary and sufficient conditions for a curve $\boldsymbol{\alpha}$ on M to be a geodesic when $\boldsymbol{\alpha}$ is parametrized by arc length. Often we need to know when a curve given in terms of some other parameter t is a geodesic [i.e., satisfies $(d^2\boldsymbol{\alpha}/ds^2)_{tan} = \mathbf{O}$ when reparametrized in terms of arc length].

Let $\boldsymbol{\alpha} = \boldsymbol{\alpha}(t)$ be a curve on M, and let $s = s(t)$ denote arc length along $\boldsymbol{\alpha}$, measured from some convenient point. Prove (a) if $\boldsymbol{\alpha}''(t)$ is normal to M for all t, then $\boldsymbol{\alpha}$ is a geodesic and has constant speed ds/dt; (b) conversely, if $\boldsymbol{\alpha}$ is a geodesic and has constant speed, then $\boldsymbol{\alpha}''(t)$ is normal to M for all t. [Hint: $d\boldsymbol{\alpha}/dt = (d\boldsymbol{\alpha}/ds)(ds/dt)$ and

$$\frac{d}{dt}\left(\frac{d\boldsymbol{\alpha}}{ds}\right) = \frac{d}{ds}\left(\frac{d\boldsymbol{\alpha}}{ds}\right)\frac{ds}{dt}$$

by the chain rule.]

6. The right circular cylinder $x^2 + y^2 = R^2$ may be parametrized as

$$\mathbf{X}(u,v) = (R \cos u, R \sin u, v)$$

A curve on this cylinder is given therefore by

$$\boldsymbol{\alpha}(t) = (R \cos u(t), R \sin u(t), v(t))$$

(a) Prove that a constant speed curve $\boldsymbol{\alpha}$ is a geodesic if and only if u and v are linear functions, i.e., $u = at + b$, $v = ct + d$, for some constants a, b, c, and d. Thus, the geodesics are the images of straight lines in the uv-plane. (Hint: Use Exercise 5.) (b) Describe all geodesics on a right circular cylinder. [Treat the three possible cases: (i) $a = 0$, (ii) $a \neq 0$ and $c = 0$, and (iii) $a \neq 0$ and $c \neq 0$. Compare the third case with Example 3.]

7. Let

$$\alpha(t) = (R \cos(at + b), R \sin(at + b), ct + d)$$

be a geodesic on the cylinder $x^2 + y^2 = R^2$ (see Exercise 6). Let $\mathbf{P} = (R, 0, z_1)$ and $\mathbf{Q} = (R, 0, z_2)$ be two points on the cylinder's element in the xz-plane. Find conditions on the constants a, b, c, and d so that $\boldsymbol{\alpha}(0) = \mathbf{P}$ and $\boldsymbol{\alpha}$ passes through $\mathbf{Q}$.

8. Determine the geodesic and normal curvatures of the helix

$$\boldsymbol{\alpha}(t) = (a \cos t, a \sin t, bt)$$

(a) as a curve on the cylinder of radius a, (b) as a curve on the helicoid of Example 9.

9. From Eqs. (31) and (29), deduce that (for $\boldsymbol{\alpha}$ parametrized in terms of arc length)

$$k_g = \sqrt{g} \det \begin{bmatrix} u^{1\prime} & u^{2\prime} \\ u^{1\prime\prime} + \Gamma^1_{ij} u^{i\prime} u^{j\prime} & u^{2\prime\prime} + \Gamma^2_{ij} u^{i\prime} u^{j\prime} \end{bmatrix}$$

(Hint: express $\mathbf{U}$ and $\boldsymbol{\alpha}'$ in terms of $\mathbf{X}_1$ and $\mathbf{X}_2$.) This result shows that the geodesic curvature of a curve $\boldsymbol{\alpha}$ on a surface is intrinsic, since, by Eq. (37), k_g is expressible in terms of the metric coefficients g_{ij} and arc length along the curve.

10. (See Exercise 3 of Section 4.) Compute the Christoffel symbols Γ^r_{ij} for

a. the helicoid (Example 9)
b. the torus (Example 12)
c. the catenoid (Exercise 2f of Section 3)

11. Verify that

$$h/r = \cos(\theta - \theta_0)$$

is a solution of Eqs. (41). [Hint: differentiate with respect to s and square. Use $\sin^2(\theta - \theta_0) = 1 - h^2/r^2$ and the left hand equality in Eq. (43).]

12. Suppose M has metric form

$$ds^2 = E\,du^2 + G\,dv^2$$

with $E_v = G_v = 0$. (These conditions are satisfied by many of our examples, including the plane in polar coordinates, the helicoid, and surfaces of revolution. The sphere is also an example provided we interchange u and v.) (a) Verify that the only non-zero Christoffel symbols Γ^r_{ij} are

$$\Gamma^1_{11} = \frac{E_u}{2E}, \quad \Gamma^2_{12} = \Gamma^2_{21} = \frac{G_u}{2G}, \quad \Gamma^1_{22} = -\frac{G_u}{2E}$$

(b) Show that a geodesic on M satisfies

$$v'' + \frac{G_u}{G} u'v' = 0$$

and integrate this equation to obtain

$$Gv' = h \text{ (a non-zero constant)}$$

(c) Mimicking calculations in Example 19, combine $Gv' = h$ with $E(u')^2 + G(v')^2 = 1$ to obtain

$$\frac{dv}{du} = \frac{\pm h\sqrt{E}}{\sqrt{G}\sqrt{(G - h^2)}} \tag{45}$$

We can drop the ± by absorbing it into h. It is easy to see that h = 0 gives a geodesic too. (All u-parameter curves are geodesic, as for example, meridians on a surface of revolution.) It can be shown conversely that a curve satisfying Eq. (45) is a geodesic. Applications are given in the following exercises.

13. Verify that for the plane with polar coordinates, Eq. (45) of Exercise 12 leads to the solution found in Example 19.

14. If M has metric form $ds^2 = E\,du^2 + G\,dv^2$, with $E_u = G_u = 0$, then a geodesic on M satisfies

$$\frac{du}{dv} = \frac{h\sqrt{G}}{\sqrt{E}\sqrt{(E - h^2)}}$$

for some constant h. (A proof is immediately obtained if in Exercise 12 we interchange u with v, E with G, and 1 with 2.) (a) Using this equation, show that a geodesic on the geographic sphere (Examples 7) satisfies

$$\frac{du}{dv} = \frac{h \sec^2 v}{\sqrt{(R^2 - h^2 \sec^2 v)}} = \frac{h \sec^2 v}{\sqrt{(R^2 - h^2 - h^2 \tan^2 v)}}$$

where h is a constant. (b) By means of the substitution $w = h \tan v$, integrate the above equation to obtain

$$\cos(u - u_0) + \gamma \tan v = 0 \tag{46}$$

where u_0 and γ are constants. [Use $\int (a^2 - x^2)^{-1/2}\,dx = -\cos^{-1}(x/a) + c$.] (c) Show that Eq. (46), when reexpressed in terms of Cartesian coordinates,

$$\begin{aligned} x &= R \cos u \cos v \\ y &= R \sin u \cos v \\ z &= R \sin v \end{aligned}$$

is a linear equation of the form $\alpha x + \beta y + \gamma z = 0$, and so represents a plane section of the sphere through the origin, i.e., a great circle.

15. In the plane with polar coordinates, consider a smooth curve

$$r = f(\theta), \quad a \leqslant \theta \leqslant b,$$

where $f(\theta) > 0$ for all θ. Let s denote arc length along the curve and assume θ increases as s increases ($d\theta/ds > 0$). Let $\mathbf{P} = \mathbf{P}(r,\theta)$ be a variable point on the curve, and let $A(\theta)$ be the area bounded by the curve, the ray $\theta = a$, and the ray $\overrightarrow{OP}$. By Exercise 14 of Section 4

$$A(\theta) = \int_a^\theta \int_0^{f(\phi)} r\,dr\,d\phi = \frac{1}{2}\int_a^\theta f(\phi)^2\,d\phi^*$$

Prove that the condition

$$r^2 \frac{d\theta}{ds} = \text{constant}$$

[cf. Eq. 42)] is equivalent to

$$\frac{dA}{ds} = \text{constant}$$

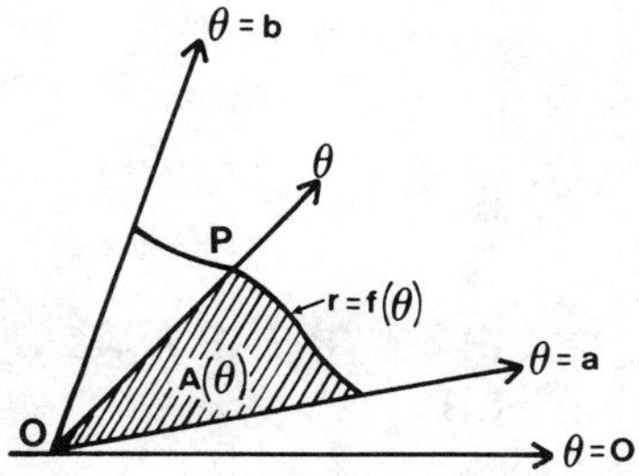

This equivalence will reappear in Chapter III as Kepler's second law of planetary motion. (Hint: by the Fundamental Theorem of Calculus, $dA/d\theta = f(\theta)^2/2$.)

16. [Alternative derivation of Eq. (44).] From Eqs. (42) and (43), we have the relations

$$\frac{dr}{ds} = \frac{dr}{d\theta}\frac{d\theta}{ds} = \frac{dr}{d\theta}\frac{h}{r^2}$$

$$1 = \frac{h^2}{r^4}\left(\frac{dr}{d\theta}\right)^2 + \frac{h^2}{r^2}$$

---

*ϕ is used as the variable of integration (instead of θ) since θ has already been used for one of the upper limits.

(a) Let $w = 1/r$ and put the latter equation into the form

$$\frac{1}{h^2} = \left(\frac{dw}{d\theta}\right)^2 + w^2$$

(b) Assuming $dw/d\theta \neq 0$, obtain

$$0 = \frac{d^2 w}{d\theta^2} + w$$

This second order differential equation has the general solution

$$\frac{1}{r} = w = \frac{1}{a}\cos(\theta - \theta_0)$$

where a and θ_0 are constants of integration. We have again arrived at the equation of a line in polar coordinates. (The calculation of gravitational orbits in Chapter III will involve similar techniques.)

8. THE CURVATURE TENSOR AND THE *THEOREMA EGREGIUM*

Here we present a proof of the *Theorema Egregium* of Gauss. The intrinsic nature of the Gauss curvature, $K = L/g$, will be established by showing that L (the determinant of the second fundamental form) is a function of the g_{ij} and their first and second derivatives.

We begin by recalling the formulas of Gauss and Weingarten [Eq. (17) in Section 5 and Eq. (28) in Section 6]:

$$\mathbf{X}_{ik} = \Gamma^h_{ik}\mathbf{X}_h + L_{ik}\mathbf{U} \quad i,k = 1,2 \tag{47}$$

$$\mathbf{U}_i = -L^j_i\mathbf{X}_j,\ i = 1,2 \tag{48}$$

Differentiating Gauss's equation (47), and denoting $\partial\mathbf{X}_{ik}/\partial u^j$ by $\mathbf{X}_{ikj}$, we have

$$\mathbf{X}_{ikj} = \frac{\partial\Gamma_{ik}}{\partial u^j}\mathbf{X}_h + \Gamma^h_{ik}\mathbf{X}_{hj} + \frac{\partial L_{ik}}{\partial u^j}\mathbf{U} + L_{ik}\mathbf{U}_j$$

If we rewrite $\mathbf{X}_{hj}$ and $\mathbf{U}_j$ by means of Eqs. (47) and (48), and then separate $\mathbf{X}_{ikj}$ into its tangent and normal components, we obtain (after some relabeling of indices)

$$\mathbf{X}_{ikj} = \left(\frac{\partial \Gamma^h_{ik}}{\partial u^j} + \Gamma^r_{ik}\Gamma^h_{rj} - L_{ik}L^h_j\right)\mathbf{X}_h + \left(\Gamma^r_{ik}L_{rj} + \frac{\partial L_{ik}}{\partial u^j}\right)\mathbf{U} \qquad (49)$$

Interchanging j and k yields

$$\mathbf{X}_{ijk} = \left(\frac{\partial \Gamma^h_{ij}}{\partial u^k} + \Gamma^r_{ij}\Gamma^h_{rk} - L_{ij}L^h_k\right)\mathbf{X}_h + \left(\Gamma^r_{ij}L_{rk} + \frac{\partial L_{ij}}{\partial u^k}\right)\mathbf{U} \qquad (50)$$

Now since the order of partial differentiation is immaterial, we must have $\mathbf{X}_{ikj} = \mathbf{X}_{ijk}$, and therefore the tangent and normal components of $\mathbf{X}_{ikj} - \mathbf{X}_{ijk}$ must both vanish. If we subtract Eq. (50) from Eq. (49) and then ignore the normal component of the result, we find that $(\mathbf{X}_{ikj} - \mathbf{X}_{ijk})_{tan} = \mathbf{O}$ is equivalent to

$$\frac{\partial \Gamma^h_{ik}}{\partial u^j} - \frac{\partial \Gamma^h_{ij}}{\partial u^k} + \Gamma^r_{ik}\Gamma^h_{rj} - \Gamma^r_{ij}\Gamma^h_{rk} = L_{ik}L^h_j - L_{ij}L^h_k \qquad (51)$$

Since the Christoffel symbols are functions of the g_{ij}, this identity establishes a relation between the coefficients of the first and second fundamental forms. We shall deduce the *Theorema Egregium* from Eq. (51).

Call the expression of the left side of Eq. (51) R^h_{ijk}:

$$R^h_{ijk} = \frac{\partial \Gamma^h_{ik}}{\partial u^j} - \frac{\partial \Gamma^h_{ij}}{\partial u^k} + \Gamma^r_{ik}\Gamma^h_{rj} - \Gamma^r_{ij}\Gamma^h_{rk} \qquad (52)$$

for h,i,j,k = 1,2. These functions obviously satisfy the identity

$$R^h_{ijk} = -R^h_{ikj} \qquad (53)$$

The R^h_{ijk} are the components of what is called the *Riemann-Christoffel curvature tensor*. Although the technical term "tensor" will not be defined precisely in this book, this section will demonstrate, at least in part, the importance of the functions defined by Eq. (52) and what they have to do with curvature. Equation (51) may be rewritten

$$R^h_{ijk} = L_{ik} L^h_j - L_{ij} L^h_k \tag{54}$$

Next, define the quantities R_{mijk} by

$$R_{mijk} = g_{mh} R^h_{ijk} \tag{55}$$

Because of Eq. (12), the inverse relations $R^r_{ijk} = g^{mr} R_{mijk}$ hold. From Eqs. (52) and (55) we see that the functions R^h_{ijk} and R_{mijk} are intrinsic, because the Christoffel symbols are functions of the g_{ij} and their first derivatives.

Multiplying Eq. (54) by g_{mh} and summing over h, we see that Eq. (51) is equivalent to the equation

$$R_{mijk} = L_{ik} L_{jm} - L_{ij} L_{km} \tag{56}$$

In particular, $R_{1212} = L_{22} L_{11} - (L_{21})^2 = \det(L_{ij}) = L$, and hence

$$K = R_{1212}/g \tag{57}$$

This establishes that K is determined by the metric form and so completes the proof of the following theorem.

Theorem I-11 (*Theorema Egregium*)

The Gauss curvature of a surface is a function of the coefficients of the metric form and their first and second derivatives. It is therefore intrinsic.

For later examples and exercises, we next derive a specific formula for K in terms of the coefficients of the first fundamental form. To do this, we need an alternative expression for R_{mijk}. From its definition [Eqs. (52) and (55)], R_{mijk} is given by

$$R_{mijk} = g_{mh} \frac{\partial \Gamma^h_{ik}}{\partial u^j} + \Gamma^r_{ik} \Gamma_{rjm} - \text{switch}(j,k)$$

where (here and in any subsequent equation) *switch(j,k)* stands for the sum of the preceding terms with j and k interchanged. From the identity

$$g_{mh} \frac{\partial \Gamma^h_{ik}}{\partial u^j} = \frac{\partial \Gamma_{ikm}}{\partial u^j} - \Gamma^h_{ik} \frac{\partial g_{hm}}{\partial u^j}$$

which results from differentiation of $g_{mh} \Gamma^h_{ik} = \Gamma_{ikm}$, we then have (replacing r by h in the term $\Gamma^r_{ik} \Gamma_{rjm}$),

$$R_{mijk} = \frac{\partial \Gamma_{ikm}}{\partial u^j} + \Gamma^h_{ik} \Gamma_{hjm} - \Gamma^h_{ik} \frac{\partial g_{hm}}{\partial u^j} - \text{switch}(j,k)$$

Making use of the identities

$$\Gamma_{ikm} = \frac{1}{2} \left(\frac{\partial g_{im}}{\partial u^k} + \frac{\partial g_{mk}}{\partial u^i} - \frac{\partial g_{ki}}{\partial u^m} \right)$$

$$\frac{\partial g_{hm}}{\partial u^j} = \Gamma_{hjm} + \Gamma_{mjh} = \Gamma_{hjm} + \Gamma^r_{mj} g_{rh}$$

[see Eqs. (35) and (36)], we next obtain

$$R_{mijk} = \frac{1}{2} \left(\frac{\partial^2 g_{im}}{\partial u^j \partial u^k} + \frac{\partial^2 g_{mk}}{\partial u^j \partial u^i} - \frac{\partial^2 g_{ki}}{\partial u^j \partial u^m} \right) + \Gamma^h_{ik} \Gamma_{hjm}$$

$$- \Gamma^h_{ik} (\Gamma_{hjm} + \Gamma^i_{mj} g_{rh}) - \text{switch}(j,k)$$

Finally, after writing out the terms of switch(j,k) and canceling where possible, we have

$$R_{mijk} = \frac{1}{2} \left(\frac{\partial^2 g_{km}}{\partial u^j \partial u^i} - \frac{\partial^2 g_{jm}}{\partial u^i \partial u^k} + \frac{\partial^2 g_{ij}}{\partial u^k \partial u^m} - \frac{\partial^2 g_{ik}}{\partial u^j \partial u^m} \right)$$

$$+ (\Gamma^h_{ij} \Gamma^r_{mk} - \Gamma^h_{ik} \Gamma^r_{mj}) g_{rh} \qquad (58)$$

From Eqs. (57) and (58) we have (in somewhat mixed notation)

$$K = \frac{R_{1212}}{g} = \frac{1}{g}\left[F_{uv} - \frac{1}{2}E_{vv} - \frac{1}{2}G_{uu} + (\Gamma^h_{12}\Gamma^r_{12} - \Gamma^h_{22}\Gamma^r_{11})g_{rh}\right]$$

Suppose the coordinates are *orthogonal*, i.e., $F = 0$. Then, substituting from Eq. (40), we obtain

$$K = \frac{1}{EG}\left[-\frac{1}{2}E_{vv} - \frac{1}{2}G_{uu} + \left(\frac{E_v^2}{4E^2} + \frac{E_u G_u}{4E^2}\right)E + \left(\frac{G_u^2}{4G^2} + \frac{E_v G_v}{4G^2}\right)G\right]$$

which, with a little manipulation, may be put in the form

$$K = -\frac{1}{2\sqrt{EG}}\left[\frac{\sqrt{EG}\,G_{uu} - \dfrac{G_u(EG_u + E_u G)}{2\sqrt{EG}}}{EG} + \frac{\sqrt{EG}\,E_{vv} - \dfrac{E_v(EG_v + E_v G)}{2\sqrt{EG}}}{EG}\right]$$

or

$$K = -\frac{1}{2\sqrt{EG}}\left[\frac{\partial}{\partial u}\left(\frac{G_u}{\sqrt{EG}}\right) + \frac{\partial}{\partial v}\left(\frac{E_v}{\sqrt{EG}}\right)\right] \tag{59}$$

This formula will be applied in the exercises. Equation (59) clearly demonstrates the sole dependence of K upon the metric form (at least for orthogonal coordinates).

Since each of the four subscripts may take on the value 1 or 2, it might appear that Eq. (56) is a system of $2^4 = 16$ equations. However, because of various symmetries, these equations are not all distinct. In fact, we can show that there is essentially only one equation contained in Eq. (56), namely, $R_{1212} = L = Kg$.

From Eq. (58) you can readily verify that the R_{mijk} satisfy the following symmetry relations:

$$R_{mijk} = - R_{imjk} \tag{60a}$$

$$R_{mijk} = - R_{mikj} \tag{60b}$$

$$R_{mijk} = R_{jkmi} \tag{60c}$$

The second one can be seen also from Eqs. (53) and (55). Thus, R_{mijk} is antisymmetric in its first pair and in its last pair of indices, but symmetric relative to an interchange of these pairs.

Because of these symmetry relations,

$$R_{11jk} = R_{22jk} = R_{mi11} = R_{mi22} = 0$$

and only the components

$$R_{1212} = R_{2121} = - R_{2112} = - R_{1221} = Kg$$

can be different from zero. Since the right hand side of Eq. (56) has the symmetry properties (60) also (check it!), set (56) contains essentially only the single equation $R_{1212} = L = Kg$. (All other choices of index values give either an equivalent equation or the equation $0 = 0$.)

In Chapter III, we shall extend many of the results of surface theory to higher-dimensional manifolds, where the concept of curvature is more complicated. The curvature tensor R^h_{ijk} will play a key role. Whereas in two dimensions this tensor consists of essentially only one component, which determines the Gauss curvature, in n dimensions it turns out that there are $n^2 (n^2 - 1)/12$ independent components.

The key result of this chapter is that curvature is an intrinsic quantity, dependent only upon the metric form. It is this fact that makes it possible to discuss the curvature of space (or of any manifold) without the need for an embedding in some higher-dimensional Euclidean space.

Exercise I-8

1. Using Eq. (59) and the results of Exercise 3 of Section 4, compute the Gauss curvature of

a. the sphere (Example 7)
b. the helicoid (Example 9)
c. the torus (Example 12)
d. the catenoid (Exercise 2f of Section 3)

2. Let

$$\mathbf{X}(u,v) = (f(u)\cos v, f(u)\sin v, g(u)$$

be a surface of revolution whose profile curve,

$$\alpha(u) = (f(u), 0, g(u))$$

has unit speed. Using Eq. (59), show that $K = -f''/f$ (as computed by a different method in Exercise 2 of Section 6).

3. Show that if $F = 0$, then

$$K = -\frac{1}{\sqrt{(EG)}}\left[\frac{\partial}{\partial u}\left(\frac{1}{\sqrt{E}}\frac{\partial\sqrt{G}}{\partial u}\right) + \frac{\partial}{\partial v}\left(\frac{1}{\sqrt{G}}\frac{\partial\sqrt{E}}{\partial v}\right)\right]$$

[Hint: see Eq. (59); show that

$$\frac{\partial}{\partial u}\left(\frac{G_u}{\sqrt{(EG)}}\right) = 2\frac{\partial}{\partial u}\left(\frac{1}{\sqrt{E}}\frac{\partial\sqrt{G}}{\partial u}\right)$$

$$\frac{\partial}{\partial v}\left(\frac{E_v}{\sqrt{(EG)}}\right) = 2\frac{\partial}{\partial v}\left(\frac{1}{\sqrt{G}}\frac{\partial\sqrt{E}}{\partial v}\right)]$$

4. Suppose a surface M has a coordinate patch in which the metric form is

$$ds^2 = du^2 + e^{2u/k}\,dv^2$$

Show that $K = -1/k^2$.

5. The *pseudosphere* (see the figure) may be represented as the surface of revolution

$$\mathbf{X}(u,v) = \left(a \sin u \cos v, a \sin u \sin v, a \left[\cos u + \ln \left(\tan\frac{u}{2}\right)\right]\right)$$

for $0 < u < \pi/2$. (Here the pseudosphere's axis is the z axis.) (a) Find the metric form of this surface. [Hint: use the identity $2 \sin(u/2) \cos(u/2) = \sin u$.] (b) Show that $K = - 1/a^2$.

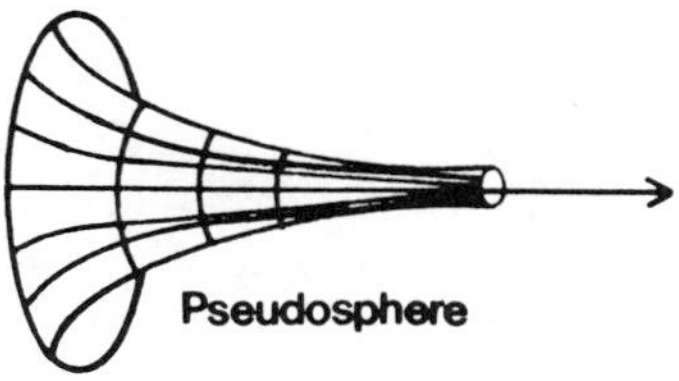

Pseudosphere

9. MANIFOLDS

Our definition of *surface* in Section 3 is deficient for several reasons. First, it is too restrictive. Many familiar objects we view as surfaces cannot be represented as the image of a *single* regular mapping $\mathbf{X}:D \to E^3$. Recall that Examples 7 and 8 gave parametrizations of only portions of the sphere. Moreover, the image of $\mathbf{X}$ could have self-intersections, violating our intuitive notion of a surface as something that looks locally (in a small neighborhood of each point) like a piece of a plane.

Second, the definition and subsequent development revolve around a given parametrization $\mathbf{X}$, whereas the truly geometric properties of a surface should be independent of a choice of parametrization. We really wish to study not $\mathbf{X}$ but the set which is its image.

Finally, since we wish to extend the intrinsic parts of our theory to higher dimensions, the most serious deficiency is the heavy dependence of all that we have done on the vector space structure and dot product of E^3.

In this section, we shall generalize our earlier definition of surface and show how the concepts of tangent vector, metric, geodesic, and curvature can be introduced without reference to an imbedding of the surface in Euclidean space. (Indeed, the extended definition will encompass examples of surfaces that are impossible to imbed in E^3.) In so doing, we shall prepare the way for the application of what we have learned in this chapter to the 4-dimensional world of relativity theory.

Let M be a non-empty set, whose elements will be called *points*. A *coordinate patch* on M is a one-to-one function $\mathbf{X}:D \to M$ from an open subset D of E^2 into M.

Definition I-12

An *abstract surface* or *2-manifold* is a set M together with a collection $\mathscr{C}$ of coordinate patches on M satisfying the following conditions:

a. M is the union of the images of the patches in $\mathscr{C}$
b. The patches of $\mathscr{C}$ *overlap smoothly*, i.e., if $\mathbf{X}:D \to M$ and $\bar{\mathbf{X}}:\bar{D} \to M$ are two patches in $\mathscr{C}$, then the composite functions $\bar{\mathbf{X}}^{-1} \circ \mathbf{X}$ and $\mathbf{X}^{-1} \circ \bar{\mathbf{X}}$ have open domains and are smooth.
c. Given two points P and P′ of M, there exist patches $X:D \to M$ and $\bar{X}:\bar{D} \to M$ in $\mathscr{C}$ such that $P \in X(D)$, $P' \in \bar{X}(\bar{D})$, and $X(D) \cap \bar{X}(\bar{D}) = \phi$.
d. The collection $\mathscr{C}$ is *maximal*, in the sense that any coordinate patch on M which overlaps smoothly with every patch of $\mathscr{C}$ is itself in $\mathscr{C}$.

Condition (c), known as the *Hausdorff property*, is added to avoid technical problems that need not concern us here. We shall always understand the word *domain* to mean the largest set on which the function in question is defined; e.g., the domain of $\bar{X}^{-1} \circ X$ is $X^{-1}[X(D) \cap \bar{X}(\bar{D})]$. A function whose domain is empty will be considered smooth by default. As usual, *smooth* means sufficiently differentiable for our purposes (in most cases, having continuous partial derivatives up to order at least three).

Examples appear in the exercises. The collection $\mathscr{C}$ is sometimes called a *differentiable structure* on M, and patches in $\mathscr{C}$ are called *admissible patches*. Any collection $\mathscr{C}'$ of patches on a set M that satisfy only (a), (b), and (c) of Definition I-12 can immediately be extended to a differentiable structure on M by adjoining to $\mathscr{C}'$ all patches that overlap smoothly with the patches of $\mathscr{C}'$. $\mathscr{C}'$ is then said to *generate* this differentiable structure.

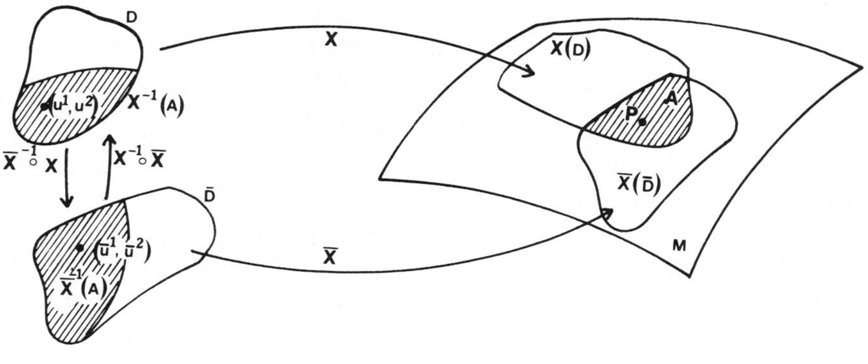

Figure I - 29.

An admissible patch $\mathbf{X}:D \to M$ associates with each point P of $\mathbf{X}(D)$ a unique ordered pair $\mathbf{X}^{-1}(P) = (u^1, u^2)$, called *local coordinates* of P. If $\bar{\mathbf{X}}:\bar{D} \to M$ is a second admissible patch, associating local coordinates $(\bar{u}^1, \bar{u}^2)$ to each point of $\bar{\mathbf{X}}(\bar{D})$, then points in the intersection $\mathbf{X}(D) \cap \bar{\mathbf{X}}(\bar{D})$ have two local coordinate pairs, (u^1, u^2) and $(\bar{u}^1, \bar{u}^2)$ and the composite functions $\bar{\mathbf{X}}^{-1} \circ \mathbf{X}$ and $\mathbf{X}^{-1} \circ \bar{\mathbf{X}}$ make $\bar{u}^1$ and $\bar{u}^2$ smooth functions of u^1 and u^2, and vice versa (see Fig. I-29). In terms of the local coordinates, $\bar{\mathbf{X}}^{-1} \circ \mathbf{X}$ is given by

$$\bar{u}^i = \bar{u}^i(u^1, u^2), \quad i = 1,2 \tag{61a}$$

and $\mathbf{X}^{-1} \circ \bar{\mathbf{X}}$ is given by

$$u^i = u^i(\bar{u}^1, \bar{u}^2), \quad i = 1,2 \tag{61b}$$

These are the equations for a *change of coordinates*.

Sometimes, instead of naming the mapping $\mathbf{X}$, we shall simply refer to (u^1, u^2) as a system of local coordinates in M. In keeping with our earlier practice, we shall occasionally label local coordinates (u,v). Of course, every parametrized surface in the sense of Section 3 is a coordinate patch and generates a differentiable structure on its image.

A subset Ω of M will be called a *neighborhood* of a point **P** of M if there exists an admissible patch $\mathbf{X}:D \to M$ such that $\mathbf{P} \in \mathbf{X}(D)$ and

$\mathbf{X}(D) \subset \Omega$. A subset of M is called *open* if it is a neighborhood of each of its points.

Definition I-13

Let Ω be an open subset of the 2-manifold M. A function $f:\Omega \to \mathcal{R}$ will be called *smooth* if $f \circ \mathbf{X}$ is smooth for every admissible patch $\mathbf{X}$ in M.

If $f:\Omega \to \mathcal{R}$ is smooth and $\mathbf{X}:D \to M$ is an admissible patch whose image meets Ω, define the functions $\partial f/\partial u^i:\mathbf{X}(D) \cap \Omega \to \mathcal{R}$, for i = 1, 2, by

$$\frac{\partial f}{\partial u^i} = \frac{\partial(f \circ \mathbf{X})}{\partial u^i} \circ \mathbf{X}^{-1} \tag{62}$$

i.e., for each P in $\mathbf{X}(D) \cap \Omega$,

$$\frac{\partial f}{\partial u^i}(P) = \frac{\partial(f \circ \mathbf{X})}{\partial u^i}(\mathbf{X}^{-1}(P))$$

This is the *partial derivative* of f with respect to u^i.

For each i, the operator $\partial/\partial u^i$ is linear and satisfies the usual product rule for derivatives,

$$\frac{\partial}{\partial u^i}(fg) = f\frac{\partial g}{\partial u^i} + g\frac{\partial f}{\partial u^i}$$

(assume f and g have the same domain here). For each **P** in $\mathbf{X}(D)$, $\partial/\partial u^i(\mathbf{P})$ denotes the partial derivative operator at **P**:

$$\frac{\partial}{\partial u^i}(P)[f] = \frac{\partial f}{\partial u^i}(\mathbf{P})$$

for f a smooth function defined in a neighborhood of **P**.

If $\bar{\mathbf{X}}:\bar{D} \to M$ is another admissible patch, then on the overlap, $\mathbf{X}(D) \cap \bar{\mathbf{X}}(\bar{D})$, we have the following operator identities*:

*Recall that for purposes of the summation convention, a denominator superscript, as in $\partial/\partial u^i$, counts as a subscript.

$$\frac{\partial}{\partial u^i} = \frac{\partial \bar{u}^j}{\partial u^i} \frac{\partial}{\partial \bar{u}^j}, \qquad i = 1,2 \tag{63a}$$

$$\frac{\partial}{\partial \bar{u}^k} = \frac{\partial u^i}{\partial \bar{u}^k} \frac{\partial}{\partial u^i}, \qquad k = 1,2 \tag{63b}$$

[These follow from Eq. (61) and the multivariable chain rule.]

Definition I-14

Let m be a positive integer and suppose $\mathcal{O}$ is an open subset of E^m. A function $f:\mathcal{O} \to M$ will be called *smooth* if $\mathbf{X}^{-1} \circ f$ is smooth for every admissible patch $\mathbf{X}$ on M.

If $\mathcal{O}$ is not open, we shall call $f:\mathcal{O} \to M$ smooth if f is the restriction to $\mathcal{O}$ of some smooth function with open domain. A *curve* in M is a smooth function from an interval into M. Consequently, if $\alpha:I \to M$ is a curve in M and $\mathbf{X}$ is an admissible patch whose image contains that of α, we have $(\mathbf{X}^{-1} \circ \alpha)(t) = (u^1(t), u^2(t))$ or

$$\alpha(t) = X(u^1(t), u^2(t))$$

with u^1 and u^2 smooth functions of t (compare Section I-3).

Our next task is to define tangent vectors for abstract surfaces. In the case of a surface M in E^3, a tangent vector was defined as the velocity vector of a curve in M (Definition I-2). Unfortunately, the definition of velocity vector given in Section 1 does not carry over to abstract surfaces, because the vector space structure of E^3 is not available to us.

However, there is an alternative way of looking at vectors. In E^3, every vector $\mathbf{v} = (a,b,c)$ has associated with it a corresponding *directional derivative* operator,

$$D_{\mathbf{v}} = a \frac{\partial}{\partial x} + b \frac{\partial}{\partial y} + c \frac{\partial}{\partial z}$$

Since this correspondence is one-to-one, we may think of vectors, not as "arrows," but as operators of a certain kind. This is the key to adapting the concept of tangent vector to abstract surfaces. First, we define the velocity vector of a curve as the directional derivative along the curve.

Definition I-15

Let $\boldsymbol{\alpha}: I \rightarrow M$ be a curve on the 2-manifold M. For each t in I, the *velocity vector* of $\boldsymbol{\alpha}$ at $\boldsymbol{\alpha}(t)$ is the operator $\boldsymbol{\alpha}'(t)$ defined by

$$\boldsymbol{\alpha}'(t)[f] = (f \circ \boldsymbol{\alpha})'(t) = \frac{d}{dt}[f(\boldsymbol{\alpha}(t)]$$

for every smooth real-valued function f defined in a neighborhood of $\alpha(t)$.

Definition I-16

Let **P** be a point of the 2-manifold M. An operator **v** which assigns a real number $\mathbf{v}[f]$ to each smooth real-valued function* f on M is called a *tangent vector* to M at **P** if there exists a curve in M which passes through **P** and has velocity vector **v** at **P**. The set of all tangent vectors to M at **P** is called the *tangent plane* of M at **P** and is denoted $T_{\mathbf{P}}M$.

The preceding two definitions have the desirable property of being independent of a choice of coordinate patch. However, it is often necessary to compute in local coordinates. Suppose $\boldsymbol{\alpha}$ is a curve on M and let $\mathbf{X}: D \rightarrow M$ be an admissible patch defining local coordinates (u^1, u^2) in a neighborhood of a point $\boldsymbol{\alpha}(t)$, for some fixed t. Then, for any smooth function f on M, we have (from the multivariable chain rule)

$$\boldsymbol{\alpha}'(t)[f] = \frac{d}{dt}[f \circ X(u^1(t), u^2(t))] = \frac{\partial (f \circ \mathbf{X})}{\partial u^i}(\mathbf{X}^{-1} \circ \boldsymbol{\alpha}(t)) \frac{du^i}{dt}$$

$$= \frac{\partial f}{\partial u^i}(\boldsymbol{\alpha}(t))\, u^{i\prime}(t)$$

Thus, $\boldsymbol{\alpha}'(t) = u^{i\prime}(t)\, \boldsymbol{\partial}/\boldsymbol{\partial u}^i(\boldsymbol{\alpha}(t))$, or, in abbreviated notation,

$$\boldsymbol{\alpha}' = u^{i\prime} \frac{\boldsymbol{\partial}}{\boldsymbol{\partial u}^i} \tag{64}$$

*When we say "function on M," it is to be understood that the domain is a subset of M that includes a neighborhood of whatever point(s) is (are) under discussion.

It follows that any tangent vector **v** at a point $\mathbf{P} = \mathbf{X}(u^1, u^2)$ is a linear combination of $\boldsymbol{\partial}/\boldsymbol{\partial}\mathbf{u}^1(\mathbf{P})$ and $\boldsymbol{\partial}/\boldsymbol{\partial}\mathbf{u}^2(\mathbf{P})$:

$$\mathbf{v} = v^i \frac{\boldsymbol{\partial}}{\boldsymbol{\partial}\mathbf{u}^i}(\mathbf{P}) \tag{65}$$

and conversely. The coefficients v^i in Eq. (65) satisfy $\mathbf{v}[u^j] = v^j$, $j = 1, 2$, because $\partial u^j/\partial u^i = \delta^j_i$. $T_\mathbf{P}M$ is thus a vector space spanned by $\boldsymbol{\partial}/\boldsymbol{\partial}\mathbf{u}^1(\mathbf{P})$ and $\boldsymbol{\partial}/\boldsymbol{\partial}\mathbf{u}^2(\mathbf{P})$. These vectors are independent (prove it!) and so form a basis of $T_\mathbf{P}M$, which therefore has dimension two. (For a parametrized surface in E^3, $\boldsymbol{\partial}/\boldsymbol{\partial}\mathbf{u}^i$ is the directional derivative in the direction of $\mathbf{X}_i$.)

Where two local coordinate systems apply, a tangent vector **v** has two coordinate representations

$$\mathbf{v} = v^i \frac{\boldsymbol{\partial}}{\boldsymbol{\partial}\mathbf{u}^i}(\mathbf{P}) = \bar{v}^j \frac{\boldsymbol{\partial}}{\boldsymbol{\partial}\bar{\mathbf{u}}^j}(\mathbf{P}). \tag{66}$$

From Eq. (63) you can readily verify that

$$\bar{v}^j = v^i \frac{\partial \bar{u}^j}{\partial u^i}(\mathbf{P}), \quad j = 1, 2 \tag{67a}$$

$$v^i = \bar{v}^j \frac{\partial u^i}{\partial \bar{u}^j}(\mathbf{P}), \quad i = 1, 2 \tag{67b}$$

—or simply apply Eq. (66), first to $\bar{u}^k$ and then to u^m and use $\partial \bar{u}^k/\partial \bar{u}^j = \delta^k_j$, $\partial u^m/\partial u^i = \delta^m_i$.

Next, we develop the concept of a metric on an abstract surface. This will enable us to define curvature and geodesics. First we need the notion of an inner product on a vector space.

Definition I-17

Let $\mathscr{V}$ be a vector space over $\mathbb{R}$. An *inner product* on $\mathscr{V}$ is a rule which assigns to each pair of vectors **v**, **w** in $\mathscr{V}$ a real number $\langle \mathbf{v}, \mathbf{w} \rangle$ such that the following properties hold (for all **v**, **v**′, **w**, **w**′ in $\mathscr{V}$ and a, a′ in $\mathbb{R}$).

a. $\langle \mathbf{v}, \mathbf{w} \rangle = \langle \mathbf{w}, \mathbf{v} \rangle$ ($\langle\ ,\ \rangle$ is *symmetric*)
b. $\langle a\mathbf{v} + a'\mathbf{v}', \mathbf{w} \rangle = a\langle \mathbf{v}, \mathbf{w} \rangle + a'\langle \mathbf{v}', \mathbf{w} \rangle$ and $\langle \mathbf{v}, a\mathbf{w} + a'\mathbf{w}' \rangle = a\langle \mathbf{v}, \mathbf{w} \rangle + a'\langle \mathbf{v}, \mathbf{w}' \rangle$ ($\langle\ ,\ \rangle$ is *bilinear*)
c. $\langle \mathbf{v}, \mathbf{v} \rangle \geqslant 0$ for all $\mathbf{v}$ in $\mathscr{V}$, and $\langle \mathbf{v}, \mathbf{v} \rangle = 0$ only if $\mathbf{v} = \mathbf{O}$. ($\langle\ ,\ \rangle$ is *positive definite*)

The ordinary dot product for vectors in E^3 is an example, of course, as is the restriction of the ordinary dot product to the tangent plane of a surface in E^3 at any point of the surface.

Definition I-18

A *Riemannian metric* (or *metric*, for short) on a 2-manifold M is an assignment of an inner product $\langle\ ,\ \rangle$ to each tangent plane of M. For each coordinate patch $\mathbf{X}:D \to M$, the functions $g_{ij}:\mathbf{X}(D) \to \mathcal{R}$, defined by

$$g_{ij}(\mathbf{P}) = \left\langle \frac{\partial}{\partial u^i}(\mathbf{P}), \frac{\partial}{\partial u^j}(\mathbf{P}) \right\rangle, \quad i,j = 1,2 \tag{68}$$

are required to be smooth. A 2-manifold furnished with a Riemannian metric is called a *Riemannian 2-manifold*.

If property c of Definition I-17 is replaced by the weaker requirement c′ below, then the preceding defines what is called a *semi-Riemannian 2-manifold*.

c′. If $\langle \mathbf{v}, \mathbf{w} \rangle = 0$ for all $\mathbf{w}$ in $\mathscr{V}$, then $\mathbf{v} = \mathbf{O}$. ($\langle\ ,\ \rangle$ is *nonsingular*.)

From Definition I-17(a), we have in each local coordinate system $g_{ij} = g_{ji}$, for all i, j. Let $g = \det(g_{ij})$.

If $(\bar{u}^1, \bar{u}^2)$ is a second local coordinate system, we have the analogous functions $\bar{g}_{ij} = \langle \partial/\partial\bar{u}^i, \partial/\partial\bar{u}^j \rangle$. Substituting from Eq. (63), we find that

$$\bar{g}_{mn} = g_{ij} \frac{\partial u^i}{\partial \bar{u}^m} \frac{\partial u^j}{\partial \bar{u}^n}, \quad m,n = 1,2 \tag{69a}$$

$$g_{ij} = \bar{g}_{mn} \frac{\partial \bar{u}^m}{\partial u^i} \frac{\partial \bar{u}^n}{\partial u^j}, \quad i,j = 1,2 \tag{69b}$$

If $\mathbf{v} = v^i\, \partial/\partial u^i$ and $\mathbf{w} = w^j\, \partial/\partial u^j$ are tangent vectors at a given point, then

$$\langle \mathbf{v},\mathbf{w}\rangle = g_{ij} v^i w^j \tag{70}$$

Since the left side of Eq. (70) is independent of a choice of coordinates, so is the right side. Therefore, the corresponding expression $\bar{g}_{ij}\bar{v}^i\bar{w}^j$ in another local coordinate system must have the same value, $\langle \mathbf{v},\mathbf{w}\rangle$. In other words, the expression $g_{ij}v^i w^j$ is an *invariant*.

For any tangent vector $\mathbf{v}$, define $\|\mathbf{v}\| = \langle \mathbf{v},\mathbf{v}\rangle^{1/2}$. If $\boldsymbol{\alpha} = \boldsymbol{\alpha}(t)$, $a \leqslant t \leqslant b$, is a curve in M, we define the length of α exactly as in Section 1, as

$$L = \int_a^b \|\boldsymbol{\alpha}'(t)\|\, dt$$

If $s = s(t)$ denotes arc length along the curve from $\boldsymbol{\alpha}(a)$ to $\boldsymbol{\alpha}(t)$, then

$$\left(\frac{ds}{dt}\right)^2 = g_{ij}\,\frac{du^i}{dt}\,\frac{du^j}{dt} \tag{71}$$

for any local coordinate system defined in a neighborhood of $\boldsymbol{\alpha}(t)$. The invariant expression on the right side of Eq. (71) is the *metric* or *fundamental form* of the 2-manifold.

In each system of local coordinates, define (g^{ij}) to be the matrix inverse of (g_{ij}), i.e., $g_{ij}g^{jk} = \delta_i^k$, for $i,k = 1,2$. We can then define the *Christoffel symbols* (in each coordinate system) by adopting relations (36) and (37) as definitions.

Definition I-19

If $\boldsymbol{\alpha} = \boldsymbol{\alpha}(s)$ is a curve in M, where s is arc length, then $\boldsymbol{\alpha}$ is called a *geodesic* if in each local coordinate system defined on part of $\boldsymbol{\alpha}$,

$$\frac{d^2 u^r}{ds^2} + \Gamma^r_{ij}\,\frac{du^i}{ds}\,\frac{du^j}{ds} = 0, \qquad r = 1,2 \tag{72}$$

[It can be shown that if Eq. (72) holds in one coordinate system, then it necessarily holds in another, where the two systems overlap.]

Theorems I-9 and I-10 carry over to Riemannian 2-manifolds with only minor changes required in their proofs. In fact, any concepts or results for parametrized surfaces that are intrinsic carry over to general Riemannian 2-manifolds. The components of the curvature tensor R^h_{ijk}, and the related functions R_{mijk} are defined by Eqs. (52) and (55). Formula (58) can be derived as in Section 8, and symmetries (60) hold as before. In dimension two, the only non-zero R_{mijk} are

$$R_{1212} = R_{2121} = -R_{2112} = -R_{1221}$$

The curvature at a point **P** of M is then defined as the value at **P** of the function

$$K = \frac{R_{1212}}{g}$$

For an orthogonal coordinate system, Eq. (59) is valid. Examples are give in the exercises.

All the basic ideas in this section generalize to higher dimensions. If the domain D of every coordinate patch $\mathbf{X}:D \to M$ is a subset of Euclidean n-space E^n, then Definition I-12 becomes the definition of an *n-manifold*. Local coordinates are then n-tuples, $(u^1, u^2, \ldots, u^n)$, and at each point **P**, the tangent "plane" is now an n-dimensional *tangent space*, spanned by $\partial/\partial u^1(\mathbf{P}), \partial/\partial u^2(\mathbf{P}), \ldots, \partial/\partial u^n(\mathbf{P})$.

All indices take on values from 1 to n, and (g_{ij}) is an $n \times n$ symmetric matrix. There are not one but $n^2(n^2 - 1)/12$ independent components of the curvature tensor R^h_{ijk}: for example, 6 in dimension 3, and 20 in dimension 4. Consequently, it take more than a single function $K(\mathbf{P})$ to describe curvature. If, for some coordinate patch $\mathbf{X}:D \to M$,

$$R^h_{ijk} = 0, \quad \text{for all} \quad h,i,j,k$$

throughout $\mathbf{X}(D)$, then $\mathbf{X}(D)$ is *flat*, i.e., locally isometric to Euclidean space.

Exercises I-9

1. Let M be the plane furnished with Cartesian coordinates, (u,v). The identity mapping of M onto itself is a coordinate patch,

and the collection consisting of this single patch obviously satisfies (a) and (b) of Definition I-12. A differentiable structure on M is obtained by adjoining to this collection all patches in M which overlap smoothly with this one.

Show that the (polar coordinate) patch

$$u = r\cos\theta, \quad v = r\sin\theta, \quad (r,\theta) \in D$$

overlaps smoothly with the identity patch, provided D is a set of the form

$$D = \{(r,\theta) \mid r > 0,\ a < \theta < b, \text{ where } b - a \leqslant 2\pi\}$$

(Hint: A one-to-one regular mapping is invertible.)

2. The mapping

$$\mathbf{X}(u,v) = (u, v, (1 - u^2 - v^2)^{1/2}), \quad u^2 + v^2 < 1$$

is a coordinate patch on the unit sphere $x^2 + y^2 + z^2 = 1$. As pointed out in Example 8, $\mathbf{X}^{-1}$ is the projection of the "northern hemisphere" $(z > 0)$ onto the open equatorial disk $(x^2 + y^2 < 1, z = 0)$, provided we identify $(x,y,0)$ with (x,y). Similarly, a patch with image the "southern hemisphere" is

$$\mathbf{Y}(u,v) = (u, v, -(1 - u^2 - v^2)^{1/2}), \quad u^2 + v^2 < 1$$

To obtain a collection of patches whose images cover the entire sphere, we add to the two above, the following, each of which covers an open hemisphere:

$$(u,v) \to (u, (1 - u^2 - v^2)^{1/2}, v)$$
$$(u,v) \to (u, -(1 - u^2 - v^2)^{1/2}, v)$$
$$(u,v) \to ((1 - u^2 - v^2)^{1/2}, u, v)$$
$$(u,v) \to (-(1 - u^2 - v^2)^{1/2}, u, v)$$

(a) Select any two of these six patches and show that they overlap smoothly. (b) Show that the geographic coordinate patch of Example 7 overlaps smoothly with the patch **X** above.

3. (*The Projective Plane*) Let S^2 be the unit sphere with the differentiable structure generated by the six patches of Exercise 2. Call two points of S^2 *antipodal* if they are the opposite end points of a diameter. The point antipodal to a point $\mathbf{P}$ will be denoted $-\mathbf{P}$. Let $\mathbb{P}^2$ be the set of all unordered pairs $\{\mathbf{P}, -\mathbf{P}\}$ of antipodal points. Define the functions $p:S^2 \to \mathbb{P}^2$ and $q:S^2 \to S^2$ by

$$p(\mathbf{P}) = \{\mathbf{P}, -\mathbf{P}\}, \quad q(\mathbf{P}) = -\mathbf{P}$$

for any $\mathbf{P}$ in S^2.

Can an admissible patch $\mathbf{X}:D \to S^2$ *small* if the chord joining any two points of $\mathbf{X}(D)$ has length less than unity. Then if $\mathbf{X}$ is a small patch in S^2, $p \circ \mathbf{X}$ is a patch in $\mathbb{P}^2$. Show that if $\mathbf{X}:D \to S^2$ and $\bar{\mathbf{X}}:\bar{D} \to S^2$ are small, then $p \circ \mathbf{X}$ and $p \circ \bar{\mathbf{X}}$ are patches in $\mathbb{P}^2$ which overlap smoothly. [Note: If the images of $p \circ \mathbf{X}$ and $p \circ \bar{\mathbf{X}}$ meet, then either $\mathbf{X}(D) \cap \bar{\mathbf{X}}(\bar{D}) \neq \phi$ or else $\mathbf{X}(D) \cap q\bar{\mathbf{X}}(\bar{D}) \neq \phi$.]

The collection $\{p \circ \mathbf{X} \mid \mathbf{X}$ is a small patch in $S^2\}$ generates a differentiable structure on $\mathbb{P}^2$. With this structure, $\mathbb{P}^2$ is the *projective plane*, an example of an abstract surface that cannot be imbedded in E^3.

4. Let M be an open subset of $\mathcal{R}^2$ with the differentiable structure generated by the identity mapping. Suppose a Riemannian metric on M is given by

$$ds^2 = \frac{1}{\gamma^2}(du^2 + dv^2)$$

where $\gamma = \gamma(u,v)$ is a smooth, positive-valued function of the Cartesian coordinates (u,v). [In other coordinate systems, the metric is obtainable from Eq. (69).] Show that M has Gauss curvature

$$K = \gamma(\gamma_{uu} + \gamma_{vv}) - (\gamma_u{}^2 + \gamma_v{}^2)$$

5. Let $M = \mathcal{R}^2$ with the metric

$$ds^2 = \frac{1}{\gamma^2}(du^2 + dv^2), \quad \gamma(u,v) = 1 + \frac{u^2 + v^2}{4R^2}$$

Show that $K = 1/R^2$. (It can be shown that $\mathcal{R}^2$ with this metric can be mapped isometrically onto the sphere of radius R with one point removed. See O'Neill [37, p. 314].)

6. (*Poincaré Upper Half-Plane*) Suppose, in Exercise 4, that

$$M = \{(u,v) \mid v > 0\}, \quad \gamma(u,v) = v/k$$

where k is a positive constant. (a) Show that $K = -1/k^2$. (b) By Exercise 14 of Section 7, a geodesic in M satisfies

$$\frac{du}{dv} = \frac{hv}{(k^2 - h^2 v^2)^{1/2}}$$

for some constant h. Show that the solutions to this differential equation are semicircles centered on the x-axis and (when $h = 0$) rays perpendicular to the x-axis.

7. (*Poincaré Disk*) Let M be the subset of $\mathcal{R}^2$

$$M = \{(u,v) \mid u^2 + v^2 < 4k^2\}$$

the interior of a disk of radius 2k. Introduce the metric

$$ds^2 = \frac{1}{\gamma^2}(du^2 + dv^2), \qquad \gamma(u,v) = 1 - \frac{u^2 + v^2}{4k^2}$$

Show that $K = -1/k^2$. (It can be shown that the geodesics of M are the diameters of the boundary circle $u^2 + v^2 = 4k^2$, together with circular arcs that meet the boundary orthogonally. See O'Neill [37, pp. 315, 334].)

8. In Example 7 of Section 3, make the substitutions $\theta = u$, $r/R = \pi/2 - v$. Then $r = R(\pi/2 - v)$ is the distance from the north pole to the point

$$X(r,\theta) = \left(R \sin\frac{r}{R} \cos\theta, \quad R \sin\frac{r}{R} \sin\theta, R \cos\frac{r}{R}\right)$$

measured along a meridian, and θ is the angle between this meridian and the meridian $\theta = 0$. (a) Show that in these coordinates the metric on the sphere is

$$ds^2 = dr^2 + R^2 \sin^2\left(\frac{r}{R}\right) d\theta^2$$

(b) In Exercise 7 above, introduce new coordinates (r, θ) satisfying

$$u = 2k \tanh \frac{r}{2k} \cos \theta$$

$$v = 2k \tanh \frac{r}{2k} \sin \theta$$

Show that the metric for the Poincaré Disk in these coordinates is

$$ds^2 = dr^2 + k^2 \sinh^2\left(\frac{r}{k}\right) d\theta^2$$

Note the analogy with (a).

II

Special Relativity: The Geometry of Flat Spacetime

Henceforth space by itself, and time by itself, are doomed to fade away into mere shadows, and only a kind of union of the two will preserve an independent reality.

—H. Minkowski

INTRODUCTION

To Sir Isaac Newton, physical space was thought of as something akin to an empty container in which was distributed all the matter of the universe. Although matter was movable and ever changing, space itself was fixed and unalterable:

> Absolute space, in its own nature, without relation to anything external, remains always similar and immovable . . .
>
> (Newton, *Principia Mathematica*)

Space was believed unbounded and infinite, and modeled by 3-dimensional Euclidean geometry. Accordingly, a stationary observer could, in principle, set up a system of rectangular coordinates in space, (x,y,z), and locate any event by giving its spatial coordinates and its time t.

Classical mechanics was based on Newton's three Laws of Motion, published in his great work, *Principia Mathematica* (1687):

1. (The *Law of Inertia*) A body at rest remains at rest and a body in motion continues in motion with constant speed and in a straight line, unless, in either case, the body is acted upon by an outside force.
2. The acceleration of an object is proportional to the force acting upon it, and is directed in the direction in which this force acts.
3. To every action there is an equal and opposite reaction.

The Second Law (tantamount to a definition of force) is expressed succinctly as "force = mass × acceleration," with both force and acceleration vector quantities: $\mathbf{F} = m\mathbf{a}$.

These three Laws of Motion, together with Newton's Law of Universal Gravitation (cf. Section III-4), also given in the *Principia*, provided the basis for a highly successful explanation for the workings of the known astronomical universe. However, toward the end of the 19th century, flaws were discovered in the classical view of the universe. Certain observed phenomena seemed inexplicable in terms of the physical laws propounded by Newton and accepted since his time as accurate descriptions of reality. Attempts to explain these phenomena led to a completely new theory that forever changed our understanding of the basic physical concepts of space, time, mass, force, and energy.

For us, the theory of relativity is of interest because of its geometric interpretation and its significance for the evolution of geometric thought. We shall begin this chapter with a discussion of the physical foundations of special relativity, and then follow with the geometric formulation (Sections 8 and 9). In Chapter III, with the analytic tools of Chapter I now at our disposal, we shall take up the general theory of relativity, where geometry comes to the forefront.

1. INERTIAL FRAMES OF REFERENCE

Imagine that the room in which you are sitting is suddenly transported deep into space, far from any celestial bodies, so that the effects of gravity are negligible. You and all the objects in the room will become weightless. If you have been holding a package and now release your grasp, the package will float motionless beside you.

If you give the package a push, it will move away from you in a straight line and with constant speed until it hits a wall or some other obstacle. Therefore, from the point of view of this room in space, the Law of Inertia seems to hold.

A frame of reference in which the Law of Inertia holds is called an *inertial frame*, or an *inertial system*. We shall refer to an observer at rest in such a system as an *inertial observer*.

Suppose S is an inertial frame, so that relative to S free test particles move *uniformly* (i.e., in straight lines with constant speed). Let S′ be a second frame of reference that is moving relative to S. As long as S′ moves uniformly relative to S, and does not accelerate, then these same test particles will follow straight paths with constant speed when observed from S′, so that S′ is inertial also (see Exercise 1).

Special relativity deals with the observation of phenomena by inertial observers, and with the comparison of the observations of two different observers in uniform relative motion. (General relativity, the subject of Chapter III, extends special relativity by treating accelerating observers and gravitation.)

In actuality, an inertial frame is only an idealization, since the effects of gravity can never be completely eliminated. However, for most experiments with light, or with particles traveling near the speed of light, the earth may be considered an inertial frame. (In the incredibly short period of time required to traverse a typical laboratory, such a particle will have fallen from its intended straight line path by a negligible distance (see Exercise 2).

Exercises II-1

1. Here is an intuitive argument that a frame S′ moving uniformly relative to an inertial frame S is itself inertial.

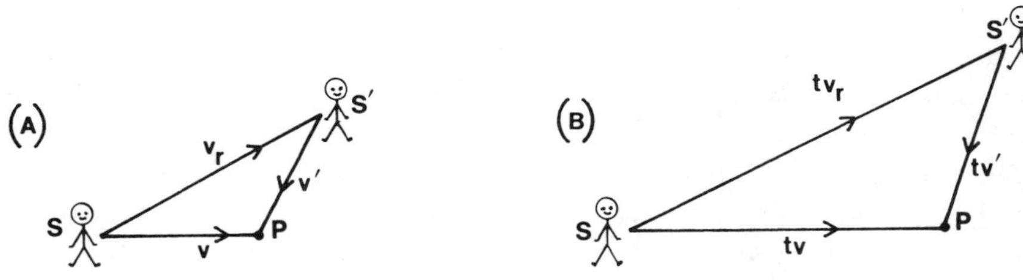

Suppose S′ moves in a straight line with constant speed v_r relative to S. Let P be a free test particle moving in a straight line with constant speed v relative to S. We assume that S, S′, and P were all at the same place initially, and that their relative positions and directions after one second are as indicated in Figure A. Figure B depicts the situation after t seconds. Explain, with the aid of these figures, why, from the point of view of S′, P moves with constant speed and direction.

2. An atomic particle traveling at 96% the speed of light passes horizontally through a chamber 100 m long. During this passage, the particle will have fallen vertically a distance $d = at^2/2$, where $a = 9.8\ m/sec^2$ is the accleration due to gravity at the earth's surface, and t is the time it takes the particle to cross the chamber. The speed of light is approximately 3×10^8 m/sec. Compute d. (By comparison, a typical wavelength for visible light is about 5.5×10^{-7} m, and the classical Bohr radius of the hydrogen atom is about 5.3×10^{-11} m.)

2. THE MICHELSON-MORLEY EXPERIMENT

Prior to the advent of relativity, it was thought that all forms of wave motion require a medium through which the waves can propagate. For example, a sound wave, which is simply a series of alternating compressions and rarefactions of the air, is propagated by the transfer of energy from one air molecule to the next in the direction of the wave's motion through the air. Under given conditions of constant atmospheric temperature, humidity, density and pressure, sound waves always travel with the same velocity (about 1100 ft. per sec.) relative to the air, regardless of the velocity of the source.

When the Scottish physicist *James Clerk Maxwell* showed in 1864 that light was an electromagnetic wave phenomenon, scientists hypothesized a medium by which the waves propagated. This was called the luminiferous (or "light-bearing") *ether* (not to be confused with the anesthetic ether). The ether presumably pervaded all of space, including the "empty" space between galaxies. Yet the ether neither acted upon nor could be affected by other matter. Its only assumed property was its role as the propagator of electromagnetic waves, which traveled through the ether with speed $c \approx 2.997925 \times 10^{10}$ cm per second (or 186,282 miles per second).

A famous experiment designed to detect the earth's motion through the ether was performed in 1881 by A. A. Michelson (the first American to receive the Nobel prize) and later, in 1887, by Michelson and Edward W. Morley, a chemist by training.

The method of detection attempted by Michelson and Morley makes use of what is called the principle of *addition of velocities*, of which the following is a commonplace illustration.

Johnny is a delivery boy who can toss a newspaper with velocity $u = 40$ feet per second. One morning, as Johnny is riding his bicycle at speed $v = 20$ feet per second directly toward Mr. Smith's front door, he hurls Mr. Smith's newspaper with his usual enthusiasm and without slowing his bicycle. The newspaper will strike the door at $u + v = 60$ feet per second (thereby alerting Mr. Smith's attack dog which has been dozing on the doorstep). In other words, the newspaper is propelled forward not only with velocity u due to Johnny's throwing arm, but also with an additional inertial velocity v acquired by virtue of the bicycle's forward motion. Relative to Mr. Smith's door the newspaper has velocity

$$u' = u + v \tag{73}$$

This is the *addition of velocities rule*.

Another example is furnished by a rower who can row with speed c in still water (c is not the speed of light here). If the river current has speed v, then the speed of the boat relative to a fixed point on the shore is $c + v$ if he is rowing downstream with the current, and $c - v$ if he is rowing upstream against it.

Before turning to the Michelson-Morley experiment, let us first look into the implications of the hypothesis that the ether exists and that light always travels with speed c relative to it. Addition of velocities will play an important role.

Imagine a space ship, S, assumed to be *at rest* in the ether. Two observers, A and B, are situated at opposite ends of S and have determined, by actual measurement with meter sticks, that they are 2L cm apart. Wishing to synchronize their watches, they arrange to have a flash bulb fired midway between A and B, with the instant of the flash to serve for them as the zero point of time. Since S is at rest in the ether, light signals will be propagated from the flash toward A and toward B with the speed c relative to them, and will reach A and

B after a time interval of $\Delta t = L/c$. Hence if both observers set their watches to Δt when they receive the light signals, their watches will be synchronized.

Now that A and B have synchronized watches as above, suppose they bring their space ship S to a certain speed v (relative to the ether) in the direction from A to B, and again fire a flash bulb midway between them. (If clock rates are temporarily affected by acceleration, both watches will be affected equally.) Because S is now moving through the ether with velocity v, which means that relative to S the ether is passing through the ship from B to A with velocity v, we would expect that the light signal approaching A would have speed $c + v$ relative to A. Similarly, we would expect that the signal approaching B would have speed $c - v$ relative to B. Here we have extended the principle of addition of velocities from everyday speeds to the speed of light, under the assumption that light travels with speed c relative to the ether, and the ether is moving past S with speed v.

The observers would find therefore that the light signals from the flash would not reach them simultaneously (according to their watches), but would be separated in time by an amount

$$\delta = L/(c - v) - L/(c + v)$$

They could in principle measure their velocity relative to the ether by solving this equation for v. Motion through the ether should therefore be detectable.

The Michelson-Morley experiment was devised to measure the expected changes in the velocity of light due to the motion of the earth through the ether. If the earth is indeed moving through the ether, an observer on the earth should be able to detect an "ether wind" whose velocity one would expect is at least on the order of the earth's orbital velocity, about 10^{-4} c ($\approx$ 30 km/sec). We say *at least* because our entire solar system is in motion relative to the center of mass of our galaxy, which in turn is in motion with respect to other galaxies. (Of course, the ether wind would vary in the course of a year because of the earth's changing speed and direction.)

The principle behind the Michelson-Morley experiment can most easily be understood through analogy with a simpler problem.

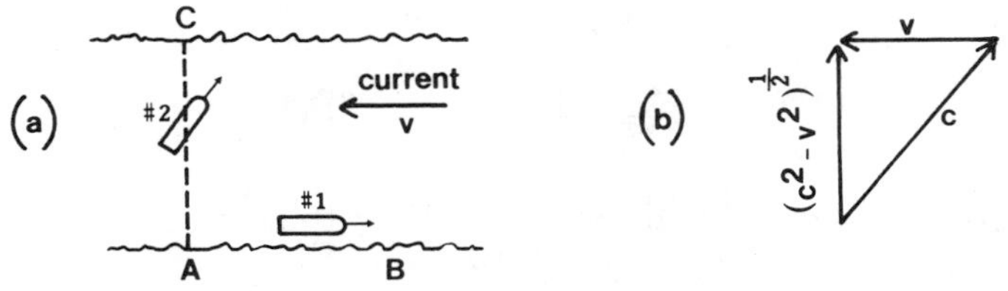

Figure II - 1.

The current on a river 1 km wide has speed v (in km per hour). Rower #1, who can row with speed c in still water, sets out from point A along the river bank and proceeds upstream 1 km to point B [Fig. II-1(a)]. At B, he turns around and rows downstream back to A. Total round trip: 2 km. (We of course assume $v < c$.)

While rowing upstream, against the current, his speed relative to the river bank is only $c - v$. The time to travel the kilometer from A to B is therefore $1/(c - v)$ (time = distance/rate). While rowing downstream, with the current, his speed relative to the river bank is $c + v$, and so the time required for the return trip from B to A is $1/(c + v)$.

Adding the two together, we see that the round trip time for rower #1 is

$$t_1 = \frac{1}{c - v} + \frac{1}{c + v} = \frac{2c}{c^2 - v^2} = \frac{2}{c}\left(1 - \frac{v^2}{c^2}\right)^{-1}$$

Now suppose that rower #2, who also can row with speed c in still water, sets out from point A and heads for point C directly across the river on the opposite shore, 1 km away. In order to compensate for the current, rower #2 must aim her boat slightly upstream, at an angle to her intended path [Fig. II-1(a)]. The resultant of her rowing speed c (relative to the water) and the speed v of the current will be a speed $(c^2 - v^2)^{1/2}$ in the direction from A to C [Fig. II-1(b)].

Accordingly, the time required for rower #2 to row the kilometer from A to C will be $1/(c^2 - v^2)^{1/2}$. Similarly, the time needed for the return trip from C to A also will be $1/(c^2 - v^2)^{1/2}$. The round trip time for rower #2 is therefore

$$t_2 = 2/(c^2 - v^2)^{1/2} = (2/c)(1 - v^2/c^2)^{-1/2}$$

(Note that t_1 and t_2 are both greater than the time $2/c$ that would be required in the absence of any current.)

To compare rowing times, we may form the ratio

$$t_1/t_2 = (1 - v^2/c^2)^{-1/2} > 1$$

The upstream-downstream time is longer than the across-stream time.

The experiment carried out by Michelson and Morley was designed to compare the travel times for light traveling in two different directions. The idea behind their apparatus is illustrated in the simplified diagram of Figure II-2. A beam of monochromatic

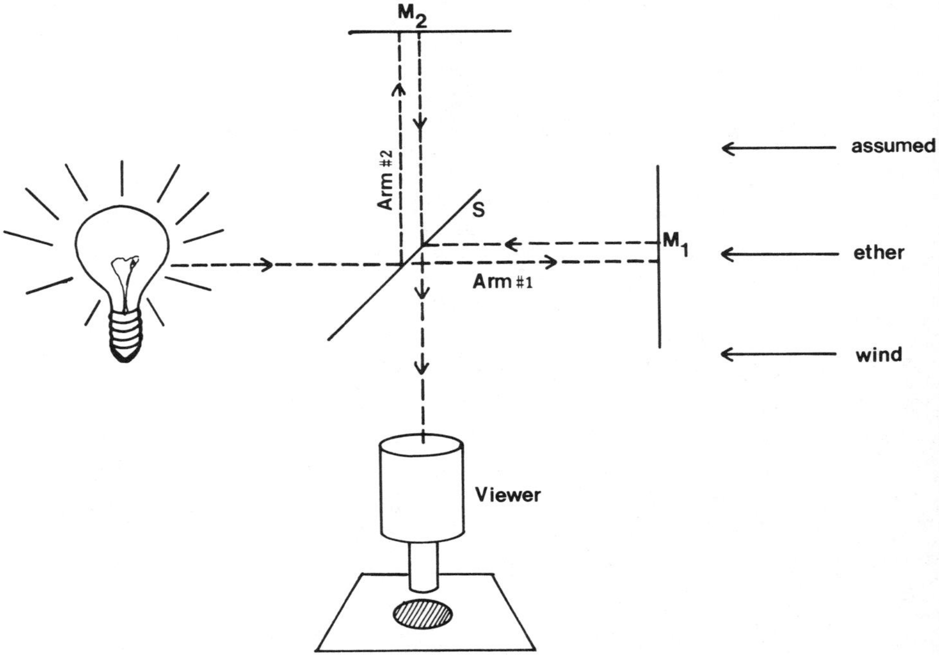

Figure II - 2.

(i.e., single frequency) light is split by a half-silvered mirror S into 2 beams which travel equal distances L from S to mirrors M_1 and M_2. There they are reflected back to mirror S, where again the beams are split, with half of each beam passing through an optical device to a viewing screen.

If the apparatus is oriented with respect to the assumed "ether wind" as in Figure II-2, then the light beam of arm #1 is the analogue of the upstream-downstream rower #1, while the light beam of arm #2 plays the role of the across-stream rower #2. Here v is the speed of the assumed ether wind (or equivalently, the speed of the earthborn laboratory relative to the ether), and c is the speed of light relative to the ether. As in the rowing example, we would expect a difference in the two round trip travel times, $t_1 > t_2$, and so the arm #1 rays should arrive at the viewing screen slightly later than (and out of phase with) the arm #2 rays. This creates a wave phenomenon known as *interference*, and a resulting pattern of alternating light and dark bands on the viewing screen.

The crucial step in the experiment is the rotation of the entire apparatus (which was floating in a pool of mercury) through an angle of 90°. This reverses the roles of the two arms, so that arm #2 is now the "upstream-downstream" arm and arm #1 is the "across-stream" arm. Michelson and Morley had calculated (from the dimensions of their apparatus and the wavelength of the light used) that when the device was rotated 90°, the interference pattern should shift by a detectable amount.

However, the expected shift was not observed! In order to rule out the possibility that the laboratory was temporarily at rest in the ether, the experiment was repeated at different times of the day and during different seasons. Always the result was the same: no shift in the interference pattern was observed.

If we consider the earth laboratory as representing different inertial frames of reference at different times of the year, the experiment seemed to suggest that the speed of light is isotropic (i.e., the same in all directions) in every inertial frame of reference. The ether wind (if it exists) cannot be detected. Apparently, then, observers A and B in our earlier space ship example, would be unable to detect their motion through the ether.

One possible explanation of the experiment's negative result is that the speed of light is an *invariant*, i.e., it has the same magnitude in all inertial frames, regardless of the motion of the observer or the source. This, however, was totally alien to the known nature of all other types of wave motion, which always travel at a fixed velocity relative to the propagating medium.

More basically, the invariance of the speed of light for all observers would contradict the principle of addition of velocities which had always seemed to be dictated by common sense.

Since the invariance of the speed of light seemed incomprehensible, other explanations were proposed to account for the negative result of the Michelson-Morley experiment. One such explanation was the suggestion that the moving earth drags some of the ether with it, so that the laboratory is at rest in the local ether. This would certainly account for the experiment's outcome. However, this possibility is ruled out by a phenomenon known as stellar *aberration*, first observed by James Bradley in 1725 (see Exercise 9).

Other theories propounded to explain the Michelson-Morley outcome hypothesized that the speed of light is constant with respect to the source, as if a moving source propels light forward just as the moving bicycle of our delivery boy given an extra boost to the toss of a newspaper. These "emission theories" are also ruled out by astronomical observations.

Historically, the most important concept used to explain the Michelson-Morley result and to retain the idea of the ether was the so-called *Lorentz-Fitzgerald Contraction*. In 1892, Fitzgerald suggested that any object moving through the ether with velocity v undergoes a shrinkage or contraction by the factor $(1 - v^2/c^2)^{1/2}$, the reciprocal of the very factor by which the "upstream-downstream" travel time exceeds the "across-stream" travel time in the Michelson-Morley experiment. This contraction applies *only* in the direction of motion. Lengths perpendicular to this direction are unaffected. Thus a ball of radius one, when put in motion through the ether, becomes an ellipsoid of revolution with semi-axes of length 1, 1, and $(1 - v^2/c^2)^{1/2}$, the latter semi-axis parallel to the direction of motion. In the Michelson-Morley apparatus, the forshortening of the "upstream-downstream" arm would exactly compensate for the longer travel time, and so the contraction hypothesis would explain why no shift is observed. Of course, the contraction could never be verified directly, since our measuring sticks would be contracted too.

In 1895 the Dutch physicist Lorentz put forward an electron theory of matter, according to which matter is composed of electric charges which generate electric and magnetic fields. According to Lorentz, these fields were affected by motion through the ether in such a way that the contraction suggested by Fitzgerald occurred. Hence the name Lorentz-Fitzgerald Contraction.

Even with the assumption of the Lorentz-Fitzgerald contraction, it turns out that if the Michelson-Morley apparatus has unequal arms, then a pattern shift could occur over a six-month period when the earth changes direction in its orbit around the sun. The earth viewed at two instances six months apart may be taken to represent two inertial frames with relative velocity twice the earth's orbital velocity: $v - v' \approx 30 - (-30) = 60$ km/sec. In 1932, Kennedy and Thorndike performed an unequal arm Michelson-Morley experiment and observed the interference pattern over many months. The degree of shift predicted on the basis of the contraction hypothesis was not observed.

Exercises II-2

1. Estimate the earth's orbital velocity. (The mean radius of the orbit is about 93 million miles.)

2. Let L_1 and L_2 be the lengths of the arms of the Michelson-Morley apparatus. Assume the speed of light is c relative to the ether. If initially arm #1 is parallel to the direction of the "ether wind," and arm #2 is perpendicular to it, then the round trip light travel times for the two arms are, respectively,

$$t_1 = \frac{2L_1}{c}\left(1 - \frac{v^2}{c^2}\right)^{-1} \approx \frac{2L_1}{c}\left(1 + \frac{v^2}{c^2}\right)$$

$$t_2 = \frac{2L_2}{c}\left(1 - \frac{v^2}{c^2}\right)^{-1/2} \approx \frac{2L_2}{c}\left(1 + \frac{v^2}{2c^2}\right)$$

[Since $v/c \approx 10^{-4}$, we are justified in using the binomial approximation $(1 - v^2/c^2)^m \approx 1 - mv^2/c^2$.] Therefore the rays which recombine at the viewer are separated by a time delay of

$$\Delta t = t_1 - t_2 \approx \frac{2}{c}\left(L_1 - L_2 + \frac{L_1 v^2}{c^2} - \frac{L_2 v^2}{2c^2}\right)$$

Now suppose the apparatus is rotated 90° so that arm #1 is now transverse to the ether wind. Let t'_1 and t'_2 denote the new round trip light travel times for the two arms, and let $\Delta t' = t'_1 - t'_2$. Compute $\Delta t'$ and show that

$$\Delta t - \Delta t' = \frac{v^2}{c^2}\left[\frac{L_1 + L_2}{c}\right]$$

This is the change in time delay produced by rotating the apparatus.

3. (See Exercise 2.) In the Michelson-Morley experiment, $L_1 = L_2 = 11$ m, and the light used had a wavelength $\lambda \approx 5.9 \times 10^{-7}$ m. The period T of this light, given by $T = \lambda/c$, is approximately 2×10^{-15} sec. Let $N = (\Delta t - \Delta t')/T$, the number of periods contained in the change in time delay. (a) Assuming the earth's speed through the ether is 3×10^4 m/sec, verify that $N \approx 0.37$ (on the basis of the ether theory), i.e., the interference pattern should shift by 37% of the distance between consecutive bands. (b) In actuality, Michelson and Morley observed a pattern shift less than 1% of the distance between consecutive bands, i.e., $N \approx 0.01$. Show this implies that the "ether wind," if it exists at all, is "blowing" at less than one-sixth the earth's orbital speed.

In the following four exercises, assume that any object moving through the ether undergoes a contraction by the factor $(1 - v^2/c^2)^{1/2}$ in the direction of its motion, as suggested by Fitzgerald. The travel times t_1 and t_2 are then

$$t_1 = \frac{2L_1}{c}\left(1 - \frac{v^2}{c^2}\right)^{-1/2}$$

$$t_2 = \frac{2L_2}{c}\left(1 - \frac{v^2}{c^2}\right)^{-1/2}$$

(There is now no predicted shift when the apparatus is rotated.)

4. Assume $\Delta L = L_1 - L_2 \neq 0$, as in the Kennedy-Thorndike experiment. Let $\Delta t = t_1 - t_2$. Show that

$$\Delta t \approx \frac{2}{c}\,\Delta L\left(1 + \frac{v^2}{2c^2}\right)$$

5. Suppose the experiment is repeated several months from now when the speed of the earth relative to the ether is v′ instead of v. This of course changes the value of Δt. Prove that this change in Δt results in an interference pattern shift of

$$\Delta N = \frac{\Delta L}{\lambda}\left(\frac{v^2 - v'^2}{c^2}\right)$$

bands, where λ (about 5.5×10^{-7} m in the Kennedy-Thorndike experiment) is the wavelength of the light used.

6. Suppose we assume that the sun is at rest in the ether. Let u ($\approx 3 \times 10^4$ m/sec) be the earth's orbital velocity, and let w (≈ 460 m/sec) be the surface velocity of the earth due to the earth's rotation on its axis. Compute the interference pattern shift ΔN that would be expected over the course of a day when the speed of the apparatus varies from $v = u + w$ to $v' = u - w$.

7. The speed of the sun relative to the galactic center is believed to be about 2.2×10^5 m/sec. Assuming the galactic center is stationary in the ether, compute the maximum pattern shift ΔN that would be expected over the course of a year, because of the earth's motion about the sun.

8. Let us now reject both the ether hypothesis and the contraction hypothesis and assume the speed of light is the same in all directions in all inertial frames. If $\Delta L = L_1 - L_2 \neq 0$, then the light rays from the two arms will arrive at the viewer separated by a time interval of $2\,\Delta L/c$ seconds. If T is the period of the light used, then the number of periods in this time interval is

$$N = \frac{2\Delta L}{cT}$$

In the Kennedy-Thorndike experiment, $\Delta L = 0.16$ m and $T \approx 1.8 \times 10^{-15}$ sec. (a) Compute N. (b) The equation for N above may be solved for c to obtain

$$c = \frac{2\Delta L}{NT}$$

Over six months of observation, N was found to vary by less than 0.003. On the assumption that ΔL and T remained constant, show that this implies a variation in the speed of light of about 1.5 m/sec. [Hint: $\Delta c \approx (dc/dN)\,\Delta N$.]

9. (Stellar Aberration) Light rays from a star directly overhead enter a telescope. Suppose the earth, in its orbit around the sun, is moving at a right angle to the incoming rays, in the direction indicated in the figure. In the time it takes a ray to travel down the barrel to the eyepiece, the telescope will have moved slightly to the right. Therefore, in order to prevent the light ray from falling on the side of the barrel rather than on the eyepiece lens, we must tilt the telescope slightly from the vertical, if we are to see the star. Consequently, the *apparent* position of the star is displaced forward somewhat from the actual position.

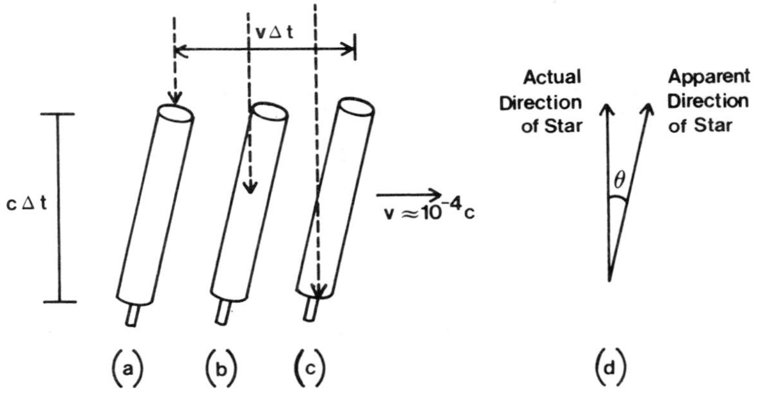

Telescope moves to right as light travels down barrel.

Show that the angle of displacement, θ, is given (in radians) by

$$\theta = \tan^{-1}(v/c) \approx v/c$$

where v is the earth's orbital velocity. Verify that $\theta \approx 20.6''$ (roughly the angle subtended by an object 0.1 mm in diameter held at arm's length).

(As the earth revolves around the sun in its nearly circular annual orbit, the apparent position of the star will trace a circle with angular radius $20.6''$. This is indeed observed. If the earth dragged a layer of ether along with it, the light rays, upon entering this layer, would acquire a horizontal velocity component matching the forward velocity v of the telescope. There would then be no aberration effect.)

3. THE POSTULATES OF RELATIVITY

In Bern, Switzerland, in the year 1905, a young patent examiner named *Albert Einstein* (1879–1955) resolved the mystery posed by relative motion and the velocity of light. (At that time Einstein was unaware of the Michelson-Morley experiment but had been studying other problems associated with relative motion.) In his famous paper *On the Electrodynamics of Moving Bodies*, Einstein stated two simple but seemingly contradictory postulates as the cornerstone of his *Special Theory of Relativity*:

> P1. All physical laws valid in one frame of reference are equally valid in any other frame moving uniformly relative to the first.
> P2. The speed of light (in a vacuum) is the same in all inertial frames of reference, regardless of the motion of the light source.

P1, which is called the *Principle of Relativity*, implies that no physical experiment carried out entirely within one inertial system can ever reveal the motion of that system with respect to any other. In practice, this means that whenever we are considering two inertial systems in uniform relative motion, we may, with equal validity and equal success, think of either one of them as at rest and the other as moving. As far as the study of just these two systems is concerned, it makes no difference whatsoever. The situation is entirely symmetric.

In fact, a more restricted form of this principle had already been given by Newton in his *Principia*:

> The motions of bodies included in a given space are the same among themselves whether that space is at rest or moves uniformly forward in a straight line.

Newton's relativity principle states that the laws of mechanics (moving bodies) are the same in any two reference frames moving uniformly relative to one another. For example, we can pour a cup of tea in a 600 mi/hr airplane (barring any turbulence) just as easily as we can in our own homes. A ball thrown across the airplane cabin will follow the same kind of parabolic trajectory as would a ball thrown across a room at home. Thus mechanical laws which are valid in one frame of reference are equally valid in any frame moving uniformly relative to the first. Einstein's great insight, characteristic of his firm belief in the unity of nature, was the extension of this principle to encompass all phenomena, electromagnetic as well as mechanical.

Accordingly, the ether, if it exists at all, can never be detected, there is no preferred frame of reference, and no such thing as an absolute state of rest. Since Einstein's theory of relativity neither requires nor presupposes the existence of the ether, we may dispense with the idea entirely.

As we have already seen, P2 contradicts the classical principle of addition of velocities, (73), which had always been accepted without question. The source of this apparent contradiction, it turns out, lies in our traditional conception of time: for example, the notion that if two events are simultaneous for one observer, then they must be so for any other observer. It had always been assumed that absolute time, like absolute space, was the same for everyone:

> Absolute, true, and mathematical time, of itself, and from its own nature, flows equably, without relation to anything external.
>
> (Newton, *Principia*)

Einstein was led to his postulates through a realization that these classical ideas concerning time and space were fundamentally wrong. According to his theory, addition of velocities is only an approximation, quite good at "everyday" speeds, but measurably inaccurate at speeds approaching the speed of light.

4. RELATIVITY OF SIMULTANEITY

In the remainder of this chapter, we shall examine the consequences of Einstein's postulates. The mathematics of the resulting *Special Theory of Relativity* is quite elementary. The difficulties you will experience are conceptual rather than mathematical. You will be asked to abandon some of the preconceived ideas of space and time that have been implanted in all our minds and reinforced over many years. Perhaps the most difficult pill to swallow is the new concept of time, to which we now turn.

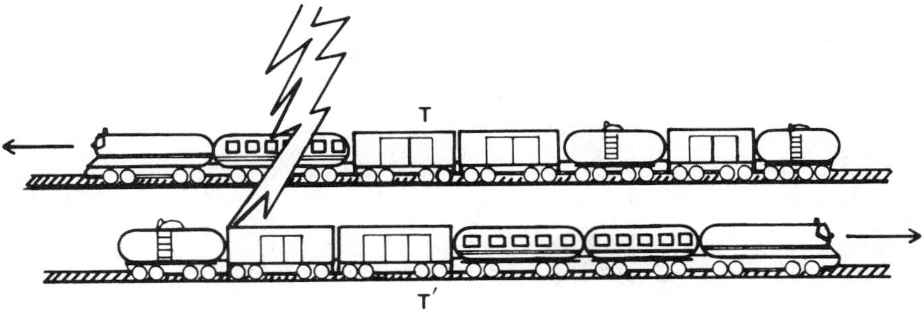

Figure II - 3.

Imagine two very long railroad trains T and T′ in relative motion along two adjacent lengths of track, as in Figure II-3. Assume that T′ is moving to the right with constant speed v relative to T. Of course, it would be equivalent to say that T is moving to the left with speed v relative to T′. (The ground will play no significant role in this discussion.) At fixed locations on these trains are two observers: O in T and O′ in T′.

Suppose that two bolts of lightning happen to strike the road bed and in so doing leave permanent marks on both trains. At this point, we make no assumption concerning the temporal order of the two lightning flashes. Let's say that one of the flashes leaves marks at point A on train T and at point A′ on train T′. Of course, since these two marks are caused by one and the same flash, the two points, A and A′, must be passing one another at the very instant (and place where) this flash occurs.

Similarly, we shall call the points on the two trains where the other flash leaves its marks B and B′ (on T and T′, respectively). These points must be opposite each other at the instant this other flash occurs.

Let us assume that O happens to be situated exactly midway between points A and B of his train, while O′ is midway between points A′ and B′ of his train. (Assume A, A′ are to the left of B, B′, respectively.)

Now suppose O *sees* the two flashes simultaneously, i.e., he receives light signals from A and from B simultaneously. Since the distances AO and BO are equal (and light travels with the same speed c in all directions), O is justified in asserting that the two bolts of lightning struck his train simultaneously. (Note that there is no need to worry about synchronization of watches here. The light rays from A and B have arrived at one and the same place, O, and simultaneity of events happening at a single location is a well-defined concept independent of time-keeping devices.) Moreover, since O′ is midway between A′ and B′, and since the two simultaneous flashes occurred when A′ was opposite A and B′ was opposite B, O can also assert that O′ was directly opposite him at the same instant. In other words, according to O, the following three events occurred *simultaneously*:

a. a lightning bolt struck at A and A′,
b. another lightning bolt struck at B and B′, and
c. O′ was opposite O.

What does O′ observe? As the light signals from the two flashes travel with finite speed toward O, O′ moves to the right relative to O (Figure II-4), toward B and away from A. Consequently, the BB′ signal will meet O′ before the AA′ signal does. Since O′ is equidistant from A′ and B′, he will naturally conclude that the flash at B′ occurred *before* the flash at A′! (Of course, O′ will explain the simultaneous meeting of both signals with O by citing the motion of O to the left: away from the earlier flash's signal and toward the latter's.)

We emphasize that it makes no difference whether T is at rest on the track and T′ is moving to the right, or T′ is at rest and T is moving to the left, or even if they are both moving on the tracks. By Einstein's postulates, all these points of view are entirely equivalent for our purposes (as long as the *relative* speed of the trains remains the same). In any case the speed of light is c according to each observer.

The above mental experiment shows that spatially separated events which are simultaneous according to one observer, need *not* be simultaneous from the point of view of another. This is the *relativity of simultaneity*.

The sequence of events from the point of view of O is represented in Figure II-4, where T is stationary and T′ moves to the right. In Figure II-5, these same events are depicted from the point of view of O′. Here T′ is viewed as stationary, while T moves to the left.

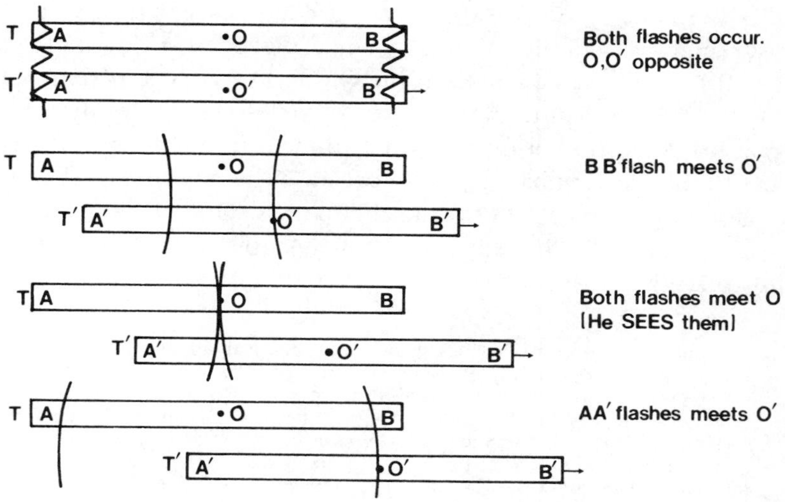

Figure II - 4. The O point of view.

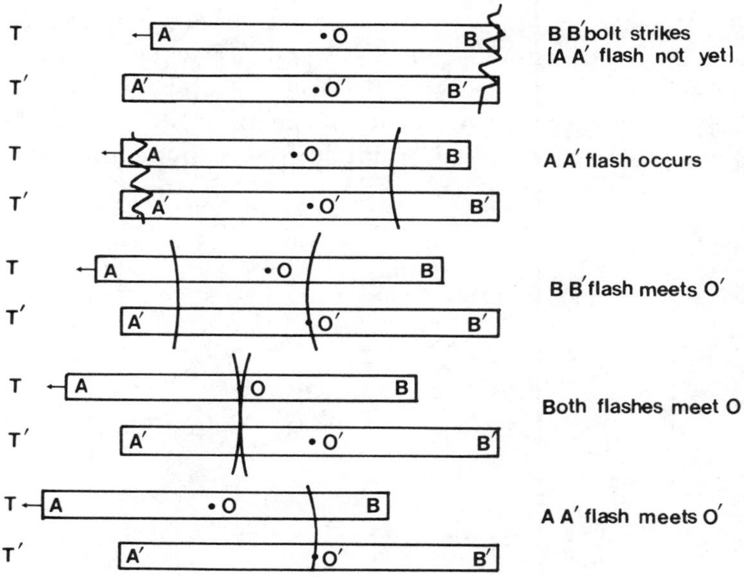

Figure II - 5. The O′ point of view.

The relativity of simultaneity is a complicating factor in the measurement of lengths. You may be perplexed at seeing that $\overline{AB}$ and $\overline{A'B'}$ have the same length in the O point of view, but that $\overline{AB}$ is shorter than $\overline{A'B'}$ in the O′ point of view. However this discrepancy is not a contradiction. In order to measure the length of any object that is moving past us, we must note the position of the front and back of the object at one and the same instant. L. R. Lieber [34] gives the following illustration of this obvious but important point (see Fig. II-6).

> . . . suppose the object is say, a fish swimming about in a tank? To measure its length while it is in motion, by placing two marks on the walls of the tank, one at the head, and the other at the tail, it would obviously be necessary to make these two marks SIMULTANEOUSLY–for, otherwise, if mark B is made at a certain time, then the fish allowed to swim in the direction indicated by the arrow, and then the mark at the head is made at some later time, when it has reached C, then you would say that the length of the fish is the distance BC, which would be a fish story indeed!

Now the discrepancy in lengths AB and A′B′ between Figures II-4 and II-5 should be less perplexing. Because two observers in relative motion will not agree on simultaneity, they will in general record different lengths for the same object. In our train example, observer O considers the distances AB and A′B′ equal because of the simultaneity of the lightning flashes. However, observer O′ considers AB shorter than A′B′ since at the instant B′ is opposite B, A has not yet reached A′ (according to his reckoning of time).

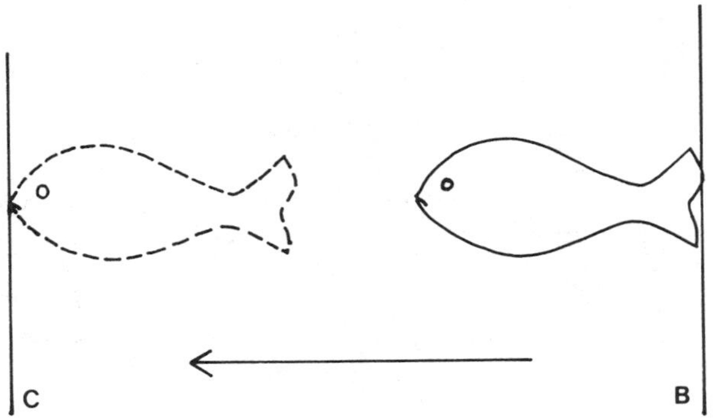

Figure II - 6.

If you are wondering whether the above asymmetry violates the Principle of Relativity, be assured that there *is* symmetry, but in the following sense: if the lightning flashes had been simultaneous for O′ (rather than O), then O′ would have considered AB and A′B′ equal, while O would have received the AA′ signal before the BB′ signal and would have considered A′B′ shorter than AB. (See Exercise 1.)

We may still ask (in either situation) whose measurements are right, those of O or those of O′. It will become clear to you shortly that they both are, because one of the key ideas of relativity theory is that *length* is not an invariant attribute of an object but depends on the observer.

Exercises II-4

1. Draw the counterparts of Figures II-4 and II-5 for the case in which O′ sees the lightning flashes simultaneously.

2. In Figure a, rocket S′ is moving to the right with velocity v relative to rocket S. At the instants the rockets pass one another, a spark jumps from S to S′ and emits a flash of light. If the speed of light were c (= 3×10^8) m/sec for all observers, then S and S′ each would claim that the light flash is moving away from him with speed c. Suppose the flash is seen by an observer at point C one second after the spark discharge (Figure b). Then according to S, AC = c

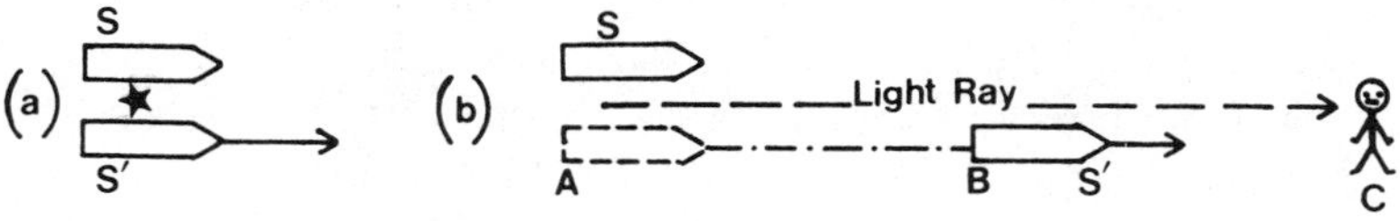

meters. On the other hand, as seen by the S′ observer, in one second the light has traveled the distance BC (the S′ observer is just as entitled to consider his rocket as the source of the flash as is the S observer). Therefore BC = c. It is obvious that AC and BC cannot both equal c. (Indeed, according to addition of velocities, if AC = c, then BC = c − v; while if BC = c, then AC = c + v.) Thus the invariance of the speed of light and the principle of addition of velocities are incompatible. Resolve this apparent paradox in the light of what you now know about the relativity of time measurements.

3. Jill is going on a train trip. Her brother Jack bids her good-bye at the station. As the train pulls away, each sees the other grow smaller and smaller as the distance between them increases. Why is it

not a contradiction for each to see the other become smaller? Is there any connection between the present situation and the discrepancy in length measurements in the train example of this section?

4. Suppose observer O′ is moving to the right with constant speed v relative to observer O. Let us assume that at the very instant O and O′ pass each other, rays of light proceed in all directions from the place where they pass. O will observe a spherical light wave front moving away from him with velocity c. After t seconds, the light will have reached all points at distance ct from him. According to O, at the instant the light has reached point A (in the figure), it has also reached point B.

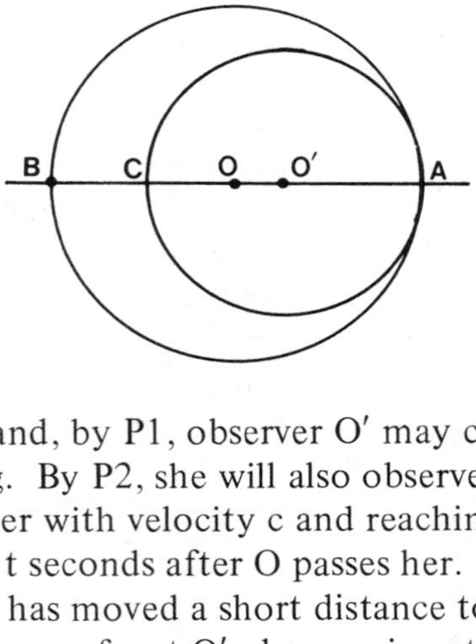

On the other hand, by P1, observer O′ may consider herself at rest and O as moving. By P2, she will also observe a spherical wave, moving away from her with velocity c and reaching all points at a distance ct from her t seconds after O passes her. However, as the light travels to A, O′ has moved a short distance to the right of O, so that the spherical wave front O′ observes is not concentric with the one observed by O. Therefore, in the view of O′, at the instant the light has reached A, it has also reached (not B yet, but) C. They cannot both be right, and yet Einstein's postulates imply they are. Resolve this apparent paradox.

5. COORDINATES

In three-dimensional Euclidean geometry, places or *points* can be located by reference to a set of mutually orthogonal coordinate axes. Once a unit of distance is selected, each point is specified by its triple of Cartesian coordinates: (x, y, z). In physics, the fundamental entities are *events* rather than points. An event determines both a time and a place, and so will be assigned four coordinates: (t, x, y, z). These coordinates will depend upon the particular observer who is doing the measuring and upon his choice of coordinate axes and the units he chooses. In general, different observers will assign different coordinates to an event. Because of the relativity of simultaneity, time as well as space coordinates will have different values for different observers. The collection of all possible events is called *spacetime*.

From a purely mathematical point of view, it would be convenient to measure all four coordinates in terms of the same units. Otherwise, we would be acting like a surveyor measuring north-south distances in meters and east-west distances in yards. It may at first seem strange to suggest measuring distances and times in the same units, but the invariant speed of light provides us with a conversion factor. One second corresponds approximately to 3×10^{10} cm of light travel (a "light-second"); or equivalently, one centimeter will be traversed by light in $(3 \times 10^{10})^{-1} \approx 3.33 \times 10^{-11}$ seconds. The latter will be called a *centimeter (of light travel time).* It is the time it takes light to travel one centimeter. From now on, we will give both distances and times in terms of *geometric units*, namely centimeters. (Actually, astronomers have been measuring distances and times in the same units for some time. A year of light travel, or a light-year is the distance light travels in one year. Hence distance as well as time can be measured in years.)

Since light travels one centimeter of distance in one centimeter of light travel time, the speed of light is one in geometric units. If v is a velocity measured in conventional units, for example cm/sec, then the corresponding value of the velocity in geometric units is

$$\beta = \frac{\text{v cm/sec}}{\text{c cm/sec}} = \frac{\text{v}}{\text{c}} \tag{74}$$

Being a ratio of two velocities, β is a "dimensionless" quantity, i.e., its numerical value is independent of the conventional units chosen, be they centimeters, inches, or miles for distance, or seconds, hours, or years for time—as long as v and c are reckoned in the same conventional units.

Suppose an observer in an inertial frame S wishes to assign coordinates to events. Imagine that he has constructed throughout his region of space a rectangular framework or lattice of rigid rods all of the same length, which we shall take as one centimeter, although a meter, millimeter, or any other convenient length would do.* At each lattice point, we imagine a clock calibrated in centimeters. To picture this think of a Cartesian coordinate system in 3-dimensional Euclidean space, with a tiny clock attached to each point whose coordinates are integers.

Suppose further that one of the lattice points is chosen as the origin of the coordinate system. Set the clock at this point to $t = 0$, and let it begin running just as a flash of light leaves the origin and spreads out in all directions. A lattice clock L cm from the origin is said to be *synchronized* with the clock at the origin if it reads L cm of time when the light signal reaches it. In principle this could be ensured by having an assistant hold the clock hand at $t = L$ and then release the pointer at the moment the flash is seen. We assume all the clocks in S have been synchronized with the origin clock in this way.

We might also mention that distances can be verified by means of light signals too. If the flash emitted from the origin at $t = 0$ is reflected back by a mirror L cm away, then the origin clock will read 2L cm when it receives the reflected signal.

We may now choose three of the rods emanating from the origin to indicate the directions of the positive x-, y-, and z-axes. The spatial coordinates of an event will be the x, y, and z coordinates of the clock at (or nearest) the place of the event, and the t coordinate of the event will be provided by the reading of that clock. Moreover, we may imagine that the clocks are recording clocks that keep a record of the events that occur at their localities. When we speak of the S observer, we really mean the collection of these recording clocks, whose records can be gathered up and examined. Since S is inertial, these records would show that any free test particle moves uniformly (equal spatial coordinate changes in equal time changes).

*Here we are assuming that physical space is modeled by Euclidean geometry. This assumption will be challenged in the following chapter.

It must be clearly understood that *observing* in the above sense, is quite different from *seeing*. When we see an object at a particular instant, our eyes receive light rays emanating from all parts of the object. Rays from the more distant parts must have left the object earlier than rays from the nearer parts, if all these rays are to enter our eyes at the same time. Hence a visual image is a composite of light signals emitted from the seen object at different times. On the other hand, when an object is *observed* in an inertial system S, the locations of all its parts are measured simultaneously (as defined by the synchronized clocks of S).

Now suppose that two inertial frames S and S′ are moving uniformly relative to each other. We may imagine that the lattices of rods and clocks of the two systems pass through one another. We wish to compare the coordinates t, x, y, and z which S assigns to an event with the coordinates t′, x′, y′, and z′ which S′ assigns to the same event.

To make things simple, we shall henceforth assume that the coordinate axes have been chosen in such a way that

i. the x- and x′-axes (and their positive directions) coincide,
ii. relative to S, S′ is moving in the positive x direction with velocity β (which means that, relative to S′, S is moving in the negative x′ direction with velocity $-\beta$),
iii. the y- and y′-axes are always parallel, and
iv. the z- and z′-axes are always parallel.

For convenience, we shall arbitrarily call S the *laboratory frame* and S′ the *rocket frame*.

First we shall show that lengths perpendicular to the direction of relative motion are the same for the observers in both frames. Imagine that there is at rest in the laboratory frame S, a circular cylinder C of radius R, as measured in S. Similarly, imagine there is at rest in the rocket frame S′ a cylinder C′ of the same radius, as measured by the S′ observer. The xx′-axis is the common axis of the two cylinders (Fig. II-7). Since we shall assume that space is *homogeneous* and *isotropic*, i.e., the same at all points and in all directions, each cylinder will be observed to be a circular cylinder by the observer in the other frame. (It could not, for instance, look like an elliptical cylinder, since that would attribute directional asymmetry

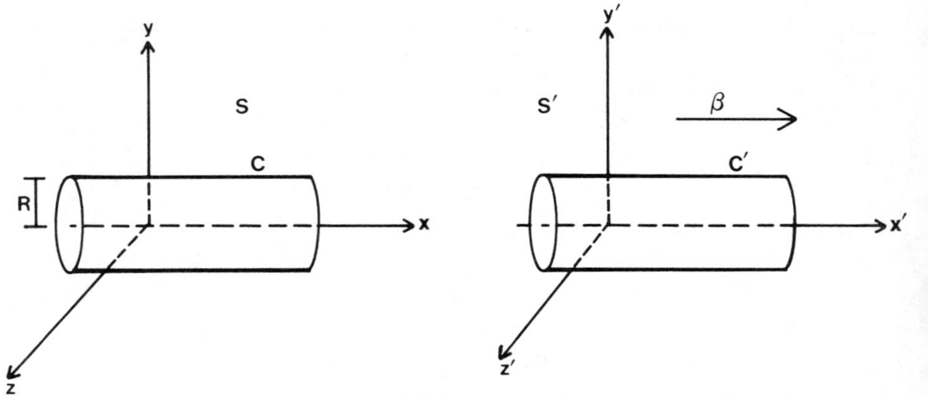

Figure II - 7.

to space.) What will the radius of the rocket cylinder be *as measured by the observer in the latoratory*? We claim that it is R, the same *as is measured by the rocket observer.*

If the lab observer measures a smaller radius, $r < R$, then he will see the rocket cylinder C′ pass through the interior of his cylinder C. S′ of course will also see this happen, since the coincidence of a point of C′ with a point in the interior of C cannot be observed by one observer without being observed by all. However, by the symmetry expressed in the Principle of Relativity, we may interchange the roles of S and S′ and conclude also that C must pass through the interior of C′. This contradiction tells us that S cannot observe the radius of C′ as being less than R—nor more than R, by completely analogous reasoning. Hence the radius of C′ is R whether S or S′ is doing the measuring (by means of his respective lattice of rods and synchronized clocks).

Another way of reaching the same conclusion is this: if S measures a smaller (or larger) radius for C′ than does S′, then there would be a way of distinguishing between two inertial frames in relative uniform motion. This violates the Principle of Relativity. Accordingly, when two inertial observers have set up coordinate systems satisfying (i), (ii), (iii), and (iv), they will agree on the values of the y and z coordinates of each event. In general, any two inertial observers in uniform relative motion will agree on the measurement of any distance perpendicular to the direction of relative motion.

The measurement of lengths parallel to this direction will be discussed in the next section.

Exercises II-5

1. Compute $\beta = v/c$ for the following.

   i. an automobile traveling 55 mph,
   ii. a satellite orbiting the earth at 8 km/sec,
   iii. the earth in its orbit around the sun (see Exercise 1 of Section II-2)

2. As pointed out in this section, when we see an object at a particular moment, light rays reaching our eyes from the most distant parts of the object must have left the object earlier than rays from the nearer parts, since all these rays enter our eyes at the same instant. Suppose the object we are seeing is a very long train approaching us at a substantial fraction of the speed of light. Would its (seen) length appear longer or shorter than if the train were seen at rest? Explain. (Be qualitative only. For a relativistic analysis of the distinction between observing and seeing, cf. [48].)

6. INVARIANCE OF THE INTERVAL

Let A and B be two points of 3-dimensional Euclidean space. Relative to a set of mutually orthogonal coordinate axes, these points have Cartesian coordinates (x_A, y_A, z_A) and (x_B, y_B, z_B), respectively. The distance d_{AB} from A to B is given by the expression

$$d_{AB} = [(\Delta x)^2 + (\Delta y)^2 + (\Delta z)^2]^{1/2} \tag{75}$$

where $\Delta x = x_B - x_A$, $\Delta y = y_B - y_A$, and $\Delta z = z_B - z_A$. If we adopt a different choice of orthogonal coordinate axes, then A and B will in general be assigned different coordinates (x'_A, y'_A, z'_A) and (x'_B, y'_B, z'_B). However, the expression

$$[(\Delta x')^2 + (\Delta y')^2 + (\Delta z')^2]^{1/2} \tag{76}$$

computed with the new coordinate differences, yields the same value, d_{AB}, for the distance. In other words, distance is *invariant* under a change of Cartesian coordinates. In this section, we shall discover an analogous quantity, called the *interval* between two events. which is invariant under a change of spacetime coordinates from one inertial system to another.

Let us compare how a simple experiment is viewed by the laboratory observer S and the rocket observer S′ (who is moving with velocity β relative to S, as in Section 5). Suppose the rocket observer, at $t' = 0$, flashes a beam of light from the origin of her coordinate system up to a mirror located at the point L cm away on the y′-axis. The light is reflected by the mirror and returns to the S′ origin after $\Delta t' = 2L$ cm of light travel time. To the rocket observer, the light has traveled a strictly vertical path, and the time interval $\Delta t'$ between emission (event A) and reception (event B) of the light is measured by means of a *single clock* located at the origin [Fig. II-8(a)].

However, the laboratory observer S sees the rocket and its mirror move to the right a distance Δx during this interval. Accordingly, for him the light has traveled the oblique path of Figure II-8(b), a longer path than that reckoned by the rocket observer. (The altitude of ΔAMB is L since lengths perpendicular to the direction of relative motion are invariant.) Because the speed of light is the same for both observers, S will record a greater time interval Δt between events A and B than will S′. (The laboratory observer will think that

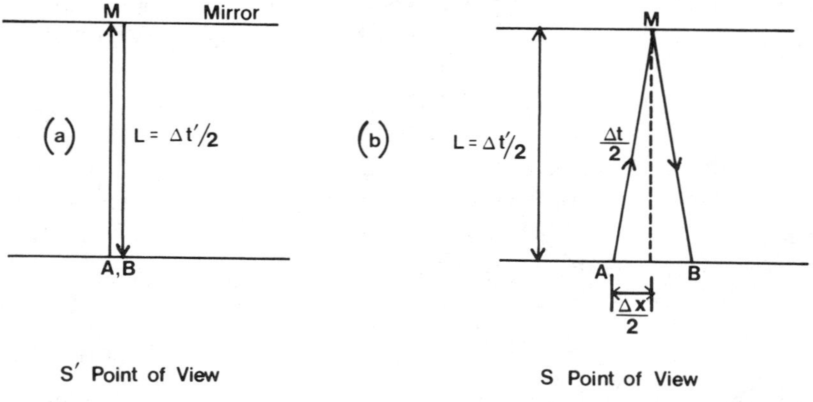

Figure II - 8.

the rocket observer's watch is running slow.) Why is this not a contradiction to the Principle of Relativity, just as it would have been a contradiction for cylinder C′ to pass inside cylinder C in the previous section? The reason is that the situation here is not symmetrical. Where S′ needs only a *single clock* to measure the time between A and B, S requires *two stationary clocks*, one at the place where A occurs in his frame and another where B occurs. The relativity of simultaneity then comes into play. According to S, the occurrence of B at $x = \Delta x$ is simultaneous with the reading $t = \Delta t$ of the origin clock (and hence of all S clocks). This simultaneity will not be observed by S′. The result here is that the observer who assigns the same spatial coordinates to the two events—so that only one clock is needed—will record the shorter time interval.

It should be noted that the same conclusion would hold for any pair of events A, B which occur at the same place in S′ (but at different places in S). This is because we could always arrange to have a light ray emitted from the place and time of event A and bounced off a suitably placed mirror so as to return to event B.

From Figure II-8(b), we obtain (in geometric units)

$$\Delta t'/2 = [(\Delta t/2)^2 - (\Delta x/2)^2]^{1/2}$$

or

$$(\Delta t')^2 = (\Delta t)^2 - (\Delta x)^2 \tag{77}$$

Since the velocity of S′ relative to S is $\beta = \Delta x/\Delta t$, we can substitute $\Delta x = \beta\, \Delta t$ and deduce

$$\Delta t' = (1 - \beta^2)^{1/2}\, \Delta t \tag{78}$$

as the quantitative relationship between the time intervals recorded by our two observers.

For "ordinary" speeds, the difference between Δt and $\Delta t'$ is exceedingly small. For instance, suppose L = 100 cm and that S′ is a train moving at $v = 2 \times 10^3$ cm/sec (about 45 mi/hr) relative to the laboratory S. Then $\beta = v/c = (2 \times 10^3)/(3 \times 10^{10})$ and $\Delta t = \Delta t'(1 - \beta^2)^{-1/2}$. Hence,

$$\Delta t - \Delta t' \approx \frac{1}{2}\beta^2 \Delta t' = 100\beta^2 \text{ cm} \approx 4.4 \times 10^{-13} \text{ cm}$$

$$\approx 1.5 \times 10^{-23} \text{ sec*}$$

In view of this tiny quantity, it is not surprising that prior to the advent of relativity theory, time intervals had always been assumed to be the same for everyone. It is only at very high speeds, such as are realized in the particle accelerators of the nuclear physicist, that the relativity of time intervals becomes readily noticeable.

Since $\Delta x' = 0$ (both events occur at $x' = 0$ in S′), (77) may be rewritten

$$(\Delta t')^2 - (\Delta x')^2 = (\Delta t)^2 - (\Delta x)^2 \tag{79}$$

The quantity

$$\Delta\tau = [(\Delta t)^2 - (\Delta x)^2]^{1/2} = [(\Delta t')^2 - (\Delta x')^2]^{1/2} \tag{80}$$

is called the *interval* between event A and event B. According to Eqs. (79) and (80), $\Delta\tau$ has the same value (2L cm in this example) whether computed in S coordinates or S′ coordinates, and this is so regardless of the value of the relative velocity β. $\Delta\tau$ is therefore an *invariant* under such a change of coordinates. In S′ coordinates, $\Delta x' = 0$, and so $\Delta\tau$ coincides with the time interval $\Delta t'$ that would be recorded by a *single clock* at rest in the rocket frame (where the two events occur at the same place). For this reason, $\Delta\tau$ is also called the *proper time* between events A and B. In any inertial frame moving relative to S′, $\Delta x \neq 0$ and so [from Eq. (80)], $\Delta t > \Delta\tau$.

As we have seen, this inequality arises because two synchronized laboratory clocks are needed to measure Δt, whereas a single rocket clock suffices to measure the proper time $\Delta t'$. Therefore, by symmetry, if two events occur at the same place in the *laboratory frame*, where they are separated in time by Δt, then their time separation $\Delta t'$ for the rocket observer (she is the one requiring two clocks this time) satisfies

$$\Delta t = (1 - \beta^2)^{1/2} \Delta t'$$

---

*For β small, $(1 - \beta^2)^{-1/2} \approx 1 + 1/2\,\beta^2$.

[compare this with Eq. (78)]. Consequently this time the rocket observer will think that the laboratory observer's watch is slow.

In brief, the time interval between two events depends not only on these events, but upon the observer as well. Time is not universal, as Newton believed. It is relative.

When the relative motion of the two systems is not along a common xx'-axis, the squared distance $(\Delta x)^2$ in formula (80) must be replaced by

$$(\Delta x)^2 + (\Delta y)^2 + (\Delta z)^2$$

in accordance with the Euclidean distance formula (75) given earlier. The general expression for the interval is therefore

$$(\Delta \tau) = [(\Delta t)^2 - (\Delta x)^2 - (\Delta y)^2 - (\Delta z)^2]^{1/2} \tag{81}$$

and so $(\text{interval})^2 = (\text{time separation})^2 - (\text{space separation})^2$. The individual coordinate differences appearing on the right side of this equation will in general have different values in different inertial frames, but the computed value of $\Delta\tau$ will always be the same, so that $\Delta\tau$ is an *invariant*. The interval is to spacetime geometry what the distance is to Euclidean geometry. When we have derived the equations for a change of coordinates in switching from one inertial observer to another (Section 7), we will be able to verify the invariance of $\Delta\tau$ algebraically.

An important consequence of Einstein's postulates is that the speed of any material object must be less than unity in every frame of reference. In the example above, we may take the view that the rocket origin clock has moved from the place of event A to the place of event B, and since $|\Delta x/\Delta t| = |\beta| < 1$, the space separation is less than the time separation. (This is so for *all* inertial observers, by the invariance of $\Delta\tau$.) Consequently, $(\Delta\tau)^2 > 0$. Such intervals, in which the time separation dominates, are called *timelike*. In general, if it is possible for a material object (or an observer) to be present at two events, then the events are separated by a timelike interval. In the coordinate system of an observer present at both events, $\Delta x = 0$ and the interval equals the time lapse between the events, as measured by a single clock: $\Delta\tau = \Delta t$.

On the other hand, if two events, A and B, are such that $(\Delta\tau)^2 < 0$ (so that $\Delta\tau$ is imaginary), then in any inertial coordinate system, the space separation will exceed the time separation. In this case, the interval is called *spacelike* and the quantity

$$\Delta\sigma = [(\Delta x)^2 + (\Delta y)^2 + (\Delta z)^2 - (\Delta t)^2]^{1/2} \qquad (82)$$

is called the *proper distance* between the events. No material particle can be present at both A and B, for to do so, it would have to travel faster than light.

Consequently, there can be no causal connection between events that are separated by a spacelike interval. Moreover, it can be shown that if the interval between A and B is spacelike then there exists an inertial observer for whom A and B are simultaneous ($\Delta t = 0$), and so the proper distance between A and B is in principle measurable by this observer's meter stick.

Finally, if a ray of light can travel from A to B, then their space and time separations are equal in magnitude for all observers, and so $\Delta\tau = 0$. Such an interval is called *lightlike*. In Figure II-8, event A and the reflection of the light beam at M are separated by a lightlike interval (although A and B, as we saw, are separated by a timelike interval).

Now let us examine how relative motion affects the measurements of lengths parallel to the direction of relative motion. Suppose train S′ passes a platform S of length L at speed β. To the observer on the platform, the time required for the front of the train to go from one end of the platform to the other is $\Delta t = L/\beta$. Of course, this observer could verify this by using two synchronized clocks at opposite ends of the platform.

On the other hand, the time of passage as recorded by the engineer on board the train is, by Eq. (78),

$$\Delta t' = \Delta t\,(1 - \beta^2)^{1/2}$$

(she can use a single clock). Since each observer measures the other's speed as $|\beta|$ (in absolute value), the length of the platform as reckoned by the engineer (the platform is moving relative to her) is

$$L' = \beta\,\Delta t' = \beta\,\Delta t\,(1 - \beta^2)^{1/2} = L\,(1 - \beta^2)^{1/2} \qquad (83)$$

We are thus led to the conclusion that the observed length of an object moving uniformly with speed β relative to the observer is $(1 - \beta^2)^{1/2}$ times the length measured by a second observer at rest relative to the object. This "contraction," which applies only in the direction of relative motion (recall that lengths perpendicular to this direction are invariant), obeys the same equations used to describe the hypothetical Lorentz-Fitzgerald contraction, but the reason for the relativistic length contraction is quite different: it is due to the relative motion of object and observer.

Exercises II-6

1. What is the speed of an object relative to an inertial reference frame in which the object is observed Lorentz-contracted to one-half its rest length?

2. Pions are subatomic particles which decay radioactively. At rest, they have a half-life of 1.8×10^{-8} sec. (After that amount of time half of any initial amount of pions will have disintegrated.) A pion beam is accelerated to $\beta = 0.99$. According to classical physics, this beam should drop to one-half its original intensity after traveling for $(0.99)(3 \times 10^8)(1.8 \times 10^{-8}) \approx 5.3$ m.

However, it is found that it drops to about one-half intensity after traveling 38 m. Explain, using either time dilation or length contraction.

3. The radius of our galaxy, the Milky Way, is about 5×10^4 light years (one light year $\approx 9.45 \times 10^{17}$ cm). Can a person, in theory, travel from the center to the edge of our galaxy in a normal lifetime? Explain, using either time dilation or length contraction.

4. μ-mesons at rest have an average lifetime of about 2.3×10^{-6} sec. These particles are produced high in the earth's atmosphere by cosmic rays. Suppose a μ-meson is created and travels downward with speed $\beta = 0.99$. How far will it travel before disintegrating?

5. At the end of this section, we deduced relativistic length contraction from time dilation. Show conversely, that if length contraction is assumed, time dilation follows.

6. (See Exercise 3.) With what speed (relative to the galaxy) must an astronaut travel in order to go from the center to the edge of our galaxy in 5 years of his own time? (Hint: use years and light-years for units. The distance (traveled at speed β for 5 years) is

$5 \times 10^4 (1 - \beta^2)^{1/2}$ as measured by the astronaut, where β is very close to unity. You will need the approximation $(1 + x)^m \approx 1 + mx$, valid for $|x|$ small.)

7. THE LORENTZ TRANSFORMATION

Here we shall derive the formulas that tell us how to translate from S coordinates to S′ coordinates and vice versa. As usual, we will make the simplifying assumption that the relative velocity of the two frames is along a common xx′-axis, that the corresponding coordinate planes remain parallel, and that the two origins O and O′ coincide when $t = t' = 0$.

For the purpose of comparison, we shall first derive the analogous formulas of Newtonian physics, where the time of any event is the same for all observers. Assuming S′ is moving with speed β relative to S, the y′z′-plane will be βt cm from the yz-plane at time t. Consequently (see Fig. II-9), a point with S′ coordinates (x', y', z') at time t will have S coordinates (x, y, z) (at time t) satisfying

$$x = x' + \beta t, \quad y = y', \quad z = z'$$

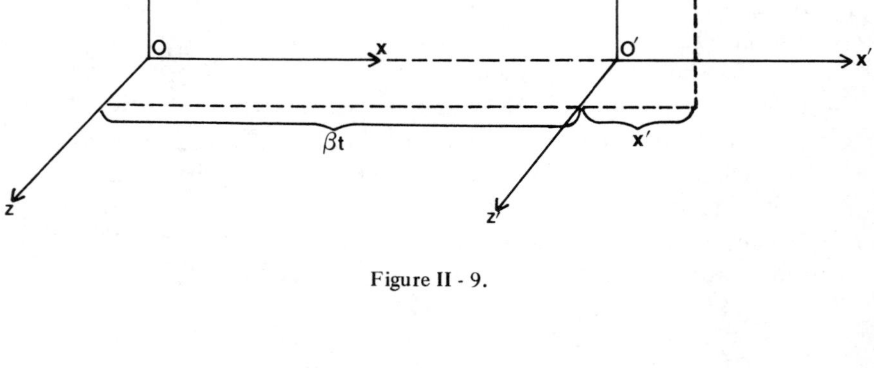

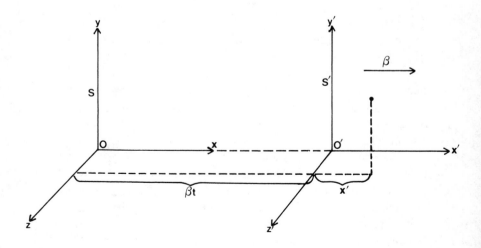

Figure II - 9.

These are the non-relativistic equations for transformation of coordinates. (A fourth equation, $t = t'$, is taken for granted in classical physics, and so is not written.)

The derivation of the corresponding relativistic formulas depends on Einstein's postulates and the physical assumption that space is homogeneous and isotropic. This means that there is no preferred point or direction in space, so that the formulas must be linear equations:

$$\begin{aligned} x &= a_{11}x' + a_{12}y' + a_{13}z' + a_{14}t' \\ y &= a_{21}x' + a_{22}y' + a_{23}z' + a_{24}t' \\ z &= a_{31}x' + a_{32}y' + a_{33}z' + a_{34}t' \\ t &= a_{41}x' + a_{42}y' + a_{43}z' + a_{44}t' \end{aligned} \tag{84}$$

Otherwise, if, say, x were given by a non-linear formula such as $x = ax'^2$, then the S length, $x_B - x_A$, of a rod which is at rest in S′ and lies along the x′-axis would depend on where the rod was located. If the ends of the rod were located at $x'_A = 0$, $x'_B = 1$, its S length would be a; but if its ends were at $x'_A = 1$, $x'_B = 2$, its S length would be 3a. This violates homogeneity. Similarly, any dependence on t must be linear, since neither a measured length nor the size of a time interval should depend on the initial numerical setting of the observer's clock. The coefficients in the above linear transformation are assumed independent of the event in question but may be expected to involve the relative velocity β. Actually, we have determined most of the coefficients already. Since lengths perpendicular to the direction of relative motion are invariant, the second and third equations reduce to

$$y = y', \quad z = z'$$

Moreover, it is not hard to see that x does not depend on y′ and z′. To convince yourself of this, picture a flat plate at rest in S′ and perpendicular to the x′-axis ($x' = $ constant). This plate should appear to the S observer as a flat plate perpendicular to the x-axis and moving away from him with speed β (for each value of t, $x = $ constant).

However, if x depended on y′ and z′ as well as on x′, then S would still observe a flat plate (because the equations are linear), but it would no longer be perpendicular to the x-axis (for each value of t, $x \neq$ constant): its normal vector would be inclined at an angle to the x-axis, thus indicating a special direction in space, which was assumed to be isotropic. Therefore $a_{12} = a_{13} = 0$. Similarly, isotropy implies $a_{42} = a_{43} = 0$. The first and last equations now may be written

$$x = a_{11} x' + a_{14} t' \tag{85}$$

$$t = a_{41} x' + a_{44} t' \tag{86}$$

Now recall the thought experiment of Figure II-8, in which a beam of light, emitted from the S′ origin at time zero (event A), bounced off a mirror, and returned to the S′ origin ($x' = 0$) at time $t' = \Delta t'$ (event B). In S, event B occurred at time $t = \Delta t$, where $t = (1 - \beta^2)^{-1/2} t'$ [cf. Eq. (78)]. (Both of Einstein's postulates contributed to this result.) Inasmuch as the same equation should be obtainable by substituting $x' = 0$ in Eq. (86), a_{44} must equal $(1 - \beta^2)^{-1/2}$. Furthermore, since the S equation for the motion of the S′ origin is $x = \beta t$ and since when $x' = 0$, $t = (1 - \beta^2)^{-1/2} t'$, we have $x = \beta (1 - \beta^2)^{-1/2} t'$, when $x' = 0$. Accordingly, $a_{14} = \beta (1 - \beta^2)^{-1/2}$ and we now have

$$x = a_{11} x' + \beta (1 - \beta^2)^{-1/2} t' \tag{87}$$

$$t = a_{41} x' + (1 - \beta^2)^{-1/2} t' \tag{88}$$

Let us apply these equations to the collection of events on a spherical light wave spreading out in all directions from a light flash emitted from the origin at $t = t' = 0$. By the invariance of the speed of light, the coordinates of every such event satisfy

$$t^2 = x^2 + y^2 + z^2$$

$$t'^2 = x'^2 + y'^2 + z'^2$$

and hence $t^2 - x^2 = t'^2 - x'^2$. Substitution from Eqs. (87) and (88) yields

$$[a_{41} x' + (1 - \beta^2)^{-1/2} t']^2 - [a_{11} x' + \beta (1 - \beta^2)^{-1/2} t']^2 = t'^2 - x'^2$$

Expanding the left side and collecting terms, we obtain

$$(a_{41}^2 - a_{11}^2) x'^2 + 2 (1 - \beta^2)^{-1/2} (a_{41} - \beta a_{11}) t'x' + t'^2$$
$$= t'^2 - x'^2$$

Both sides are polynomials in the variables t' and x'. These polynomials cannot have the same value at each of the events that comprise the spreading wave front, unless they are actually the same polynomial. This means that the coefficients of t'^2, $t'x'$, and x'^2 on the left side of the equation must equal the corresponding coefficients on the right side. This gives us

$$a_{41} - \beta a_{11} = 0, \quad a_{41}^2 - a_{11}^2 = -1$$

These have the solution

$$a_{11} = (1 - \beta^2)^{-1/2}, \quad a_{41} = \beta (1 - \beta^2)^{-1/2}$$

[We must choose the positive roots to ensure that Eq. (87) reduces to $x = x'$ when $\beta = 0$.]

Having found all the coefficients in Eq. (84) we may now write down the equations for the transformation of coordinates from S′ to S in the form

$$x = (x' + \beta t') (1 - \beta^2)^{-1/2} \tag{89a}$$

$$y = y' \tag{89b}$$

$$z = z' \tag{89c}$$

$$t = (\beta x' + t') (1 - \beta^2)^{-1/2} \tag{89d}$$

Because these are the same equations originally given by Lorentz in connection with his contraction hypothesis, they are called the *Lorentz Transformation*.

[Einstein was the first to show that time is relative. Lorentz, as did everyone else before Einstein, took it for granted that time was the same for all observers. The variable t′ that appears in Eqs. (89) was considered by Lorentz the result of a correction that had to be applied to true measured time in order to make his contraction hypothesis consistent with experimental facts other than the Michelson-Morley experiment. In other words, he viewed t′ as a sort of artificial time, a mathematical construct devoid of obvious physical meaning, but needed to fit theory to experimental fact. Later physicists considered that a clock's running rate might be affected by motion through the ether, and so gave t′ a physical meaning. It turns out that this effect on clock rates, in addition to the contraction hypothesis, would account for the null result of the Kennedy-Thorndike experiment, but this is a very complicated way of explaining a result which is easily obtained from the invariance of the speed of light in Einstein's theory.]

We shall usually omit the writing of Eqs. (89b) and (89c).

It is important to realize that for speeds which are small compared to the speed of light, Eqs. (89) reduce to the classical transformation derived at the beginning of this section:

$$x \approx x' + \beta t'$$

$$t \approx t'$$

[In geometric units, and for low speeds, x and x′ are small compared to t and t′. Thus $\beta x'$ in (89d) can be neglected, but not $\beta t'$ in (89a). For example, for a satellite traving 8 km/sec, $\beta \approx 2.7 \times 10^{-5}$ and $(1 - \beta^2)^{-1/2} \approx 1 + 3.6 \times 10^{-10} \approx 1$.]

To find the inverse transformation expressing S′ coordinates in terms of S coordinates, we could solve the system Eqs. (89) for x′, y′, z′, and t′, but there is an easier way: By symmetry (the Principle of Relativity), the inverse transformation should have the same form, but with primed coordinates on the left instead of on the right, and with β replaced by $-\beta$ (since, relative to S′, S moves in the negative x′ direction). We then obtain

$$x' = (x - \beta t)(1 - \beta^2)^{-1/2} \tag{90a}$$

$$t' = (-\beta x + t)(1 - \beta^2)^{-1/2} \tag{90b}$$

When dealing with pairs of events, it will often be convenient to work with the coordinate differences, Δx, Δy, Δz, and Δt, rather than with the coordinates themselves. Since the equations of Eqs. (89) are linear, the equations for these differences will have the same coefficients:

$$\Delta x = (\Delta x' + \beta \Delta t')(1 - \beta^2)^{-1/2} \tag{91a}$$

$$\Delta t = (\beta \Delta x' + \Delta t')(1 - \beta^2)^{-1/2} \tag{91b}$$

Using Eqs. (91), you can now verify the invariance of the interval *algebraically* by substituting for Δt and Δx in the formula for $\Delta \tau$ (see Exercise 14).

To derive time dilation [Eq. (78)], set $\Delta x' = 0$ in Eq. (91b).

To establish length contraction algebraically, consider a rod at rest in S and lying along the x-axis. Denote its S length (Δx) by L. To measure the rod's length $L' = \Delta x'$ in the rocket frame, S′ must observe the rod's ends simultaneously ($\Delta t' = 0$). Accordingly, substitute $\Delta t' = 0$ in Eq. (91a) and obtain $L = L'(1 - \beta^2)^{-1/2}$, or

$$L' = L(1 - \beta^2)^{1/2}$$

[Of course, if the rod were at rest in S′ instead of in S, we would have $L = L'(1 - \beta^2)^{1/2}$, which could be obtained algebraically from the inverse of Eqs. (91).]

Exercises II-7

1. Consider, if you will, the following statement:

> "Two events occurring at different times at the same place of one reference system will be separated by a definite space interval from the point of view of another system (moving uniformly relative to the first)."

(a) Describe an everyday illustration of this commonplace fact to show you understand what it means. (b) Now, in the quoted sentence above, interchange the words denoting spatial concepts ("place," "space") with the word "time." What statement results? [Although the statement above is "commonplace," that of (b) is strange, though equally true according to the theory of relativity. It is illustrated in the next exercise. This *space-time duality* is one aspect of the equivalence of space and time in relativity. Note that duality is also evident in the symmetrical form of Eqs. (89). Interchanging x with t and x′ with t′ leaves the equations invariant.]

2. Observer S′ seated at the center of a railroad car observes two men, seated at opposite ends of the car, light cigarettes simultaneously ($\Delta t' = 0$). However for S, an observer on the station platform, these events are not simultaneous ($\Delta t \neq 0$). If the length of the railroad car is $\Delta x' = 25$ m and the speed of the car relative to the platform is 20 m/sec ($\beta = 20/3 \times 10^8$), find Δt and convert your answer to seconds. (The extreme smallness of your answer shows why the statement in Exercise 1b is unfamiliar: only at much higher speeds is the relativity of simultaneity apparent.)

3. The observer in frame S finds that a certain event A occurs at the origin of his coordinate system, and that a second event B occurs 2×10^{-8} sec later at the point x = 1200 cm, y = z = 0 cm. (a) Is there an inertial observer S′ for whom these two events are simultaneous? (b) If so, what is the speed and direction of motion of S′ relative to S? (c) Verify that the proper distance $\Delta\sigma$ between A and B is the same in both coordinate systems. (d) Repeat part (a) for the case in which events A and B are only 400 cm apart in the S coordinate system.

4. The S coordinates of two events are:

event A: $x = 0, \quad y = 0, \quad z = 0, \quad t = 5$

event B: $x = 4, \quad y = 0, \quad z = 0, \quad t = 8.2$

(distance and time measured in the same units). An observer in frame S′, moving uniformly relative to S, observes the two events simultaneously. What is the velocity of S′ relative to S?

5. Observer S′ moves relative to observer S with speed $\beta = 0.8$. Assume their coordinate axes are oriented as in Section 5. Suppose event A occurs at $t = 3 \times 10^{-6}$ sec at x = 2 km, and event B occurs at $t = 5 \times 10^{-6}$ sec at x = 1 km. What is the time interval between these events as measured by S′?

6. We now derive the *relativistic addition of velocities* rule that replaces the classical law invalidated by Einstein's postulates. Here the relative velocity of S′ with respect to S will be denoted β_r. Suppose a missile is fired from rocket S′ in a direction parallel to

the x′ axis with velocity β' relative to S′. If the missile travels $\Delta x'$ cm in $\Delta t'$ cm of light travel time (as recorded by the rocket clocks), then $\beta' = \Delta x'/\Delta t'$. Similarly,the speed of the missile as measured with laboratory rods and clocks is $\beta = \Delta x/\Delta t$. Using Eqs. (91), show that

$$\beta = \frac{\beta' + \beta_r}{1 + \beta_r\beta'}$$

(Note that for $|\beta'| \ll 1$ and $|\beta_r| \ll 1$, the relativistic formula is approximated by the classical rule $\beta = \beta' + \beta_r$.)

7. (See Exercise 6.) Show that if $\beta' = \pm 1$, then $\beta = \pm 1$ also. [If the missile fired from the rocket is a light photon, then both observers will measure the photon's speed as unity. The invariance of the speed of light (Einstein's P2) is therefore embodied in the relativistic law for the addition of velocities.]

8. Show that if $|\beta'| \leqslant 1$ and $|\beta_r| < 1$, then

$$\left|\frac{\beta' + \beta_r}{1 + \beta_r\beta'}\right| \leqslant 1$$

(Hint: Fix β_r and let β' vary. Use Calculus.)

9. Since $-1 < \beta_r < +1$, there is a unique real number θ_r such that $\beta_r = \tanh\theta_r$. θ_r is called the *velocity parameter*. Similarly, we define θ and θ' by $\beta = \tanh\theta$, $\beta' = \tanh\theta'$. If $\beta = 1$ or -1, then we take $\theta = +\infty$ or $-\infty$, respectively, and similarly, for β' and θ'. Hence the velocity parameter for the speed of light is infinite.

Show that in terms of velocity parameters, the Lorentz transformation, Eqs. (89) takes the form

$$x = x'\cosh\theta_r + t'\sinh\theta_r$$

$$t = x'\sinh\theta_r + t'\cosh\theta_r$$

10. Rewrite the relativistic addition of velocities rule,

$$\beta = \frac{\beta_r + \beta'}{1 + \beta_r\beta'}$$

in terms of velocity parameters, and deduce that $\theta = \theta_r + \theta'$. (Since $\tanh\theta \approx \theta$ for θ small, the classical addition of velocities formula,

$\beta = \beta_r + \beta'$, may be viewed as a low speed approximation to $\theta = \theta_r + \theta'$, which is correct for all speeds.)

11. (See Exercise 9.) Define the imaginery-valued variable w by $w = it$. Using the identities $\cosh\theta = \cosh(i\theta)$ and $\sinh\theta = -\, i\sin(i\theta)$ (cf. Appendix B), show that Eqs. (89) may be expressed as

$$x = x'\cos(i\theta_r) - w'\sin(i\theta_r)$$

$$w = x'\sin(i\theta_r) + w'\cos(i\theta_r)$$

[Since these equations resemble those for a rotation of coordinates in the Cartesian plane, namely,

$$x = x'\cos(\theta) - y'\sin(\theta)$$

$$y = x'\sin(\theta) + y'\cos(\theta)$$

the Lorentz transformation is sometimes described as an "imaginary rotation." Such transformations leave invariant the expression $dw^2 + dx^2 + dy^2 + dz^2$. Note that the addition of velocities rule, $\theta = \theta_r + \theta'$, may be viewed as corresponding to the addition of angles when two (imaginary) rotations are composed.]

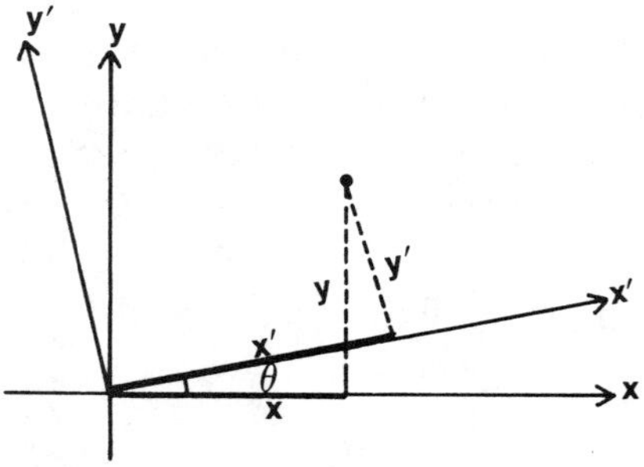

rotation of coordinates

12. (See Exercises 9 and 11.) If the angle (between the x- and x'-axes) for a rotation of Cartesian coordinate systems is the analogue of the velocity parameter for a Lorentz transformation, what is the analogue of the velocity itself?

13. (a) Rewrite the relativistic addition of velocities rule in classical notation, in terms of conventional velocities ($v = \beta c$). (b) Rewrite the Lorentz transformation, Eqs. (89) and (90), in classical notation.

14. Substitute the transformation Eq. (91) into the formula for the interval and verify that

$$(\Delta t)^2 - (\Delta x)^2 - (\Delta y)^2 - (\Delta z)^2 = (\Delta t')^2 - (\Delta x')^2 - (\Delta y')^2 - (\Delta z')^2$$

15. Suppose, in contradiction to relativity theory, it were possible for two particles to travel in the same direction with speeds $1 + \epsilon$ and $1 - \epsilon$ relative to some inertial observer (ϵ is a small positive number). Show that their speed relative to each other would be $2/\epsilon$.

8. SPACETIME DIAGRAMS

It is not possible to represent all of spacetime graphically, since this would require four dimensions. We can, however, construct such a representation when the objects we wish to study are confined to move in a plane or along a straight line. Where motion is along the x-axis, we can depict the movement of any particle by plotting x against t (y and z remain zero).

In Figure II-10(a) we see a particle traveling along the x-axis with constant speed β, so that $x = \beta t$. In Figure II-10(b), the motion

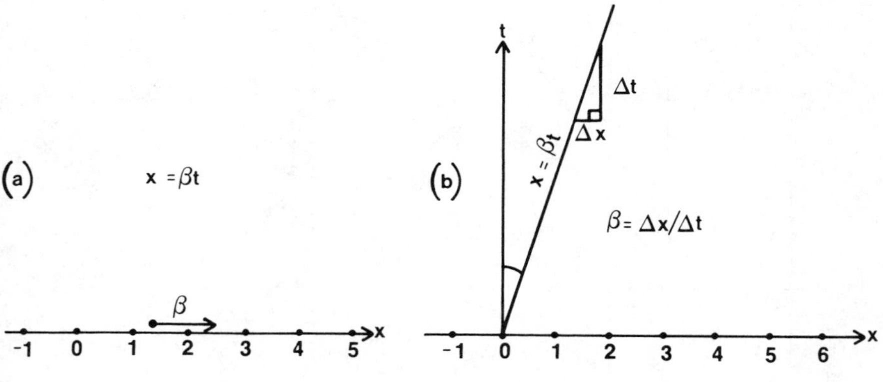

Figure II - 10.

is represented in a *spacetime diagram* as a straight line whose slope with respect to the t-axis (not the x-axis as in freshman calculus) is $\Delta x/\Delta t = \beta$. This line is called the *world-line* of the particle. Since we have chosen one centimeter as the unit for both time and distance, the absolute value of this slope will be less than one for any material particle and equal to one for a light ray.

Figure II-11 shows how the "histories" of other particles would be represented. Line a is the world-line of a particle traveling to the left ($\beta = -0.2$). Line b represents a faster particle traveling to the right ($\beta = +0.7$), and the dotted line c is the world-line of a light ray. The world-line of a particle that is stationary in this reference system is a vertical line such as d, since, as time passes, x remains constant. Curve e is the world-line of an initially stationary particle which starts toward the right from x = 6 at t = 1, picks up speed for a while, then slows down and reverses direction, and finally returns to a halt at its original position at about t = 2.3.

A collision is indicated by the meeting of two world-lines, such as f and g. From the slopes, we might conjecture that f is the world-line of a heavy particle that was only slightly slowed by its collision

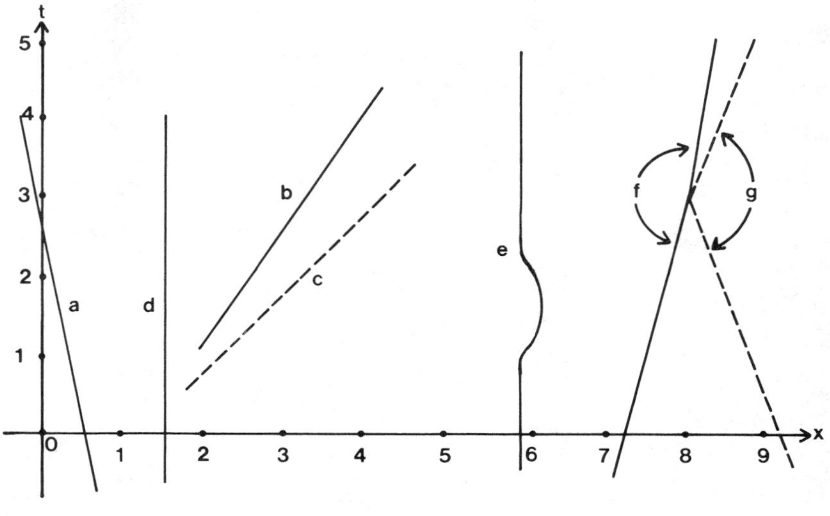

Figure II - 11.

with a lighter particle (whose motion was almost completely reversed).

The use of spacetime diagrams and the term "world-line" were first introduced by H. Minkowski in a 1908 address entitled "Space and Time":

> The whole universe is seen to resolve itself into similar world-lines, and I would fain anticipate myself by saying that in my opinion physical laws might find their most perfect expression as reciprocal relations between these world-lines.

To portray planar motion, we may represent space two-dimensionally and indicate time on a third axis. As an illustration, the planar orbit of the earth around the sun is shown schematically in Figure II-12. As the earth traces out an ellipse in space, time passes.

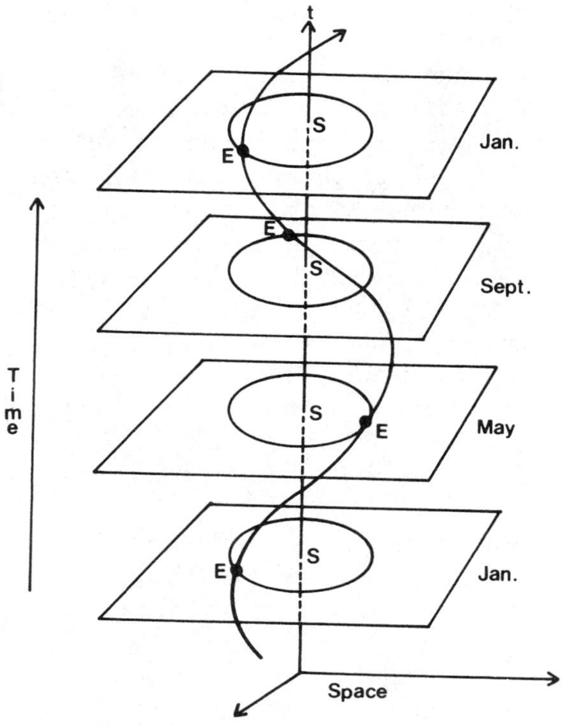

Figure II - 12.

Consequently, the spacetime diagram shows an elliptical coil as the earth's world-line. Slices perpendicular to the t-axis give the earth's orbital plane at different times of the year. This particular diagram is not drawn to scale. The diameter of the earth's orbit is about 3×10^{13} cm, whereas the length of a year (January to January on the time axis) is about 9.5×10^{17} cm. Therefore, if we were to use the same scale on all three axes, the coil would appear so stretched in the vertical direction as to look nearly straight (as we shall discover in Chapter III, this is a very significant remark).

More generally, the world-line of a particle moving in space would be represented as a curve on a 4-dimensional spacetime diagram. We cannot draw such a diagram, but we can deal with it mathematically. The spacetime point of view represents a moving particle's history not as a vector function of time, $(x(t), y(t), z(t))$—this casts t in the special role of parameter—but as a locus in txyz-space. This approach is essentially the transfer of emphasis from a function to its graph [compare Figs. II-10(a) and 10(b)]. The result is that the coordinates t, x, y, and z are put on an equal footing. We shall see in the next section how the invariance of the interval leads us to consider spacetime as a semi-Riemannian manifold.

Returning now to our laboratory and rocket frames, we shall now show how to represent the S′ coordinate system in the spacetime diagram of S. In Figure II-13, the line labeled $x = \beta t$ is the world-line of the rocket origin clock. Since all events on this world-line occur at $x' = 0$, this line may be taken as the t′-axis. To calibrate this axis, we have drawn one branch of the hyperbola $t^2 - x^2 = 1$. Points on this curve also must satisfy $t'^2 - x'^2 = 1$, by invariance of the interval. Therefore, the intersection of this hyperbola with the t′-axis (along which $x' = 0$) is the point $x' = 0$, $t' = 1$. This intersection, therefore, gives us the unit for measurement along the t′-axis.

How do we draw the x′-axis? This consists of all events for which $t' = 0$. Substituting this in Eq. (90b) gives $t = \beta x$ as the equation for the x′-axis. Notice that the x′-axis has the same slope with respect to the x-axis as the t′ axis has with respect to the t-axis. This may have been anticipated in view of space-time duality ($t = \beta x$ dual to $x = \beta t$) or from the necessity for the speed of light to equal one in S′ as it is for S. You can easily check, by drawing the hyperbola $t^2 - x^2 = -1$, that the x′- and t′-axis are calibrated by means of

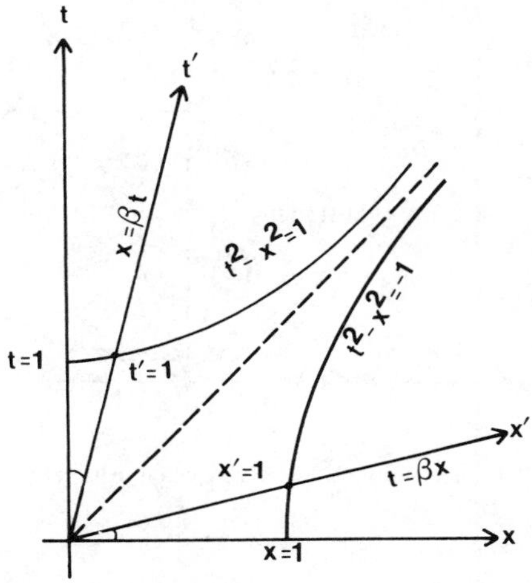

Figure II - 13.

the same unit. Some of the lines x′ = constant and t′ = constant are drawn in Figure II-14. The S′ coordinate system is oblique when represented on the spacetime diagram of S.

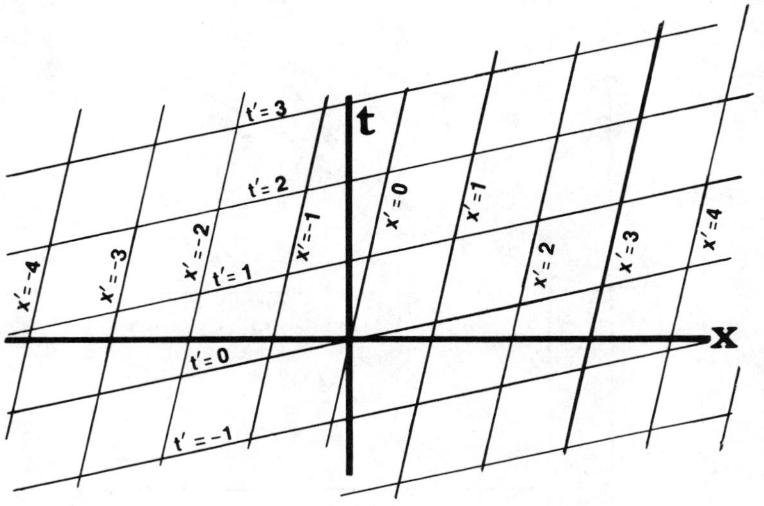

Figure II - 14.

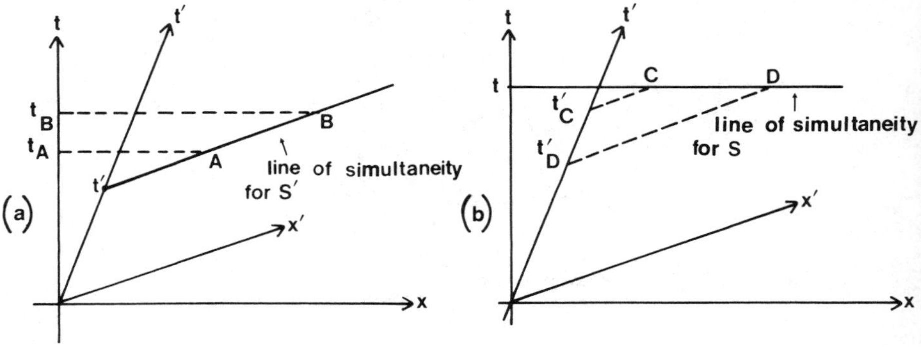

Figure II - 15. Relativity of simultaneity.

Spacetime diagrams provide further insight into the relativity of time and space intervals. In Figure II-15 (a) events A and B occur simultaneously, at time t', in the rocket frame. To S, however, they are not simultaneous, occurring at times t_A and t_B. Dually, two events C and D which occur simultaneously in S [Fig. 11-15(b)] are not simultaneous in S'.

In Figure II-16, we see clearly how each observer considers the other's clock to be running behind (relativistic time dilation). When

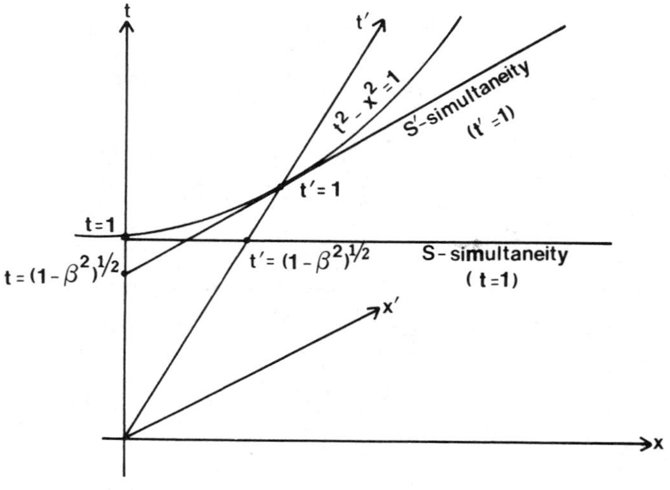

Figure II - 16. Time dilation.

the laboratory clock reads t = 1, S observes that the rocket clock reads less than one. On the other hand, when the rocket clock reads t′ = 1, S′ observes that the lab clock reads less than one. Each observer thinks the other one has a slow-running clock. This is not a contradiction. There is nothing the matter with either clock. We must remember that time and distance measurements are made with one system of rigid rods and synchronized clocks by S and with another system of rigid rods and synchronized clocks by S′. (Re-read distinction between observing and seeing, Section 5.) Notice the "lines of simultaneity" in Figures II-15 and II-16.

We may illustrate length contraction too with the aid of space-time diagrams. Figure II-17(a) shows the world-lines of the ends of a meter stick which is at rest in S. The S-length of this stick is the amount of space separation between two events on these world-lines (such as A and B) which are simultaneous in S: $x_B - x_A = 1$ m. For S′ to measure the length of this stick, which to him is moving to the left with speed β, he must subtract the x′ coordinates of the two ends simultaneously in his system, e.g., at two events such as A and C. $x'_C - x'_A < 1$ here—even though $\overline{AC}$ is the hypotenuse of ΔABC—because of the calibration of the x′-axis. You should carry out a similar analysis for the case where a meter stick is at rest in S′ [Fig. II-17(b)].

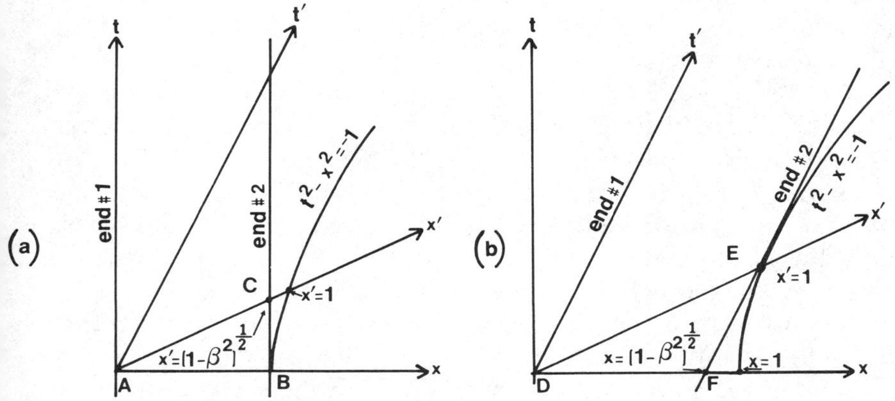

Figure II - 17. Length contraction.

Exercises II-8

1. In diagrams II-15, II-16, and II-17, the S′ coordinate axes are shown on a spacetime diagram of S. Draw the S coordinate axes on a spacetime diagram of S′, i.e., one in which the x′- and t′-axes are perpendicular. Calibrate the x- and t-axes by drawing the appropriate hyperbolas.

2. (See Exercise 1.) Using the spacetime diagram of the S′ observer, demonstrate (a) relativity of simultaneity, (b) time dilation, and (c) length contraction, as was done in this section with the spacetime diagram of the S observer.

3. An athlete carrying a pole 16 m long runs toward the front door of a barn so rapidly that an observer in the barn measures the pole's length as only 8 m, which is exactly the length of the barn. Therefore at some instant the pole will be observeed entirely contained within the barn. However, from the athlete's viewpoint, it is the barn that is Lorentz-contracted, with an observed length of 4 m, and the 16 m pole could not possibly fit within the barn. (a) What is the relative speed of pole and barn? (b) Sketch the spacetime diagram of the barn observer S and indicate on it the world-lines of the front and back of the pole and the world-lines of the front and back doors of the barn. If the front end of the pole enters the barn at $t = 0$, at what time is the pole contained within the barn? (Assume the pole moves along the x-axis and that both barn doors are on this axis.) (c) Sketch the spacetime diagram of the athlete S′ and indicate on it the four world-lines requested in (b). Assume the front of the pole is met by the front of the barn at $t' = 0$.

4. Suppose, in the thought experiment of the preceding exercise, that the barn observer closes both the front and back doors of the barn at the instant he observes the pole entirely contained by the barn. What will the athlete observe? Assume the doors are made of some impenetrable substance. (Hint: since in relativity no signal or influence can travel faster than light, no material object, such as a pole, can be perfectly rigid and incompressible, since otherwise it would instantaneously transmit an impulse received at one end to the opposite end, with infinite speed.)

5. Using a diagram similar to Figure II-15, show that (a) at time $t = 0$ in the laboratory frame, the rocket clocks that lie along the positive x-axis are observed by S to be set behind the laboratory

clocks, with the clocks further from the origin set further behind, and that (b) at time $t' = 0$ in the rocket frame, the laboratory clocks that lie along the positive x'-axis are observed by S' to be set ahead of the rocket clocks, with the clocks further from the origin set further ahead.

6. Deduce (and make quantitative) the results of the previous exercise by means of the Lorentz transformation. (The difference in sign between the two parts is not an asymmetry that might serve to distinguish between two frames in relative motion. It is due to an asymmetry in the choice of positive x direction. If each observer had chosen his positive x-axis so as to point in the direction of the other's motion, then the results would have been symmetric.)

7. Verify that in Figure II-17, the line $\overleftrightarrow{EF}$ is tangent to the hyperbola $t^2 - x^2 = -1$ at E.

8. Observers A and B are situated at opposite ends of a train moving with speed β relative to two stationary observers C and D on the tracks:

Suppose that, to the observers C and D, the passing of C by A (call this event AC) is simultaneous with the passing of D by B (event BD). By the relativity of simultaneity, AC and BD are not simultaneous in the train reference frame. (a) Use a spacetime diagram of observer A to determine which event is earlier in the train reference frame. (b) Show that the same conclusion can be drawn from a spacetime diagram of observer C.

9. LORENTZ GEOMETRY

In the Euclidean plane, the length of a curve α is obtained as the integral of the element of arc length, ds, along the curve:

$$L(\alpha) = \int_\alpha ds = \int_\alpha [(dx)^2 + (dy)^2]^{1/2}$$

as in Example 10 of Chapter I. Similarly, the length of a curve in E^3 is given by

$$L(\alpha) = \int_\alpha ds = \int_\alpha [(dx)^2 + (dy)^2 + (dz)^2]^{1/2} \qquad (92)$$

We often think of this integral as a limit of sums, as follows. Partition the curve α into a number of small pieces, each of which is approximated by the straight line segment which connects its end points and which has length given by an expression of the form

$$[(\Delta x)^2 + (\Delta y)^2 + (\Delta z)^2]^{1/2}$$

in accordance with the Euclidean distance formula at the beginning of Section 6. If we sum up the lengths of all the approximating segments and then pass to the limit as the number of these segments goes to infinity and their lengths go to zero, we obtain Eq. (92) in the limit.

If we wish to assign a type of length or measure to paths in spacetime geometry—which is also known as *Lorentz geometry*—we must replace distance by interval or proper time. (We shall be interested mainly in *timelike curves*, i.e., those for which $(d\tau/dt)^2 > 0$ according to every inertial observer.)

The total lapse of proper time along α is found by summing the proper time contributions $\Delta\tau$ from all the (approximately straight) segments in a suitable partition of α, and then taking the limit as the number of these segments goes to infinity and the sizes (i.e., the proper times) of these segments go to zero. In other words, the proper time of α, also called the *spacetime length* of α, is given by the integral

$$L(\alpha) = \int_\alpha d\tau = \int_\alpha [(dt)^2 - (dx)^2 - (dy)^2 - (dz)^2]^{1/2}$$

Except for the minus signs, which distinguish Lorentz geometry from Euclidean geometry, this is completely analogous to the concept of length of a curve in (4-dimensional) Euclidean space.

Since $\Delta\tau$ is an invariant, so is $L(\alpha)$. Its value will be the same in whatever inertial frame it is computed. $L(\alpha)$ may also be viewed as the actual passage of time that would be recorded by a clock whose world-line is α. (We assume the clock is infinitesimally small and sturdy enough to withstand the accelerations along α.) This interpretation will be exploited in Section 10 to explain the famous "twin paradox."

Thus proper time in Lorentz geometry is the analogue of arc length in Euclidean geometry, and the quadratic form

$$(d\tau)^2 = (dt)^2 - (dx)^2 - (dy)^2 - (dz)^2 \qquad (93)$$

plays the role of a metric (with the important difference that $(d\tau)^2$ may be zero or even negative). $\mathcal{R}^4$ endowed with this (semi-Riemannian) metric is sometimes called *Minkowski space*, after H. Minkowski, who in 1908 [35] developed the geometric interpretation of Einstein's special relativity. Changing coordinates from those of one inertial observer to those of another is analogous to a change of rectangular Cartesian coordinates in Euclidean geometry.

Now just as it proved convenient to parametrize curves in E^3 in terms of arc length, it is convenient to parametrize timelike curves in spacetime in terms of proper time. For example, according to the Law of Inertia, a free particle travels with constant speed and direction, so that

$$\frac{dx}{dt} = a, \quad \frac{dy}{dt} = b, \quad \frac{dz}{dt} = c$$

for some constants a, b, and c. Since t, x, y, and z are functions of proper time along the particle's world-line, we may consider this world-line as a parametrized curve in spacetime, with τ the parameter. Let $\beta = (a^2 + b^2 + c^2)^{1/2}$, the particle's speed. From Eq. (93), we have

$$\left(\frac{d\tau}{dt}\right)^2 = 1 - \left(\frac{dx}{dt}\right)^2 - \left(\frac{dy}{dt}\right)^2 - \left(\frac{dz}{dt}\right)^2 = 1 - \beta^2$$

and so

$$\frac{dt}{d\tau} = \frac{1}{(1 - \beta^2)^{1/2}}$$

$$\frac{dx}{d\tau} = \frac{dx}{dt}\frac{dt}{d\tau} = \frac{a}{(1 - \beta^2)^{1/2}}$$

$$\frac{dy}{d\tau} = \frac{dy}{dt}\frac{dt}{d\tau} = \frac{b}{(1 - \beta^2)^{1/2}}$$

$$\frac{dz}{d\tau} = \frac{dz}{dt}\frac{dt}{d\tau} = \frac{c}{(1 - \beta^2)^{1/2}}$$

Therefore, since these four derivatives are constant, the world-line of a free particle is a straight line in spacetime, as expected. Differentiating again, we may put the equations of a free particle's motion in the simpler form

$$\frac{d^2 t}{d\tau^2} = \frac{d^2 x}{d\tau^2} = \frac{d^2 y}{d\tau^2} = \frac{d^2 z}{d\tau^2} = 0 \tag{94}$$

(For the world-line of a light ray, the lapse of proper time is zero, and so τ cannot be used as a parameter. This case will be detailed in Chapter III.)

Up to now, we have been using the coordinate systems of inertial observers, as defined in terms of distance measurements and synchronized clock readings (cf. Section 5). We shall call such coordinate systems *Lorentz coordinates*. From a purely mathematical point of view, there is no rule saying we cannot change to an arbitrary system of coordinates, i.e., replace t, x, y, z by u^0, u^1, u^2, u^3, where the transformation

$$\begin{aligned} u^0 &= u^0(t, x, y, z) \\ u^1 &= u^1(t, x, y, z) \\ u^2 &= u^2(t, x, y, z) \\ u^3 &= u^3(t, x, y, z) \end{aligned} \tag{95}$$

is smooth with a non-singular Jacobian matrix, as in Section I-9.* In these arbitrary coordinates, the interval $d\tau$ would not be given by Eq. (93), but rather by an expression of the more general form

$$d\tau^2 = g_{ij} du^i du^j \tag{96}$$

In each such system of coordinates, we could define Christoffel symbols and the components of the curvature tensor just as in Chapter I. A curve $u^i(\tau)$, i = 0, 1, 2, 3, is called a *geodesic* if

*Following many authors, we shall let the range of all spacetime indices be from 0 to 3 rather than from 1 to 4.

$$\frac{d^2 u^r}{d\tau^2} + \Gamma^r_{ij} \frac{du^i}{d\tau} \frac{du^j}{d\tau} = 0, \quad r = 0,1,2,3 \tag{97}$$

In other words, we may view spacetime as a 4-dimensional manifold with semi-Riemannian metric (96). In the special case of a Lorentz coordinate system (each g_{ij} is constant), the Christoffel symbols and hence the curvature tensor vanish, which shows that the spacetime of special relativity is *flat*. Free particles follow geodesics because Eq. (95) reduces to Eq. (94) in Lorentz coordinates. The geodesics of flat spacetime are straight lines.

Although arbitrary (non-Lorentz) coordinates are perfectly acceptable for the description of events, they do not in general represent the distance measurements and clock readings of a particular observer. (Indeed it might not be obvious which of the new coordinates u^i, if any, play the role of "time.") For this reason, Lorentz coordinates are preferable in the flat spacetime of special relativity. (By analogy, for most applications in the Euclidean plane, rectangular Cartesian coordinates are preferable, although curvilinear coordinates, such as polar coordinates, can legitimately be used.)

In the curved spacetime of general relativity (Chapter III), we shall be forced to consider arbitrary coordinate systems, just as curvilinear coordinates must be used on a curved surface. We defer any further discussion of general coordinates to that chapter.

Exercise II-9

Substitute the Lorentz transformation (89) into $d\tau^2$ and verify that

$$dt^2 - dx^2 - dy^2 - dz^2 = dt'^2 - dx'^2 - dy'^2 - dz'^2$$

10. THE TWIN PARADOX

Everyone knows that in Euclidean geometry, "a straight line is the shortest distance between two points." In Figure II-19(a), the straight path α_0 between points A and B is obviously shorter than the curvilinear path α. This is simply because the length

$$\Delta s \approx [(\Delta x)^2 + (\Delta y)^2]^{1/2}$$

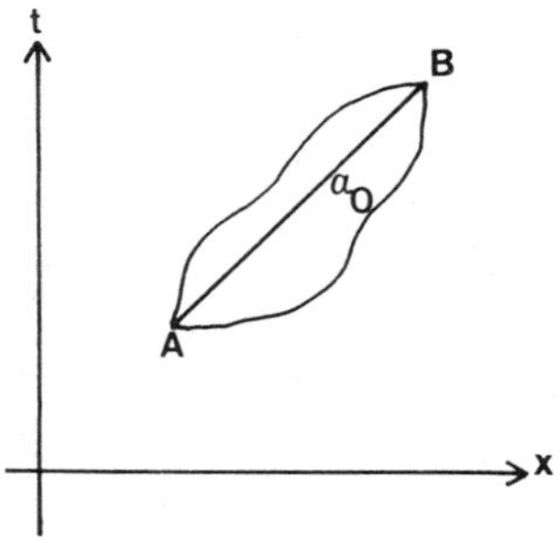

Figure II - 18. The spacetime diagram of some inertial observer. The straight line $\overline{AB}$ is the world line α_0 of an inertial observer present at events A and B.

of a small segment of α is longer than the length Δy of the corresponding segment of the straight path. In detail, the length of α is

$$L(\alpha) = \int_\alpha ds = \int_\alpha [(dx)^2 + (dy)^2]^{1/2} > \int_A^B dy = y_B - y_A$$

Let us investigate whether the straight line enjoys a similar property in Lorentz geometry. Let A and B be two events separated by a timelike interval.

There are many possible world-lines connecting A to B, each representing the possible history of a particle or observer which is present at both events (Fig. II-18). Of particular interest is the world-line of an *inertial* observer S present at both events. His world-line, α_0, would appear straight on a space-time diagram drawn by *any* inertial observer, in particular, on one drawn by S himself. Since A and B occur at the same place for S, he may choose this place as the origin of space, so that his world-line is a portion of his t-axis. In Figure II-19(b), we have drawn the spacetime diagram of S and have given an example of a curved world-line α from A to B, such as would be realized by a non-uniformly moving particle or observer (one that accelerates). For simplicity, we have assumed this particle moves back and forth along the x-axis of S, so that $y = z = 0$ along α. The numbers alongside each path in Figure II-19(b) give the lapse of proper time, i.e., the time that would be registered on a clock following that path.

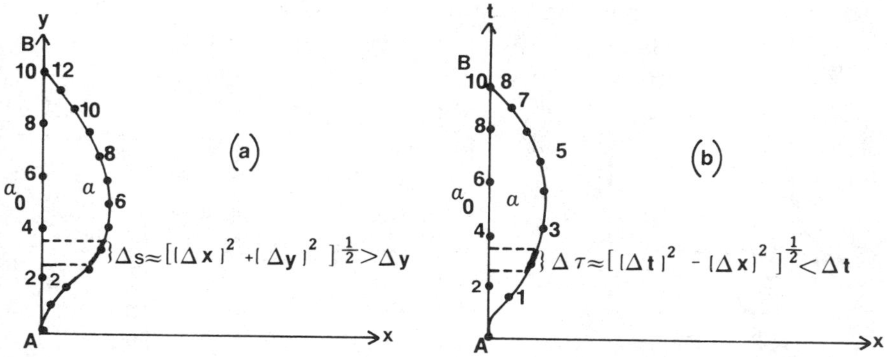

Figure II - 19. (a) Euclidean geometry; (b) Lorentz geometry.

For the straight world-line, the proper time is

$$L(\alpha_0) = \int_{\alpha_0} [(dt)^2 - (dx)^2]^{1/2} = \int_{\alpha_0} [(dt)^2 - 0^2]^{1/2}$$

$$= \int_A^B dt = t_B - t_A$$

the latter being the time between A and B as recorded by a clock at rest in S. However, along α, $(dx)^2$ is positive, and so

$$d\tau = [(dt)^2 - (dx)^2]^{1/2} < dt$$

Consequently, $L(\alpha) < L(\alpha_0)$. In other words, the elapsed time recored by the *interial* observer traveling from A to B is *greater* than the elapsed time recorded by any non-inertial (accelerating) observer traveling from A to B.* In brief, in Lorentz geometry a straight line gives the *longest* (spacetime) distance between two points.

The discrepancy in proper time means that the non-inertial traveler from A to B will age less than his inertial companion, since the clock that each carries could just as well be a biological clock such as heart rate. In this section, we shall quantitatively examine

*The same conclusion would hold if y or z varied along the curve α.

this difference in aging and resolve the famous "twin" or "clock paradox." If we have gained any relativistic intuition thus far, we should not be surprised to learn that time, like distance, is a route-dependent quantity. If two drivers follow different routes from town A to town B, their odometers will most likely show different distances for their trips. This is the obvious route-dependence of distance. It is equally true in relativity theory that two travelers who are present at two events A and B, but who traveled different spacetime routes in going from A to B will in general record different time intervals between these events. This is the route-dependence of time. (Of course, as with all relativistic effect, the time difference will be readily observable only when very high speeds are involved.)

Let us imagine that Jack is the occupant of a laboratory floating freely in intergalactic space. He can be considered at the origin of an inertial frame of reference. His twin sister, Jill, fires the engines in her rocket, initially alongside Jack's space laboratory. Jill's rocket is accelerated to a speed of 0.8 relative to Jack and then travels at that speed for three years of Jill's time. At the end of that time, Jill fires powerful reversing engines that turn her rocket around and head it back toward Jack's laboratory at the same speed, 0.8. After another three-year period, Jill returns to Jack and slows to a halt beside her brother. Jill is then six years older.

We can simplify the analysis by assuming that the three periods of acceleration are so brief as to be negligible. The error introduced is not important, since by making Jill's journey sufficiently long and far, without changing the acceleration intervals, we could make the fraction of time spent in acceleration as small as we wish.

Assume Jill travels along Jack's x-axis. In Figure II-20, Jill's world-line is represented on Jack's spacetime diagram. It consists of two straight line segments inclined to the t-axis with slopes + 0.8 and − 0.8, respectively. For convenience, we are using units of years for time and light-years for distance. (The speed of light is still one in these units.)

During each half of her voyage, Jill is an inertial observer. However, because of her direction reversal at the half-way point, there is no single inertial frame in which she is at rest for the entire trip. She must, therefore, use one (imaginary) lattice of rigid rods and recording clocks for the outbound half of her journey, and a second lattice for the inbound half. Her world-line for the outbound

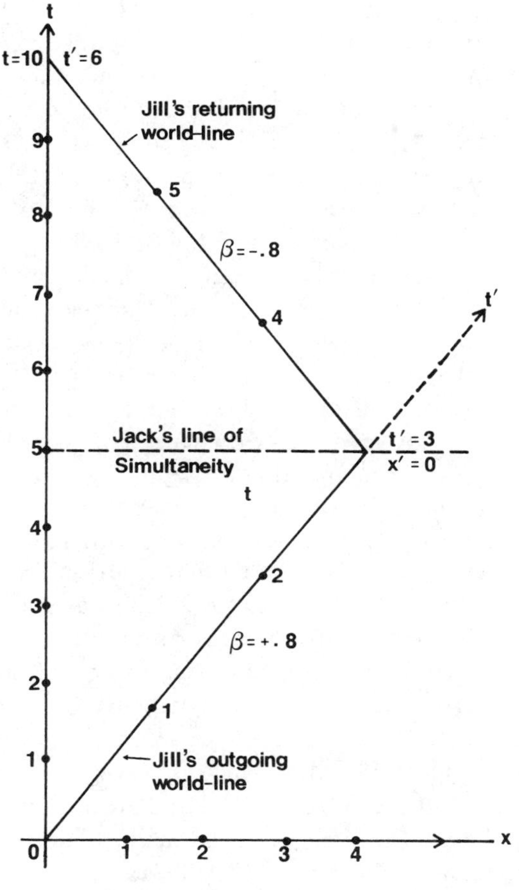

Figure II - 20.

half may be taken as the t'-axis for the first reference frame, and is calibrated as in Section 8. Jill's coordinates for the turn-around point are therefore t' = 3, x' = 0. Substituting these in Eq. (89d), we obtain t = 5 [since $(1 - \beta^2)^{1/2} = 0.6$]. This means that in Jack's reckoning, Jill turned back after five years, not three. By symmetry, it will take another five years in Jack's reckoning for Jill to return. Therefore, Jack will be ten years older (four years older than his sister) when the twins are reunited.

If Jack knows relativity theory, this may not be surprising, since he will no doubt recall the dictum "a moving clock runs slow." Because of time dilation, Jack will observe that the five year interval registered on his clock for each half of Jill's voyage will register as a three-year interval on Jill's clock [see Eq. (78)].

On the other hand, from Jill's viewpoint, Jack is the one who traveled away and returned. Consequently, by the Principle of Relativity, Jill should observe a slowing of Jack's clock, so that Jack (and not Jill) would be the younger twin at journey's end. It is obviously impossible for each to be younger than the other. Where is the flaw in the above reasoning? Have we not found an inconsistency in relativity theory?

The solution to this mystery (the so-called twin paradox) is the lack of symmetry between the two twins. Jack is an inertial observer. Floating freely, far from any gravitating masses, he finds that the law of inertia holds in his laboratory. However, as we have already noted, there is no single inertial system in which Jill remains at rest. A pail of water at rest inside her rocket during the outgoing trip will crash into the wall of her cabin as soon as she reverses her direction. Jill must be considered as one inertial observer for the first half of her trip and another inertial observer for the second half.

In Figures II-21 (a) and II-21 (b), we have indicated the coordinate axes for these two observers—"outgoing Jull" and "incoming Jill"—on Jack's spacetime diagram. The origins have been chosen so that in both of these systems, the turn-around point has coordinates $t' = 3$, $x' = 0$. From Figure II-21 (a), we see that just as Jill arrives at the turn-around point, she observes that Jack's clock reads $t = 1.8$ years (remember that "observing" is not "seeing": cf. Section 5). So it *is* true that Jack's clock seems slow to Jill (whose clock reads three years). However, as soon as Jill reverses direction, she switches to a new inertial frame, a new lattice of rods and clocks, and a new standard of simultaneity. In fact, as can be seen in Figure II-21 (b), as Jill beings her return trip, she observes that Jack's clock reads $t = 8.2$ years. Moreover, she observes that over the next three years of her own time, Jack will age only 1.8 years—again illustrating that Jack's clock seems to her to be running slow.

What Jill has overlooked is clear from Figure II-22, where we have drawn the two "lines of simultaneity" from Figures II-21 (a) and II-21 (b). The gap of 6.4 years compensates for the difference in standards of simultaneity between outgoing Jill and incoming Jill.

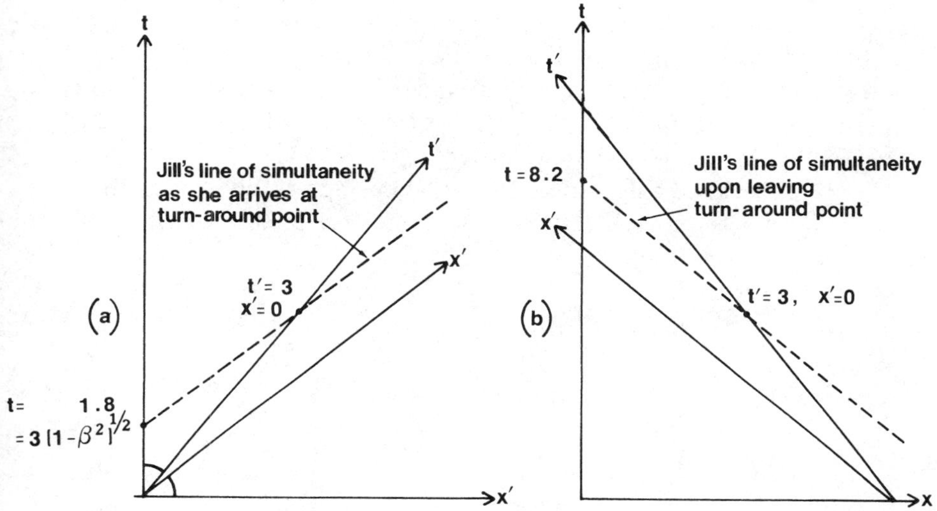

Figure II - 21. Jill's outgoing inertial frame; (b) Jill's incoming inertial frame.

Because of this gap, which did not show up in the records of Jill's two sets of recording clocks, the slowing of Jack's clocks as observed by Jill was insufficient to prevent Jack from ending up older than his twin.

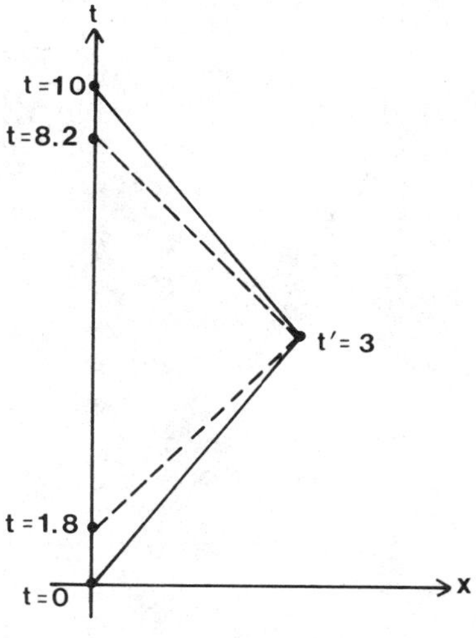

Figure II - 22.

As a check, let us analyze the problem in another way. Suppose Jill flashes a light signal to Jack at the end of each of the six years of her voyage. These flashes are represented by the dotted 45° world-lines in Figure II-23. (The flash at $t' = 6$ is received instantaneously by Jack, since the twins are together then.)

First we shall calculate the time when Jack receives Jill's first flash. This is the time coordinate of event C in Jack's coordinates (see Fig. II-23):

$$t_C = t_A + AC = t_A + AB = t_B + x_B$$

Substituting $t'_B = 1$, $x'_B = 0$ in Eqs. (89), we obtain

$$t_B = (1 - \beta^2)^{-1/2}, \qquad x_B = \beta(1 - \beta^2)^{-1/2}$$

from which it follows that

$$t_C = (1 + \beta)(1 - \beta^2)^{-1/2} = \left(\frac{1 + \beta}{1 - \beta}\right)^{1/2} = 3 \text{ years}$$

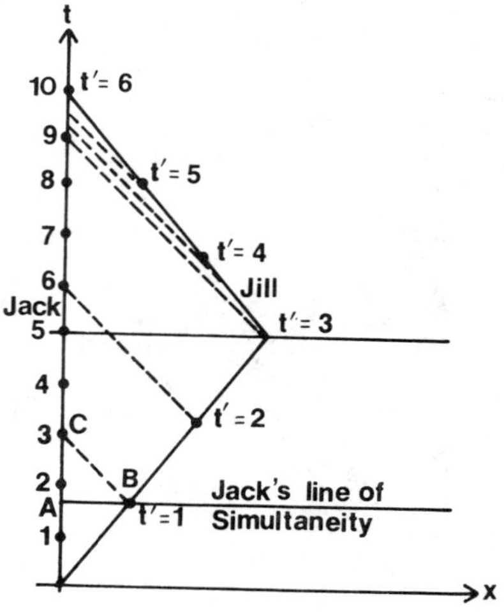

Figure II - 23.

Since the world-lines of the light flashes are mutually parallel, we can conclude that Jack will receive Jill's first three flashes at t = 3, t = 6, and t = 9 years. More generally, if two inertial observers are moving away from each other with relative speed β and if one of them emits a light signal every T units of her own time, then her colleague will receive these signals every kT units of his time where

$$k = \left(\frac{1+\beta}{1-\beta}\right)^{1/2} \qquad (98)$$

This change in frequency, known as the *Doppler effect* arises because each succeeding signal has further to go to reach the receiver than its predecessors. The phenomenon is analogous to the change we notice in the pitch or frequency of a train whistle as the train roars past us. However, because sound travels at a fixed speed relative to the air, this effect is quantitatively not the same when the source is moving through the air and the receiver is at rest as when the receiver is moving and the source is at rest (see Bondi [5], pp. 41-43).

When two observers approach each other (with speed $-\beta$), the succeeding signals have shorter distances to travel and so the time between receptions is less than the time between emissions. In fact, replacing β by $-\beta$ in the formula above, we see that k is replaced by its reciprocal. In the present example, this means that Jill's last three flashes will be received at 1/3 year intervals by Jack, i.e., at t = 9 1/3, t = 9 2/3, and t = 10 years. Notice that altogether six signals were sent and six were received.

You are urged to carry out a similar analysis for the case in which Jack sends signals, one at the end of each of the ten years that Jill is away (see Exercise 1).

In Chapter III, we shall analyze the twin paradox still another time, in the context of general relativity. (See Exercises 9 and 10 of Section III-8.)

Exercises II-10

1. Suppose Jack sends light signals to Jill, one at the end of each of the ten years (of Jack's time) that Jill is away. Indicate the light signals on a spacetime diagram like Figure II-23. When, in Jill's time reckoning, will the signals be received by her?

2. Suppose Jill sets off from earth in her rocket with acceleration that is constant in the following sense. A weight scale has been taken along on the voyage and is placed against the wall of her ship that is opposite the direction of acceleration. (This wall is her "floor.") The rocket engines are automatically and continuously adjusted so that at any time during the flight, if Jill steps on the scale, her correct earth weight of 120 pounds will be indicated.

Because of the difference between Jack's earth time and Jill's proper time, her rocket does not have constant acceleration relative to Jack. We can say, however, that at any moment Jill is accelerating at 980 cm/sec^2 with respect to a hypothetical inertial observer who happens to be instantaneously at rest relative to her. Call this observer the instantaneous inertial observer (for the given instant).

Let us express the acceleration in geometric units, as $g = 980/(3 \times 10^{10})^2$ cm per cm of time per cm of time. Suppose, during an interval $\Delta\tau$ of Jill's time, her velocity relative to the instantaneous observer (for the start of the interval) increases from 0 to $\Delta\beta$. Let $\Delta\theta$ be the corresponding increase in the velocity parameter (see Exercises 9, 10 of Section 7). Then for small $\Delta\tau$, $\Delta\beta/\Delta\tau \approx g$, and so

$$\Delta\theta \approx \tanh(\Delta\theta) = \Delta\beta \approx g\Delta\tau$$

Since velocity parameters add, during the same interval of Jill's time, her velocity parameter relative to Jack increases from its initial value θ to $\theta + \Delta\theta$. Thus the lapse of Jill's proper time, $\Delta\tau$, and the increase in her velocity parameter relative to Jack, $\Delta\theta$, are related by $\Delta\theta/\Delta\tau \approx g$, with equality holding in the limit as $\Delta\tau$ goes to zero:

$$\frac{d\theta}{d\tau} = g$$

Assuming $\theta = 0$ when $\tau = 0$, we then have $\theta = g\tau$.

Now let x and t be the distance and time of Jill's rocket as measured by Jack. Then $dx/dt = \beta = \tanh\theta$ = Jill's (varying) speed, and the time dilation is expressed as $dt/d\tau = (1 - \beta^2)^{-1/2} = \cosh\theta$, where $\theta = g\tau$. (a) Find x and t as functions of τ. (b) Eliminate τ from these equations and find an equation relating x and t. Sketch the graph of this equation on Jack's spacetime diagram. (It is a branch of a hyperbola. Use $\cosh^2\theta - \sinh^2\theta = 1$.) (c) As Jill travels farther and farther from Jack, her world-line approaches an asymptote (on Jack's spacetime diagram). What is this asymptote's slope, and what is the physical significance of this asymptotic behavior? (d) Compute the coordinate acceleration d^2x/dt^2.

11. TEMPORAL ORDER AND CAUSALITY*

In Figure II-24 the lines $x = \pm t$ are the world-lines of two light signals which pass through the laboratory (S) origin at $t = 0$. One signal is moving in the positive x direction; the other, in the negative x direction. Points of the spacetime diagram not on these lines fall into three classes, which are labeled F, P, and E in the diagram.

Events in F, such as A, are separated from event O by a timelike interval. A material particle could in theory travel from O to A with speed less than the speed of light. Accordingly, O could physically influence what happens at A; i.e., O could be, as we say, *causally connected* to A. Moreover, there is an inertial frame S′ moving relative to S in such a way that events O and A occur at the same place of S′ and are separated in time only. In fact, if A has S coordinates (x_A, t_A), then an observer S′ moving relative to S with velocity

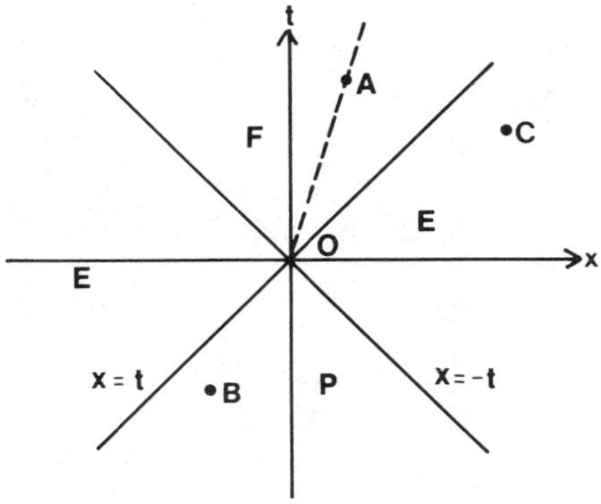

Figure II - 24.

*This section is not required for Chapter III and may be omitted on a first reading.

$\beta = x_A/t_A < 1$ assigns the spatial coordinate $x' = 0$ to all events on line $\overleftrightarrow{OA}$ [simply substitute $x = \beta t$ in Eq. (90a)]. In other words, $\overleftrightarrow{OA}$ is a possible t′ axis. (More generally, given any timelike interval $\Delta\tau$, we can, by transforming to a suitable inertial frame S′ measure this interval by means of a single clock: $\Delta\tau = \Delta t'$. This justifies the use of the name "timelike.")

In addition, the S′ time coordinate of A is positive. Since O and A occur at the same place of S′ and A occurs later than O, A is *absolutely in the future* relative to O. This means that *all* observers will agree on the temporal order of these two events. We should certainly expect this on the basis of our beliefs about causality: if A is an effect caused by O, then no observer could observe A happen before O. By drawing the coordinate axes of an arbitrary inertial observer on the spacetime diagram of Figure II-24, you can easily show geometrically that O will precede A according to every inertial observer. For this reason, region F is called the *abolute future* (relative to event O).

Analogously, events in region P, such as B, are also separated from O by timelike intervals, but these events are *absolutely in the past* relative to O: each event of P will be observed earlier than O in every inertial frame. P is called the *absolute past* (relative to O). An event of P could physically influence what happens at O, since a particle could travel from that event to O with speed less than that of light.

On the other hand, events in E, such as C, are separated from O by a spacelike interval. No physical agent could travel from O to C, for to do so would require a speed greater than that of light. However, there is an inertial frame S″ in which O and C occur at the same time and are separated only in space: substituting the relative speed $\beta = t_C/x_C < 1$ in Eq. (90b) gives $t'' = 0$ for all events on line $\overleftrightarrow{OC}$. In other words, $\overleftrightarrow{OC}$ is a possible x″ axis. In general, given any spacelike interval $\Delta\sigma$, we can, by transforming to a suitable inertial frame S″, measure this interval by means of a single meter stick $\Delta\sigma = \Delta x''$ (hence the term "spacelike").

However, there are observers for whom O occurs earlier than C (e.g., observer S), and other observers for whom O occurs later (e.g., an observer for whom $t_C/x_C < \beta < 1$), as you can check on the spacetime diagram. Consequently, there is no absolute temporal order between O and an event of E. However, this does not violate the principle of causality, inasmuch as two events which are separated by a spacelike interval can never be causally related anyway: no signal or physical agent can travel faster than light. We shall refer to E as *elsewhere*. Some authors label it the "relative present," since O and C are simultaneous for *some* observers. We might also call it the "unreachable" region.

To depict events that occur in a plane rather than on a line, we can employ a three-dimensional spacetime diagram, as in Figure II-25. The cone $t^2 = x^2 + y^2$ is made up of the world lines of all possible light signals which pass through the S origin at $t = 0$. It is called the *light cone*. The upper half (the future light cone) can be viewed as representing a circular wave front spreading out in all directions from a light flash emitted at the origin at $t = 0$. Planar cross sections perpendicular to the t axis portray the wave front at various instants of time. The lower half (the past light cone) might be visualized as the history of light rays approaching the origin from all directions in the plane and arriving there at $t = 0$.

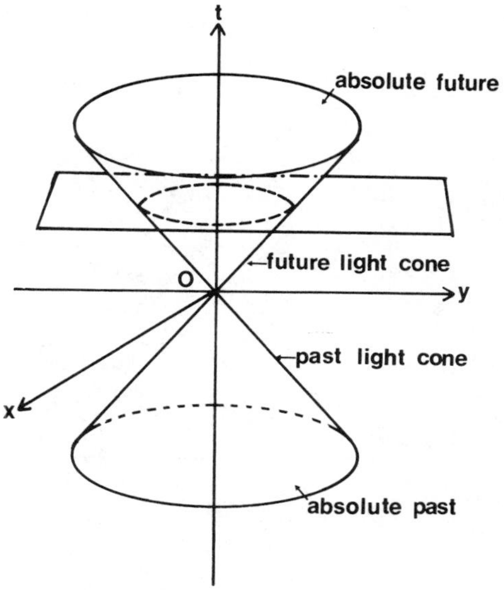

Figure II - 25.

The absolute future (relative to O) consists of all events inside the future light cone; the absolute past, the events inside the past light cone. Elsewhere consists of all events outside the light cone. As before, these events cannot be causally connected to O.

Actually there is nothing sacred about event O. In fact, for every event, there is a light cone with vertex at that event. If event A has coordinates x_A, y_A, and t_A, then the light cone at A has equation

$$(t - t_A)^2 - (x - x_A)^2 - (y - y_A)^2 = 0$$

By invariance of the interval, the light cone at any event is invariant, i.e., it exists independently of a choice of coordinates. Therefore, the causal "connectability" of two given events is invariant also (see Fig. II-26). For example, if according to one observer event A caused a later event B, then any other observer will agree on the time order of these two events.

Although we cannot graphically depict spacetime of four dimensions, the ideas above extend quite naturally. The analog of the light cone with vertex O is the collection of all events satisfying $t^2 - x^2 - y^2 - z^2 = 0$. This is still called the light cone even though it is a three-dimensional set of points (a "hypercone").

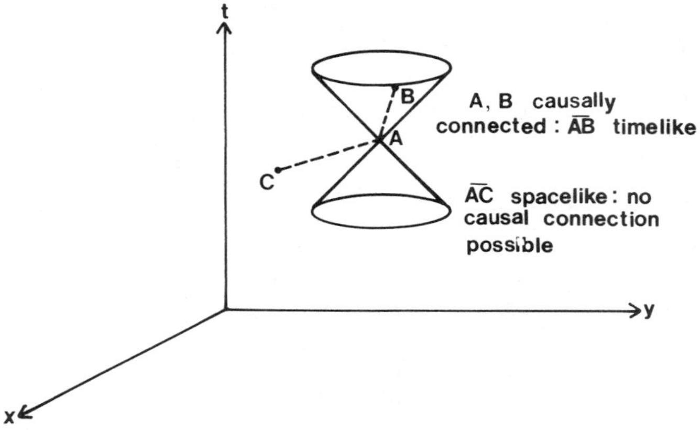

Figure II - 26.

By analogy with Figure II-25, it can be shown that a cross section of the future light cone by a three-dimensional hyperplane perpendicular to the t-axis is a sphere and represents a spreading spherical wave front at a particular instant. The division of space-time into absolute future, absolute past, and elsewhere is completely analogous to the lower-dimensional cases.

One of the consequences of the theory of relativity is that no material object can travel faster than light. This was suggested by the relativistic addition of velocities formula (Exercises 8, 9, and 10 of Section 7). Equivalently, the sum of velocity parameters with finite values is finite. This limiting property of the speed of light is also implicit in the equivalence of mass and energy discovered by Einstein (but which we have not discussed). It turns out that an infinite amount of energy would be needed to accelerate any mass up to the speed of light.

If we accept the invariance of causal connectability, we are forced to conclude that no signal or physical agent can travel faster than light. Here is why. Suppose some sort of object or signal having speed $u = \Delta x/\Delta t > 1$ travels between event P, with S coordinates x, t, and event Q, with S coordinates $x + \Delta x$, $t + \Delta t$. Suppose S′ is an inertial observer moving with speed β relative to S. Using the difference equation corresponding to Eq. (90b), we have

$$\Delta t' = (\Delta t - \beta \Delta x)(1 - \beta^2)^{-1/2} = \Delta t\,(1 - \beta u)(1 - \beta^2)^{-1/2}$$

Now if $\beta < 1/u$, then $\beta u < 1$, and $\Delta t'$ and Δt will agree in sign. Therefore, S and S′ will agree on which event preceded the other. However, if $1/u < \beta < 1$, then $\beta u > 1$, and Δt and $\Delta t'$ will differ in sign. Hence, in this case, if S believes that event P caused event Q, then S′ will find that event P occurred after Q, and so would be likely to believe that Q caused P. Most of us would assume that no one can observe an effect before a cause, and so we have additional corroboration of the limiting property of the velocity of light.

III

General Relativity: The Geometry of Curved Spacetime

The mathematician, carried along on his flood of symbols, dealing apparently with purely formal truths, may still reach results of endless importance for our description of the physical universe.
—Karl Pearson

The Great Architect of the Universe now begins to appear as a pure mathematician.
—J. H. Jeans

INTRODUCTION

Special relativity theory relates the descriptions of events as measured by two different inertial observers in relative uniform motion. According to this theory, we cannot speak of the state of motion of an inertial reference frame except as motion relative to another such frame. Moreover, no experiment carried out entirely within one inertial frame can ever reveal the motion of that frame with respect to another.

Accelerated motion, however, is quite a different matter, since its effects are readily felt. The occupants of a vehicle that is rapidly accelerating feel forces acting upon them. One of the wilder rides at any amusement park will quickly convince you of this. Thus, while uniform motion is relative, accelerated motions appears to be absolute. This distinction between uniform and accelerated motion was disturbing to Einstein, who sought to modify the special theory so as to put all reference frames on an equal footing.

In his 1916 *The Foundation of the General Theory of Relativity* [35] Einstein extended his special theory to arbitrary frames of reference, and at the same time proposed a completely new theory of gravitation. In the next few sections, we shall describe the principal ideas of general relativity and their geometric nature. Later sections will present some of the details, which involve the mathematical tools developed in Chapter I.

1. THE PRINCIPLE OF EQUIVALENCE

Under the influence of gravity, all bodies fall with the same acceleration (at least when air resistance is negligible): a baseball and a bowling ball, dropped from a tall building at the same instant, will strike the ground together. This fact was known to Galileo (1564-1642) and is a simple consequence of the laws of mechanics and gravitation later formulated by Sir Isaac Newton, born the year Galileo died. More recently (1969) the first astronaut to walk on the moon, Neil Armstrong, performed the experiment of dropping a feather and a hammer from equal heights above the lunar surface. In the absence of air resistance, both objects landed simultaneously.

Now, according to Newton's Second Law of Motion, the force needed to impart a given magnitude of acceleration to a given object is proportional to the object's mass. (It takes more force to move a bowling ball than a baseball.) All bodies fall with the same acceleration because, according to Newton's Law of Universal Gravitation, the force exerted by gravity on a given object is likewise proportional to the object's mass. It's as if the earth, somehow "sensing" that a bowling ball has a relatively larger mass, exerts a correspondingly larger gravitational attraction in order to give the bowling ball the same acceleration as a less massive baseball.

Einstein was puzzled by this feature of gravity, because it exhibits two aspects of mass whose relationship was never satisfactorily explained. The "mass" appearing in Newton's Second Law is *inertial mass*, a measure of an object's resistance to a change in motion i.e., to acceleration. The "mass" appearing in the Law of Gravitation is *gravitational mass*, a measure of an object's response to gravitational attraction. Einstein recognized that in using the single word "mass" we are really identifying two different physical attributes of matter. Einstein resolved this mystery of the dual nature of mass in a far reaching principle that became a cornerstone of his general theory of relativity. We can best convey his great idea by means of a simple "thought experiment."

Consider a tall building in which an elevator is suddenly cut loose from its cables and begins to fall. (We shall neglect any effects of air resistance or friction.) Since the elevator and all the objects within it fall with the same acceleration relative to the ground, they do not have any acceleration *relative to one another*. Therefore, if one of the passengers had been holding a package and now releases it, the package will appear to float alongside its owner as he and it plummet together toward the ground. If he pushes the package away, it will move away from him in a straight line and with constant speed relative to him and the elevator. (Of course, *relative to the ground*, the passenger and his package are both accelerating.)

Consequently, the elevator in free fall appears to be an inertial frame of reference. If the passengers did not know where they were, they might very well believe they were in a space station floating weightlessly in interstellar space.

According to Einstein's *Principle of Equivalence*, which we shall state in a moment, the falling elevator and the floating station are in fact entirely equivalent, in the sense that if you were suddenly placed within one of these two vehicles, without being told which one, then no internal experiment you could perform could ever tell you in which of the two you were located. Since it is impossible in the above example for you to determine whether your weightless state is produced by free fall (accelerated motion) in a gravitational field, or by (uniform motion in) the absence of a gravitational field, then gravity and acceleration no longer have any absolute meaning, but depend on the choice of observer (or, in mathematical terms, on the choice of coordinate system). A second thought experiment should clarify this further.

Imagine a group of scientists inside a windowless rocket in deep space, far from the gravitational influence of any celestial bodies. The engines have been fired, and the rocket is moving with constant acceleration (relative to any nearby inertial observer).

Because of this acceleration, the scientists will feel themselves pressed against the cabin wall opposite the direction of acceleration. This wall will become for them the "floor," and the opposite wall, the "ceiling." If one of them had been holding a package and now releases it, the package will "fall" to the floor. This is because, by the Law of Inertia, the package remains, relative to an inertial observer, in a state of uniform motion with the velocity it had at the instant it was released; but the rocket and its passenger standing on the floor continue to gain speed, so that the package lags behind the hand of the scientist who dropped it. The mass of the package is irrelevant here: relative to the rocket, all objects "fall" with the same acceleration (equal in magnitude but opposite in direction to the acceleration of the rocket relative to an inertial frame).

Therefore, if the rocket's acceleration has the same magnitude as the acceleration of the freely falling elevator (gaining in speed about 9.8 meters per second per second), then the scientists might easily be fooled into believing they were in a windowless laboratory on earth, at rest in the earth's gravitational field. Indeed, the Principle of Equivalence asserts that the restults of any experiment carried out inside the accelerating rocket will be identical to the results of the same experiment carried out in the laboratory. (Of course, the experimenters must not be allowed to exchange signals with observers in other reference frames. We are speaking here of wholly internal experiments.)

The Principle of Equivalence in essence states that there is no way to distinguish between the effects of acceleration and the effects of gravity: acceleration and gravity are *equivalent*, and this equivalence accounts for the identification of inertial and gravitational mass. If a particular observer finds that free particles do not move uniformly, but accelerate relative to him, he is equally justified in attributing their motion either to an acceleration of his frame of reference, or to the presence of a gravitational field. Consequently, motion, whether uniform or not, has no absolute meaning, but exists only in relation to some frame of reference.

An important consequence of the Principle of Equivalence is that light rays are "bent" by a gravitational field. To see this, suppose that the accelerating rocket of our earlier example has a window through which a ray of light enters. Assume the direction of the ray is perpendicular to the direction of the rocket's acceleration.

To the outside inertial observer, the light travels a straight path to the opposite wall of the cabin. However, in the small amount of time it takes light to traverse the craft's interior, the rocket will have moved a small but non-zero distance. Consequently, the ray will strike the opposite wall a short distance below the spot it would have struck had the rocket not been accelerating.

Therefore, to the scientists inside the rocket, the light ray follows a curved path (see Exercise 1). By the Principle of Equivalence, a light ray crossing an earthbound laboratory must also follow a curved path, similar to the parabolic path of a ball thrown across the room, but of course of much smaller curvature. (Over the dimensions of a typical laboratory, the deviation from straightness is too small to be measured. This is why the effect was unknown before Einstein. We shall see later how the effect of gravity on light is observable on a much larger scale.)

Here is a mathematical demonstration of the Principle of Equivalence in a simple case where the relative acceleration of two observers has constant magnitude and direction. We consider a collection of particles with masses $m_1, m_2, \ldots, m_n$, interacting in such a way that each particle exerts a force (e.g., electrostatic or gravitational) on each of the other particles. For $i \neq j$, let $\mathbf{F}_{ij}$ be the force exerted by the ith particle on the jth particle. To make the analysis simple, we shall assume this force is directed along the line between the two particles and is a function of the distance between them. $\mathbf{F}_{ij} = -\mathbf{F}_{ji}$, for all i, j.

Let observer #1 measure events in terms of coordinates x, y, z, t, and observer #2, in terms of coordinates x′, y′, z′, t′. We will assume the velocities involved are small enough so that a non-relativistic analysis will suffice. We therefore take $t = t'$, and for convenience use the notation

$$\mathbf{X} = (x, y, z), \qquad \mathbf{X}' = (x', y', z')$$

for the spatial coordinate vectors. Label the location of the ith particle $\mathbf{X}_i$ in observer #1's coordinates and $\mathbf{X}'_i$ in observer #2's coordinates. (These are vector functions of t.)

If the first observer believes he is in the presence of a uniform gravitational field, so that relative to him all free particles fall with the same acceleration, given by a vector g with constant components, then the equations of motion are, for observer #1,

$$m_i \frac{d^2 \mathbf{X}_i}{dt^2} = m_i g + \sum_{j \neq i} \mathbf{F}_{ji}, \quad i = 1, 2, \ldots, n \tag{99}$$

(in accordance with Newton's Second Law of Motion: mass × acceleration = total force = gravitational force + interactive forces.)

Now suppose observer #2 is moving relative to observer #1 in such a way that

$$\mathbf{X}' = \mathbf{X} - \frac{1}{2} g t^2 \tag{100}$$

This means that frame #2 is accelerating relative to frame #1 and vice versa. Then (differentiate twice) for each i,

$$\frac{d^2 \mathbf{X}_i}{dt^2} = \frac{d^2 \mathbf{X}'_i}{dt^2} + g$$

and so, from Eq. (99), the equations of motion in terms of observer #2's coordinates are

$$m_i \frac{d^2 \mathbf{X}'_i}{dt^2} = \sum_{j \neq i} \mathbf{F}_{ji} \tag{101}$$

(Since $\mathbf{F}_{ji}$ is a function of the $\|\mathbf{X}_i - \mathbf{X}_j\|$ and since $\|\mathbf{X}'_i - \mathbf{X}'_j\| = \|\mathbf{X}_i - \mathbf{X}_j\|$, $\mathbf{F}_{ji}$ is the same in both coordinate systems.)

The gravitational field has been "transformed" away! To observer #2, there is no gravitational field present. Observer #1 would account for this by saying observer #2 is in free fall, and only thinks there is no gravitational field present because of the weightless state created by free fall. On the other hand, #2 might say, "No, my friend, there really is no gravitational field present. I am an inertial observer, and your acceleration relative to me makes you think you are in a gravitational field." Who is right? They both are, and the two points of view are equally valid for the description of physical events—mechanical, electromagnetic, or whatever.* Thus, the Principle of Equivalence extends the Principle of Relativity by placing all frames, inertial or not, on an equal footing.

In the following section, a more careful analysis will point out the need for a modification of the equivalence principle as stated up to now; but rather than weaken the equivalence principle, this modification will enrich the physical theory by drawing geometry into it.

Exercises III-1

1. Assume rocket S′ accelerates at a constant rate g relative to an inertial observer S. The relative motion is along a common yy′-axis, as in the figure, so that t seconds after the S′ origin passes the S origin, S and S′ coordinates are related by

$$x' = x, \quad y' = y - \frac{1}{2}gt^2, \quad z' = z$$

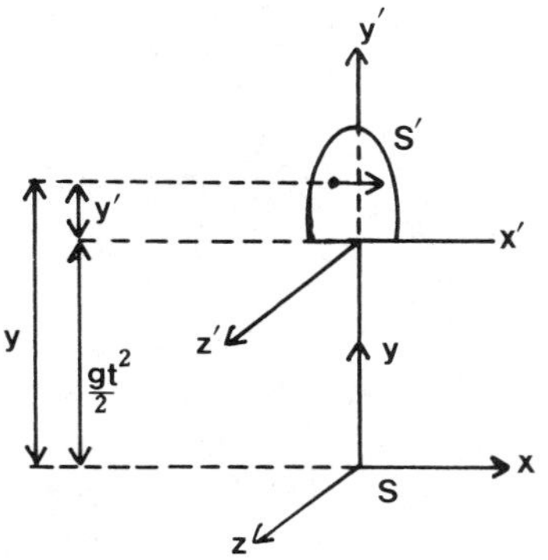

(Assume non-relativistic speeds, so that t = t′.) Show that a free particle or light ray moving at constant speed parallel to the x axis is observed by S′ to follow a parabolic trajectory, as in a room on earth.

2. Carry out an analysis of the falling elevator similar to that of the accelerating rocket in Exercise 1. Verify that if the elevator falls with acceleration −g, then a free particle (which accelerates

*Of course, observer #2, being inertial, may adopt a coordinate system in which the interval takes the simple form of Lorentz geometry.

relative to the ground: $d^2y/dt^2 = -g$) travels uniformaly in S' ($d^2y'/dt^2 = 0$).

2. GRAVITY AS SPACETIME CURVATURE

In the thought experiments of the preceding section, we have assumed that the gravitational field is *uniform*, by which is meant that, relative to an inertial observer, the acceleration of all falling objects has constant magnitude and direction. This is an idealization that is only approximately true. In reality, the natural acceleration due to gravity varies from point to point.

For example, think of two test particles initially at rest relative to a space capsule freely falling to the earth's surface. Suppose the line of separation of these particles is horizontal, i.e., perpendicular to the direction from the capsule to the earth's center [Fig. III-1(a)]. Since both particles are heading toward the center of the earth, the distance between them will necessarily decrease, in fact, at a gradually increasing rate (cf. Exercise 2). Consequently, we cannot say that both particles remain at rest relative to the space capsule.

Likewise, if the separation between the particles is vertical, the one nearer the earth (where the pull of gravity is slightly stronger) will undergo a slightly greater acceleration, and so the particles will gradually draw farther apart [Fig. III-1 (b)].

These effects, which are called *tidal effects* of gravity, are caused by nonuniformity in the earth's gravitational field—the pull of gravity varies from point to point. On a global scale, tidal effects are responsible for the oceanic tides; but on the scale of our space

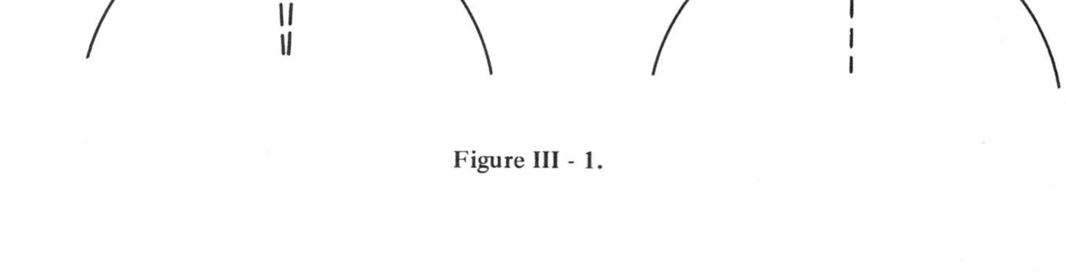

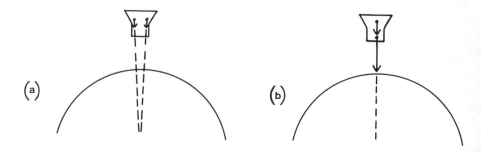

Figure III - 1.

capsule, these effects are small. Consequently, if the two particles are initially fairly close, and if we study their motion for but a short period of time, their relative acceleration will be too small for detection by our measuring devices (Exercises 1, 2) and the gravitational field will not be detectable. In other words, a sufficiently small, freely falling space capsule, over a sufficiently short span of time, is, to a first approximation, an inertial frame.

The Principle of Equivalence now takes the following form:

> For each spacetime point (i.e., event), and for a given degree of accuracy, there exists a frame of reference in which, in a certain region of space and for a certain interval of time (i.e., in a sufficiently small spacetime neighborhood of the event), the effects of gravity are negligible, and the frame is inertial to the degree of accuracy specified.

This is the frame of a freely falling observer.

Such a reference frame will be called a *locally inertial frame* (at the event), and the freely falling observer at rest in such a frame will be called a *locally inertial observer*. Any frame moving uniformly relative to a locally inertial frame is also a locally inertial frame in a neighborhood of the spacetime point under discussion.

Let us look a little more closely at the tidal effects (the nonuniformity) of gravity. As we have seen, in the presence of a gravitational field—for example, in the vicinity of any large mass, such as the earth, the sun, or any star—two test particles, initially moving with the same speed along two nearby parallel paths in space, will in general, not remain on parallel paths, but will gradually accelerate with respect to each other, because of the nonuniformity of the field.

This relative acceleration of freely falling test particles has a purely geometric analogue, which we can visualize quite easily on a sphere. Imagine two travelers setting out on separate journeys from two nearby points on the earth's equator (A and B in Fig. III-2). Assume they both travel northward, along great circles, so that their paths are initially parallel. As is clear from Figure III-2, these paths do not remain parallel, but draw closer and eventually meet at the north pole. This is because the sphere is a curved surface.

To express this relationship between two nearby geodesics (great circles) in mathematical terms, let $\alpha(s)$ be the parametrization of one of the two geodesics in terms of arc length—$\alpha(0)$ is on the equator—and let $v(s)$ be the distance from $\alpha(s)$ to the other geodesic,

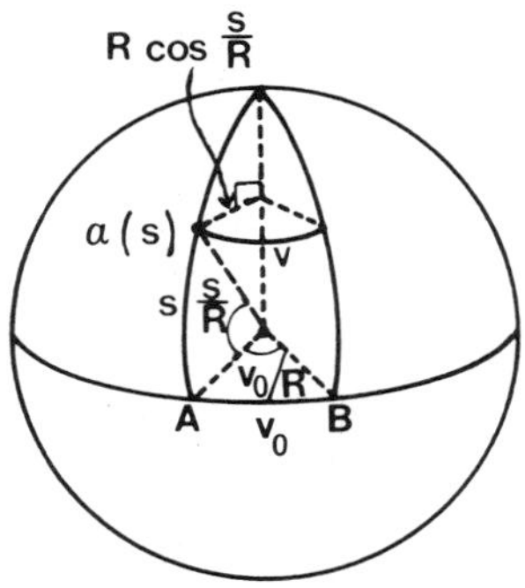

Figure III - 2.

as measured along the parallel (circle of constant latitude) through α(s), i.e., the "east-west distance." Then it can be shown that

$$\frac{d^2 v}{ds^2} + \frac{v}{R^2} = 0$$

where R is the sphere's radius (Exercise 4). Recalling that $1/R^2$ is the Gauss curvature of the sphere, we can rewrite this equation as

$$\frac{d^2 v}{ds^2} + Kv = 0 \tag{102}$$

It turns out that this differential equation, known as *Jacobi's equation*, or the equation of *geodesic deviation*, holds equally true on an arbitrary surface (with the appropriate definition of v), where the curvature varies from point to point, and K in Eq. (102) is a function of arc length along α: K = K(α(s)) (cf. Laugwitz [31, p. 132], Stoker [45, p. 250]). On a surface of negative curvature, for example, initially parallel geodesics draw apart from one another (see Fig. III-3). (We have to be careful about what "initially parallel" means here. In the present context, it means that at the initial point, $v > 0$ and $dv/ds = 0$.)

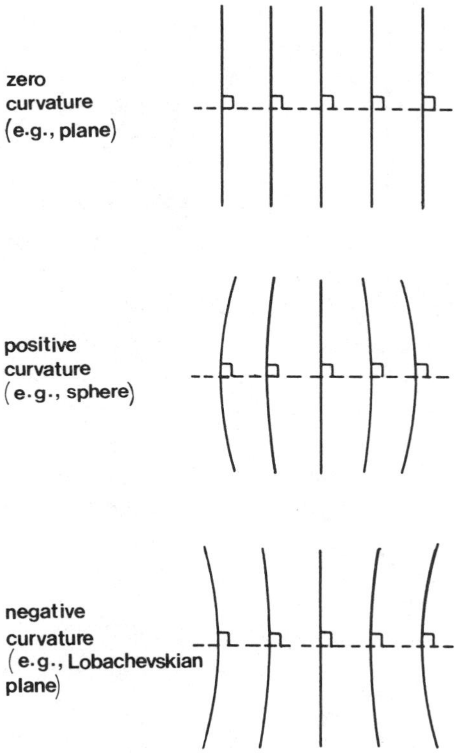

Figure III - 3.

The similarity between (a) the relative acceleration of free particles in a nonuniform gravitational field, and (b) the varying separation of nearby geodesics on a curved surface is more than a vague resemblance. Reasoning from his Principle of Equivalence, Einstein concluded that gravity is not a force, as Newton believed, but rather the curvature of spacetime. The source of this curvature is matter itself. Just as a magnet sets up a magnetic field in the surrounding space, so does a material object create a gravitational field which distorts or "curves" the surrounding spacetime.

Generalizing the Law of Inertia, which says that free particles follow geodesics (straight lines) in the flat spacetime of special relativity (no gravity = no curvature), Einstein hypothesized that free particles must follow geodesics in the curved spacetime of a gravitational field. Thus all material objects, from an apple falling from a tree to an orbiting planet, move along geodesics in spacetime, unless prevented from doing so by some force or obstacle. According to Einstein, light rays also pursue geodesics.

It is important to realize that these geodesics are curves in four dimensions. We are not speaking about spatial trajectories. In Section II-8, we sketched a spacetime diagram of the earth's orbit (Fig. II-12). Since the orbit lies in a plane, we were able to ignore one of the spatial coordinates (but not the time coordinate). In discussing the earth's world-line, which is (to a very close approximation) an elliptical helix (coil) we pointed out that if the diagram had been drawn to scale, with both time and distance measured in the same units, e.g., centimeters, then the helix would have to be stretched in the direction of its axis by a huge factor (on the order of 10^4). It would then be practically indistinguishable from a straight line. It is not at all surprising, therefore, to be told that this world-line is a geodesic—after all, the stretched helix would appear hardly curved at all.

Actually this curve is completely straight, in the sense of being a geodesic: it is the straightest possible spacetime path between any (nearby) two of its points. It may *look* curved on the diagram, but it is the surrounding region of spacetime, and not the world-line, that has curvature. (Indeed, our spacetime diagram, even if drawn to scale, is somewhat misleading, because rectilinear axes cannot be set up in a curved region of spacetime, any more than Cartesian coordinates can be set up on a curved surface, except as a local approximation. Curvilinear coordinates are required in the large.) Likewise, although we usually describe the effect of a gravitational field on light as a "bending" of light, such language, in the strictest sense, is inappropriate, since the light follows a geodesic (straight path), while the spacetime it traverses is curved.

We conclude this section with a more mathematical formulation of the term *locally inertial frame*. For convenience, introduce the symbols

$$\eta_{ij} = \begin{cases} 1 & \text{if } i = j = 0 \\ -1 & \text{if } i = j = 1, 2, \text{ or } 3 \\ 0 & \text{if } i \neq j \end{cases}$$

Then (η_{ij}) is the matrix of the Lorentz metric.

Since the effects of gravity are negligible in a locally inertial frame, a locally inertial observer may set up a system of coordinates in which the interval is given approximately by

$$d\tau^2 = \eta_{ij} du^i du^j$$

$$= (du^0)^2 - (du^1)^2 - (du^2)^2 - (du^3)^2 \qquad (103)$$

the form taken in special relativity.* (Within the accuracy of the observer's clocks and measuring rods, spacetime is flat in a neighborhood of the event being considered, and a Lorentz coordinate system, $t = u^0$, $x = u^1$, $y = u^2$, $z = u^3$, as described in Section II-9 is possible.)

More precisely, in a locally inertial frame at an event **P**, it is possible to set up a coordinate system (u^i) such that

$$d\tau^2 = g_{ij} du^i du^j,$$

and the functions g_{ij} satisfy

$$g_{ij}(\mathbf{P}) = \eta_{ij} \qquad (104)$$

$$\frac{\partial g_{ij}}{\partial u^k}(\mathbf{P}) = 0 \qquad (105)$$

for all i,j,k = 0,1,2,3 (see Exercise 3 for an analytic proof). The first condition stipulates that the metric is Lorentzian at **P** exactly. The second implies (by continuity) that the g_{ij} have small rates of change in a small neighborhood of **P** and so differ very little from the respective η_{ij}. In other words,

$$d\tau^2 \approx \eta_{ij} du^i du^j,$$

and the difference can be made arbitrarily small by restricting the size of the spacetime neighborhood. We shall call a system of coordinates satisfying Eqs. (104) and (105), a *locally Lorentz coordinate system* at **P**.

---

*The *Einstein summation convention* (Section I-5) is in effect throughout this chapter.

The Principle of Equivalence is a physical analogue of the fact that a sufficiently small piece of a 2-dimensional surface looks approximately like a piece of the flat Euclidean plane and may be parametrized in such a way that the metric form is approximately $ds^2 = (du^1)^2 + (du^2)^2$ (locally Cartesian coordinates). The extent to which $d\tau^2$ departs from the Lorentz metric $\eta_{ij}du^i du^j$ as we move away from **P** (in time and space) corresponds physically to the nonuniformity of gravity. Mathematically, it corresponds to the curvature of spacetime, as we shall see shortly.

Exercises III-2

1. Two test particles are dropped from points 10 m apart on the roof of an office building 240 m high. (a) Neglecting air resistance, how long will it take the particles to reach the street below? (b) The earth's mean radius is $R_\oplus \approx 6.37 \times 10^6$ m. Using similar triangles, prove that in the time it takes the two particles to reach the street, the distance between them will have decreased by about 0.4 mm.

2. In the previous exercise, suppose the initial distance between the particles is Δx. (a) Show that the relative speed v_r of the particles, after they have fallen a distance h is

$$v_r \approx \frac{-\Delta x}{R_\oplus}(2gh)^{1/2}$$

(See also Exercise 6 of Section 4.) (b) Estimate v_r for $\Delta x = 10$ m, and a fall of one kilometer near the earth's surface.

3. (Existence of locally Lorentz coordinates) Let u^0, u^1, u^2, u^3 be an arbitrary coordinate system in a neighborhood of an event **P**. After making a suitable spatial translation if necessary, we may assume **P** is the origin: $u^i(\mathbf{P}) = 0$, all i. Let new coordinates $\bar{u}^0, \bar{u}^1, \bar{u}^2, \bar{u}^3$ be introduced satisfying

$$u^i = a^i_h \bar{u}^h - \frac{1}{2}(\Gamma^i_{jk})_o a^j_m a^k_n \bar{u}^m \bar{u}^n$$

for i = 0, 1, 2, 3, where a subscript 'o' indicates evaluation at the origin, and (a^i_j) is a non-singular matrix of constants. (a) Show that, at the origin, we have

$$\left(\frac{\partial u^i}{\partial \bar{u}^h}\right)_o = a^i_h$$

$$\left(\frac{\partial^2 u^i}{\partial \bar{u}^m \partial \bar{u}^n}\right)_o = -\left[\Gamma^i_{jk} \frac{\partial u^j}{\partial \bar{u}^m} \frac{\partial u^k}{\partial \bar{u}^n}\right]_o$$

(b) Assume the following fact, proved below: the Christoffel symbols for any two coordinate systems u^i and $\bar{u}^i$ are related by the identities

$$(*) \quad \bar{\Gamma}^h_{mn} = \left(\frac{\partial^2 u^i}{\partial \bar{u}^m \partial \bar{u}^n} + \Gamma^i_{jk} \frac{\partial u^j}{\partial \bar{u}^m} \frac{\partial u^k}{\partial \bar{u}^n}\right) \frac{\partial \bar{u}^h}{\partial u^i}$$

Show that in the new system $(\partial \bar{g}_{ij}/\partial \bar{u}^k)_o = 0$, for all i, j, k, i.e., Eq. (105) holds. Finally, by standard methods of linear algebra, we can follow the transformation $u^i \rightarrow \bar{u}^i$ by a linear transformation whose matrix, when multiplied by the matrix $(\partial \bar{u}^i/\partial u^j)_o$—the inverse of (a^i_j)—yields (η_{ij}). Under a linear transformation Eq. (105) is preserved, and now Eq. (104) holds as well.

Proof of (*): For an arbitrary geodesic,

$$0 = \frac{d^2 u^i}{d\tau^2} + \Gamma^i_{jk} \frac{du^j}{d\tau} \frac{du^k}{d\tau} = \frac{d}{d\tau}\left(\frac{\partial u^i}{\partial \bar{u}^m} \frac{d\bar{u}^m}{d\tau}\right)$$

$$+ \Gamma^i_{jk} \frac{\partial u^j}{\partial \bar{u}^m} \frac{d\bar{u}^m}{d\tau} \frac{\partial u^k}{\partial \bar{u}^n} \frac{d\bar{u}^n}{d\tau}$$

$$= \frac{\partial u^i}{\partial \bar{u}^m} \frac{d^2 \bar{u}^m}{d\tau^2} + \frac{\partial^2 u^i}{\partial \bar{u}^m \partial \bar{u}^n} \frac{d\bar{u}^m}{d\tau} \frac{d\bar{u}^n}{d\tau}$$

$$+ \Gamma^i_{jk} \frac{\partial u^j}{\partial \bar{u}^m} \frac{\partial u^k}{\partial \bar{u}^n} \frac{d\bar{u}^m}{d\tau} \frac{d\bar{u}^n}{d\tau}$$

Multiply by $\partial \bar{u}^h/\partial u^i$ and sum over i. We obtain, using

$$\frac{\partial u^i}{\partial \bar{u}^m} \frac{\partial \bar{u}^h}{\partial u^i} = \delta^h_m$$

$$0 = \frac{d^2\bar{u}^h}{d\tau^2} + \left[\frac{\partial^2 u^i}{\partial \bar{u}^m \partial \bar{u}^n} + \Gamma^i_{jk}\frac{\partial u^j}{\partial \bar{u}^m}\frac{\partial u^k}{\partial \bar{u}^n}\right]\frac{\partial \bar{u}^h}{\partial u^i}\frac{d\bar{u}^m}{d\tau}\frac{d\bar{u}^n}{d\tau}$$

However, the geodesic equations are invariant under a change of coordinates, and so we must have also

$$0 = \frac{d^2\bar{u}^h}{d\tau^2} + \bar{\Gamma}^h_{mn}\frac{d\bar{u}^m}{d\tau}\frac{d\bar{u}^n}{d\tau}$$

The equivalence of the last two equations for all possible geodesics necessitates the validity of (*).

4. Using Figure III-2, show that $v(s) = v_0 \cos(s/R)$ and thereby verify Jacobi's equation for the sphere.

5. In Exercise 2 of Section II-10, Jill sets out with "constant acceleration" (as defined in that exercise) relative to Jack, who is an inertial observer: $d\tau^2 = dt^2 - dx^2 - dy^2 - dz^2$, in his coordinates. Assuming Jill travels along the x axis, starting from x_0, we found that Jill's world line is described by

$$t = \frac{1}{g}\sinh(gt')$$

$$x = \frac{1}{g}[\cosh(gt') - 1] + x_0$$

$$y = z = 0$$

where t' is Jill's proper time. Define a new coordinate system t', x', y', z' by

$$t = \left(\frac{1}{g} + x'\right)\sinh(gt')$$

$$x = \left(\frac{1}{g} + x'\right)\cosh(gt') + x_0 - \frac{1}{g}$$

$$y = y', \quad z = z'$$

Show that in the primed coordinate system

i. Jill remains at rest at the origin ($x' = y' = z' = 0$)
ii. $d\tau^2 = (1 + gx')^2 dt'^2 - dx'^2 - dy'^2 - dz'^2$

[Since this reduces to the Lorentz metric at Jill's position—$x' = 0$—this coordinate system approximates a Lorentz coordinate system in her immediate vicinity. However, Eq. (105) does not hold. Notice that x', y', and z' are analogous to Euclidean spatial coordinates, since when t' is held constant, $d\sigma^2 = - d\tau^2 = dx'^2 + dy'^2 + dz'^2$.]

6. (a) The shaded area in figure A is a segment of a circle in the Euclidean plane.

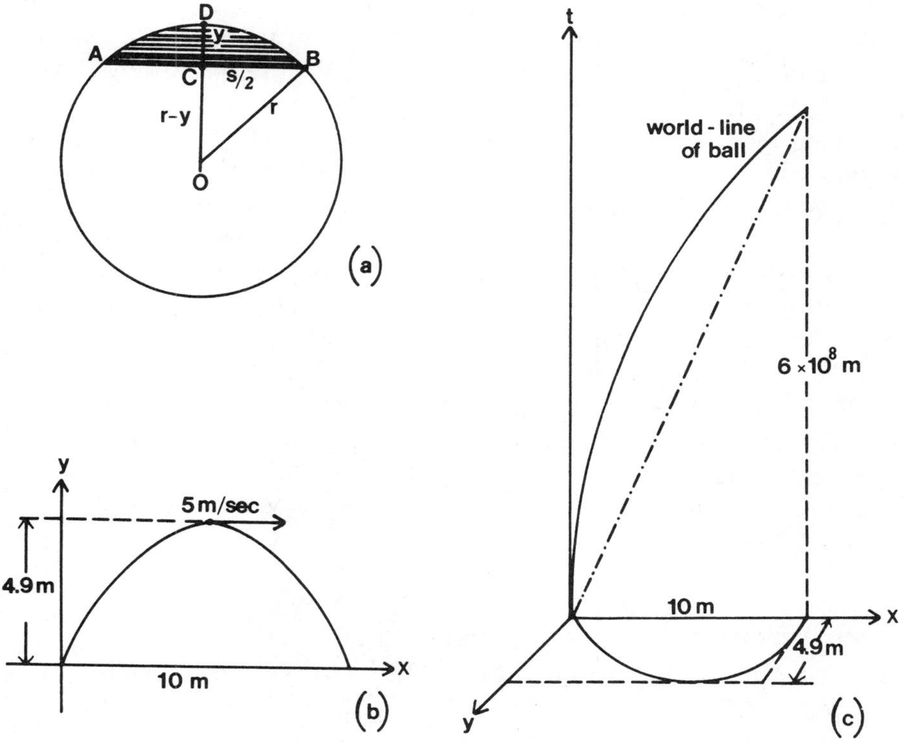

Let s = AB be the base of the segment and let y = CD be the height. Show that the radius of the circle is given by

$$r = \frac{s^2}{8y} + \frac{y}{2}$$

Show that if ∡COB is small, the second term on the right is negligible compared to the first, and so

$$r \approx \frac{s^2}{8y}$$

(b) Figure B shows the parabolic path of a ball, thrown with such an initial velocity and angle of elevation that it lands 10 m down-range 2 sec (6×10^8 m) after it is thrown. Show that its maximum height is 4.9 m. (c) Figure C depicts the world-line of the ball on a spacetime diagram. Note that the scale used on the t axis is different from that used for the spatial axes. Assuming this world-line is approximated by a circular arc, use the approximation formula of (a) to estimate the radius of this arc. (It is roughly one light-year.) (d) Consider now the parabolic path of a bullet which travels 10 m horizontally in 0.02 sec. What is the bullet's maximum height? (e) As in (c), sketch the bullet's world-line and estimate the radius of this world-line (assuming it is roughly circular).

Although the path of a ball and the path of a bullet appear to have quite different curvatures *in space*, their world-lines *in spacetime* appear to have essentially the same curvature, which is very small: $(1 \text{ lt-yr})^{-1}$. (Gravity is very weak.) Actually, the spacetime diagrams are misleading (just as was Fig. II-12), because it is spacetime that is curved rather than the world-lines, which are geodesics and therefore "straight."

3. THE CONSEQUENCES OF EINSTEIN'S THEORY

As pointed out in Section 1, the path of a light ray is affected by a gravitational field: gravity bends light! For instance, if a light ray from a distant star passes near the sun on its way to the earth, the sun's gravitational field will cause the ray to bend in toward the sun so that the apparent position of the star as seen by an earth observer, is displaced outward from the sun, away from what would be the star's position if the sun were in some other part of the sky (see Fig. III-4).

By computing the curvature of spacetime around the sun, Einstein predicted that if stars near the edge of the sun's disk were observed during a solar eclipse, the bending ($\measuredangle SPS'$ in Fig. III-4) would amount to about 1.74″. Plans were made to test Einstein's theory during the eclipse of May 29, 1919. From Britain, a committee of the Royal Society and the Royal Astronomical Society dispatched scientific teams to two remote corners of the globe where the eclipse might be observed in its totality. Although bad weather conditions at one of the locations and some technical problems at the other limited the amount of good data, Einstein's theory was supported (see Eddington [12]).

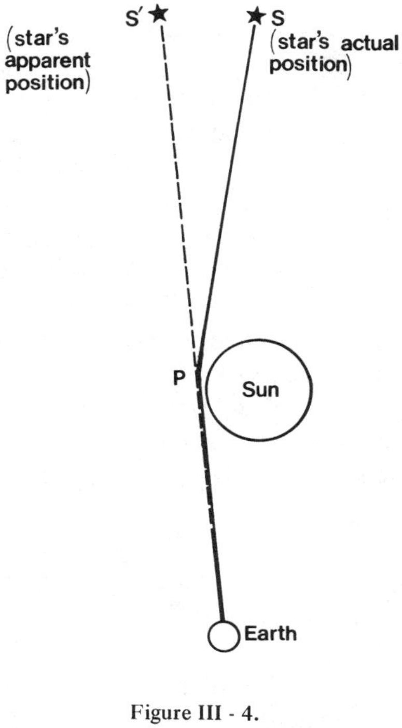

Figure III - 4.

The experiment has been repeated during later eclipses, with the observed deflection in good agreement with general relativity.

More recently, similar experiments have been performed using not light waves but radio waves emitted by quasars ("quasi-stellar" radio sources), extremely distant objects whose exact nature is presently a mystery, but which emit vast amounts of energy, some at radio frequencies. Once a year, the quasar known as 3C279 is occulted by the sun. At such times scientists can observe the changes in the angle between 3C279 and another reference object (quasar 3C273) due to the deflection of the radio waves by the sun. Here too, the observations support Einstein's theory.

Actually, if one postulates that a light ray is a beam of tiny particles, then the classical Newtonian theory predicts a gravitational bending of light also. A particle of light passing by the sun is in effect a very small high speed comet traveling along a hyperbolic orbit. The amount of deflection is the acute angle between the orbit's asymptotes. However, the amount of deflection calculated from Newton's Law of Gravity is only half that predicted by general relativity (see Section 10). The experimental results have therefore lent strong support to Einstein's theory as a more accurate theory of gravitation than that of Newton.

Another consequence of general relativity has to do with the orbit of Mercury, closest planet to the sun. According to Newtonian physics, the orbit of each planet is an ellipse with the sun at one focus. However, in the case of Mercury the orbit is not precisely elliptical. The orbit's perihelion, or closest point to the sun, slowly advances (see Fig. III-5), so that after one revolution (perihelion to perihelion), the perihelion advances a very small fraction of a revolution. This advance or *precession*, as it is often called, is so slow that it takes about 3,000,000 years for the perihelion to go around the sun once. This amount to around 43 seconds of arc in a century, a very small angle indeed, but nevertheless discernible, and the source of a long standing mystery.

Actually, these figures (3,000,000 as well as 43) refer to a perihelion advance in addition to a precession produced by perturbing effects of the other planets on Mercury's orbit (see Weinberg [50, p. 199] and Misner, Thorne and Wheeler [36, p. 1113]. The total precession amounts to about 5600″ per year. The other planets have perihelion advances also, but these are much smaller because of the greater distance from the sun.

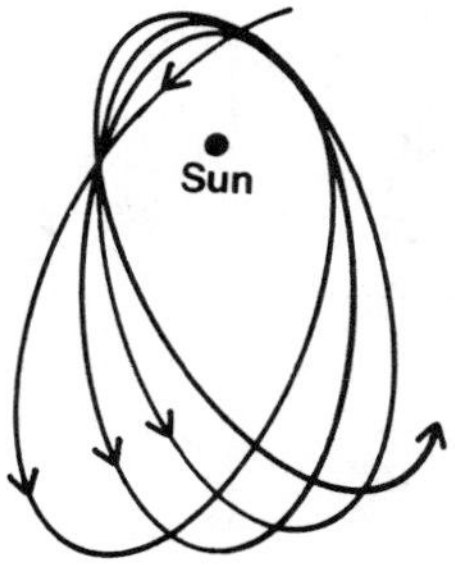

Perihelial
Precession
(greatly exaggerated)

Figure III - 5.

General relativity does indeed predict the amount of perihelion advance observed. In a little while we shall go through the detailed relativistic calculations for both the bending of light and the precession of perihelion. For the sake of comparison, we shall perform the corresponding calculations in the context of Newtonian theory as well. To this end, the next section will briefly describe Newton's theory and its predictions.

Another prediction of Einstein's theory is the *gravitational redshift*. This is discussed in the exercises of Sections 7 and 8.

4. THE UNIVERSAL LAW OF GRAVITATION

Imagine a ball fired horizontally from a cannon atop a high mountain. If we ignore air resistance, the ball follows an approximately parabolic path down to the ground. If we repeat the experiment with greater and greater muzzle velocities, the ball lands farther and farther from the base of the mountain.

For very high velocities the curvature of the earth's surface becomes significant, and we can imagine that a sufficiently great muzzle velocity would send the ball into orbit about the earth, similar to the moon's orbit except for the greater size of the latter (see Fig. III-6).

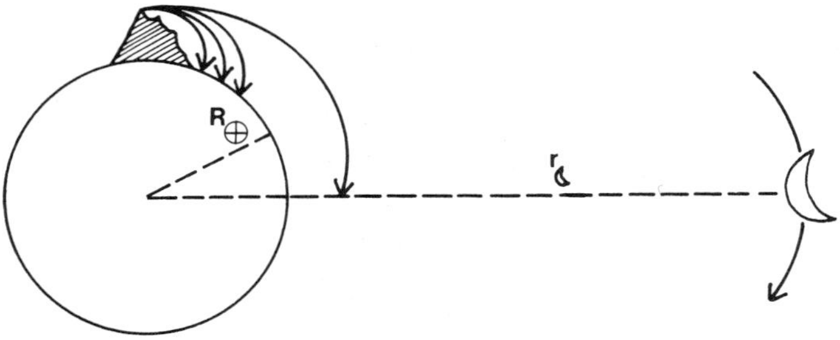

Figure III - 6.

No doubt through similar reasoning, Sir Isaac Newton (1642-1727) realized that the force which causes the cannon ball to follow a curved trajectory or an apple to fall from a tree is of the same nature as the force which maintains the moon in its orbit about the earth, or the planets in their orbits about the sun.

In reasoning toward a mathematical formulation of this universality of gravity, Newton was influenced by the three laws of planetary motion discovered by Johannes Kepler (1571-1630). It was Kepler who made the final break with the entrenched idea that only circles or spheres could be used to describe astronomical motions. Stated in modern terms, Kepler's laws are

a. A planet revolves about the sun in a (planar) elliptical orbit with the sun at one focus.*
b. The radius vector drawn from the sun to the planet sweeps out area at a constant rate.
c. The period of the planet is proportional to the 3/2 power of the orbit's major axis.

Newton proclaimed his Law of Universal Gravitation in his *Principia*:

> Every particle in the universe attracts every other particle in such a way that the force between the two is directed along the line between them and has magnitude proportional to the product of their masses and inversely proportional to the square of the distance between them.

*The other focus is simply a mathematical point in space, with no physical significance.

Thus the force between two particles of masses M and m separated by a distance r has magnitude

$$F = \frac{GMm}{r^2} \tag{106}$$

where the constant of proportionality G is called the *gravitational constant*. In terms of conventional cgs units (centimeters, grams, seconds), G has the approximate value 6.67×10^{-8} $cm^3/(g\ sec^2)$.

In the next section, we shall see how Kepler's Laws can be derived from the Law of Universal Gravitation coupled with Newton's Second Law of Motion (cf. the Introduction to Chapter II).

Before continuing, however, we shall make a few conventions concerning units of measurement. In Chapter II, we saw how time as well as distance can be reckoned in centimeters, using the speed of light ($c \approx 2.997925 \times 10^{10}$ cm/sec) as a conversion factor. Surprisingly, mass can also be reckoned in centimeters, rather than in grams. (Would you like to know your mass in centimeters? It has nothing to do with how tall you are.)

Here is how. The gravitational constant G is approximately 6.67×10^{-8} $cm^3/(g\ sec^2)$. As you can check, when G is multiplied by mM/r^2 the resulting expression has the units of a force (mass $\times$ acceleration), and this is the gravitational force according to Newton's law:

$$\frac{GmM}{r^2}\left(\frac{cm^3}{g\ sec^2}\ \frac{g^2}{cm^2}\right) = F\left(\frac{g\ cm}{sec^2}\right)$$

Symbols like cm and g can be canceled as in algebra! This explains why G has the units $cm^3/(g\ sec^2)$. When G is divided by c^2, the result has the units cm/g (check it!) and so

$$G/c^2 \approx 7.425 \times 10^{-29}\ cm/g$$

can be used as a conversion factor from grams to centimeters.

For example, the mass of the sun in conventional units is $M_{conv} \approx 1.99 \times 10^{33}$ g. In centimeters, this mass is $M_\odot = GM_{conv}/c^2 \approx 1.48 \times 10^5$ cm. In the remaining sections, we shall always use

Table III - 1.

Time	$t_{conv}(\text{sec}) \times c(\text{cm sec}^{-1}) = t(\text{cm})$
Velocity	$v_{conv}(\text{cm sec}^{-1}) \times \frac{1}{c}(\text{cm}^{-1}\text{sec}) = v(\text{dimensionless})$*
Acceleration	$a_{conv}(\text{cm sec}^{-2}) \times \frac{1}{c^2}(\text{cm}^{-2}\text{sec}^2) = a(\text{cm}^{-1})$
Mass	$m_{conv}(\text{g}) \times \frac{G}{c^2}(\text{cm g}^{-1}) = m(\text{cm})$
Force	$F_{conv}(\text{cm g sec}^{-2}) \times \frac{G}{c^4}(\text{cm}^{-1}\text{g}^{-1}\text{sec}^2) = F(\text{dimensionless})$
Energy	$E_{conv}(\text{cm}^2\text{g sec}^{-2}) \times \frac{G}{c^4}(\text{cm}^{-1}\text{g}^{-1}\text{sec}^2) = E(\text{cm})$

*Denoted β in Chapter II.

geometric units, i.e., we shall measure distance, time, and mass in centimeters, and define all other quantities in terms of these. The relationships between conventional (cgs) measurements and measurements in geometric units are presented in Table III-1. (The subscript "conv" indicates conventional units.) In Exercise 7, you will show that Newton's

$$F_{conv} = \frac{GM_{conv}m_{conv}}{r^2}$$

becomes

$$F = Mm/r^2$$

in geometric units. Thus the effect of using geometric units is that the gravitational constant as well as the speed of light are unity in these units.

Exercises III-4

1. According to Newton's theory, at each instant, the magnitude of the gravitational force exerted by one particle on another is inversely proportional to the square of the distance between them *at that instant*. What difficulties does this concept of "action at a distance" create vis-a-vis special relativity?

2. Suppose a new planet is discovered orbiting the sun in a nearly circular orbit. Observations indicate the planet's period is 8.6 years. How far is the planet from the sun? (Hint: the mean earth-to-sun distance is approximately 1.5×10^{13} cm.)

3. (a) Show that the acceleration of a freely falling particle at distance r from a point mass M has magnitude

$$a_{conv} = \frac{GM_{conv}}{r^2}$$

(b) Show that in geometric units, the equivalent formula is

$$a = \frac{M}{r^2}$$

(c) Show that $|\Delta a/a| \approx 2\,|\Delta r/r|$ for a change in distance Δr that is negligible compared to r itself.

4. Using the formula of Exercise 3a, compute the acceleration g_{conv} of a freely falling particle near the earth's surface. $[M_\oplus)_{conv} \approx 5.98 \times 10^{27}$ g, $R_\oplus \approx 6.37 \times 10^8$ cm.]

5. At what altitude above sea level will the acceleration a of a falling object be less than g (i) by 1% (ii) by 0.1%. (See Exercise 3c.)

6. Suppose two test particles are released from rest above the earth's surface. Assume the separation Δr between them is vertical, as in Figure III-1 (b). (a) Show that after they have fallen a distance h, their relative velocity v_r is given by

$$v_r \approx \frac{2\Delta r}{R_\oplus} (2gh)^{1/2}$$

(b) Estimate v_r (in conventional units) for $\Delta r = 10$ m and $h = 1$ km.

7. Compute the masses of the following, in centimeters:

i. yourself
ii. a 1500 kg automobile
iii. the earth $[(M_\oplus)_{conv} \approx 5.98 \times 10^{27}$ g]

8. Re-express each of the following formulas in geometric units:

i. $F_{conv} = GM_{conv}m_{conv}/r^2$ (Newton's Law of Gravitation)
ii. $F_{conv} = m_{conv}a_{conv}$ (Newton's Second Law)
iii. $E_{conv} = m_{conv}c^2$ (Einstein's Equivalence of Energy and Mass)

5. ORBITS IN NEWTON'S THEORY

Let us now apply Newton's Law of Gravity to the case of a planet orbiting the sun. (We shall ignore the influence of planets other than the one under discussion.) As Newton had shown, a spherically symmetric mass acts gravitationally like a "point mass," i.e., as if all its matter were concentrated at its center. Therefore in Eq. (106), we may take r to be the distance between the centers of the sun and the planet. Since the sun is many times more massive than the planet, we may, to a close approximation, take the sun as fixed at the origin of a rectangular coordinate system (although strictly speaking both sun and planet move about their common center of gravity).

Let $\mathbf{X}(t) = (x^1(t), x^2(t), x^3(t))$ be the position of the planet as a function of time t, and let M and m denote the respective masses of sun and planet. First, we show that the orbit is planar. Since the force on the planet, $\mathbf{F} = m\mathbf{X}''$, is radial (directed toward the origin), $\mathbf{X}''$ and $\mathbf{X}$ are proportional. Therefore,

$$\mathbf{O} = \mathbf{X} \times \mathbf{X}'' = (\mathbf{X} \times \mathbf{X}'') + (\mathbf{X}' \times \mathbf{X}') = \frac{d}{dt}(\mathbf{X} \times \mathbf{X}')$$

Consequently the vector $\mathbf{A} = \mathbf{X} \times \mathbf{X}'$ has constant components. Moreover,

$$\mathbf{X} \cdot \mathbf{A} = \mathbf{X} \cdot (\mathbf{X} \times \mathbf{X}') = \mathbf{X}' \cdot (\mathbf{X} \times \mathbf{X}) = \mathbf{O}$$

which implies that $\mathbf{X}$ is restricted to the plane through the origin with normal vector $\mathbf{A}$.

We can therefore orient our coordinate axes so the planet moves in the $x^1 x^2$-plane, with the sun at the origin. Let r, θ be the corresponding polar coordinates, so that $x^1 = r \cos\theta$, $x^2 = r \sin\theta$. Then r and θ are functions of t along the orbit. Let $\mathbf{u}_r$ and $\mathbf{u}_\theta$ be the unit vectors defined (in Cartesian coordinates) by

$$\mathbf{u}_r = (\cos\theta, \sin\theta)$$

$$\mathbf{u}_\theta = (-\sin\theta, \cos\theta)$$

These are the unit vectors in the radial and transverse directions. They are orthogonal and satisfy

$$\frac{d\mathbf{u}_r}{d\theta} = \mathbf{u}_\theta, \qquad \frac{d\mathbf{u}_\theta}{d\theta} = -\mathbf{u}_r$$

We can now write $\mathbf{X}(t) = r\mathbf{u}_r$. Differentiating by the product rule, and using the relations

$$\frac{d\mathbf{u}_r}{dt} = \frac{d\mathbf{u}_r}{d\theta}\frac{d\theta}{dt} = \frac{d\theta}{dt}\mathbf{u}_\theta$$

$$\frac{d\mathbf{u}_\theta}{dt} = \frac{d\mathbf{u}_\theta}{d\theta}\frac{d\theta}{dt} = -\frac{d\theta}{dt}\mathbf{u}_r$$

we have

$$\mathbf{X}'(t) = \frac{dr}{dt}\mathbf{u}_r + r\frac{d\theta}{dt}\mathbf{u}_\theta \qquad (107)$$

and

$$\mathbf{X}''(t) = \frac{d^2 r}{dt^2}\mathbf{u}_r + \frac{dr}{dt}\frac{d\theta}{dt}\mathbf{u}_\theta + \frac{dr}{dt}\frac{d\theta}{dt}\mathbf{u}_\theta + r\frac{d^2\theta}{dt^2}\mathbf{u}_\theta - r\left(\frac{d\theta}{dt}\right)^2\mathbf{u}_r$$

or

$$\mathbf{X}''(t) = \left[\frac{d^2 r}{dt^2} - r\left(\frac{d\theta}{dt}\right)^2\right]\mathbf{u}_r + \left[r\frac{d^2\theta}{dt^2} + 2\frac{dr}{dt}\frac{d\theta}{dt}\right]\mathbf{u}_\theta \qquad (108)$$

Since the force is radial, the coefficient of $\mathbf{u}_\theta$, which can be rewritten

$$r \frac{d^2\theta}{dt^2} + 2 \frac{dr}{dt} \frac{d\theta}{dt} = \frac{1}{r}\left(\frac{d}{dt}\left(r^2 \frac{d\theta}{dt}\right)\right)$$

must vanish. Therefore,

$$r^2 \frac{d\theta}{dt} = h \quad \text{(constant).} \tag{109}$$

By the results of Exercise 15 of Section I-7, (1/2) r^2 $d\theta/dt$ is the rate at which the radius vector X sweeps out area as the planet traverses its orbit. We have therefore established Kepler's Second Law.

Now, according to Newton's Law of Gravity, the force on the planet is

$$\mathbf{F} = -\frac{Mm}{r^2}\mathbf{u}_r \tag{110}$$

which, together with Newton's Second Law of Motion, $\mathbf{F} = m\mathbf{X}''$, gives

$$\mathbf{X}'' = -\frac{M}{r^2}\mathbf{u}_r \tag{111}$$

Equating this to the radial component of Eq. (108), we have

$$-\frac{M}{r^2} = \frac{d^2 r}{dt^2} - r\left(\frac{d\theta}{dt}\right)^2 \tag{112}$$

We shall deduce Kepler's First and Third Laws from this equation.

From Eq. (109), we have $d\theta/dt = h/r^2$. Substituting this into Eq. (112) and multiplying the resulting equation by r^2/h^2, we obtain

$$-\frac{r^2}{h^2}\frac{d^2r}{dt^2} + \frac{1}{r} = \frac{M}{h^2} \tag{113}$$

At this point it is convenient to introduce a new variable, u = 1/r. It satisfies

$$\frac{du}{d\theta} = -\frac{1}{r^2}\frac{dr}{d\theta} = -\frac{1}{r^2}\frac{dr}{dt}\Big/\frac{d\theta}{dt} = -\frac{1}{h}\frac{dr}{dt}$$

$$\frac{d^2u}{d\theta^2} = -\frac{1}{h}\frac{d}{dt}\left(\frac{dr}{dt}\right)\Big/\frac{d\theta}{dt} = -\frac{r^2}{h^2}\frac{d^2r}{dt^2}$$

The orbital equation (113) now becomes

$$\frac{d^2u}{d\theta^2} + u = \frac{M}{h^2} \tag{114}$$

The constant function $u = M/h^2$ is one obvious solution to Eq. (114). According to the theory of linear differential equations, the general solution to Eq. (114) is then obtained by adding to this "particular solution" the general solution of the corresponding "homogeneous equation,"

$$\frac{d^2u}{d\theta^2} + u = 0 \tag{115}$$

Since the solutions of the latter (apart from the trivial solution u = 0) are all functions expressible in the form

$$u = \frac{1}{d}\cos(\theta - \theta_o)$$

where θ_o and d are constants of integration, we see that the general solution of Eq. (114) is

$$u = \frac{1}{d}\cos(\theta - \theta_o) + \frac{M}{h^2}$$

By reorienting our coordinate axes if necessary, we can assume $d > 0$ and $\theta_0 = 0$. The orbital equation can now be written

$$u = \frac{M}{h^2}(1 + e\cos\theta) \qquad (116)$$

where $e = h^2/(Md)$. Recalling that $u = 1/r$ and noting that $h^2/M = ed$, we have also

$$ed = r(1 + e\cos\theta) \qquad (117)$$

or

$$r = e(d - r\cos\theta) \qquad (118)$$

We can now show that Eq. (118) is the polar coordinate equation of a conic section. In Figure III-7, F is the origin, P is the planet's location, and D is the foot of the perpendicular dropped from P to the line $x^1 = d$. Since $x^1 = r\cos\theta$, and $d - x_1 = PD$, then Eq. (118) is equivalent to

$$PF = e\,PD,$$

the defining equation of a conic section with a focus at F and eccentricity e. The orbit is therefore an ellipse if $e < 1$, a parabola if $e = 1$, or a hyperbola if $e > 1$.

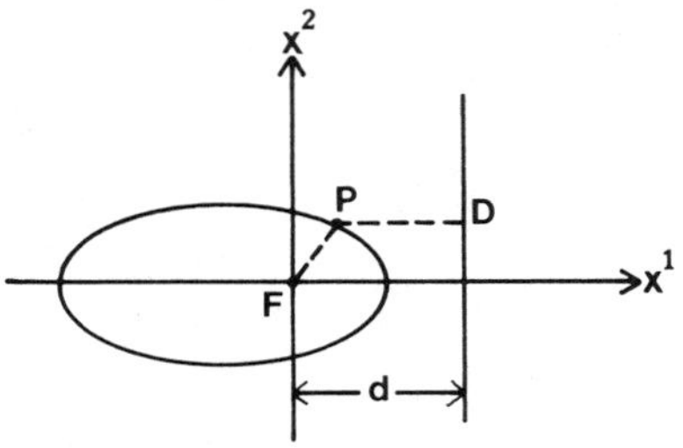

Figure III - 7.

Of course, for a planetary orbit, which is a closed curve, we must have $e < 1$. The orbit of a comet could have $e < 1$ or $e \geqslant 1$ depending upon whether or not it had sufficient energy to escape permanently from the sun's gravitational attraction.

In Exercise 9, you will transform Eq. (117) to Cartesian coordinates, $x = x^1 = r \cos\theta$, $y = x^2 = r \sin\theta$, and obtain, for $e < 1$,

$$\frac{(x+c)^2}{a^2} + \frac{y^2}{b^2} = 1$$

(an ellipse with semi-axes a and b and one focus at the origin), where

$$a = \frac{ed}{1 - e^2}, \qquad b = a(1 - e^2)^{1/2}, \qquad c = ae \tag{119}$$

To establish Kepler's Third Law, recall that the area of an ellipse with semi-axes a and b is

$$A = \pi ab = \pi a^2 (1 - e^2)^{1/2}$$

from Eq. (119). Dividing the area by the constant rate $dA/dt = h/2$, we obtain the planet's period T:

$$T = \frac{2A}{h} = \frac{2\pi a^2 (1 - e^2)^{1/2}}{h}$$

Since $a(1 - e^2) = ed = h^2/M$, then $(1 - e^2)^{1/2} = h/(Ma)^{1/2}$, and so

$$T = \frac{2\pi a^2}{h} \cdot \frac{h}{(Ma)^{1/2}} = \frac{2\pi a^{3/2}}{M^{1/2}}$$

in agreement with Kepler's Third Law.

Exercises III-5

1. Show that a planet whose orbit is circular travels with constant speed.

2. Show that an object moving with constant speed in a circular orbit of radius r has speed $v = r\omega$ and acceleration of magnitude $r\omega^2 = v^2/r$, where $\omega = d\theta/dt$.

3. In Figure III-6, assume the height of the mountain is negligible compared to $R_\oplus \approx 6.37 \times 10^6$ m. If air resistance is neglected, what muzzle velocity is required to send the cannon ball into a circular orbit about the earth?

4. Show that for a planet moving about M in a circular orbit the speed v and radius r are related by $v^2 = M/r$.

5. A satellite is said to be in *synchronous orbit* if it continually remains above the same spot on the surface of the rotating earth. For a circular orbit, this means that the satellite's angular velocity $\omega = d\theta/dt$ (with respect to, say, a locally inertial observer in free fall relative to the sun) matches the earth's own angular velocity,

$$\Omega = 2\pi/(24 \times 3600 \times 3 \times 10^{10}) \text{ radians/cm of time}$$

What is the altitude of such a satellite? ($M_\oplus \approx 0.44$ cm, $R_\oplus \approx 6.37 \times 10^8$ cm; Ω has the units cm^{-1}, since radians are dimensionless.)

6. To a reasonable approximation, the moon travels about the earth in a circular orbit of radius $r \approx 3.84 \times 10^{10}$ cm and period of about 27 1/3 days. Compute the magnitude a of the moon's acceleration relative to the earth and show that

$$\frac{a}{g} \approx \frac{1/r^2}{1/R_\oplus{}^2}$$

with an error of less than 1% (under our assumptions).

7. Let g(r) denote the magnitude of the acceleration due to the sun's gravitational field at a distance r from the sun's center. Prove that Kepler's Third Law (T^2 proportional to r^3) implies that g(r) is proportional to $1/r^2$.

8. Estimate the mass of the earth (in grams), using Exercise 3 of Section 4 and the moon's acceleration computed in Exercise 6.

9. Using the conversion formulas

$$x = r\cos\theta, \quad y = r\sin\theta, \quad x^2 + y^2 = r^2$$

convert Eq. (117) to Cartesian coordinates and show that if $e < 1$ the equation of an ellipse,

$$\frac{(x+c)^2}{a^2} + \frac{y^2}{b^2} = 1$$

results, where a, b, and c are defined by Eq. (119). Show that $c^2 = a^2 - b^2$. ($e = c/a$ is the ellipse's eccentricity.)

10. Let r_{min} and r_{max} denote, respectively, the smallest and largest values of r in a Newtonian orbit, Eqs. (116), (117). Show that

$$r_{min} = a(1-e), \quad r_{max} = a(1+e)$$

11. Show that the Newtonian orbit, Eq. (117), may be expressed in the equivalent form

$$r = \frac{r_{min}(1+e)}{1 + e\cos\theta}$$

6. GEODESICS

As we saw in Section 2, general relativity interprets the motion of a planet or other free object not as a response to a force, but as natural unconstrained motion along a geodesic in four dimensions. Space-time is viewed as a semi-Riemannian 4-manifold, with metric form

$$(d\tau)^2 = g_{\mu\nu}dx^\mu dx^\nu$$

in each coordinate system (x^0, x^1, x^2, x^3). (In order to agree with most modern texts on relativity theory, we henceforth use Greek letters for all indices which take on the values 0, 1, 2, 3. We also denote general coordinates by x^0, x^1, x^2, x^3, instead of u^0, u^1, u^2, u^3, used in Section II-9.)

A vector $\mathbf{v} = v^\mu \, \partial/\partial x^\mu$ will be called *timelike, lightlike,* or *spacelike* if the number

$$\langle \mathbf{v},\mathbf{v} \rangle = g_{\mu\nu} v^\mu v^\nu$$

is, respectively, positive, zero, or negative. In this chapter, we shall adopt the following definition of *geodesic*.

Definition III-1

A spacetime curve $\boldsymbol{\alpha}$ is a geodesic if it has a parametrization $x^\lambda(\rho)$ satisfying

$$\frac{d^2 x^\lambda}{d\rho^2} + \Gamma^\lambda_{\mu\nu} \frac{dx^\mu}{d\rho} \frac{dx^\nu}{d\rho} = 0, \qquad \lambda = 0, 1, 2, 3 \tag{120}$$

It can be shown that this definition is independent of a choice of coordinate system. Equations (120) imply that the function

$$\langle \boldsymbol{\alpha}', \boldsymbol{\alpha}' \rangle = \left(\frac{d\tau}{d\rho}\right)^2 = g_{\mu\nu} \frac{dx^\mu}{d\rho} \frac{dx^\nu}{d\rho}$$

is a constant (see the last part of the proof of Theorem I-10). Call this constant C^2. We call α *timelike, lightlike,* or *spacelike* according as C^2 is positive, zero, or negative, respectively.

If $\boldsymbol{\alpha}$ is timelike, we may integrate $d\tau/d\rho = \pm C$, solve for ρ, and deduce $\rho = a\tau + b$ for some real numbers a and b. We shall assume $a > 0$, so that $\boldsymbol{\alpha}$ is what is called "future-directed."* It is then a simple matter to reparametrize $\boldsymbol{\alpha}$ in terms of proper time, and then Eq. (120) holds with ρ replaced by τ.

If $\boldsymbol{\alpha}$ is lightlike, τ is constant along $\boldsymbol{\alpha}$ and $\langle \boldsymbol{\alpha}', \boldsymbol{\alpha}' \rangle = 0$. In this case proper time cannot be used as a parameter.

If $\boldsymbol{\alpha}$ is spacelike, $d\tau/d\rho$ is imaginary and the curve may be reparametrized in terms of proper distance (cf. Section II-6): $\rho = a\sigma + b$. We shall not need this case, since no signal or material object can travel a spacelike path.

---

*Otherwise, we would have a path corresponding to a particle traveling backward in time.

A curve $\boldsymbol{\alpha}$ (not necessarily a geodesic) is called *timelike* if $\langle \boldsymbol{\alpha}', \boldsymbol{\alpha}' \rangle > 0$ at each of its points. Theorems I-9 and I-10 have the following analogues for spacetime.

Theorem III-2

Let $\boldsymbol{\alpha}$ be a timelike curve which extremizes the spacetime distance (i.e., the lapse of proper time) between its two end points. Then $\boldsymbol{\alpha}$ is a geodesic.

Proof: Parametrize in terms of τ and mimic the proof of Theorem I-9. (See Exercise 1.)

Theorem III-3

Given an event **P** and a non-zero vector **v** at **P**, there exists a unique geodesic $\boldsymbol{\alpha} = \boldsymbol{\alpha}(\rho)$ such that $\boldsymbol{\alpha}(0) = \mathbf{P}$ and $\boldsymbol{\alpha}'(0) = \mathbf{v}$.

For **v** timelike, this theorem embodies the principle that all particles fall with the same acceleration in the same gravitational field, since the spacetime path depends only on the initial conditions $\boldsymbol{\alpha}(0)$ and $\boldsymbol{\alpha}'(0)$.

Einstein postulated that the world-line of a free particle is a timelike geodesic in spacetime. Moreover, the world-line of a light photon is a lightlike geodesic. The coefficients $\Gamma^{\lambda}_{\mu\nu}$ in Eq. (120) determine the deviation from the "straight" path $d^2 x^{\lambda}/d\rho^2 = 0$ followed in the absence of the gravitational field. We may therefore think of the Christoffel symbols as the components of the gravitational field (relative to the given coordinate system). In the next section, we will see how the metric coefficients $g_{\mu\nu}$ are analogous to the gravitational potential function of Newton's theory.

Exercise III-6

A curve $\boldsymbol{\alpha}$ (not necessarily a geodesic) is *timelike* if it has a parametrization $x^{\lambda}(\rho)$ for which

$$g_{\mu\nu} \frac{dx^{\mu}}{d\rho} \frac{dx^{\nu}}{d\rho} > 0$$

for all ρ. Show that a timelike curve can be reparametrized in terms of arc length, $x^{\lambda}(\tau)$, and that $g_{\mu\nu}(dx^{\mu}/d\tau)(dx^{\nu}/d\tau)$ is then constant. What is this constant?

7. THE FIELD EQUATIONS

In stating that particles follow spacetime geodesics, we are describing how the geometry of spacetime influences matter. This is only half the story. To complete the theory, it is necessary to describe also how matter determines the geometry: we need a set of equations relating the metric coefficients $g_{\mu\nu}$ to the distribution of matter. These are the famous *field equations* of Einstein, first given in his 1916 paper, *The Foundation of the General Theory of Relativity* [35]. In this section we shall outline some of the reasoning which led Einstein to these equations.

We shall consider in detail only the case of the gravitational field outside an isolated spherically symmetric mass M, as, e.g., the field produced by the sun (we will ignore the perturbing effects of the planets on this field). We shall be able to solve the field equations in this case and compute the orbit of a planet and the path of a light ray. In order to gain some insight into why Einstein chose the particular equations he did, let us first take another look at Newton's Law of Gravitation.

Consider a point mass M at the origin of a 3-dimensional Cartesian coordinate system, x, y, z or x^1, x^2, x^3. (Both notations are used interchangeably below.) Let $\mathbf{X} = (x, y, z)$ and

$$r = (x^2 + y^2 + z^2)^{1/2} = \|\mathbf{X}\|$$

Let $\mathbf{u}_r$ be the unit "radial vector," $(1/r)\mathbf{X}$. By Newton's law, the force $\mathbf{F}$ on a particle m located at $\mathbf{X}$ is (in geometric units)

$$\mathbf{F} = -\frac{Mm}{r^2}\mathbf{u}_r$$

Combining this with Newton's Second Law,

$$\mathbf{F} = m\frac{d^2\mathbf{X}}{dt^2}$$

we have

$$\frac{d^2\mathbf{X}}{dt^2} = -\frac{M}{r^2}\mathbf{u}_r$$

Define the *potential function* $\Phi = \Phi(r)$ by

$$\Phi = -\frac{M}{r}, \quad r > 0 \tag{121}$$

Since

$$\frac{\partial r}{\partial x^i} = \frac{\partial}{\partial x^i}(\mathbf{X}\cdot\mathbf{X})^{1/2} = \frac{2x^i}{2(\mathbf{X}\cdot\mathbf{X})^{1/2}} = \frac{x^i}{r}, \quad i = 1,2,3$$

and $\partial\Phi/\partial x^i = (\partial\Phi/\partial r)(\partial r/\partial x^i)$, we have

$$-\nabla\Phi = -\left(\frac{\partial\Phi}{\partial x}, \frac{\partial\Phi}{\partial y}, \frac{\partial\Phi}{\partial z}\right) = -\frac{M}{r^2}\left(\frac{x}{r}, \frac{y}{r}, \frac{z}{r}\right)$$

$$= -\frac{M}{r^2}\mathbf{u}_r = \frac{d^2\mathbf{X}}{dt^2}$$

or, in coordinates,

$$\frac{d^2x^i}{dt^2} = -\frac{\partial\Phi}{\partial x^i}, \quad i = 1,2,3 \tag{122}$$

Differentiating the relation

$$\frac{\partial\Phi}{\partial x^i} = \frac{Mx^i}{r^3}$$

we obtain

$$\frac{\partial^2\Phi}{(\partial x^i)^2} = M\left[\frac{r^3 - 3(x^i)^2 r}{r^6}\right] = \frac{M}{r^5}(r^2 - 3(x^i)^2)$$

and then summing, we obtain Laplace's equation,

$$\nabla^2\Phi = \frac{\partial^2\Phi}{\partial x^2} + \frac{\partial^2\Phi}{\partial y^2} + \frac{\partial^2\Phi}{\partial z^2} = 0 \tag{123}$$

valid everywhere but at the origin (where the mass is located). This is the equation for the potential function in the "empty space" surrounding an isolated point mass (or outside a spherically symmetric homogeneous mass distribution).

In the case of a finite number of point masses, Eqs. (122) and (123) are still valid in the region between the masses, but the potential function Φ is now a sum of terms, each contributed by one of the masses. For a continuous distribution of matter throughout a region of space, it can be shown that Laplace's equation is replaced by Poisson's equation

$$\nabla^2 \Phi = 4\pi\rho, \tag{124}$$

where $\rho = \rho(\mathbf{X})$ is the density of matter at $\mathbf{X}$ (see Harris [25]). Once Eq. (124) is solved for the potential Φ—a difficult task except in special cases—the acceleration of any free particle, and hence the gravitational force, can be found from Eq. (122).

Equations (122), which contain the first partial derivatives of the potential function Φ, are replaced in general relativity by the equations

$$\frac{d^2 x^\lambda}{d\tau^2} + \Gamma^\lambda_{\mu\nu} \frac{dx^\mu}{d\tau} \frac{dx^\nu}{d\tau} = 0, \quad \lambda = 0,1,2,3 \tag{125}$$

which contain the first partial derivatives of the metric coefficients $g_{\mu\nu}$, since

$$\Gamma^\lambda_{\mu\nu} = \frac{1}{2} g^{\lambda\beta} \left(\frac{\partial g_{\mu\beta}}{\partial x^\nu} + \frac{\partial g_{\nu\beta}}{\partial x^\mu} - \frac{\partial g_{\mu\nu}}{\partial x^\beta} \right) \tag{126}$$

In a sense then, the metric coefficients play the role of gravitational potential functions in Einstein's theory. The field equations, which relate these potential functions to the distribution of matter, correspond to Eq. (124)—or Eq. (123) in the case of the empty space outside a given mass distribution.

In keeping with this analogy, we might expect the field equations in empty space to be a system of equations of the form

$$G = 0 \tag{127}$$

where G is some expression involving the second partial derivatives of the "potentials" $g_{\mu\nu}$. In view of the discussion in Section 2, we would expect also that G would have something to do with curvature. Another important consideration is that Eq. (127), like Eq. (125), should have the same form in every coordinate system. [In technical terms, Eq. (127) should be a tensor equation.] Now it turns out that the only tensors that are constructible from the metric coefficients $g_{\mu\nu}$ and their first and second derivatives are those that are functions of the $g_{\mu\nu}$ and the components $R^{\lambda}_{\mu\nu\sigma}$ of the curvature tensor, introduced in Section I-8:

$$R^{\lambda}_{\mu\nu\sigma} = \frac{\partial \Gamma^{\lambda}_{\mu\sigma}}{\partial x^{\nu}} - \frac{\partial \Gamma^{\lambda}_{\mu\nu}}{\partial x^{\sigma}} + \Gamma^{\beta}_{\mu\sigma}\,\Gamma^{\lambda}_{\nu\beta} - \Gamma^{\beta}_{\mu\nu}\,\Gamma^{\lambda}_{\beta\sigma} \tag{128}$$

In addition, the field equations should have as one possible solution, the flat spacetime of special relativity. Here, in suitable coordinates (e.g., Lorentz coordinates) all the $g_{\mu\nu}$ are constants, and therefore, from Eqs. (126) and (128),

$$R^{\lambda}_{\mu\nu\sigma} = 0, \qquad \lambda,\mu,\nu,\sigma = 0,1,2,3 \tag{129}$$

(in fact in every coordinate system).

System (129) might therefore be a good candidate for the field equations. However, this condition is too restrictive since it can be shown (conversely) that (129) implies the existence of a coordinate system in which the $g_{\mu\nu}$ are constants. This in turn leads to the flat spacetime of special relativity. To allow for essential gravitational fields that cannot be transformed away (curved spacetime), we need a less restrictive condition.

If we set $\sigma = \lambda$ in Eq. (129) and then sum over λ, we obtain the components of the so-called *Ricci tensor,*

$$R_{\mu\nu} = R^{\lambda}_{\mu\nu\lambda} = \frac{\partial \Gamma^{\lambda}_{\mu\lambda}}{\partial x^{\nu}} - \frac{\partial \Gamma^{\lambda}_{\mu\nu}}{\partial x^{\lambda}} + \Gamma^{\beta}_{\mu\lambda}\,\Gamma^{\lambda}_{\nu\beta} - \Gamma^{\beta}_{\mu\nu}\,\Gamma^{\lambda}_{\beta\lambda} \tag{130}$$

Einstein chose, as his field equations for the gravitational field in empty space (outside matter),

$$R_{\mu\nu} = 0, \qquad \mu,\nu = 0,1,2,3 \tag{131}$$

These are second order partial differential equations to be solved for the $g_{\mu\nu}$. [In the case of a continuous distribution of matter throughout space, the right hand side of Eq. (131) is replaced by the components of a tensor having to do with the energy, density, and pressure of matter. Thus, the field equations really do relate the geometry to the distribution of matter.]

The choice of Eq. (131) no doubt involved some guesswork, but not as much as one would infer from the preceding sketch. There were other considerations, both mathematical and physical, that greatly limited the number of possibilities. For example, Einstein had reason to believe that G in Eq. (127) had two subscripts and was a symmetric tensor ($G_{\mu\nu} = G_{\nu\mu}$). The proof that the Ricci tensor is symmetric will be given later. Moreover, in the continuous mass distribution case, the requirement that the field equations must reduce to Eq. (123) in the limit of a very weak field and low velocities gave an additional clue. We cannot go into the details here. Of course, the ultimate justification for Eq. (131) is the extent of agreement between the theoretical predictions and observed phenomena.

For a slow moving particle in a weak static gravitational field, the predictions of general relativity must agree with those of classical physics to a first approximation. Let us see what this agreement says about the metric coefficients.

Let (x^0, x^1, x^2, x^3) be a locally Lorentz coordinate system in a neighborhood of an event on a moving particle's world-line. For convenience, relabel the time coordinate $t = x^0$.

If the particle's velocity is sufficiently small ($|dx^i/dt| \ll 1$ for $i = 1,2,3$), we may neglect $dx^i/d\tau$ compared to $dt/d\tau$ and write the geodesic equations as

$$\frac{d^2 x^\lambda}{d\tau^2} = - \Gamma^\lambda_{\mu\nu} \frac{dx^\mu}{d\tau} \frac{dx^\nu}{d\tau} \approx - \Gamma^\lambda_{00} \left(\frac{dt}{d\tau}\right)^2 \tag{132}$$

Since we are assuming the gravitational field is *static* (time derivatives of the metric coefficients vanish), formula (126) for the Christoffel symbols gives

$$\Gamma^\lambda_{00} = - \frac{1}{2} g^{\beta\lambda} \frac{\partial g_{00}}{\partial x^\beta}$$

Since the field is weak, we may choose locally Lorentz coordinates satisfying

$$g_{\mu\nu} = \eta_{\mu\nu} + h_{\mu\nu}$$

where

$$(\eta_{\mu\nu}) = \begin{pmatrix} 1 & 0 & 0 & 0 \\ 0 & -1 & 0 & 0 \\ 0 & 0 & -1 & 0 \\ 0 & 0 & 0 & -1 \end{pmatrix}$$

and the $h_{\mu\nu}$ are small compared to unity. Then the matrix inverse of $(g_{\mu\nu})$ has entries of the form $g^{\mu\nu} = \eta^{\mu\nu} + k^{\mu\nu}$, where $\eta^{\mu\nu} = \eta_{\mu\nu}$—the matrix $(\eta_{\mu\nu})$ is its own inverse—and the $k_{\mu\nu}$ are small compared to unity (they are power series in the $h_{\mu\nu}$ with zero constant term). Accordingly, we have

$$\Gamma^{\lambda}_{00} \approx -\frac{1}{2}\eta^{\beta\lambda}\frac{\partial h_{00}}{\partial x^{\beta}}, \tag{133}$$

to first order in the $h_{\mu\nu}$.

Substituting Eq. (133) into (132), we obtain (to a close approximation)

$$\frac{d^2 x^{\lambda}}{d\tau^2} = \frac{1}{2}\left(\frac{dt}{d\tau}\right)^2 \eta^{\alpha\lambda}\frac{\partial h_{00}}{\partial x^{\alpha}} \tag{134}$$

For $\lambda = 0$, this yields $d^2 t/d\tau^2 = 0$, or $dt/d\tau$ = constant. Consequently,

$$\frac{dx^i}{dt} = \frac{dx^i}{d\tau}\bigg/\frac{dt}{d\tau}, \quad \frac{d^2 x^i}{dt^2} = \frac{d^2 x^i}{d\tau^2}\bigg/\left(\frac{dt}{d\tau}\right)^2$$

and from Eq. (134) with $\lambda = i \neq 0$, we then obtain

$$\frac{d^2 x^i}{dt^2} = \frac{1}{2}\eta^{\alpha i}\frac{\partial h_{00}}{\partial x^{\alpha}} = -\frac{1}{2}\frac{\partial h_{00}}{\partial x^i}, \quad i = 1,2,3$$

If we compare the latter with the Newtonian

$$\frac{d^2 x^i}{dt^2} = -\frac{\partial \Phi}{\partial x^i}$$

we see that the two theories will agree provided

$$\frac{\partial h_{00}}{\partial x^i} = 2\frac{\partial \Phi}{\partial x^i}, \qquad i = 1,2,3$$

or

$$h_{00} = 2\Phi + C$$

for some constant C.

Now the Newtonian potential function Φ is determined up to an arbitrary additive constant. We have followed the convention of choosing this constant so as to make Φ vanish at infinity. Since $h_{\mu\nu}$ should also vanish at infinity (spacetime flat far away from the source of gravity), we must have $C = 0$, and so $h_{00} = 2\Phi$, and

$$g_{00} = 1 + 2\Phi \tag{135}$$

In brief then, the low speed, weak field agreement between Newtonian theory and general relativity requires that Eq. (135) hold in locally Lorentz coordinates.

We conclude this section with a lemma that will allow us to write the Ricci tensor in a more convenient form.*

Lemma III-4

For each μ,

$$g^{\lambda\beta}\frac{\partial g_{\lambda\beta}}{\partial x^\mu} = \frac{1}{g}\frac{\partial g}{\partial x^\mu} = \frac{\partial}{\partial x^\mu}\ln|g|$$

*Those wishing to omit the details may accept Eq. (140) on faith and move on to Section 8.

Proof. Since $g_{\lambda\beta}g^{\beta\sigma} = \delta^{\sigma}_{\lambda} = 1$ if $\sigma = \lambda$, we have (summing over β only)

$$\sum_{\beta} g_{\lambda\beta}gg^{\beta\lambda} = g \tag{136}$$

for each fixed λ. This equation is actually the expansion of the determinant g along the λth row, and $gg^{\beta\lambda}$ is the cofactor of $g_{\beta\lambda}$.

Now there is another expression for the determinant, namely,

$$g = \sum_{\pi} \operatorname{sign}(\pi)\, g_{0\pi(0)}\, g_{1\pi(1)}\, g_{2\pi(2)}\, g_{3\pi(3)} \tag{137}$$

where π ranges over all permutations of $\{0, 1, 2, 3\}$. Differentiation of the latter with respect to x^{μ} gives

$$\frac{\partial g}{\partial x^{\mu}} = \sum_{\pi} \operatorname{sign}(\pi)\, \frac{\partial g_{0\pi(0)}}{\partial x^{\mu}}\, g_{1\pi(1)}\, g_{2\pi(2)}\, g_{3\pi(3)} + \cdots$$

$$+ \sum_{\pi} \operatorname{sign}(\pi)\, g_{0\pi(0)}\, g_{1\pi(1)}\, g_{2\pi(2)}\, \frac{\partial g_{3\pi(3)}}{\partial x^{\mu}} \tag{138}$$

The first term in this sum is similar in form to the right side of Eq. (137), and is in fact the determinant of $(g_{\mu\nu})$ with the first (or zeroth) row replaced by

$$\frac{\partial g_{00}}{\partial x^{\mu}}, \quad \frac{\partial g_{01}}{\partial x^{\mu}}, \quad \frac{\partial g_{02}}{\partial x^{\mu}}, \quad \frac{\partial g_{03}}{\partial x^{\mu}}$$

Applying Eq. (136) (with $\lambda = 0$) to this modified matrix [each of whose zeroth row cofactors is the same as those of $(g_{\mu\nu})$], we conclude that

$$\sum_{\pi} \operatorname{sign}(\pi)\, \frac{\partial g_{0\pi(0)}}{\partial x^{\mu}}\, g_{1\pi(1)}\, g_{2\pi(2)}\, g_{3\pi(3)} = \frac{\partial g_{0\beta}}{\partial x^{\mu}}\, gg^{0\beta}$$

Similarly, the second, third, and fourth terms in Eq. (138) are

$$\frac{\partial g_{1\beta}}{\partial x^\mu}\, g g^{1\beta}, \qquad \frac{\partial g_{2\beta}}{\partial x^\mu}\, g g^{2\beta}, \qquad \frac{\partial g_{3\beta}}{\partial x^\mu}\, g g^{3\beta}$$

and so the entire sum is

$$\frac{\partial g}{\partial x^\mu} = \frac{\partial g_{\lambda\beta}}{\partial x^\mu}\, g g^{\lambda\beta} \tag{139}$$

from which the statement of the lemma is immediate.

We can apply this lemma to rewrite formula (130) for the Ricci tensor. Setting $\nu = \lambda$ in Eq. (126) and summing over λ, we obtain

$$\Gamma^\lambda_{\mu\lambda} = \frac{1}{2} g^{\lambda\beta} \left(\frac{\partial g_{\mu\beta}}{\partial x^\lambda} + \frac{\partial g_{\lambda\beta}}{\partial x^\mu} - \frac{\partial g_{\mu\lambda}}{\partial x^\beta} \right) = \frac{1}{2} g^{\lambda\beta} \frac{\partial g_{\lambda\beta}}{\partial x^\mu}$$

$$= \frac{1}{2} \frac{\partial}{\partial x^\mu} \ln|g| = \frac{\partial}{\partial x^\mu} \ln|g|^{1/2}$$

since the first and third terms in parentheses cancel in the sum (we can interchange λ and β in one of these terms because $g^{\lambda\beta} = g^{\beta\lambda}$). Substituting this result into Eq. (130) gives

$$R_{\mu\nu} = \frac{\partial^2}{\partial x^\mu \partial x^\nu} \ln|g|^{1/2} - \frac{\partial \Gamma^\lambda_{\mu\nu}}{\partial x^\mu} + \Gamma^\beta_{\mu\lambda} \Gamma^\lambda_{\nu\beta}$$

$$- \Gamma^\beta_{\mu\nu} \frac{\partial}{\partial x^\beta} \ln|g|^{1/2} \tag{140}$$

It is clear from this formula that $R_{\mu\nu} = R_{\nu\mu}$, as claimed earlier.

Exercises III-7

1. Deduce from Eq. (122) that a Newtonian orbit satisfies

$$\frac{v^2}{2} + \Phi = \text{constant}$$

where

$$v = \left[\sum_{i=1}^{3} (dx^i/dt)^2 \right]^{1/2}$$

[Note: according to classical physics, an orbiting particle of mass m has kinetic energy $mv^2/2$ and potential energy $m\Phi$. The above result is the classical Law of Conservation of Energy: kinetic energy plus potential energy (i.e., total energy) is constant.]

2. Show that in the case of a circular orbit of radius r, the constant of the previous exercise equals $-M/(2r)$.

3. Show that a negative total energy (see Exercise 1) signifies a bound orbit: the planet can never escape the attracting mass M.

4. (a) A particle of mass m falls freely in a uniform gravitational field, so that it undergoes constant acceleration g relative to an observer initially at rest relative to the particle. After time t, the particle's speed and distance fallen are given by $v = gt$, $h = gt^2/2$. By Conservation of Energy (cf. Exercise 1), the gain in kinetic energy, $mv^2/2$, equals the loss, $-\Delta\Phi$, in potential energy. Show that

$$\Delta\Phi = -gh$$

(b) Suppose a particle is in free fall above the earth's surface. The potential is now given by $\Phi = -M_\oplus/r$, where r is the distance from the earth's center. Show that if the particle falls a distance $\Delta r = -h$ near the earth's surface ($r \approx R_\oplus$), then the change in potential satisfies

$$\Delta\Phi \approx -gh$$

in close agreement with the result of (a), which is exact when the gravitational field is uniform.

5. (*Gravitational Redshift*) Consider once again the windowless rocket of Section 1. Suppose that, as the rocket begins accelerating (with constant acceleration with respect to a nearby inertial observer), a scientist on the rocket floor sends a light signal to another scientist on the ceiling. Let Δt_f be the period of the emitted light (time between successive wave crests), as measured by the floor observer. If the height of the rocket is h, then the light travel time from ceiling to floor is approximately $t = h$, in geometric

units. During this time, the rocket will have acquired a speed $v = gt = gh$ relative to the floor observer's initial position at the time the light was emitted (i.e., relative to an inertial observer initially at rest relative to the floor observer). By the Doppler effect [formula (98) of Section II-10], the time Δt_c between successive wave crest receptions, as measured by the ceiling observer, is

$$\Delta t_c = \left(\frac{1 + gh}{1 - gh}\right)^{1/2} \Delta t_f$$

Using the binomial approximation $(1 \pm x)^{1/2} \approx 1 \pm x/2$, for x small, show that

$$\Delta t_c \approx (1 + gh)\,\Delta t_f$$

The light is thus observed to have a longer period and hence a longer wavelength, which means a shift toward the red end of the visible light spectrum. (Since the ceiling and floor observers are not inertial observers in relative uniform motion, the above derivation is not very rigorous. An alternative derivation will appear in the exercises of Section 8.)

By the Equivalence Principle, a similar redshift should be observed in a laboratory on earth when light is beamed from the floor to the ceiling. For this situation, we can use the results of Exercise 4 to write the relationship above as

$$\Delta t_c \approx (1 + \Delta\Phi)\,\Delta t_f$$

where $\Delta\phi = \phi(\text{ceiling}) - \phi(\text{floor}) = gh$. If the light were sent downward, from ceiling to floor, the effect would be a fall in potential $(\Delta\Phi = -gh)$ and a resulting decrease in wavelength or a blueshift.

6. Show that for locally Lorentz coordinates in a weak, static field

$$\Gamma^i_{00} = \frac{\partial\Phi}{\partial x^i}, \qquad R^i_{0j0} = \frac{\partial^2\Phi}{\partial x^i \partial x^j}$$

[Hint: in locally Lorentz coordinates, Eq. (105) and continuity imply that the $g_{\mu\nu}$ have small derivatives. Therefore, products such as $\Gamma^\beta_{\mu\nu}\,\Gamma^\lambda_{\nu\beta}$ may be neglected.]

8. THE SCHWARZSCHILD SOLUTION

Here we shall solve Einstein's equations,

$$R_{\mu\nu} = 0, \quad \mu,\nu = 0,1,2,3 \tag{141}$$

for the gravitational field outside an isolated spherically symmetric mass M, assumed permanently at rest at the origin of our coordinate system.

If the mass is a typical star, like our sun, the departure from flatness is not great, and we may suppose there is a coordinate system in which the metric differs only slightly from the special relativity form

$$d\tau^2 = dt^2 - dx^2 - dy^2 - dz^2 \tag{142}$$

It will be convenient to replace the spatial Cartesian coordinates x, y, and z by spherical coordinates ρ, ϕ, and θ. These are related to Cartesian coordinates by the transformation

$$\begin{aligned} x &= \rho \sin\phi \cos\theta \\ y &= \rho \sin\phi \sin\theta \\ z &= \rho \cos\phi \end{aligned} \tag{143}$$

[The geometric significance of these coordinates is indicated in Fig. III-8: if we denote (x, y, z) by **X**, then $\rho = \|\mathbf{X}\|$, ϕ is the angle between **X** and the positive z axis, and θ is the angle between the vector (x, y, 0) and the positive x axis.]

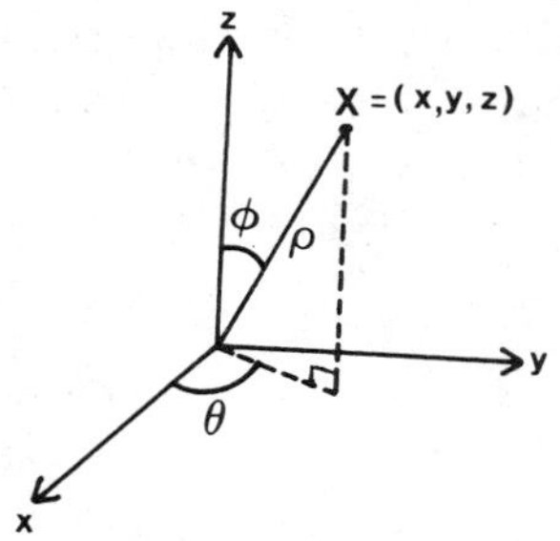

Figure III - 8.

When dx, dy, and dz are computed from Eq. (143) and substituted into Eq. (142), the result is

$$d\tau^2 = dt^2 - d\rho^2 - \rho^2 d\phi^2 - \rho^2 \sin^2\phi \, d\theta^2 \tag{144}$$

This is still, of course, the metric of *flat* spacetime. Our approach will be to modify Eq. (144) so as to produce the metric of curved spacetime surrounding our mass M. The derivation that follows is not entirely rigorous, but it does not have to be—as long as the resulting metric form is a solution to the field equations. The ultimate test of course is how well the predictions of the theory are borne out by experiment.

Since the metric we seek should be static (i.e., with coefficients independent of t) and spherically symmetric, it is reasonable to assume the interval is given by an expression of the form

$$d\tau^2 = U(\rho)dt^2 - V(\rho)d\rho^2 - W(\rho)(\rho^2 d\phi^2 + \rho^2 \sin^2\phi \, d\theta^2) \tag{145}$$

where U, V, and W are functions of ρ only and differ slightly from unity. If we introduce the variable $r = \rho W(\rho)^{1/2}$, then we can put the metric in the form

$$d\tau^2 = A(r)dt^2 - B(r)dr^2 - r^2 d\phi^2 - r^2 \sin^2\phi \, d\theta^2 \tag{146}$$

This form contains only two unknown functions for us to determine.

Note: since the functions U, V, W, A, and B all differ slightly from unity, both r in Eq. (146) and ρ in Eq. (145) are approximate counterparts of the radial distance, ρ in Eq. (144). However, since our spacetime is curved, no variable—neither r nor ρ—can have exactly the properties of distance from the origin in Euclidean space. Therefore, we may choose to view r rather than ρ, as the analog of the radial distance of Newtonian theory.

Finally, define the functions m = m(r) and n = n(r) by

$$e^{2m} = A, \qquad e^{2n} = B$$

We then have

$$d\tau^2 = e^{2m} dt^2 - e^{2n} dr^2 - r^2 d\phi^2 - r^2 \sin^2\phi \, d\theta^2 \tag{147}$$

Comparing this with the general expression

$$d\tau^2 = g_{\mu\nu}\, dx^\mu\, dx^\nu$$

and labeling

$$x^0 = t, \quad x^1 = r, \quad x^2 = \phi, \quad x^3 = \theta$$

we have

$$(g_{\mu\nu}) = \begin{pmatrix} e^{2m} & 0 & 0 & 0 \\ 0 & -e^{2n} & 0 & 0 \\ 0 & 0 & -r^2 & 0 \\ 0 & 0 & 0 & -r^2 \sin^2 \phi \end{pmatrix} \tag{148}$$

and determinant

$$g = -\, e^{2m+2n}\, r^4 \sin^2 \phi$$

Our task is to determine the functions m and n.

To this end, we next compute the Christoffel symbols, Eq. (126), in terms of m and n. Since our coordinate system is orthogonal ($g_{\mu\nu} = 0$ for $\mu \neq \nu$), we have $g^{\mu\mu} = 1/g_{\mu\mu}$ for all μ, and $g^{\mu\nu} = 0$ if $\mu \neq \nu$. Equation (126) then reduces to

$$\Gamma^\lambda_{\mu\nu} = \frac{1}{2g_{\lambda\lambda}} \left(\frac{\partial g_{\mu\lambda}}{\partial x^\nu} + \frac{\partial g_{\nu\lambda}}{\partial x^\mu} - \frac{\partial g_{\mu\nu}}{\partial x^\lambda} \right) \quad \text{(no sum)} \tag{149}$$

As you can check, there are now only three cases to consider [compare with Eq. (39) in Section I-7]:

Case 1: For $\lambda = \nu$,

$$\Gamma^\nu_{\mu\nu} = \frac{1}{2} \frac{\partial}{\partial x^\mu} \ln |g_{\nu\nu}| \quad \text{(no sum)} \tag{150}$$

Case 2: For $\mu = \nu \neq \lambda$,

$$\Gamma^\lambda_{\mu\mu} = -\frac{1}{2g_{\lambda\lambda}} \frac{\partial g_{\mu\mu}}{\partial x^\lambda} \quad \text{(no sum)} \tag{151}$$

Case 3: For μ, ν, λ all distinct,

$$\Gamma^{\lambda}_{\mu\nu} = 0 \tag{152}$$

It is now simple (though tedious) to verify that the only non-zero Christoffel symbols are the following:

$$\begin{aligned}
&\Gamma^0_{10} = \Gamma^0_{01} = m', && \Gamma^1_{00} = m'e^{2m-2n} \\
&\Gamma^1_{11} = n', && \Gamma^1_{22} = -re^{-2n} \\
&\Gamma^2_{12} = \Gamma^2_{21} = 1/r, && \Gamma^1_{33} = -re^{-2n}\sin^2\phi \\
&\Gamma^3_{13} = \Gamma^3_{31} = 1/r, && \Gamma^3_{23} = \Gamma^3_{32} = \cot\phi \\
& && \Gamma^2_{33} = -\sin\phi\cos\phi
\end{aligned} \tag{153}$$

These expressions, together with the formula

$$\ln|g|^{1/2} = \frac{1}{2}\ln(e^{2m+2n}r^4\sin^2\phi) = m + n + 2\ln r + \ln\sin\phi$$

are to be substituted into Einstein's field equations

$$R_{\mu\nu} = \frac{\partial^2}{\partial x^\mu \partial x^\nu}\ln|g|^{1/2} - \frac{\partial\Gamma^\lambda_{\mu\nu}}{\partial x^\lambda} + \Gamma^\beta_{\mu\lambda}\Gamma^\lambda_{\nu\beta} - \Gamma^\beta_{\mu\nu}\frac{\partial}{\partial x^\beta}\ln|g|^{1/2} = 0 \tag{154}$$

For example, for $\mu = \nu = 0$, we have (using primes to denote differentiation with respect to r)

$$\begin{aligned}
R_{00} &= \frac{\partial^2}{\partial t^2}\ln|g|^{1/2} - \frac{\partial\Gamma^\lambda_{00}}{\partial x^\lambda} + \Gamma^\beta_{\lambda 0}\Gamma^\lambda_{\beta 0} - \Gamma^\beta_{00}\frac{\partial}{\partial x^\beta}\ln|g|^{1/2} \\
&= 0 - \frac{\partial}{\partial r}(m'e^{2m-2n}) + (\Gamma^0_{10}\Gamma^1_{00} + \Gamma^1_{00}\Gamma^0_{10}) - \Gamma^1_{00}\frac{\partial}{\partial r}\ln|g|^{1/2} \\
&= [-m'' - m'(2m' - 2n')]e^{2m-2n} + 2m'^2 e^{2m-2n} \\
&\qquad - m'e^{2m-2n}\left(m' + n' + \frac{2}{r}\right) \\
&= \left(-m'' + m'n' - m'^2 - \frac{2m'}{r}\right)e^{2m-2n} = 0
\end{aligned}$$

The rest of the calculations are similar, and we list only the results.

$$R_{00}/e^{2m-2n} = \left(-m'' + m'n' - m'^2 - \frac{2m'}{r}\right) = 0 \tag{155a}$$

$$R_{11} = m'' - m'n' + m'^2 - \frac{2n'}{r} = 0 \tag{155b}$$

$$R_{22} = e^{-2n}(1 + rm' - rn') - 1 = 0 \tag{155c}$$

$$R_{33} = R_{22} \sin^2 \phi = 0 \tag{155d}$$

All other $R_{\mu\nu}$ are identically zero.

Now adding Eqs. (155a) and (155b), we find $m' + n' = 0$, and so $m + n = b$, a constant. However, both m and n must vanish as $r \to \infty$, since the metric (147) must approach the Lorentz metric at great distances from the mass M. Consequently, $b = 0$ and $n = - m$. Equation (155c) then becomes

$$1 = (1 + 2rm')e^{2m} = (re^{2m})'$$

Hence

$$re^{2m} = r + C$$

for some constant C, or

$$g_{00} = e^{2m} = 1 + C/r$$

However, we saw in the previous section that far from the mass, where the field is weak and Newton's theory holds, we must have $g_{00} = 1 - 2M/r$. We therefore have $C = - 2M$. We have finally arrived at the solution

$$d\tau^2 = \left(1 - \frac{2M}{r}\right)dt^2 - \left(1 - \frac{2M}{r}\right)^{-1} dr^2 - r^2\, d\phi^2 - r^2 \sin^2\phi\, d\theta^2 \tag{156}$$

originally derived by Schwarzschild in 1916, just a few months after the publication of Einstein's paper on the general theory.

Exercises III-8

1. Derive Eq. (144) from Eqs. (142) and (143).

2. Compute any three of the Christoffel symbols in Eq. (153).

3. Compute R_{11} and R_{22} in the Schwarzschild metric [see Eq. (155)].

4. Assume the interval is given by Eq. (156). By setting $d\tau = 0$ and integrating, compute the time required for a light photon to travel radially (i.e., with ϕ and θ constant) from $r = r_1$ to $r = r_2$.

5. (*Redshift in Schwarzschild metric*) Adopt Schwarzschild coordinates for the field produced by an isolated mass M. Suppose a light source located at $\mathbf{P}_1(r_1, \phi_1, \theta_1)$ sends light of a single frequency to the observer at $\mathbf{P}_2(r_2, \phi_2, \theta_2)$. (Both source and observer are at rest relative to M.) Since metric (156) is static, each wave crest requires the same amount of coordinate time to travel from $\mathbf{P}_1$ to $\mathbf{P}_2$ as does any other wave crest (cf. Exercise 4 for the case $\phi_2 = \phi_1$, $\theta_2 = \theta_1$). Therefore, the coordinate time between the reception of successive wave crests at $\mathbf{P}_2$ is the same as the coordinate time between their emission at $\mathbf{P}_1$. Call this time interval Δt.

However, good clocks (such as atomic clocks) are assumed to measure proper time, not coordinate time. By Eq. (156), when r, ϕ, and θ are constant, proper time is related to coordinate time by

$$d\tau = \left(1 - \frac{2M}{r}\right)^{1/2} dt \approx \left(1 - \frac{M}{r}\right) dt$$

Therefore, the ratio of the wavelength of the received light at $\mathbf{P}_2$ to the wavelength of the same light when emitted at $\mathbf{P}_1$ is

$$\frac{\lambda_{rec}}{\lambda_{em}} = \frac{\Delta\tau_{rec}}{\Delta\tau_{em}} = \frac{1 - M/r_2}{1 - M/r_1} \approx 1 + \left[-\frac{M}{r_2} - \left(-\frac{M}{r_1}\right)\right]$$

$$= 1 + \Delta\phi$$

where $\Delta\phi = \phi(r_2) - \phi(r_1)$, the difference in potential between $\mathbf{P}_2$ and $\mathbf{P}_1$. If $r_2 > r_1$, $\lambda_{rec} > \lambda_{em}$, and the received light is redder than when emitted. The *spectral shift* is defined to be the ratio

$$\frac{\Delta\lambda}{\lambda} = \frac{\lambda_{rec} - \lambda_{em}}{\lambda_{em}} = \frac{\lambda_{rec}}{\lambda_{em}} - 1 = \Delta\phi$$

Light emitted from the surface of a star of radius R and mass M, when observed far away from the star ($r_2 \approx \infty$) should be observed redshifted by the factor

$$\frac{\Delta\lambda}{\lambda} = \frac{M}{R}$$

Compute $\Delta\lambda/\lambda$ for light emitted from the surface of the sun ($M_\odot \approx 1.48 \times 10^5$ cm, $R_\odot \approx 1.50 \times 10^{13}$ cm) and observed on earth. (As you can check, the potential due to the earth's own field and the potential of the sun at the earth's surface are practically negligible.) Observations agree with this theoretical prediction to within about 5%.

6. The experiments of Pound and Rebka (1960) and Pound and Snider (1965) [39] increased the agreement between the theoretical and observed redshift to about 1 percent. They examined γ-rays rising a distance $h \approx 22.6$ m in a laboratory on earth. From the results of the previous exercise, the expected spectral shift is

$$\frac{\Delta\lambda}{\lambda} = \Delta\Phi \approx \left.\frac{\partial\Phi}{\partial r}\right|_{r=R_\oplus} \Delta r = \frac{M_\oplus}{R_\oplus{}^2}\, h = gh = g_{conv} h/c^2$$

Compute the predicted shift. (The details of how such a tiny quantity was observed can be found in Weinberg [50, pp. 82, 83] and Misner, Thorne, and Wheeler [36, pp. 1056, 1057].

7. (*Redshift, alternative derivation*) In the experiment of the previous exercise, we may assume the gravitational field is essentially uniform, causing all free particles to fall with constant acceleration g, or equivalently, that the building accelerates upward with acceleration g relative to a freely falling inertial observer.

By the results of Exercise 5 of Section 2, the observer at rest in the building may choose coordinates t, x, y, z in which he is at rest and the metric is

$$d\tau^2 = (1 + gz)^2\, dt^2 - dx^2 - dy^2 - dz^2$$

(The γ-rays travel in a positive z direction.) Imitating the argument of Exercise 5 above, show that the light undergoes a spectral shift

$$\frac{\Delta\lambda}{\lambda} \approx gh$$

where h is the height of the building.

8. Let t, r, ϕ, and θ be Schwarzschild coordinates. Introduce the new radial variable ρ defined by

$$\rho = \frac{1}{2}\left[r - M + (r^2 - 2Mr)^{1/2}\right]$$

Verify the following:

(a) $$r = \rho\left(1 + \frac{M}{2\rho}\right)^2$$

(b) $$dr = \left(1 - \frac{M}{4\rho^2}\right) d\rho \quad \text{and} \quad 1 - \frac{2M}{r} = \frac{\left(1 - \frac{M}{2\rho}\right)^2}{\left(1 + \frac{M}{2\rho}\right)^2}$$

(c) $$d\tau^2 = \frac{\left(1 - \frac{M}{2\rho}\right)^2}{\left(1 + \frac{M}{2\rho}\right)^2} dt^2 - \left(1 + \frac{M}{2\rho}\right)^4 (d\rho^2 + \rho^2\, d\theta^2 + \rho^2 \sin^2\phi\, d\theta^2)$$

Define x, y, and z by

$$x = \rho \sin\phi \cos\theta$$

$$y = \rho \sin\phi \cos\theta$$

$$z = \rho \cos\phi$$

Then

$$d\tau^2 = \frac{\left(1 - \frac{M}{2\rho}\right)^2}{\left(1 + \frac{M}{2\rho}\right)^2} dt^2 - \left(1 + \frac{M}{2\rho}\right)^4 (dx^2 + dy^2 + dz^2)$$

(d) Show that at any point in space the speed of light, in the coordinate system t, x, y, z is

$$\frac{1 - \dfrac{M}{2\rho}}{\left(1 + \dfrac{M}{2\rho}\right)^3}$$

in all directions. (In Schwarzschild coordinates, the speed of light at any point varies with direction.)

9. (*The Twin Paradox revisited*) As in Section II-10, Jack is an inertial observer in a space laboratory. This time however, Jill sets out from the laboratory with speed $\beta = 0.8$ on a circular path of circumference 8 light-years.

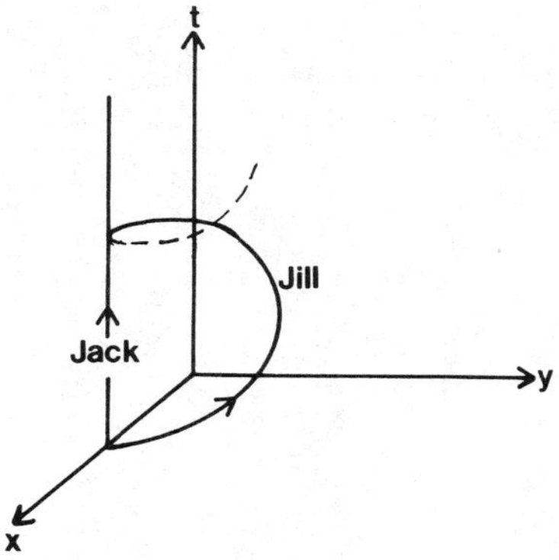

Therefore, after 10 years of Jack's time, Jill will have completed the circuit and returned to her brother. We shall orient Jack's coordinate axes so that Jill is confined to the xy-plane and Jack's lab is located on the positive x-axis. The figure shows the world-lines of Jack and Jill in this coordinate system (z is identically zero and so is suppressed).

Since Jack is an inertial observer, the interval is given in his coordinates by

$$d\tau^2 = dt^2 - dx^2 - dy^2$$

(We assume the experiment is performed far from any celestial bodies, so that spacetime is flat.) Along his world-line (where $dx = dy = 0$), proper time equals coordinate time:

$$d\tau_{\text{Jack}} = dt$$

For Jill however, her proper time along her world-line is given by

$$d\tau_{\text{Jill}}^2 = dt^2 - \left[\left(\frac{dx}{dt}\right)^2 + \left(\frac{dy}{dt}\right)^2\right] dt^2 = dt^2\,(1 - \beta^2)$$

or

$$d\tau_{\text{Jill}} = (1 - \beta^2)^{1/2}\, dt = 0.6\, dt$$

Consequently, after 10 years of Jack's time, Jill will have aged only 6 years.

Now let us analyze the situation from Jill's point of view and show that the same conclusion follows. First, make the following change of coordinates:

$$x' = x \cos \omega t + y \sin \omega t$$

$$y' = -x \sin \omega t + y \cos \omega t$$

$$t' = t$$

where $\omega = 2\pi/10$ is Jill's angular speed relative to Jack, in radians per year. The inverse transformation is

$$x = x' \cos \omega t' - y' \sin \omega t'$$

$$y = x' \sin \omega t' + y' \cos \omega t'$$

$$t = t'$$

(a) Show that Jill is at rest in the primed coordinate system, i.e., x' constant, y' constant. (b) Let $r = (x^2 + y^2)^{1/2} = (x'^2 + y'^2)^{1/2}$. Then $\beta = r(d\theta/dt) = r\omega$. Show that

$$d\tau^2 = (1 - \omega^2 r^2)\, dt'^2 - dx'^2 - dy'^2 + 2\omega y' dx' dt' - 2\omega x' dy' dt'$$

Now by (a), along Jill's world-line $dx' = dy' = 0$, and so

$$d\tau_{\text{Jill}} = (1 - \omega^2 r^2)^{1/2}\, dt' = (1 - \beta^2)^{1/2}\, dt = 0.6\, dt$$

as obtained before with Jack's coordinates. On the other hand, on Jack's world-line, $x = x_0$ (constant), $y = 0$, or

$$x' = x_0 \cos \omega t, \qquad y' = -x_0 \sin \omega t$$

(c) Deduce (from the interval formula in primed coordinates) that along Jack's world-line, $d\tau_{\text{Jack}} = dt$.

Computation with either coordinate system yields the same conclusion: when the twins are reunited, Jack will have aged 10 years, but Jill only 6 years.

10. (*Experimental Verification of the Twin Paradox*) Consider two identically constructed clocks initially together on earth. One of them remains on earth, in the laboratory. The other clock is taken aboard a jet airliner, flown around the world at altitude H, and then brought down to earth and reunited with its twin clock. Let us compare the proper times of the two clocks, using the Schwarzschild metric for the gravitational field produced by the earth ($M = M_\oplus$ and the origin is at the earth's center). For simplicity, we can assume the laboratory and jet move in the equatorial plane ($\phi = \pi/2$). Let Ω denote the angular velocity of the earth lab relative to a local inertial observer (the angular speed $d\theta/dt$ with which the earth turns on its axis).

(a) For the earth-bound clock, $r = R_\oplus$. Show that the total proper time for this clock is

$$\Delta\tau_E = \int d\tau_E = t\left[1 - \frac{2M}{R_\oplus} - R_\oplus^2\Omega^2\right]^{1/2} \approx t\left[1 - \frac{M}{R_\oplus} - \frac{R_\oplus^2\Omega^2}{2}\right]$$

where t is the coordinate time when the clocks are reunited.

(b) For the traveling clock, we may neglect the effects of ascending to and descending from altitude H, since, by having the plane circle the earth sufficiently many times, the fraction of time spent in changing altitudes can be made negligible. We therefore have $r = R_\oplus + H$, and the clock's coordinate speed is

$$(R_\oplus + H)\frac{d\theta}{dt} \approx (R_\oplus + H)\,\Omega + v$$

(The first term on the right is the speed of an object at altitude H and at rest relative to the ground below; the second term is the plane's speed relative to the ground. The equation, being an application of the classical, non-relativistic addition of velocity law, is a good approximation.)

Show that the total proper time for the clock traveling at altitude H is

$$\Delta\tau_H = \int d\tau_H = t\left[1 - \frac{2M_\oplus}{R_\oplus + H} - ((R_\oplus + H)\,\Omega + v)^2\right]^{1/2}$$

(c) Assume the following: $v = 10^{-6}$ (about 670 mph), $H = 10$ km $= 10^6$ cm, $R_\oplus \approx 6.37 \times 10^8$ cm, and $M_\oplus \approx 0.44$ cm. Verify that $\Omega \approx 2.4 \times 10^{-15}$ cm^{-1}, $2M_\oplus/(R_\oplus + H) \approx 1.4 \times 10^{-9}$, and $[(R_\oplus + H)\,\Omega + v]^2 \approx 6 \times 10^{-12}$.

We may therefore drop H from the last expression and write

$$\Delta\tau_H \approx t\left[1 - \frac{2M_\oplus}{R_\oplus + H} - R_\oplus^2\,\Omega^2 - 2R_\oplus\Omega v - v^2\right]^{1/2}$$

$$\approx t\left[1 - \frac{M_\oplus}{R_\oplus + H} - \frac{R_\oplus^2\,\Omega^2}{2} - R_\oplus\Omega v - \frac{v^2}{2}\right]$$

[We do not replace $R_\oplus + H$ by $R_\oplus$ in the larger term $2M_\oplus/(R_\oplus + H)$, since this would produce a more significant error.]

(d) Let ϵ be the dimensionless quantity $(\Delta\tau_H - \Delta\tau_E)/\Delta\tau_E$, the proportional difference in proper time between the traveling clock and the earth-bound clock. Since, from (a), $\Delta\tau_E/t \approx 1$, with error of order 10^{-9}, we have

$$\epsilon \approx \frac{\Delta\tau_H - \Delta\tau_E}{t}$$

Show that

$$\epsilon \approx gH - R_{\oplus}\Omega v - v^2/2$$

where $g = M_{\oplus}/R_{\oplus}^2$.

(e) Show that for a clock flown eastward ($v = +10^{-6}$), $\epsilon \approx -9.5 \times 10^{-13}$; while for a clock flown westward ($v = -10^{-6}$), $\epsilon \approx 2.1 \times 10^{-12}$.

(Similar theoretical predictions were confirmed in experiments carried out in 1971 by Hafele and Keating to compare the proper times of atomic clocks transported on commercial jets eastward and westward around the world with the proper times of similar clocks on the ground. The above discussion is another version of the twin paradox, analyzed this time in the curved spacetime of the earth's gravitational field.)

9. ORBITS IN GENERAL RELATIVITY

Having calculated the orbit of a planet on the assumption of Newton's Law of Gravitation, let us now perform the same feat under the assumption of general relativity. These two theories are so disparate, so seemingly unrelated, that we would hardly expect to arrive at an orbital equation even remotely resembling Eq. (116). Nevertheless, the predictions of the two theories are amazingly close, with observational data overwhelmingly in favor of Einstein.

We begin with the Schwarzschild metric,

$$d\tau^2 = \left(1 - \frac{2M}{r}\right) dt^2 - \left(1 - \frac{2M}{r}\right)^{-1} dr^2 - r^2\, d\phi^2 - r^2 \sin^2\phi\, d\theta^2 \qquad (157)$$

for the spacetime geometry outside a spherically symmetric mass M. (As in the Newtonian case, we can treat the sun as a point mass and neglect the effects on the orbit by other planets.) According to Einstein, a planet follows a timelike geodesic in this geometry:

$$\frac{d^2 x^\lambda}{d\tau^2} + \Gamma^\lambda_{\mu\nu} \frac{dx^\mu}{d\tau} \frac{dx^\nu}{d\tau} = 0, \qquad \lambda = 0, 1, 2, 3 \qquad (158)$$

where $x^0 = t$, $x^1 = r$, $x^2 = \phi$, and $x^3 = \theta$. The Christoffel symbols are obtained from Eq. (153), where primes denote differentiation with respect to r, n = - m, and

$$e^{2m} = 1 - 2M/r$$

Choosing first $\lambda = 2$, we obtain

$$\frac{d^2\phi}{d\tau^2} + \frac{2}{r}\frac{dr}{d\tau}\frac{d\phi}{d\tau} - \sin\phi\cos\phi\left(\frac{d\phi}{d\tau}\right)^2 = 0$$

By performing a spatial rotation if necessary—such a change preserves the form of Eq. (157)—we may assume that when $\tau = 0$, $\phi = \pi/2$, and $d\phi/d\tau = 0$. This means that the planet is initially in the plane $\phi = \pi/2$. Because of the spherical symmetry of the field, the planet must remain in this plane, since to do otherwise would give preference to one of the plane's half-spaces over the other. We therefore take $\phi = \pi/2$ identically. Substituting this into the remaining geodesic equations, we have, as our orbital equations (three equations in three unknowns),

$$\frac{d^2 t}{d\tau^2} + 2m'\frac{dr}{d\tau}\frac{dt}{d\tau} = 0 \tag{159a}$$

$$\frac{d^2 r}{d\tau^2} + m'e^{2m-2n}\left(\frac{dt}{d\tau}\right)^2 + n'\left(\frac{dr}{d\tau}\right)^2 - re^{-2m}\left(\frac{d\theta}{d\tau}\right)^2 = 0 \tag{159b}$$

$$\frac{d^2\theta}{d\tau^2} + \frac{2}{r}\frac{dr}{d\tau}\frac{d\theta}{d\tau} = 0 \tag{159c}$$

For convenience later, let

$$\gamma = 1 - \frac{2M}{r} = e^{2m}$$

Since $m'dr/d\tau = (dm/dr)\,(dr/d\tau) = dm/d\tau$, we may divide Eq. (159a) by $dt/d\tau$ and rewrite it as

$$\frac{d}{d\tau}\left(\ln \frac{dt}{d\tau}\right) = -\,2\,\frac{dm}{d\tau}$$

which can be immediately integrated and then exponentiated to give

$$\frac{dt}{d\tau} = be^{-2m} = \frac{b}{\gamma} \tag{160}$$

for some positive constant b.

Equation (159c) also can be integrated easily [as was Eq. (41b) in Section I-7] to obtain

$$r^2\,\frac{d\theta}{d\tau} = h \tag{161}$$

a positive constant.

It is interesting to compare our present calculations with those involved in Example 19 and Exercise 16 in Section I-7. Proper time τ here plays the role of arc length s there. The resemblance is perhaps not surprising, since in both cases we are computing geodesics in terms of "polar coordinates and arc length."

Rather than attempting to integrate Eq. (159b), we can obtain an alternative equation from the metric form. From Eq. (157) with $\phi = \pi/2$, we have

$$\begin{aligned} 1 &= \gamma\left(\frac{dt}{d\tau}\right)^2 - \gamma^{-1}\left(\frac{dr}{d\tau}\right)^2 - r^2\left(\frac{d\theta}{d\tau}\right)^2 \\ &= \gamma\left(\frac{b}{\gamma}\right)^2 - \gamma^{-1}\left(\frac{dr}{d\theta}\,\frac{h}{r^2}\right)^2 - r^2\left(\frac{h}{r^2}\right)^2 \end{aligned} \tag{162}$$

by Eqs. (160 and (161). Multiplying by $\gamma = 1 - 2M/r$ and rearranging, we have

$$\left(\frac{h}{r^2}\ \frac{dr}{d\theta}\right)^2 + \frac{h^2}{r^2} = b^2 - 1 + \frac{2M}{r} + \frac{2M}{r}\ \frac{h^2}{r^2}$$

Let $u = 1/r$, so that $du/d\theta = -(1/r^2)dr/d\theta$. Then the previous equation becomes, after division by h^2,

$$\left(\frac{du}{d\theta}\right)^2 + u^2 = \frac{b^2 - 1}{h^2} + \frac{2Mu}{h^2} + 2Mu^3$$

Finally, differentiate with respect to θ and divide by $2\ du/d\theta$ to obtain

$$\frac{d^2u}{d\theta^2} + u = \frac{M}{h^2} + 3Mu^2 \tag{163}$$

where $u = 1/r$ and $h = r^2\, d\theta/d\tau$, a constant.

This is similar to the Newtonian equation (114) except for the additional "relativistic" term $3Mu^2$, which is usually quite small compared to M/h^2. For instance, for the planet Mercury, the ratio of these terms is about 7.4×10^{-8}. The orbit therefore should differ only slightly from that predicted from Newton's theory, with the discrepancy becoming apparent only after many revolutions. We may therefore give, as a first approximation to a solution to Eq. (163), the function $u_1 = u_1(\theta)$ given by

$$u_1 = \frac{M}{h^2}(1 + e\cos\theta) \tag{164}$$

where e is a constant [cf. Eq. (116)].

To obtain a better approximation, substitute u_1 into the relativistic term $3Mu^2$. This gives

$$\frac{d^2u}{d\theta^2} + u = \frac{M}{h^2} + \frac{3M^3}{h^4}(1 + 2e\cos\theta + e^2\cos^2\theta)$$

or

$$\frac{d^2 u}{d\theta^2} + u = \frac{M}{h^2} + \frac{3M^3}{h^4} + \frac{6M^3 e}{h^4} \cos\theta + \frac{3M^3 e^2}{2h^4} + \frac{3M^3 e^2}{2h^4} \cos 2\theta \qquad (165)$$

the last term being obtained via the identity $\cos^2\theta = (1 + \cos 2\theta)/2$.

We can solve Eq. (165) with the aid of the following easily verified lemma.

Lemma III-5

Let A be a real number. Then

$u = A$ is a solution of $d^2u/d\theta^2 + u = A$

$u = \frac{1}{2}\,\theta \sin\theta$ is a solution of $d^2u/d\theta^2 + u = A\cos\theta$

$u = -\frac{A}{3}\cos(2\theta)$ is a solution of $d^2u/d\theta^2 + u = A\cos 2\theta$

Apply this lemma to each of the four additional terms of Eq. (165), i.e., for each term T, solve $d^2u/d\theta^2 + u = T$, and then add the sum of these four solutions to u_1. After factoring out M/h^2 and collecting terms, the result is

$$u = \frac{M}{h^2}\left[1 + \frac{3M^2}{h^2}\left(1 + \frac{e^2}{2}\right) + e\cos\theta + \frac{3M^2 e}{h^2}\,\theta\sin\theta - \frac{M^2 e^2}{2h^2}\cos 2\theta\right] \qquad (166)$$

where e is a constant.

Let us now determine how the extra terms in the brackets modify the previous elliptical orbit, Eq. (116). The added constant term, $3M^2(1 + e^2/2)/h^2$, is quite small compared to unity (e.g., 8×10^{-8} for Mercury). By itself, this term has a negligible effect on the planet's orbit. In fact, if we were to set

$$\alpha = 1 + \frac{3M^2}{h^2}\left(1 + \frac{e^2}{2}\right)$$

and define $e' = e/\alpha$, we could rewrite the first few terms of Eq. (166) as

$$\frac{M\alpha}{h^2}\ (1 + e' \cos\theta)$$

a function similar in form to Eq. (164). Since $\alpha \approx 1$, the effect of the added constant term is therefore seen to be only a negligible change in the orbit's eccentricity and perihelial distance. There is no precessional effect here.

The final term in the brackets, the one involving $\cos 2\theta$, is likewise very small and produces a negligible periodic variation in the position of perihelion.

The crucial term is the one involving $\theta \sin\theta$. Since θ appears on its own here, rather than as the argument of a periodic function, the term will have a *cumulative* effect over many revolutions, as θ becomes large. This effect is the observed perihelial advance. Consequently, we may, to a close approximation, write the relativistic orbital equation in the form

$$u \approx \frac{M}{h^2}\left(1 + e\cos\theta + \frac{3M^2 e}{h^2}\ \theta\sin\theta\right)$$

Using the approximations

$$\cos\left(\frac{3M^2\theta}{h^2}\right) \approx 1, \qquad \sin\left(\frac{3M^2\theta}{h^2}\right) \approx \frac{3M^2\theta}{h^2}$$

valid since $M^2/h^2 \approx 10^{-8}$, we can write the orbital equation as

$$u \approx \frac{M}{h^2}\left[1 + e\cos\left(\theta - \frac{3M^2}{h^2}\ \theta\right)\right] \tag{167}$$

Now perihelion occurs when u = 1/r achieves its maximum value (r minimum). From Eq. (167), we can see that two successive perihelia will occur at $\theta = 0$ and at

$$\theta = \frac{2\pi}{1 - \frac{3M^2}{h^2}} \approx 2\pi\left(1 + \frac{3M^2}{h^2}\right)$$

[since for small x, $(1 - x)^{-1} \approx 1 + x$]. Thus the direction of perihelion advances $\Delta\theta \approx 6\pi M^2/h^2$ radians per revolution (perihelion to perihelion), and this precession is in the direction of the planet's revolution (see Fig. III-9) where the advance is greatly exaggerated.

Let n be the number of revolutions of the planet per (earth) century. Then, the amount of precession or perihelion advance in 100 years, $\Delta\theta_{cent}$ is

$$\Delta\theta_{cent} = n\Delta\theta = \frac{6\pi M^2 n}{h^2} = \frac{6\pi M n}{a(1 - e^2)}$$

radians, since $h^2/M = ed = a(1 - e^2)$, by Eq. (119).* This is then converted to seconds of arc per century upon multiplication by $3600 \cdot 180/\pi$.

The predictions and observational data for Mercury, Venus, Earth, and the asteroid Icarus are given in Table III-2 (see [4]).

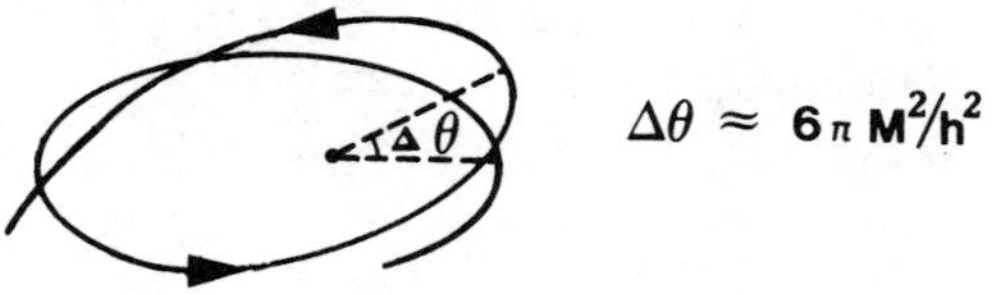

Figure III - 9.

*Since the disagreement with Newtonian theory is slight, the formulas of Section 5 are applicable to our solar system.

Table III - 2.

Planet	a ($\div 10^{11}$ cm)	e	n	$\Delta\theta_{cent}$ (" per century) Gen. Rel.	Observed
Mercury	57.91	0.2056	415	43.03	43.11 ± 0.45
Venus	108.21	0.0068	149	8.6	8.4 ± 4.8
Earth	149.60	0.0167	100	3.8	5.0 ± 1.2
Icarus	161.0	0.827	89	10.3	9.8 ± 0.8

(The tiny eccentricity of Venus's nearly circular orbit makes it difficult to locate perihelion precisely, hence the large uncertainty in the observations for that planet.) Overall, the results strongly confirm the predictions of Einstein's theory.

Exercises III-9

1. (a) Show that

$$\frac{3Mu^2}{M/h^2} = 3\left(r\frac{d\theta}{d\tau}\right)^2$$

(b) Show that if $e \ll 1$,

$$\frac{3Mu^2}{M/h^2} = \frac{3Ma(1-e^2)}{r^2} \approx \frac{3M}{a}$$

2. (See Exercise 1.) Estimate the ratio of $3M_\odot u^2$ to $M_\odot/h^2$ (i) for the planet Mercury; (ii) for the earth.

3. Verify the values of $\Delta\theta_{cent}$ given in Table III-2 for Mercury and for Earth.

10. THE BENDING OF LIGHT

According to Einstein, a ray of light follows a geodesic also, but now τ cannot be used as the parameter since the lapse of proper time along a lightlike curve is zero. The equations for a light ray have the form

$$\frac{d^2 x^\lambda}{d\rho^2} + \Gamma^\lambda_{\mu\nu} \frac{dx^\mu}{d\rho} \frac{dx^\nu}{d\rho} = 0, \quad \lambda = 0,1,2,3$$

where $d\tau/d\rho = 0$. As before, we may orient our coordinate system so that the path lies in the plane $\phi = \pi/2$, and the equations take the form (159), but with τ replaced by ρ. The derivation proceeds as for planetary orbits, except that the left side of Eq. (162) is zero instead of unity, since $d\tau/d\rho = 0$ replaces $d\tau/d\tau = 1$. The calculations yield

$$\frac{d^2 u}{d\theta^2} + u = 3Mu^2 \tag{168}$$

where $u = 1/r$.

Let R be the minimum distance from the mass M to the path of the light ray. We may assume our coordinate system oriented so that the point of closest approach is on the ray $\theta = 0$. In the absence of the mass ($M = 0$), the solution to Eq. (168) would be

$$u = \frac{1}{r} = \frac{1}{R} \cos\theta \tag{169}$$

or

$$x = r\cos\theta = R$$

a straight line. The term $3Mu^2$ therefore represents the deviation from the classical rectilinear path, and in most cases this term is small compared to R. For a light ray grazing the sun, $R = R_\odot \approx 6.96 \times 10^{10}$ cm, $M = M_\odot \approx 1.48 \times 10^5$ cm, and $3Mu^2 \approx 9.2 \times 10^{-17}$. We therefore expect the solution to Eq. (168) to be approximately

$$u \approx \frac{1}{R} \cos\theta$$

Substituting this back into the relativistic term $3Mu^2$ in Eq. (168), we have

$$\frac{d^2 u}{d\theta^2} + u \approx \frac{3M}{R^2} \cos^2 \theta = \frac{3M}{2R^2} (1 + \cos 2\theta)$$

By Lemma III-5, a particular solution of the latter is

$$u_1 = \frac{3M}{2R^2} \left(1 - \frac{1}{3} \cos 2\theta\right)$$

With the aid of the identity $\cos 2\theta = 2 \cos^2 \theta - 1$, we can rewrite this as

$$u_1 = \frac{M}{R^2} (2 - \cos^2 \theta)$$

Adding u_1 to the homogeneous solution (169), we have that the solution to Eq. (168) is, to a close approximation

$$u = \frac{1}{r} = \frac{1}{R} \cos \theta + \frac{M}{R^2} (2 - \cos^2 \theta) \tag{170}$$

Now, far away from M in either direction ($r \to \infty$) spacetime is practically flat and the path becomes nearly straight. Thus the path has asymptotic lines, and the angle $\Delta\theta$ between these lines is the amount of light deflection predicted by general relativity. From Figure III-10, we can see that as $r \to \infty$, θ will approach $\pm (\pi/2 + \Delta\theta/2)$. Since $\Delta\theta$ is very small, the $\cos^2 \theta$ term in Eq. (170) is negligible compared to the other terms ($\cos \theta \approx \cos \pi/2 = 0$). Passing to the limit of Eq. (170) as $r \to \infty$, we therefore have

$$0 = \frac{1}{R} \cos\left(\frac{\pi}{2} + \frac{\Delta\theta}{2}\right) + \frac{2M}{R^2}$$

Consequently,

$$\frac{2M}{R} = -\cos\left(\frac{\pi}{2} + \frac{\Delta\theta}{2}\right) = \sin \frac{\Delta\theta}{2} \approx \frac{\Delta\theta}{2}$$

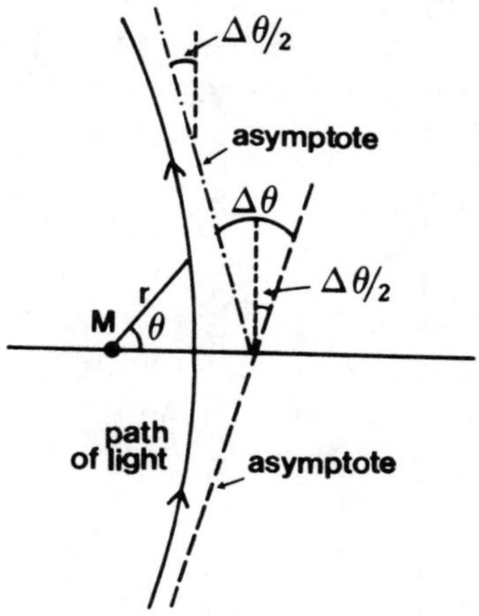

Figure III - 10.

or

$$\Delta\theta \approx 4M/R$$

For $M = M_{\odot}$, $R = R_{\odot}$, $\Delta\theta \approx 8.51 \times 10^{-6}$ radians $\approx 1.75''$.

Actually, Newtonian mechanics predicts a bending of light also, provided the assumption is made that light photons are influenced by gravity as are material particles. The path of such a photon is then one branch of a hyperbola (a tiny high speed comet!) and is given by Eq. (116) with $e > 1$:

$$\frac{1}{r} = \frac{M}{h^2}(1 + e\cos\theta) \tag{171}$$

where

$$h = r^2 \frac{d\theta}{dt} \tag{172}$$

Consider the case of a ray just grazing the sun when $\theta = 0$. At the instant of closest approach ($r = R_{\odot}$), the radial component of the photon's velocity is zero and the speed, from Eq. (107), is

$r\,d\theta/dt = 1$ (the speed of light). Equation (172) then gives $h = r(r\,d\theta/dt) = R_\odot$. Substituting $\theta = 0$, $M = M_\odot$, and $r = R_\odot = h$ into Eq. (171) and solving for e, we obtain

$$e = \frac{R_\odot}{M_\odot} - 1 \approx \frac{R_\odot}{M_\odot} \approx 4.7 \times 10^5$$

Let $\Delta\theta_N$ be the acute angle between the light path's asymptotes. As $r \to \infty$ in Eq. (171), $\theta \to \pm(\pi/2 + \Delta\theta_N/2)$. We therefore have, in the limit,

$$0 = \frac{M}{R^2}\left[1 + e\cos\left(\frac{\pi}{2} + \frac{\Delta\theta_N}{2}\right)\right]$$

or

$$\frac{1}{e} = -\cos\left(\frac{\pi}{2} + \frac{\Delta\theta_N}{2}\right) = \sin\frac{\Delta\theta_N}{2} \approx \frac{\Delta\theta_N}{2}$$

since the largeness of e implies $\Delta\theta$ is small. Hence the Newtonian deflection is

$$\Delta\theta_N \approx \frac{2}{e} = \frac{2M_\odot}{R_\odot}$$

exactly half that predicted by general relativity.

As described in Section 3, measurements of apparent star positions during solar eclipses since 1919 and more recent observations of radio sources have overwhelmingly confirmed Einstein's theory over Newton's.

Exercises III-10

1. Supply the missing details in the derivation of Eq. (168).
2. Compute the angle of deflection $\Delta\theta = 4M_\oplus/R_\oplus$ for a light ray grazing the earth (in radians and then in seconds of arc).

Appendix A

Vector Geometry and Analysis

Intuitively speaking, a vector is an entity that has both magnitude and direction, and so is often represented as a directed line segment or "arrow." Arrows with the same length and direction are considered to be equivalent and represent the same vector. Accordingly, many authors define a vector to be an equivalence class of directed line segments, i.e., a set consisting of an arrow together with all arrows having the same length and direction as it.

In E^3, if a vector **V** is represented by an arrow with initial point (x,y,z) and terminal point (x′,y′,z′), then the coordinate differences,

$$\Delta x = x' - x, \quad \Delta y = y' - y, \quad \Delta z = z' - z$$

are called the *components* of **V**. It is easy to see that equivalent arrows have the same components, so the choice of representing arrow is immaterial.

Throughout this appendix, we treat vectors in 3-dimensional Euclidean space, E^3. However, with the exception of the cross product, which is defined in three dimensions only, analogous definitions and results pertain in two dimensions: simply delete the third coordinate of every point and the third component of every vector.

For each point, we can associate the vector represented by the arrow from the origin to that point. Conversely, each vector has a representing arrow emanating from the origin and may be associated with the terminal point of that arrow. The coordinates of the point are exactly the components of the vector. Since points and vectors are thus in one-to-one correspondence, we shall use ordered triples to label vectors as well as points, e.g., $\mathbf{V} = (a, b, c)$. The context will usually make clear which is intended, but in many instances we shall want to think of the same object as a point at certain times and a vector at others.

Definition A-1

If $\mathbf{V}_1 = (a_1, a_2, a_3)$ and $\mathbf{V}_2 = (b_1, b_2, b_3)$ are vectors and k is any real number, we define

$$\mathbf{V}_1 + \mathbf{V}_2 = (a_1 + b_1, a_2 + b_2, a_3 + b_3)$$

$$k\mathbf{V}_1 = (ka_1, ka_2, ka_3)$$

Geometrically, vector addition is accomplished by choosing representatives for the vectors for which the terminal point of the first vector coincides with the initial point of the second. The sum is then the arrow from the initial point of the first to the terminal point of the second. This is illustrated in Figure A-1 for the 2-dimensional case.

Vector addition is obviously commutative and satifies

$$(k_1 + k_2)\mathbf{V} = k_1\mathbf{V} + k_2\mathbf{V}$$

$$k(\mathbf{V}_1 + \mathbf{V}_2) = k\mathbf{V}_1 + k\mathbf{V}_2$$

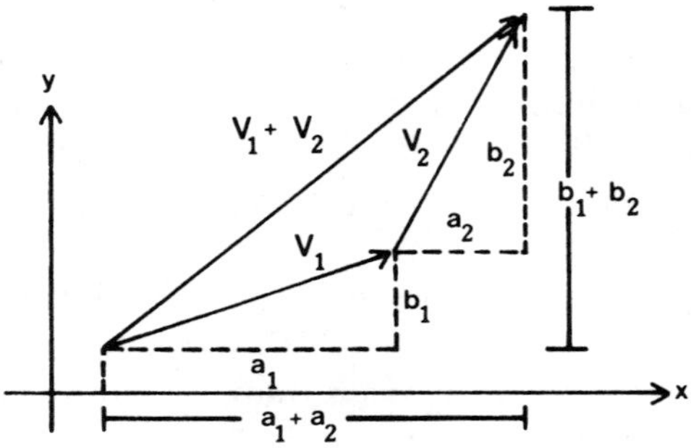

Figure A - 1.

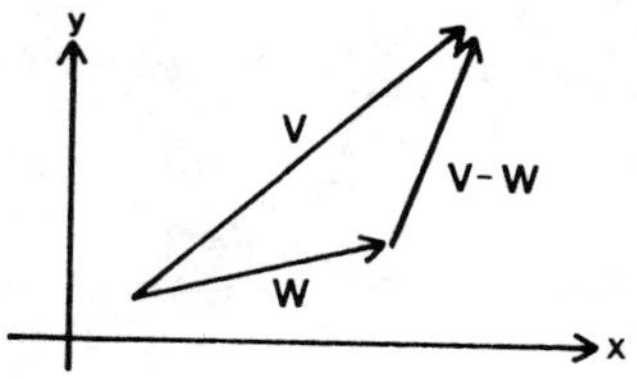

Figure A - 2.

We define **V** - **W** to mean **V** + (- 1)**W**. The geometric significance of vector substraction is shown in Figure A-2. The vector whose components are all zero is denoted **O**.

Definition A-2

The length of a vector **V** = (a,b,c) is the non-negative number **V** given by

$$\|\mathbf{V}\| = (a^2 + b^2 + c^2)^{1/2}$$

It is the distance between the initial and terminal points of any arrow representing **V**.

The length $\|\mathbf{V}\|$ is zero if and only if **V** = **O**. From the geometric interpretation of vector addition (cf. Fig. A-1), we have the *triangle inequality*,

$$\|\mathbf{V} + \mathbf{W}\| \leqslant \|\mathbf{V}\| + \|\mathbf{W}\|$$

for all **V** and **W**. For any vector **V** and real number k,

$$\|k\mathbf{V}\| = |k| \, \|\mathbf{V}\|$$

As a consequence, if $\mathbf{V} \neq \mathbf{O}$, the vector $(1/\|\mathbf{V}\|)\mathbf{V}$, abbreviated $\mathbf{V}/\|\mathbf{V}\|$, is a *unit vector*, i.e., a vector of length one, having the same direction as **V**.

Definition A-3

The *scalar product* or *dot product* of two vectors $\mathbf{V} = (a_1, b_1, c_1)$ and $\mathbf{W} = (a_2, b_2, c_2)$ is the real number $\mathbf{V} \cdot \mathbf{W}$ defined by

$$\mathbf{V} \cdot \mathbf{W} = a_1 a_2 + b_1 b_2 + c_1 c_2$$

Scalar multiplication is commutative and distributive with respect to vector addition. For any **V**, $\mathbf{V} \cdot \mathbf{V} = \|\mathbf{V}\|^2$.

Theorem A-4

If $\mathbf{V} = (a_1, b_1, c_1)$ and $\mathbf{W} = (a_2, b_2, c_2)$ then

$$\mathbf{V} \cdot \mathbf{W} = \|\mathbf{V}\| \; \|\mathbf{W}\| \cos\theta$$

where θ is the angle between **V** and **W**.

Proof. We give the proof in two dimensions, the proof for three being entirely similar. Applying the Law of Cosines to ΔOPQ in Figure A-3, we obtain

$$(a_2 - a_1)^2 + (b_2 - b_1)^2 = (PQ)^2 = \|\mathbf{V}\|^2 + \|\mathbf{W}\|^2 - 2\|\mathbf{V}\| \; \|\mathbf{W}\| \cos\theta$$

$$= a_1{}^2 + b_1{}^2 + a_2{}^2 + b_2{}^2 - 2\|\mathbf{V}\| \; \|\mathbf{W}\| \cos\theta$$

or $2\|\mathbf{V}\| \; \|\mathbf{W}\| \cos\theta = 2(a_1 a_2 + b_1 b_2) = 2(\mathbf{V} \cdot \mathbf{W})$.

Corollary A-5

Two vectors **V** and **W** are *orthogonal* (i.e., perpendicular when drawn emanating from the same point) if and only if $\mathbf{V} \cdot \mathbf{W} = \mathbf{O}$. (The zero vector, **O**, is, by convention, orthogonal to every vector.)

Definition A-6

The *vector product* or *cross product* of two vectors $\mathbf{V} = (a_1, b_1, c_1)$ and $\mathbf{W} = (a_2, b_2, c_2)$ is the vector $\mathbf{V} \times \mathbf{W}$ given by

$$\mathbf{V} \times \mathbf{W} = (b_1 c_2 - b_2 c_1, \; c_1 a_2 - c_2 a_1, \; a_1 b_2 - a_2 b_1)$$

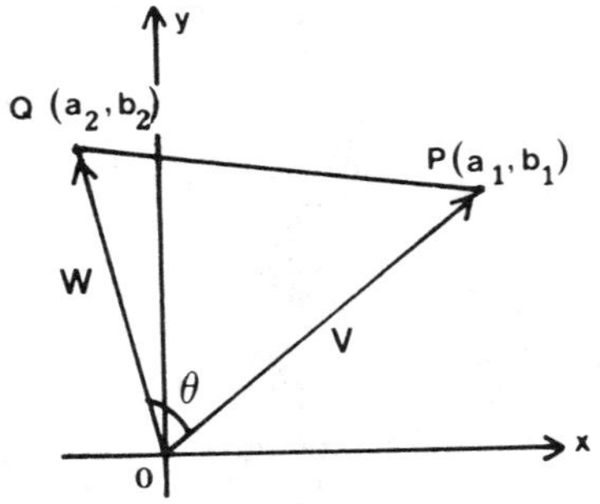

Figure A - 3.

The cross product can be computed by formally evaluating the determinant of the 3×3 matrix

$$\begin{pmatrix} \mathbf{i} & \mathbf{j} & \mathbf{k} \\ a_1 & b_1 & c_1 \\ a_2 & b_2 & c_2 \end{pmatrix}$$

where $\mathbf{i} = (1,0,0)$, $\mathbf{j} = (0,1,0)$, and $\mathbf{k} = (0,0,1)$. Thus, expanding this determinant along the first row, we have

$$\begin{aligned} \mathbf{V} \times \mathbf{W} &= \begin{vmatrix} b_1 & c_1 \\ b_2 & c_2 \end{vmatrix} \mathbf{i} - \begin{vmatrix} a_1 & c_1 \\ a_2 & c_2 \end{vmatrix} \mathbf{j} + \begin{vmatrix} a_1 & b_1 \\ a_2 & b_2 \end{vmatrix} \mathbf{k} \\ &= (b_1 c_2 - b_2 c_1, c_1 a_2 - c_2 a_1, a_1 b_2 - a_2 b_1) \end{aligned}$$

The following identities can be verified by computation with components.

$$\mathbf{V} \times \mathbf{W} = -(\mathbf{W} \times \mathbf{V})$$

$$\mathbf{V} \times \mathbf{V} = \mathbf{O}$$

$$\mathbf{V} \times (\mathbf{W}_1 + \mathbf{W}_2) = (\mathbf{V} \times \mathbf{W}_1) + (\mathbf{V} \times \mathbf{W}_2)$$

$$k_1 \mathbf{V}_1 \times k_2 \mathbf{V}_2 = k_1 k_2 (\mathbf{V}_1 \times \mathbf{V}_2)$$

$$\mathbf{U} \cdot (\mathbf{V} \times \mathbf{W}) = \mathbf{V} \cdot (\mathbf{W} \times \mathbf{U}) = \mathbf{W} \cdot (\mathbf{U} \times \mathbf{V})$$

The quantity $\mathbf{U} \cdot (\mathbf{V} \times \mathbf{W})$ is called the *triple scalar product* of $\mathbf{U}$, $\mathbf{V}$ and $\mathbf{W}$, and the last property above is often called the *cyclic property* of the triple scalar product. We usually omit the parentheses in $\mathbf{U} \cdot (\mathbf{V} \times \mathbf{W})$ since there is only one possible way of interpreting $\mathbf{U} \cdot \mathbf{V} \times \mathbf{W}$—$(\mathbf{U} \cdot \mathbf{V}) \times \mathbf{W}$ makes no sense, since we cannot cross a number and a vector.

If $\mathbf{V}_i = (a_i, b_i, c_i)$, for $i = 1,2,3$, then $\mathbf{V}_1 \cdot \mathbf{V}_2 \times \mathbf{V}_3$ is the determinant

$$\mathbf{V}_1 \cdot \mathbf{V}_2 \times \mathbf{V}_3 = \begin{vmatrix} a_1 & b_1 & c_1 \\ a_2 & b_2 & c_2 \\ a_3 & b_3 & c_3 \end{vmatrix}$$

Theorem A-7

Let **V** and **W** be non-zero vectors and let θ be the angle between **V** and **W** ($0 \leqslant \theta \leqslant \pi$). Then (a) $\mathbf{V} \times \mathbf{W}$ is orthogonal to both **V** and **W**; (b) $\|\mathbf{V} \times \mathbf{W}\| = \|\mathbf{V}\| \, \|\mathbf{W}\| \sin\theta$.

Proof. (a) can be verified by computing $\mathbf{V} \cdot \mathbf{V} \times \mathbf{W}$ and $\mathbf{W} \cdot \mathbf{V} \times \mathbf{W}$ in components, or by using the fact that the determinant of a matrix with two equal rows is zero. For (b),

$$\|\mathbf{V}\|^2 \, \|\mathbf{W}\|^2 \sin^2\theta = \|\mathbf{V}\|^2 \, \|\mathbf{W}\| \, (1 - \cos^2\theta)$$

$$= \|\mathbf{V}\|^2 \, \|\mathbf{W}\|^2 - (\mathbf{V} \cdot \mathbf{W})^2$$

If the latter is computed in components, it will be found to agree with $\|\mathbf{V} \times \mathbf{W}\|^2$ expressed in components.

The length of the cross product, $\|\mathbf{V} \times \mathbf{W}\|$, can be interpreted geometrically as the area of a parallelogram in which two adjacent side are arrows representing **V** and **W**. In Figure A-4, area = base $\times$ height = $\|\mathbf{V}\| \, \|\mathbf{W}\| \sin\theta$. If $\mathbf{V} \times \mathbf{W} \neq \mathbf{O}$, the direction of $\mathbf{V} \times \mathbf{W}$ is such that **V**, **W**, and $\mathbf{V} \times \mathbf{W}$ form a "right-handed system": if you extend the index and middle fingers and thumb of your right hand so that (a) your index finger points in the direction of **V**, (b) your middle finger points in the direction of **W**, and (c) your thumb is perpendicular to both fingers, then your thumb will point in the direction of $\mathbf{V} \times \mathbf{W}$.

Occasionally, we deal with *vector-valued functions* of a real variable t, for example,

$$\mathbf{V}(t) = (a(t), b(t), c(t))$$

where a, b, and c are differentiable real-valued functions. The derivative is defined componentwise:

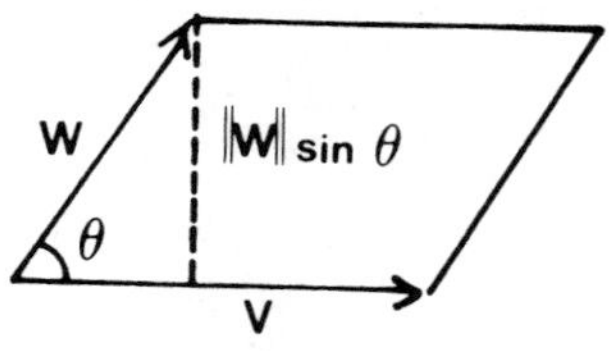

Figure A - 4.

$$\frac{d\mathbf{V}}{dt} = \left(\frac{da}{dt}, \frac{db}{dt}, \frac{dc}{dt}\right)$$

The following are immediate consequences of the sum, product, and quotient rules for real-valued functions:

$$\frac{d}{dt}(\mathbf{V} + \mathbf{W}) = \frac{d\mathbf{V}}{dt} + \frac{d\mathbf{W}}{dt}$$

$$\frac{d}{dt}(\mathbf{V} \cdot \mathbf{W}) = \mathbf{V} \cdot \frac{d\mathbf{W}}{dt} + \frac{d\mathbf{V}}{dt} \cdot \mathbf{W}$$

$$\frac{d}{dt}(f\mathbf{V}) = f\frac{d\mathbf{V}}{dt} + \frac{df}{dt}\mathbf{V}$$

$$\frac{d}{dt}\left(\frac{1}{f}\mathbf{V}\right) = \frac{f\frac{d\mathbf{V}}{dt} - \frac{df}{dt}\mathbf{V}}{f^2}$$

$$\frac{d}{dt}(\mathbf{V} \times \mathbf{W}) = \mathbf{V} \times \frac{d\mathbf{W}}{dt} + \frac{d\mathbf{V}}{dt} \times \mathbf{W}$$

for $\mathbf{V}$ and $\mathbf{W}$ vector functions and f a real-valued function. We shall interchangeably write $f\mathbf{V}$ or $\mathbf{V}f$. If $\mathbf{V} = \mathbf{V}(s)$ and $s = f(t)$, then the identity

$$\frac{d\mathbf{V}}{dt} = \frac{d\mathbf{V}}{ds}\frac{ds}{dt}$$

follows from the ordinary chain rule for real-valued functions. If $\mathbf{W}$ is a vector function of two real variables. $\mathbf{W} = \mathbf{W}(u,v)$ and u and v in turn are functions of a variable s, then

$$\frac{d\mathbf{W}}{ds} = \frac{\partial \mathbf{W}}{\partial u}\frac{du}{ds} + \frac{\partial \mathbf{W}}{\partial v}\frac{dv}{ds}$$

(All functions are assumed differentiable.)

Appendix B

Hyperbolic Functions

Definition B-1

For any real number x, we define

$$\sinh x = \frac{1}{2}(e^x - e^{-x}), \qquad \cosh x = \frac{1}{2}(e^x + e^{-x})$$

$$\tanh x = \frac{\sinh x}{\cosh x} = \frac{e^x - e^{-x}}{e^x + e^{-x}}, \qquad \operatorname{sech} x = \frac{1}{\cosh x} = \frac{2}{e^x + e^{-x}}$$

The graphs of the first three of these functions are sketched in Figure B-1. The identities below follow readily from the definitions. As with the circular (ordinary trigonometric) functions, $\cosh^2 x$, $\sinh^2 x$, etc., denote $(\sinh x)^2$, $(\cosh x)^2$, etc.:

$$\cosh^2 x - \sinh^2 x = 1$$

$$1 - \tanh^2 x = \operatorname{sech}^2 x$$

$$\sinh(x \pm y) = \sinh x \cosh y \pm \cosh x \sinh y$$

$$\cosh(x \pm y) = \cosh x \cosh y \pm \sinh x \sinh y$$

$$\tanh(x \pm y) = \frac{\tanh x \pm \tanh y}{1 \pm \tanh x \tanh y}$$

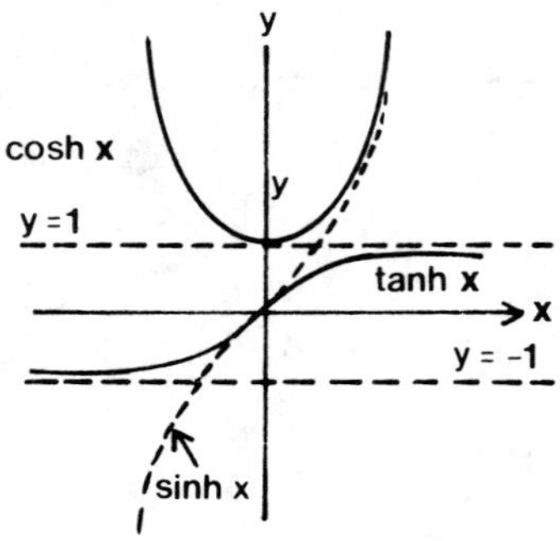

Figure B - 1.

$$\sinh 2x = 2 \sinh x \cosh x$$

$$\cosh 2x = \cosh^2 x + \sinh^2 x = 2 \cosh^2 x - 1$$

$$\frac{d}{dx} \sinh x = \cosh x, \quad \frac{d}{dx} \cosh x = \sinh x$$

$$\frac{d}{dx} \tanh x = \operatorname{sech}^2 x, \quad \frac{d}{dx} \operatorname{sech} x = - \operatorname{sech} x \tanh x$$

From the power series for e^x and e^{-x},

$$e^x = 1 + x + \frac{x^2}{2!} + \frac{x^3}{3!} + \frac{x^4}{4!} + \cdots$$

$$e^{-x} = 1 - x + \frac{x^2}{2!} - \frac{x^3}{3!} + \frac{x^4}{4!} + \cdots$$

and the definitions above, we have

$$\sinh x = x + \frac{x^3}{3!} + \frac{x^5}{5!} + \frac{x^7}{7!} + \cdots$$

$$\cosh x = 1 + \frac{x^2}{2!} + \frac{x^4}{4!} + \frac{x^6}{6!} + \cdots$$

$$\tanh x = x - \frac{x^3}{3} + \frac{2x^5}{15} - \frac{17x^7}{315} + \cdots, \quad |x| < \frac{\pi}{2}$$

Except for the differences in signs, these resemble the series for the circular functions:

$$\sin x = x - \frac{x^3}{3!} + \frac{x^5}{5!} - \frac{x^7}{7!} + \cdots$$

$$\cos x = 1 - \frac{x^2}{2!} + \frac{x^4}{4!} - \frac{x^6}{6!} + \cdots$$

$$\tan x = x + \frac{x^3}{3!} + \frac{2x^5}{15} + \frac{17x^7}{315} + \cdots, \quad |x| < \frac{\pi}{2}$$

In fact, there is a close connection between the circular and hyperbolic functions. These functions, as well as the exponential function, e^x, can be extended to functions of a complex variable by taking their power series (with the variable real or complex) as definitions. For example, sin z, for z complex, is simply defined to be the sum of the convergent series

$$z - \frac{z^3}{3!} + \frac{z^5}{5!} - \frac{z^7}{7!} + \cdots$$

If $i = (-1)^{1/2}$, then $i^2 = -1$, $i^3 = -i$, $i^4 = 1$, and in general, $i^{n+4} = i^n$, for any integer n. Replacing x by ix in the power series for sinh(x), we have

$$\sinh(ix) = (ix) + \frac{(ix)^3}{3!} + \frac{(ix)^5}{5!} + \cdots$$

$$= ix - i\frac{x^3}{3!} + i\frac{x^5}{5!} - \cdots$$

$$= i\left(x - \frac{x^3}{3!} + \frac{x^5}{5!} + \cdots\right)$$

$$= i \sin x$$

and so the identity

$$\sinh ix = i \sin x$$

(valid for all x, real or complex). In a similar fashion, we can derive

$$\cosh ix = \cos x$$

$$\tanh ix = i \tan x$$

We can reverse the process as well. Replacing x by ix in the series for the circular functions yields the dual relations

$$\sin ix = i \sinh x$$

$$\cos ix = \cosh x$$

$$\tan ix = i \tanh x$$

Bibliography

1. Abbot, Edwin A., *Flatland*, 6th ed., Dover Publications, New York, 1952
2. Barnett, Lincoln, *The Universe and Dr. Einstein*, Bantam Books, Inc., New York, 1968
3. Bergmann, P. G., *The Riddle of Gravitation*, Charles Scribner's Sons, New York, 1968
4. Berry, Michael, *Principles of Cosmology and Gravitation*, Cambridge University Press, Cambridge, England, 1976
5. Bondi, Hermann, *Relativity and Common Sense*, Doubleday & Co., Garden City, 1964
6. Bronowski, J., "The Clock Paradox," *Scientific American*, February, 1963
7. Callahan, J. J., "The Curvature of Space in a Finite Universe," *Scientific American*, 235 (1976), 90-100
8. Chiu, H. and W. Hoffman, eds., *Gravitation and Relativity*, Benjamin, New York, 1964
9. Cole, R. H., *Theory of Ordinary Differential Equations*, Appleton-Century-Crofts, New York, 1968
10. Dirac, P. A. M., "The Evolution of the Physicist's Picture of Nature," *Scientific American*, May, 1963
11. Dirac, P. A. M., *General Theory of Relativity*, John Wiley & Sons, New York, 1975
12. Eddington, A. S., *Space, Time and Gravitation*, Cambridge University Press, 1929
13. Eddington, A. S., *The Mathematical Theory of Relativity*, 3rd ed., reprinted, Chelsea Publishing Co., New York, 1975
14. Einstein, A., "On the Generalized Theory," *Scientific American*, April, 1950
15. Einstein, A., *The Meaning of Relativity*, 4th ed., Princeton University Press, Princeton, 1953
16. French, A. P., *Special Relativity*, W. W. Norton & Co., Inc., New York, 1968
17. Gamow, G., "Gravity," *Scientific American*, March 1961

18. Gamow, G., "The Evolutionary Universe," *Scientific American*, September, 1956

19. Gamow, G., *Mr. Tompkins in Paperback*, Cambridge University Press, Cambridge, England, reprinted, 1971

20. Gauss, C. F., *General Investigations of Curved Surfaces*, tr. by J. C. Morehead and A. M. Hiltebeitel, reprinted, Raven Press, New York, 1965

21. Geller, M., "The Large Scale Structure of the Universe," *American Scientist*, 66 (1978), 176 ff.

22. Gingerich, O., "Copernicus and Tycho," *Scientific American*, 226 (1973), p. 86

23. Goetz, Abraham, *Introduction to Differential Geometry*, Addison-Wesley Pub. Co., Reading, MA, 1968

24. Hall, Tord, *Carl Friedrich Gauss, A Biography*, tr. by A. Froderberg, MIT Press, Cambridge, 1970

25. Harris, E. G., *Introduction to Modern Theoretical Physics*, Vol. 1, John Wiley & Sons, New York, 1975

26. Holton, Gerald, "Einstein, Michelson, and the Crucial Experiment," *ISIS*, 60 (1969), 132-197

27. Hoyle, Fred, *Astronomy*, Rathbone Books, Ltd., London, 1962

28. Hoyle, Fred, *From Stonehenge to Modern Cosmology*, W. H. Freeman and Co., San Francisco, 1972

29. Kaufmann, W. J., *Relativity and Cosmology*, Harper & Row, New York, 1973

30. Lanczos, C., *The Einstein Decade*, Academic Press, New York, 1974

31. Laugwitz, D., *Differential and Riemannian Geometry*, tr. by F. Steinhardt, Academic Press, New York, 1965

32. Layzer, David, "The Arrow of Time," *Scientific American*, (233), 1975

33. Levi-Civita, T., "Geodesic Deviation" (in French), *Mathematische Annalen*, 97 (1927), 291-320

34. Lieber, L., and H. Lieber, *The Einstein Theory of Relativity*, Rinehart & Co., New York, 1945

35. Lorentz, H., A. Einstein et al., *The Principle of Relativity*, tr. by W. Perrett and G. Jeffery, Dover Publications, New York, 1952

36. Misner, C. W., K. S. Thorne, and J. A. Wheeler, *Gravitation*, W. H. Freeman and Co., San Francisco, 1973

37. O'Neill, Barrett, *Elementary Differential Geometry*, Academic Press, New York, 1966

38. Pontryagin, L. *Ordinary Differential Equations*, tr. by W. Counts, Addison-Wesley Pub. Co., Reading, MA, 1962

39. Pound, R. V., and J. L. Snider, "Effect of Gravity on Gamma Radiation," *Physical Review*, B140 (1965), 788-803

40. Resnick, Robert, *Introduction to Special Relativity*, John Wiley & Sons, New York, 1968
41. Rucker, Rudolf v. B., *Geometry, Relativity and the Fourth Dimension*, Dover Publications, New York, 1977
42. Sciama, D. W., *The Physical Foundations of General Relativity*, Doubleday & Co., Garden City, 1969
43. Shapiro, I., "The Universe, Open or Closed?" *Technology Review*, Dec., 1975, p. 64 ff.
44. Stewart, Ian, "Gauss," *Scientific American*, 237 (1977), 122-131
45. Stoker, J. J., *Differential Geometry*, John Wiley & Sons, New York, 1969
46. Struik, D. J., *Lectures on Classical Differential Geometry*, Addison-Wesley Press, Cambridge, 1950
47. Taylor, E. F., and J. A. Wheeler, *Spacetime Physics,* W. H. Freeman and Co., San Francisco, 1966
48. Terrell, J., "Invisibility of the Lorentz Contraction," *Physical Review*, 116 (1959), 1041-45
49. Tolman, R. C., *Relativity, Thermodynamics and Cosmology*, Oxford University Press, Oxford, England, 1934
50. Weinberg, S., *Gravitation and Cosmology*, John Wiley and Sons, New York, 1972
51. Weingard, R., "On Traveling Backward in Time," in *Space, Time, and Geometry*, ed. P. Suppes, D. Reidel Publishing Co., Dordrecht, Holland
52. Wheeler, J. A., "Our Universe: The Known and the Unknown," *American Scientist,* Spring 1968
53. White, A. J., *Real Analysis: An Introduction*, Addison-Wesley Publishing Co., London, 1968
54. Zwart, P. J., "The Flow of Time," in *Space, Time, and Geometry*, ed. by P. Suppes, D. Reidel Publishing Co., Dordrecht, Holland

Index